Unive

Subject

http:/

NITRIC OXIDE

ADVANCES IN EXPERIMENTAL BIOLOGY

NITRIC OXIDE

Edited by

Professor Bruno Tota

Department of Cellular Biology
University of Calabria
Arcavacata di Rende, Italy

and

Professor Barry Trimmer

Department of Biology
Tufts University
Medford, MA, USA

ELSEVIER

Amsterdam – Boston – Heidelberg – London – New York – Oxford
Paris – San Diego – San Francisco – Singapore – Sydney – Tokyo

9007575632

Elsevier
Radarweg 29, PO Box 211, 1000 AE Amsterdam, The Netherlands
Linacre House, Jordan Hill, Oxford OX2 8DP, UK

First edition 2007

Library of Congress Cataloging-in-Publication Data
A catalog record for this book is available from the Library of Congress

British Library Cataloguing in Publication Data
A catalogue record for this book is available from the British Library

ISBN: 978-0-444-53119-3
ISSN: 1872-2423

For information on all Elsevier publications
visit our website at books.elsevier.com

Printed and bound in The Netherlands

07 08 09 10 11   10 9 8 7 6 5 4 3 2 1

Working together to grow
libraries in developing countries
www.elsevier.com | www.bookaid.org | www.sabre.org
ELSEVIER   BOOK AID International   Sabre Foundation

Information About the Society for Experimental Biology (SEB)

The Society for Experimental Biology (SEB) is Europe's leading, not-for-profit organisation embracing all disciplines of experimental biology. Through its large membership and passion for science the Society supports and promotes experimental biology, from molecular to ecological, to benefit both the scientific community and the general public. The Society was founded in 1923 at Birkbeck College and is now well established with a current worldwide membership of over 1,900 biological researchers, teachers and students. The SEB is somewhat unusual in catering for both plant and animal biologists.

The Animal Section of the SEB launched this serial "Advances in Experimental Biology" (AEB) in 2006, with the first volume published in 2007. The aim of this new serial is to provide state-of-the-art review volumes on timely issues that have international topicality within the field of Comparative and Integrative Biology. The series as a whole will therefore cover a rather broad range of topics from the role of individual molecules (e.g. nitric oxide) to system-level approaches applied to a particular research discipline (e.g. toxicogenomics).

Each volume will contain approximately ten chapters each providing a detailed review of current understanding within a sub-topic of the volume title. In some volumes the chapters will be organised according to the level at which the research is focused (e.g. molecular, biochemical, physiological, behavioural, ecological); in others the chapters may be organised on a more taxonomic basis. Each chapter is written by leading authorities in science that have been invited, based on their international reputation, to provide their perspective on the current status and recent developments within the field. All chapters are peer-reviewed by at least two independent referees prior to acceptance for publication.

The series thus aims to provide an excellent, up-to-date resource for a global research audience within each of the volume topics.

Information about the Series Editors

Professor Mike Thorndyke

Professor Mike Thorndyke is the Director and Chair of Experimental Marine Biology at the Royal Swedish Academy of Sciences, Kristineberg Marine Research Station, Sweden. His laboratory specialises in understanding the diversity of cellular and genetic processes in marine organisms, and their importance in terms of evolution, adaptation and ecology, with particular reference to development and adult regeneration. He has over 25 years experience working with marine invertebrates. His laboratory has been the focus for the characterisation of neural complexity in invertebrate Deuterostomes including echinoderms, tunicates and the Xenoturbellids, recently confirmed as a new Deuterostome phylum. Most recently, Professor Thorndyke has become one of the leaders of the neural group in the sea urchin genome annotation consortium.

Dr Rod Wilson

Dr Rod Wilson began his academic career reading Biological Sciences at the University of Birmingham, where he subsequently completed a PhD in Fish Physiology and Ecotoxicology. After completing postdoctoral training at the University of Birmingham, McMaster University in Canada and the University of Manchester, Dr Wilson moved to the University of Exeter where he is currently a senior lecturer. His area of expertise covers comparative and integrative physiology, ranging from studies on water absorption in the mammalian kidney to the behaviour and physiology of fish in the wild. Dr Wilson's work currently focuses upon fish (marine and freshwater), he believes it is important to approach research from a multidisciplinary angle in order to guarantee a holistic understanding of homoeostasis in animals. He is particularly interested in how multiple physiological systems (*e.g.*, respiratory/cardiovascular, osmoregulation, acid-base balance, nitrogenous waste excretion) respond in an integrated manner to maintain whole animal homoeostasis in the face of environmental changes (both natural and anthropogenic). His work utilises studies at the molecular, cellular, tissue and whole animal levels. Furthermore, he is increasingly integrating this approach with behavioural studies that help link physiological mechanisms with social behaviour in fish, both in the laboratory and in the wild.

Information about the Volume Editors

Professor Bruno Tota

Bruno Tota is a full professor of general physiology, specialising in comparative cardiovascular physiology. He graduated in Medicine and Surgery from the University "Federico II" in Naples, where he then completed a master's degree in cardiology. Professor Tota is currently head of the Department of Cell Biology at the University of Calabria, Italy. His expertise lies in comparative cardiovascular physiology with particular interests in the evolutionary morpho-functional design of the heart, cardiac hormones, nitric oxide and signalling, cardiocirculatory adaptations extreme habitats and Antarctic fish biology.

Professor Barry Trimmer

Barry Trimmer is an invertebrate neurobiologist with interests in synaptic signalling (in particular acetylcholine and nitric oxide) and neuromechanics. He received his undergraduate and PhD degrees from the University of Cambridge in England and received postdoctoral training at Harvard Medical School, the University of California, Berkeley and the University of Oregon, Eugene. He has been a professor of biology at Tufts University since 1990 and was awarded an endowed chair (the *Henry Bromfield Pearson Professor of Natural Science*) in 2005. Most recently, he founded the Tufts Biomimetic Devices Laboratory to study the interactions of neural and mechanical control systems for use in novel, biologically inspired devices.

Preface

Since its initial discovery as a modulator of vascular function in mammals, nitric oxide (NO) has been found to be a ubiquitous signaling molecule of profound importance in a wide range of roles. There has been a growing appreciation of the complexity of NO production, dispersal and targeting which in turn has helped to focus attention on physical and chemical processes that were often neglected in the study of cell signaling. Our understanding of these processes has been further enhanced by comparing different functional roles in different model systems. As one might expect, some of the apparent peculiarities (or controversies) about NO signaling actually reflect the special conditions found in each tissue compartment or in the life history of the model species. Many of these complexities can only be resolved by taking a broad and comparative perspective.

The collection of reviews in this volume reflects this philosophy and helps to highlight both the commonalities and the specializations of NO signaling. The model organisms discussed in these chapters include mammals, with their clear clinical relevance, and a concentration of papers on fish, molluscs and insects that provide revealing insights into new NO functions and signaling mechanisms. NO signaling is discussed at the biochemical and molecular level (*e.g.,* Korneev and O'Shea, Morton and Vermehren, Mitchell *et al.*), at the tissue and systemic level (*e.g.,* Agnisola and Pellegrino, Fago and Jensen, Tota *et al.*), in its effects on whole animal physiology (*e.g.,* Panula *et al.*) and its possible evolutionary origins (*e.g.,* Moroz and Kohn). Functionally, there is a significant body of work on NO's role in vascular control but added to this are chapters on development (*e.g.,* Holmqvist *et al.*, Pelster), immunity and cell defenses (*e.g.,* Luckhart *et al.*, Dezfulian *et al.*, Cena *et al.*) epithelial transport (*e.g.,* Davies) and behavior (*e.g.,* Heinrich and Ganter).

Together these reviews by leading experts in their respective fields provide a unique overview of NO signaling in animal systems. These chapters also help to point out key areas for more detailed research. Often by comparing conflicting generalizations from different approaches it is possible to gain a new and more profound understanding of the important issues. It is also hoped that some of the lessons learned from NO research can be used to open up new avenues of research elsewhere. For example, NO appears to have a special place in its sensitivity to the cellular and extracellular milieu. However, it would serve us well to remember that such sensitivity is not limited to free radical signaling molecules and that classical transmitters are subject to functional variations that are not easily revealed through focused studies on a limited number of organisms.

Another aspect to emerge from the comparative perspective of these reviews is the ubiquitous pleiotropic and versatile nature of NO. It appears well suited not only as a messenger between local component sub-systems (according to Shannon's classical concept of information based on communication of a message between a source and a destination), but also as an organizer through connection-integration processes. In other words, NO has biological significance to the communication or regulation of a message and also to the configuration of networks. When placed into a wider biology scenario, this concept helps to underscore the potential of NO signaling in the evolution of emergent highly integrated networks, *i.e.,* networks possessing more information than the sets of their components.

In conclusion, by gathering together many findings, concepts and predictions centered on NO signaling, we hope that the breadth of topics covered in this Volume will provide the reader with new perspectives and help to stimulate research into major unanswered questions. It is also our hope that these questions will lead to entirely new avenues of investigation.

Finally, we would like to express our profound thanks to all the contributing authors and chapter reviewers; their hard work has made this volume an important compendium and waypoint for NO research. We are also most grateful to our managing editor, Dr Suzanne Brockhouse who handled the difficult organizational aspects of this volume with great efficiency and good humor.

Bruno Tota and Barry Trimmer

List of Contributors

Claudio Agnisola, Dipartimento delle Scienze Biologiche, Università di Napoli Federico II, Napoli, Italy.

Per Alm, Department of Pathology, Lund University, Lund, Sweden.

Marcello Canonaco, Comparative Neuroanatomy and Cytology Laboratory, Ecology Department, University of Calabria, Arcavacata di Rende (Cosenza), Italy.

Jonathan Cena, Department of Pharmacology, Cardiovascular Research Group, University of Alberta, Edmonton, Canada.

Ava K. Chow, Department of Pediatrics, Cardiovascular Research Group, University of Alberta, Edmonton, Canada.

Shireen-A. Davies, Division of Molecular Genetics, Institute of Biomedical and Life Sciences, University of Glasgow, Glasgow, UK.

Cameron Dezfulian, Vascular Medicine Branch, National Heart Lung Blood Institute, National Institutes of Health, Bethesda, Maryland, USA.

Critical Care Medicine Department, Clinical Center, National Institutes of Health, Bethesda, Maryland, USA.

Division of Pediatric Anesthesia and Critical Care Medicine, Johns Hopkins Hospital, Baltimore, Maryland, USA.

Marco d'Ischia, Department of Organic Chemistry and Biochemistry, University of Naples 'Federico II', Naples, Italy.

Lars Ebbesson, Department of Biology, Bergen University, Thormøhlensgaten 55, Bergen, Norway.

Rosa Maria Facciolo, Comparative Neuroanatomy and Cytology Laboratory, Ecology Department, University of Calabria, Arcavacata di Rende (Cosenza), Italy.

Angela Fago, Department of Biological Sciences, University of Aarhus, Aarhus C, Denmark.

Geoffrey K. Ganter, Department of Biological Sciences, University of New England, Biddeford, Maine, USA.

Alfonsina Gattuso, Department of Cell Biology, University of Calabria, Arcavacata (CS), Italy.

Cecilia Giulivi, Department of Molecular Biosciences, School of Veterinary Medicine, University of California at Davis, Davis, California, USA.

Giuseppina Giusi, Comparative Neuroanatomy and Cytology Laboratory, Ecology Department, University of Calabria, Arcavacata di Rende (Cosenza), Italy.

Mark T. Gladwin, Vascular Medicine Branch, National Heart Lung Blood Institute, National Institutes of Health, Bethesda, Maryland, USA.

Critical Care Medicine Department, Clinical Center, National Institutes of Health, Bethesda, Maryland, USA.

Ralf Heinrich, Department of Neurobiology, Institute for Zoology, Göttingen, Germany.

Bo Holmqvist, Department of Pathology, Lund University, Lund, Sweden.

Sandra Imbrogno, Department of Cell Biology, University of Calabria, Arcavacata (CS), Italy.

Frank B. Jensen, Institute of Biology, University of Southern Denmark, Odense M, Denmark.

Kazunobu Kato, Department of Molecular Biosciences, School of Veterinary Medicine, University of California at Davis, Davis, California, USA.

Andrea B. Kohn, The Whitney Laboratory for Marine Bioscience, University of Florida, St Augustine, Florida, USA.

Sergei Korneev, Sussex Centre for Neuroscience, University of Sussex, Brighton, UK.

Minamaija Lintunen, Department of Biology, Abo Akademi University, Turku, Finland.

Shirley Luckhart, Department of Medical Microbiology and Immunology, School of Medicine, University of California at Davis, Davis, California, USA.

Michael A. Marletta, Department of Molecular and Cell Biology, University of California, Berkeley, California, USA.

Department of Chemistry, University of California, Berkeley, California, USA.

Division of Physical Biosciences, Lawrence Berkeley National Laboratory, University of California, Berkeley, California, USA.

Rosa Mazza, Department of Cell Biology, University of Calabria, Arcavacata (CS), Italy.

Thomas Michel, Cardiovascular Division, Brigham and Women's Hospital, Harvard Medical School, Boston, Massachusetts, USA.

Veterans Affairs Boston Healthcare System, West Roxbury, Massachusetts, USA.

Douglas A. Mitchell, Department of Chemistry, University of California, Berkeley, California, USA.

Leonid L. Moroz, The Whitney Laboratory for Marine Bioscience, University of Florida, St Augustine, Florida, USA.

Department of Neuroscience, University of Florida, Gainesville, Florida, USA.

David B. Morton, Department of Integrative Biosciences, Oregon Health & Science University, Portland, Oregon, USA.

Michael O'Shea, Sussex Centre for Neuroscience, University of Sussex, Brighton, UK.

Anna Palumbo, Zoological Station Anton Dohrn, Naples, Italy.

Pertti Panula, Neuroscience Center, Institute of Biomedicine/Anatomy, Biomedicum Helsinki, University of Helsinki, Helsinki, Finland.

Daniela Pellegrino, Dipartimento Farmaco-Biologico, Università della Calabria, Cosenza, Italy.

Bernd Pelster, Institut für Zoologie, Leopold-Franzens-Universität Innsbruck, Innsbruck, Austria.

Tina Sallmen, Department of Biology, Abo Akademi University, Turku, Finland.

Richard Schulz, Department of Pediatrics, Cardiovascular Research Group, University of Alberta, Edmonton, Canada.

Department of Pharmacology, Cardiovascular Research Group, University of Alberta, Edmonton, Canada.

Sruti Shiva, Vascular Medicine Branch, National Heart Lung Blood Institute, National Institutes of Health, Bethesda, Maryland, USA.

Bruno Tota, Department of Cell Biology, University of Calabria, Arcavacata (CS), Italy.

Anke Vermehren, Department of Integrative Biosciences, Oregon Health & Science University, Portland, Oregon, USA.

Contents

xviii

Note: the colour plate section appears after page 291

On the comparative biology of Nitric Oxide (NO) synthetic pathways: Parallel evolution of NO-mediated signaling

Leonid L. Moroz[1,2,*] and Andrea B. Kohn[1]

[1]The Whitney Laboratory for Marine Bioscience, University of Florida, 9505 Ocean Shore Blvd., St. Augustine, FL 32080-8623, USA
[2]Department of Neuroscience, University of Florida, Gainesville, FL 32611, USA

Abstract. Nitric oxide (NO) is one of the smallest and most diffusible signal molecules known. It can be synthesized in virtually any cell of our body and found in nearly every major group of organisms on our planet. Here, we will discuss several enzymatic and nonenzymatic pathways of NO synthesis in both prokaryotes and eukaryotes including protists, plants and animals. Many of these synthetic mechanisms can coexist within the same cell or cell population. We will also briefly review comparative aspects of NO signaling with a focus on the diversity of NO synthases in invertebrate animals and nonanimal groups. NO-related regulatory mechanisms may be as old as cellular organization itself, so that "ancestral" functions of NO in prokaryotes and basal eukaryotes are likely well preserved across billions of years of biological evolution and can be essential for biomedical studies and clinical applications. On the other hand, NO synthetic pathways might represent examples of parallel evolution in different lineages of organisms.

Keywords: nitric oxide synthesis; evolution; nitric oxide synthase; nonenzymatic NO formation; nitrites; neurons; mollusca; gastropoda; cnidaria; *Aplysia*; arthropoda; insects; bacteria; plant NOS; ascorbate; cambrian; carbon monoxide; hydrogen sulfide; nitrogen fixation; archean; origin of life; denitrification; mitochondria; xanthine oxidoreductase; phylogeny; abiotic NO formation; nitrite photolysis; sea urchin; vertebrates; *Arabidopsis*.

Introduction

Nitric oxide (NO) is one of the smallest and most diffusible signal molecules known. It can be synthesized in virtually any cell of our body and can be found in nearly every major group of organisms on our planet. There have been several enzymatic and nonenzymatic pathways of NO synthesis discovered in both prokaryotes and eukaryotes including protists, plants and animals. Many of these synthetic mechanisms can coexist within the same cell or cell population. Considering the free radical nature of NO and its complex redox chemistry, which might modify practically all classes of biomolecules, it is not surprising that NO can be implicated in almost all biological functions of an organism. This

Corresponding author: Fax: [1]-904-461-4052.
E-mail: moroz@whitney.ufl.edu (L.L. Moroz).

ADVANCES IN EXPERIMENTAL BIOLOGY
VOLUME 01   ISSN 1872-2423
DOI: 10.1016/S1872-2423(07)01001-0

situation is well reflected by the presence of nearly 80,000 publications on NO during the last 20 years with no tendency to decline. Yet, ~99% of the publications within this rapidly growing field of modern biology were obtained from mammalian models. As a result, certain aspects of synthesis, regulation and even classification of NO-dependent mechanisms, although generalized, can be only applied to representatives of one class of vertebrate animals. On the other hand, a number of mechanisms widely distributed across various bacteria, plants and invertebrate animals (*e.g.*, both nonenzymatic and enzymatic NO syntheses from nitrites) have received relatively little attention in mammalian-oriented papers.

In this chapter we will briefly review comparative aspects of NO synthesis and signaling, with a focus on the distribution and function of NO in invertebrate animals and nonanimal groups. Historically, studies of relatively simple organisms have revealed the fundamental principles in more complex systems. This should be true for NO signaling as well. NO-related regulatory mechanisms may be as old as cellular organization itself, so that "ancestral" functions of NO in prokaryotes and basal eukaryotes are likely well preserved across billions of years of biological evolution and can be essential for biomedical studies and clinical applications. On the other hand, NO synthetic pathways might represent examples of parallel evolution in different lineages of organisms.

Gaseous messengers in animals

If we exclude the obvious regulatory functions of oxygen and carbon dioxide, one might ask how many gaseous molecules, in general, are confirmed to be both endogenously produced in cells and tissues and able to perform signaling functions? How different and unique are animals (vs. plants or other eukaryotic groups) with respect to the presence and distribution of these gaseous messengers? And, finally, how unique are vertebrates and mammals in particular (vs. various invertebrate phyla) with respect to gaseous signaling mechanisms?

The concept that gases can be endogenous intercellular messengers in living systems can be traced back to the mid-1900s (Thimann, 1974). It was shown that gaseous ethylene inhibits both growth and geotropism in plants, and ethylene is endogenously produced by fruits (reviewed by Somerville, 2000; Thimann, 1974). Further observations not only confirmed the crucial role of ethylene in plant development, but also provided a detailed analysis of its transduction mechanisms (Benavente and Alonso, 2006; Chang and Bleecker, 2004; Chang and Shockey, 1999; Chow and

McCourt, 2006; Hedrick *et al.*, 2005; van Loon *et al.*, 2006). Surprisingly, there is evidence that in one of the basal metazoan groups – sponges – not only is ethylene produced (Krasko *et al.*, 1999), but it is also involved in the regulation of calcium homeostasis (Perovic *et al.*, 2001; Seack *et al.*, 2001) and can interact with NO-dependent pathways (Muller *et al.*, 2006). Thus, these recent findings might radically change the commonly accepted opinion that signaling functions of ethylene are restricted only to the plant kingdom.

Two other candidates, carbon monoxide (CO) and hydrogen sulfide (H_2S), have also been proposed as gaseous messengers in animal tissues. CO is enzymatically produced from heme by (i) the constitutively expressed enzyme heme oxygenase-2 (HO-2), which is highly expressed in the brain, (ii) the inducible enzyme HO-1 and (iii) as a by-product of lipid perox-idation. CO might act on the same targets as NO (*e.g.*, soluble guanylyl cyclase). Therefore, its functions in neuronal and circulatory systems as well as in other peripheral tissues may be similar but not identical to those of NO because of different regulatory mechanisms and different biological half-lives (see recent reviews in Boehning and Snyder, 2003; Gelperin *et al.*, 2000; Kim *et al.*, 2006; Leffler *et al.*, 2006; Ryter and Otterbein, 2004; Ryter *et al.*, 2004, 2006).

H_2S is produced endogenously in mammalian tissues from L-cysteine metabolism mainly by three enzymes: cystathionine beta-synthetase (CBS), cystathionine gamma-lyase (CSE) and 3-mercaptosulfurtransferase (MST) (Stipanuk *et al.*, 2006). H_2S was first reported as an endogenous neuronal modulator (Abe and Kimura, 1996; Eto *et al.*, 2002; Hosoki *et al.*, 1997; Kimura, 2002; Kimura *et al.*, 2005; Nagai *et al.*, 2004). Later it was implicated in the regulation of vascular tone (as a dilator), in the func-tioning of the gastrointestinal tract and liver, and in the pathogenesis of various cardiovascular diseases and toxicity (Bhatia, 2005; Chahl, 2004; Ebrahimkhani *et al.*, 2005; Fiorucci *et al.*, 2006; Kimura, 2002; Leffler *et al.*, 2006; Li *et al.*, 2006; Tang *et al.*, 2006).

However, the roles and even the presence of biologically active concen-trations of CO and H_2S are still under investigation. Many mechanisms of their action, and their interactions with other regulatory systems, remain open for exploration. On the other hand, the information related to NO signaling is overwhelming and, apart from ethylene in plants, NO is still the major confirmed player in the "class" of gaseous messengers in animals. It should be noted that in many physiologically oriented studies the term NO is used to refer to three forms: (i) the nitrosyl radical ($\bullet$NO itself) and (ii) its nitroxyl (NO^-) and (iii) nitrosonium (NO^+) ions.

The concept of NO signaling in animal physiology

From a historical point of view, the discovery that the radical NO is the major dilatory agent in the circulatory system of humans (Ignarro *et al.*, 1987; Moncada *et al.*, 1991; Palmer *et al.*, 1987) generated a burst of interest in the concept that living systems are able to produce and utilize gaseous species as genuine intra- and extracellular signal molecules.

It was instantly realized by animal physiologists that, in addition to various groups of hormones, signaling peptides and classical neurotransmitters, gaseous molecules are a new and distinct class of chemical messengers with potentially unique types of signaling mechanisms (Boehning and Snyder, 2003; Crawford, 2006; Garthwaite, 2005; Ignarro, 2000; Jacklet, 1997; Lowenstein and Snyder, 1992; Moncada *et al.*, 1991; Moroz, 2000b, 2001; Mur *et al.*, 2006; Nathan, 1992; Palumbo, 2005; see also other chapters in this volume). Specifically, the synthesis of NO does not require any special storage or delivery mechanisms. NO can be released at the place of its initial synthesis and can diffuse across membrane barriers in three-dimensional space, potentially affecting targets that are far from its origin (Lancaster, 1994, 1996, 1997; Thomas *et al.*, 2001; Wood and Garthwaite, 1994). NO primarily exerts its effects through direct covalent binding to target molecules (Bartberger *et al.*, 2002; Donzelli *et al.*, 2006; Gorren and Mayer, 2007; Mancardi *et al.*, 2004; Miranda *et al.*, 2005; Roy and Garthwaite, 2006; Stamler *et al.*, 1992; Thomas *et al.*, 2002; Wink *et al.*, 1996c), resulting in their direct chemical modification [*e.g.*, forming *S*-nitrosothiols or nitrosotyrosine residues in its target proteins (Fukuto *et al.*, 2000; Hess *et al.*, 2005; Miranda *et al.*, 2000; Singel and Stamler, 2005; Stamler *et al.*, 2001)]. It appears that due to its physical and chemical properties, the concept of a "NO microenvironment" with transient NO gradients is a more accurate way to describe the actual situation in living tissues. Finally, being a radical NO is chemically very active; NO gradients are not constant, and they can be very dynamic and restricted to highly localized compartments even within small cells. Endogenous NO concentrations (ranging from 10^{-12} M to 10^{-6} M) are dramatically affected by both the intracellular and the extracellular redox states. Therefore, the biological half-life of NO can be extremely variable (from milliseconds to minutes or even hours) depending on many environmental factors (such as the NO concentration itself, pO_2, the presence of thiols, heme groups or various endogenous scavengers, etc.).

For example, NO oxidation in aqueous solutions is not linear with respect to NO concentration and can be summarized as the following reaction with NO_2^- as the predominant reaction product (Ford *et al.*, 1993; Fukuto *et al.*, 2000; Ignarro *et al.*, 1993; Lewis and Deen, 1994):

$$4NO + O_2 + 2H_2O \rightarrow 4NO_2^- + 4H^+$$

The loss of NO from the reaction is described as:

$$\frac{-d[NO]}{dt} = 4k[NO]^2[O_2]$$

where $k = 2 \times 10^6 \, M^{-2} s^{-1}$ (*i.e.*, this reaction is second order in NO concentration and first order in O_2 concentration). Thus, with $[O_2] = 200 \, \mu M$ it will take $\sim 1 \, min$ for $10 \, \mu M$ NO (a concentration which can be reached locally following inducible NO synthase (iNOS; see below) activation in macrophages) to degrade to $5 \, \mu M$ (or half of the starting level). If the initial concentration of NO is $10 \, nM$ (a level that can be achieved following activation of constitutive NOSs) it will take over $70 \, h$ to degrade to $5 \, nM$ (Fukuto *et al.*, 2000). The maximal NO concentration in solution in equilibrium with headspace gas of pure NO (at 1 atm pressure and 25°C) will be $\sim 1.9 \, mM$ and its degradation to half of its original concentration will take seconds (Shaw and Vosper, 1977). Consequently, physiologically relevant NO concentrations within the range of $100 \, pM$ to $1 \, nM$ will be very little affected by oxidation in solutions and can be maintained for many days.

However, the solubility of NO in hydrophobic solvents is approximately nine times greater than that in water (Shaw and Vosper, 1977) and that of O_2 is approximately threefold greater. As a result, NO might be nine times more concentrated in cell membranes, lipid inclusions and lipoprotein complexes. Therefore, the calculated rate of NO autoxidation in the membrane will be 243 times faster than that in the aqueous phase because of the concentration effect of the reactants (Lancaster, 2000). As a second outcome of this process, NO reacts with various lipid radicals in membranes and can act as a potential inhibitor of lipid peroxidation and low-density lipoprotein oxidation. Surprisingly, nitrated fatty acids not only represent one of the largest pools of nitrogen oxides in vasculature (Baker *et al.*, 2004; Botti *et al.*, 2005; Lima *et al.*, 2005), but they are also recognized as a novel class of bioactive cell-signaling molecules (Baker *et al.*, 2004, 2005; Kalyanaraman, 2004; Schopfer *et al.*, 2005; Wright *et al.*, 2006).

It is evident that in biological tissues NO will react reasonably well with many other biomolecules, such as thiols, metals and their heme-containing proteins and complexes, that can act both as a "NO sink" and/or as NO carriers depending on the cellular microenvironment (Miranda *et al.*,

2000). Furthermore, the reactivity of NO with radicals such as O_2^- is a diffusion-limited process. Additional details about diffusion range and modeling of NO gradients and concentrations are well described elsewhere (Lancaster, 1994, 1996, 1997; Thomas *et al.*, 2001). Importantly, even ubiquitous products of NO-dependent posttranslational modifications (via N-nitrosation, heme-nitrosylation and S-nitrosation) are also highly dynamic. They have fairly short halflives and are linked to tissue oxygenation and redox state; for example, hypoxia induces profound changes in nitrosylation within 1–5 min (Bryan *et al.*, 2004).

Both on the scale of biological evolution and at the levels of an organism, tissue, cell or cell compartment, there is a strong reciprocal, concentration-dependent influence of O_2 and NO on each other's physiological actions. As indicated by Nathan (2004), "NO exerts more control when the concentration of oxygen ($[O_2]$) falls." As a result, NO and reactive nitrogen intermediates [RNI, such as nitrites, $\bullet NO_2$, N_2O_3, N_2O_4, S-nitrothiols, peroxynitrite ($OONO^-$) and dinitrosyl-iron complexes] and their reciprocal relationship with oxygen-reactive intermediates (ROI) provide perfectly tuned and highly localized cellular redox domains and gradients. In this dynamic NO-controlled microenvironment, redox signaling provides a wide-range mechanism that integrates and functionally links different cell compartments. NO "tiers the cell's different commitments to its metabolic budget" (Nathan, 2004). In other words, the apparently nonspecific and widespread chemistry of RNIs and ROIs acts as the universal signaling that integrates the complex cell biochemical machinery and cell energetics. Recent data linking NO functions to the biogenesis of mitochondria and cellular respiration reveals a novel systemic level of integration (Moncada and Bolanos, 2006; Nisoli *et al.*, 2004, 2005; Quintero *et al.*, 2006).

In summary, NO is an extremely dynamic molecule and its behavior in living cells, tissues and organisms must be considered in the context of a specific chemical microenvironment and time (*i.e.*, the biological half-life of NO can be as short as a few milliseconds or as long as days). In some ways NO biology has its own "parallels" with the real-estate business: location, location, location, ... and oxygen.

NO in comparative and evolutionary contexts

NO has a widespread distribution and has been found among practically all animal groups investigated so far (Moroz, 2000b, 2001) as well as in plants, diatoms, slime molds and prokaryotes, where it is involved in countless biological phenomena. Still, the comparative physiology and biochemistry of NO synthetic pathways and NO-mediated transduction pathways even within major invertebrate lineages is poorly described.

For example, among more than 30 extant animal phyla with ~100 classes (see references in Brusca and Brusca, 2003), functional and biochemical data about NO signaling are mostly limited to the classes Mammalia (1 of 12 classes comprising the phylum Chordata) and Insecta (1 of the 13 classes of Arthropoda). As a result, the conditions that caused this versatile and yet potentially toxic molecule to evolve as a modulator of physiological processes are not yet apparent. Even now, evolutionary links between NO synthetic enzymes and specific signaling functions in animal and nonanimal groups are not clear. The fact that different classes of NOSs (genetically unrelated to mammalian NOSs) were recently found in plants (Guo *et al.*, 2003; Zemojtel *et al.*, 2004), and that NO itself was "officially rediscovered" by animal physiologists in prokaryotes, where it not only acts as a transient intermediate in denitrification pathways, but also operates as a potential endogenous regulator of gene expression (Cutruzzola, 1999; Spiro, 2006; Zumft, 2002), suggests that there are deep phylogenetic roots for the development of numerous signaling pathways.

Taken together, these data point out that NO-related regulatory systems may be as old as cellular organization itself, tracing their origin to the very dawn of biological evolution ~3.8–3.5 billion years ago (Gya). Furthermore, the evidence that conditions were favorable for NO synthesis and accumulation on the anoxic primitive Earth (during the Hadean and Archean eras, from 4.5 Gya to 2.5 Gya) in the ancient atmosphere and early oceans makes NO an important environmental factor contributing to the origin of life itself (Nna-Mvondo *et al.*, 2001, 2005). It is suggested that NO was a crucial intermediate in the utilization of chemically inert molecular nitrogen (N_2). Indeed, nitrogen is an essential element for life and, although the Earth's atmosphere is the major reservoir of dinitrogen, most organisms are not able to use it directly because of the high activation energy ($948 \, kJ \, mol^{-1}$) required to dissociate the nitrogen triple bond ($N \equiv N$). Thus, in order to be used by various organisms, nitrogen must be "fixed" in either reduced (ammonia, cyanide, acetonitrite) or oxidized (N_2O, NO, nitrites or nitrates) form. In fact, NO was the principal form of fixed nitrogen in the early Earth. It was produced highly efficiently in a neutral Hadean and Archean atmosphere [composed primarily of CO_2 and N_2 (Kasting, 1992), under which conditions HCN and related compounds are not synthesized] by lightning and other electrical discharges associated with volcanic eruptions, meteorite impacts and thunderstorms, with an estimated accumulated annual production rate of the order of $\sim 10^{13}$ g per year [the global production of NO by lighting at the present time is estimated as $\sim 10^{12}$ g per year (Wang *et al.*, 1998)]. This NO would have been converted into nitric and nitrous acids and delivered to the ancient ocean and early lithosphere as acid rain. NO and related species were then reduced by various minerals [primarily Fe(II), which was present in earlier oceans at much higher concentrations than today] to produce NH_4^+ (Brandes et al., 1998; Summers and Chang, 1993), thereby providing a prebiotic mechanism for nitrogen fixation on the early Earth and, possibly, for the accumulation of ammonia in localized areas.

These steps are required for further synthesis and buildup of more complex organic nitrogen-containing molecules and should be a factor in the origin of life on Earth. At later times, during the Archean eon (3.8–2.5 Gya), NO production by lighting discharge decreased (due to the decrease in CO_2 concentration) by two orders of magnitude until ~2.2 Gya. After this time, the rise in oxygen (or methane) concentrations probably initiated other abiotic sources of nitrogen (see also Kasting and Howard, 2006; Kasting and Ono, 2006; Kasting and Siefert, 2001). It was hypothesized (Navarro-Gonzalez et al., 2001) that although the temporal Archean reduction in NO production may have lasted for only 100 Myr (10^8 years) or less, this was potentially long enough to cause the ecological crisis that triggered the development of biological nitrogen fixation (again with NO as an important intermediate, and possible signal molecule). These new enzymatic mechanisms (nitrite/nitrate reductases) had emerged in some prokaryotic cells, providing the foundation for the global nitrogen cycle in the biosphere and representing ~70% of the total quantity of fixed nitrogen today. Thus, as stated by Zumft (1993), "NO is not an obscure chemical and certainly no newcomer to the life sciences, as often stated in hyperbole. Early in evolution NO took its role as a central player in bacterial bioenergetics and the global N cycle vital to all organisms."

At this point, we would like to outline different mechanisms of NO synthesis in biological systems, including multiple nonenzymatic and enzymatic pathways suggesting parallel and, possibly, convergent evolution, as well as the recruitment of NO synthetic pathways and different classes of NOSs.

Multiplicity of NO synthetic pathways

At least six pathways of enzymatic NO synthesis can be found in living systems:

- (i) classical multi-domain NOS-type enzymes (Ghosh and Salerno, 2003; Griffith and Stuehr, 1995) in animals and slime molds;
- (ii) prokaryotic "truncated" NOS found in a majority of bacteria and some archaea (Zemojtel et al., 2003);
- (iii) plant-type NOS distantly related to GTPases (Zemojtel et al., 2004);

(iv) prokaryotic and eukaryotic enzymes involved in the reduction of nitrites as part of the nitrogen cycle (Philippot, 2002) (Fig. 1);
(v) nitrite reduction by mitochondrial enzymes (Kozlov et al., 1999; Nohl et al., 2005; Reutov, 2002; Tischner et al., 2004);
(vi) xanthine oxidoreductase reduction of nitrites (Doel et al., 2001; Li et al., 2003, 2005; Millar et al., 1998; Zhang et al., 1998).

Table 1 summarizes the current status of molecular cloning and identification of major NOS-related proteins and Figs. 2, 4–6 illustrate phylogenetic relationships between different classes of NOSs.

In addition to these six pathways, NO can be produced nonenzymatically from nitrites in various cells and tissues under certain chemical conditions using various redox mechanisms. Deoxy-hemoglobin/myoglobin (Cosby et al., 2003; Nagababu et al., 2003) and even endothelial NOS (eNOS) (Gautier et al., 2006) can also reduce nitrites and recycle bioactive NO under hypoxic/anoxic conditions. Below, we will primarily discuss nonenzymatic NO formation.

Abiotic reduction of nitrites results in nonenzymatic NO formation

The requirement of an enzyme for NO synthesis in biological systems is not absolute. The fact that NO can be generated nonenzymatically (without NOS or a denitrification pathway; see below) from a nitrite solution is crucial for any physiological, comparative and evolutionary analysis of NO synthesis and signaling. In reality, abiotic NO formation

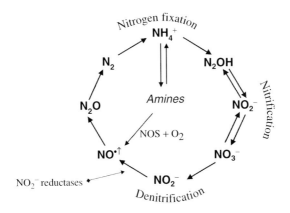

Fig. 1. Nitrogen cycle: enzymatic NOS-independent synthesis.

Table 1. Molecular identification of nitric oxide synthase (NOS) related proteins in the animal kingdom, non-animal eukaryotes and prokaryotes[a].

Taxon species	Abbreviation	GenBank accession no.	Length	Comments	Exon no.	Genomic information	Reference
Deuterostomes							
Phylum Chordata							
Class Mammalia							
Homo sapiens				3 NOS genes			Hall *et al.* 1994
	hNOSn	NP_000611	1434 aa	nNOS: constitutive; Ca-dependent, NOS1	29	12q24.2-12q24.31; GeneID 4842	
H. sapiens	hNOSe	NP_000594	1203 aa	eNOS: constitutive; Ca-dependent, NOS3	27	7q36: GeneID 4846	
H. sapiens	hNOSi	NP_000616	1153 aa	iNOS:Ca-independent NOS2A	27	17q11.2-17q12; GeneID 4843	
Pan troglodytes	PtNOSn	XP_522539	1433 aa	nNOS, NOS1	42	Chromosome 12; GeneID 467139	
P. troglodytes	PtNOSe	XP_519525	1323 aa	eNOS, NOS3	22	Chromosome 7; GeneID 463893	
P. troglodytes	PtNOSi	XP_511794	652 aa	iNOS: partial sequence, NOS2A		Chromosome 17 (not complete)	
Class Aves (birds)				2 NOS genes			
Gallus gallus	GgNOSn	XP_425296	1609 aa	nNOS, NOS1	29	Chromosome 15; GeneID 427721	
G. gallus	GgNOSi	NP_990292	1136 aa	iNOS, NOS2	28	Chromosome 19; GeneID 395807	
Class Reptilia							
Aspidoscelis uniparens (lizard)	AuNOSn	AAZ76558	718 aa	nNOS-like; partial sequence; EST			
Class Amphibia				1-3 NOS genes			
Xenopus tropicalis	XtNOSn	ENSXETG0000022354	1416 aa	nNOS, NOS1	27	GeneID 373705	Scheinker *et al.* 2003
X. tropicalis	XtNOSe	ENSXETT0000025059	1138 aa	eNOS, NOS3	26		
X. laevis	XlNOSn	AAD55136	1419 aa	nNOS, NOS1			
X. laevis NOS3	XlNOSe	AW765292, AW764664	ESTs	eNOS-like , NOS3			
Class Actinopterygii				1-3 NOS genes			
Danio rerio	DrNOSn	NP_571735	1431 aa	nNOS, NOS1	18	GeneID 60658,	Poon *et al.* 2003; Holmqvist *et al.* 2000
D. rerio	DrNOSai	XP_692454	1145 aa	iNOS, NOS2A	28	Chromosome 15; GeneID 564002	

Taxon	Abbreviation	Accession	Length	NOS type	No.	Notes	Reference
D. rerio	DrNOSbi	XP_692103	1130 aa	iNOS, NOS2B	26	Chromosome 15. GeneID 563654	
Tetraodon nigroviridis	TnNOSn	CAG08158	1429 aa	nNOS, NOS1		Chromosome 12: 1 NOS gene in genome	Wilson *et al.*
Takifugu rubripes	TrNOSn	AAL82736	1418 aa	nNOS, NOS1			Yamamoto and Suzuki. 2004
Takifugu poecilonotus	TpNOSn	AAM46138	1418 aa	nNOS, NOS1			
Oryzias latipes	OlNOSn	BAD11808	1424 aa.	nNOS, NOS1			Rose *et al.* 2005
Fundulus heteroclitus	FhNOSn	AAS21300	1420 aa	nNOS, NOS1			Reddick *et al.* 2005
Carassius auratus	CaNOSi	AAX85387	1127 aa	iNOS, NOS2a			Reddick *et al.* 2005
Carassius auratus	CaNOSi	AAX85386	1126 aa	iNOS, NOS2b			Saeij *et al.* 2000
Cyprinus carpio	CcNOSi	CAB60197	1137 aa	iNOS, NOS2			Wang *et al.* 2001
Oncorhynchus mykiss		CAC83069	1083 aa	iNOS			Wang *et al.* 2001
O. mykiss		CAC82808	1100 aa	iNOS			McNeill and Perry (2006)
O. mykiss		ABG24205	761 aa	nNOS; partial sequence			
Class Chondrichthyes							
Scyliorhinus canicula	ScNOSi	AAX85385	1125 aa	iNOS, NOS2			Reddick *et al.* 2005
Order Petromyzontiformes							
Petromyzon marinus	PmNOSn	EB720322, EB083238	ESTs	nNOS-like, NOS1			
Class Cephalochordata							
Branchiostoma floridae	BfNOS	AAQ02989	1332 aa	nNOS-like			Panchin, Sudreyev, and Moroz. 2001
Branchiostoma floridae				NOS-like, partial sequence			Panchin and Moroz. unpublished
Class Urochordata (tunicates)							
Ciona savignyi	CsNOS	ENSCSAVG0000009725	1128 aa	NOS-like	25		
C. intestinalis	CiNOS	ENSCING0000002710	1379 aa	NOS-like	32	10q31	
Phylum Hemichordata							
Balanoglossus sp.				nNOS-like, partial			Panchin *et al.* unpublished
Phylum Echinodermata							
	SpNOS	XP_781703	1488 aa		23		Hibino *et al.* 2006

Table 1 (*Continued*)

Taxon species	Abbreviation	GenBank accession no.	Length	Comments	Exon no.	Genomic information	Reference
Strongylocentrotus purpuratus				NNOS-like All sequences for sea urchin obtained at: http://annotation.hgsc.bcm.tmc.edu/Urchin cgibin pubSearchGene.cgi		GeneID: SPU_002328/025118	
S. purpuratus	SpNOS	XM_001179342	1385 aa	nNOS-like		GeneID:587111	Hibino *et al.*, 2006
S. purpuratus	SpNOS	XM_792802	442 aa	Truncated NOS-like Flavodoxin and oxygenase domains		SPU_013373	
S. purpuratus	SpNOS	XM_792378	539 aa	Truncated NOS-like.Oxidoreductase NAD-binding and FAD binding domains			
		SPU_019970	Hibino *et al.*, 2006				
Arbacia punctulata	ApNOS	AF191751 AF191750	ESTs	nNOS-like			Cox *et al.*, 2001
Phylum Arthropoda **Class Crustacea (Decapoda)**							
Gecarcinus lateralis (crab)	GlNOS	AAT46681	1199 aa				Kim *et al.*, 2004
Homarus americanus (lobster)	HaNOS	CN853572	EST	NOS-like			
Class Insecta							
Tribolium castaneum[a]	TcNOS	XP_967195	1105 aa	NOS-like	13	Chromosome LG9: GeneID 655549	
Rhodnius prolixus	RpNOS	Q26240	1174 aa	NOS-like			Yuda *et al.*, 1996
Bombyx mori	BmNOS	BAB85836	1209 aa	NOS-like			Imamura *et al.*, 2002
Manduca sexta	MsNOS	AAC61262	1206 aa	NOS-like			Nighorn *et al.*, 1998
Apis mellifera	AmNOS	NP_001012980	1143 aa	NOS-like	25	Chromosome LG3: GeneID 503861	Watanabe *et al.*, 2005
Drosophila melanogaster	DmNOS	NP_523541	1349 aa	NOS-like	19	Chromosome 2L: Location 32B1: GeneID 34495	Regulski and Tully, 1995: Stasiv *et al.*, 2001
D. pseudoobscura	DpNOS	EAL33128	1348 aa	NOS-like			

Species	Gene	Accession	Size	Description		Notes	References
Anopheles gambiae	AgNOS	XP_317213	1113 aa	NOS-like		Chromosome 3R: GeneID 1277726	Luckhart and Rosenberg, 1999
Anopheles stephensi	AsNOS	O61608	1247 aa	NOS-like			
Aedes aegypti	AaNOS	EAT38354	1112 aa	NOS-like			
Acheta domesticus	AdNOS	AAR88326	210 aa	Partial sequence			
Gryllus bimaculatu	GbNOS	BAE66755	124 aa	Partial sequence			
Glossina morsitans	GmNOS	AY152725, DV603147	429 bp	NOS-like; ESTs, Partial sequence			
Phylum Mollusca							
Class Gastropoda							
Aplysia californica	AcNOS1	AAK83069	1387 aa			Genome not completed	Sadreyev et al., 2001; Moroz et al., unpublished
A. californica	AcNOS2	AAK92211	1175 aa		>20	Genome not completed	Sadreyev et al., 2003; Moroz et al., unpublished
Lymnaea stagnalis	LsNOS1	O61309	1153 aa	NOS-like (2 genes)			Korneev et al., 1998
L. stagnalis	LsNOS2	AAW88577	1218 aa				Korneev et al., 2005
Lymnaea sp.		AAM21319	397 aa	NOS-related protein; partial sequence			Korneev and O'Shea, 2002
Limax marginatus	LmNOS	BAC80150	209 aa	NOS-like; partial sequence			
Ilyanassa obsoleta	IoNOS	AAV31753	77 aa	NOS-like; partial sequence			
Class Cephalopoda							
Sepia officinalis	SoNOSa	AAS93626	1133 aa	NOSa-like			Scheinker et al., 2005
S. officinalis	SoNOSb	AAS93627	1139 aa	NOSb-like			
Phylum Nematoda							
Caenorhabditis elegans				Not found in genome		NOS gene loss in this lineage	
C. briggsae				Not found in genome		NOS gene loss in this lineage	
Phylum Platyhelminthes (Flatworms)		No data		Histochemical NADPHd labeling of putative NOS			
Phylum Cnidaria							
Discosoma striata	DsNOS	AAK61379	1115 aa	NOS-like			Panchin, Sadreyev and Moroz, 2001; unpublished

Table 1 (*Continued*)

Taxon/species	Abbreviation	GenBank accession no.	Length	Comments	Exon no.	Genomic information	Reference
Hydra vulgaris	HvNOS	DY449349	EST	NOS-like			
Hydra magnipapillata	HmNOS	DT608891, DT607441, DR434830, CN776807	ESTs	NOS-like: 2 distinct genes. Partial sequence			
Phylum Porifera (sponges)							
Reniera sp.		No data		EST; genome			
Phylum Placozoa							
Trichoplax adhaerens	TaNOS	Partial clone					Moroz, unpublished
Other Eukaryotes							
Mycetozoa							
Physarum polycephalum	PpNOSa	AAK43730	1055 aa	NOSa-like			Golderer *et al.*, 2001
P. polycephalum	PpNOSb	AAK43729	1046 aa	NOSb-like			Golderer *et al.*, 2001
Fungi							
Aspergillus oryzae	AoOx	BAE64541	178 aa	Not found NOS-like		NOS oxygenase domain, 5'-end	
Prokaryotes: bacteria							
Staphylococcus epidermidis	SeOx	NP_765153	355 aa	Prokaryotic NOS-like		NOS oxygenase domain	
S. saprophyticus	SsOx	YP_300967	354 aa	Prokaryotic NOS-like		NOS oxygenase domain	
S. aureus	SaOx	P0A092	358 aa	Prokaryotic NOS-like		NOS oxygenase domain	
Oceanobacillus iheyensis	OiOx	NP_693612	369 aa	Prokaryotic NOS-like		NOS oxygenase domain	
Deinococcus radiodurans	DrOx	Q9RR97	356 aa	Prokaryotic NOS-like		NOS oxygenase domain	
D. geothermalis	DgOx	YP_603740	375 aa	Prokaryotic NOS-like		NOS oxygenase domain	
Bacillus subtilis	BsOx	O34453	336 aa	Prokaryotic NOS-like		NOS oxygenase domain	
B. halodurans	BhOx	NP_241689	366 aa	Prokaryotic NOS-like		NOS oxygenase domain	

B. anthracis	BaOx	ZP_00390385	356 aa	Prokaryotic NOS-like	NOS oxygenase domain
B. thuringiensis	BtOx	ZP_00741647	215 aa	Prokaryotic NOS-like	NOS oxygenase domain
B. thuringiensis	BtOx2	YP_039435	356 aa	Prokaryotic NOS-like	NOS oxygenase domain
B. clausii	BcOx	YP_174766	363 aa	Prokaryotic NOS-like	NOS oxygenase domain
B. licheniformis	BlOx	YP_090413	365 aa	Prokaryotic NOS-like	NOS oxygenase domain
B. cereus	BceOx	AAU20271	440 aa	Prokaryotic NOS-like	NOS oxygenase domain
Geobacillus kaustophilus	GkOx	YP_147529	440 aa	Prokaryotic NOS-like	NOS oxygenase domain
Exiguobacterium sibiricum	EsOx	ZP_00539087	366 aa	Prokaryotic NOS-like	NOS oxygenase domain
Streptomyces avermitilis	SavOx	NP_82706	516 aa	Prokaryotic NOS-like	NOS oxygenase domain
S. turgidiscabies	StOx	AAW49313	400 aa	Prokaryotic NOS-like	NOS oxygenase domain
S. scabiei	SscOx	AAO53225	400 aa	Prokaryotic NOS-like	NOS oxygenase domain
Archaea (halobacteria)					
Natronomonas pharaonis	NpOx	CAI49045	378 aa	Prokaryotic NOS-like	NOS oxygenase domain
Different class of putative NOS-like proteins					
Class Mammalia					
Homo sapiens	Human	BAC05262	623 aa	Novel NOS	
Mus musculus	Mouse	Q99LH1	693 aa	Novel NOS	
Class *Actinopterygii*					
Danio rerio	Danio	Xp_693129	702 aa	Novel NOS	
Phylum Nematoda					
Caenorhabditis elegans	Caenorhabditis	NP_001045614	388 aa	Novel NOS	
Phylum					
Dictyostelium discoideum	Dictyostelium	XP_640529	650 aa	Novel NOS	
Class Insecta					
Tribolium castaneum	Tribolium	XP_967515	709 aa	Novel NOS	

16

Table 1 (*Continued*)

Taxon/species	Abbreviation	GenBank accession no.	Length	Comments	Exon no.	Genomic information	Reference
Apis mellifera	Apis	XP_396974	533 aa	Novel NOS			
Drosophila melanogaster	Drosophila	NP_611297	624 aa	Novel NOS			
Anopheles gambiae	Anopheles	XP_319462	536 aa				
Aedes aegypti	Aedes	EAT40167	689 aa	Novel NOS			
Phylum Echinodermata							
Strongylocentrotus purpuratus	Strongylocentrotus			XP_782008	564 aa	Novel NOS	
Phylum Mollusca							
Class Gastropoda							
Aplysia californica	Aplysia	EF043280	712 aa	Novel NOS			
Helix pomatia	Helix	CAA65719	433 aa	Novel NOS			
Planta							
Arabidopsis thaliana	Arabidopsis	NP_190329	561 aa	Novel NOS			
Oryza sativa	Oryza	NP_001045614	547 aa	Novel NOS			
Prokaryotes: bacteria							
Lactobacillus plantarum WCFS1	Lactobacillus	NP_785132|	378 aa	Novel NOS			
Pediococcus pentosaceus ATCC 25745	Pediococcus	ZP_00323186	370 aa	Novel NOS			
Bacillus anthracis str. A2012	Bacillus	ZP_00394666	368 aa	Novel NOS			
Staphylococcus aureus subsp. *aureus* NCTC	Staphylococcus		366 aa	YP_500210		Novel NOS	
Streptococcus pneumoniae R6	Streptococcus	NP_359186	368 aa	Novel NOS			

aa, amino acids. bp, base pairs. EST, expressed sequence tag.
Abbreviations and GenBank accession numbers for all trees.
[a]The most up-to-date references about individual NOS sequences can be found at http://www.ncbi.nlm.nih.gov using GenBank accession numbers.

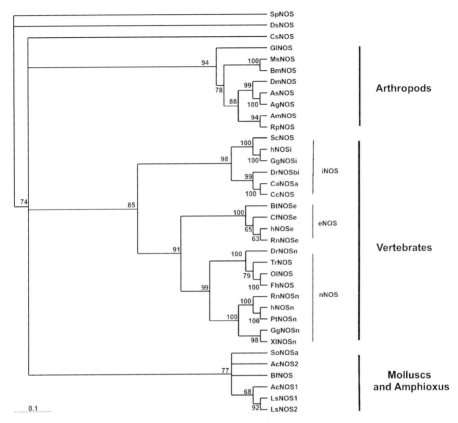

Fig. 2. Rooted tree for animal NOSs (see Table 1). Full-length amino acid sequences were aligned in "ClustalX" version 1.83 using default parameters (Jeanmougin *et al.*, 1998); all gaps were manually removed in "GeneDoc" (Nicholas *et al.*, 1997). The phylogentic tree was generated using default parameters and 10,000 iterations of the maximum likelihood algorithm implemented in the program "TREE-PUZZLE" (http://www.tree-puzzle.de). Numbers at branches represent bootstrap values for 10,000 iterations. Branch-length scale bar represents 0.1 amino acid substitutions per site. The graphic output was generated using "Treeview" (Page, 1996). See abbreviations for species and gene bank accession numbers for NOSs in Table 1; iNOS, vertebrate inducible like NOS; eNOS, vertebrate endothelial like NOS; nNOS, vertebrate neuronal-like NOS.

is well known from the chemistry of NO_x species, but until recently it was not considered to be an endogenous source of NO in animals (Lundberg and Govoni, 2004; Lundberg and Weitzberg, 2005; Lundberg *et al.*, 2004; Suschek *et al.*, 2006; Weitzberg and Lundberg, 1998; Zweier *et al.*, 1999).

The chemistry of NO oxidation is very complex, with many transient nitrogen/oxygen species (Kharitonov *et al.*, 1994, 1995; Saran and Bors, 1994; Wink and Mitchell, 1998; Wink *et al.*, 1996a,b), where nitrites (NO_2^-) and nitrates (NO_3^-) are major sequential products of NO oxidation. In tissues, hemoproteins convert NO and NO_2^- to NO_3^- (Grisham *et al.*, 1996; Ignarro *et al.*, 1993). However, NO_2^- is the only stable product formed by the spontaneous oxidation of NO in oxygenated solutions (Ignarro *et al.*, 1993; Kharitonov *et al.*, 1994; Lewis and Deen, 1994).

In an acidic environment nitrites are easily converted to NO according to the equation:

$$NO_2^- + 2H^+ \rightarrow NO\bullet + H_2O$$

Again, the actual mechanisms are more complicated (Butler *et al.*, 1995; Feelisch and Stamler, 1996; Kharitonov *et al.*, 1994; Lewis and Deen, 1994; Saran and Bors, 1994; Wink and Mitchell, 1998), but without describing secondary pathways, the sequence can be presented as follows:

$$2NO_2^- + 2H^+ \rightleftharpoons 2HONO\ (pK_a \sim 3.2 - 3.4) \rightarrow H_2O + N_2O_3$$
$$\rightarrow NO\bullet + NO_2 + H_2O$$

Ascorbate (Asc) (Archer, 1993; Dahn *et al.*, 1960; Mirvish *et al.*, 1972; Yamasaki, 2000) and some reducing compounds, such as NADPH, L-cysteine, reduced glutathione and other thiols (Feelisch, 1993; Feelisch and Kelm, 1991; Feelisch and Noack, 1987; Scorza and Minetti, 1998; Scorza *et al.*, 1997), have been reported to stimulate NO formation from nitrite. For example:

$$2HNO_2 + Asc \rightarrow 2NO + DHAsc + 2H_2O$$

where DHAsc is dehydroascorbate.

Nonenzymatic NO production from dietary nitrates *in vivo* was originally demonstrated in the gut (Benjamin *et al.*, 1994; Lundberg *et al.*, 1994) and in the human oral cavity (Duncan *et al.*, 1995). In both cases NO concentrations were sufficient to be involved in primary antimicrobial, nonimmune defense reactions (Allaker *et al.*, 2001; Anyim *et al.*, 2005; Dykhuizen *et al.*, 1996, 1998; Lundberg *et al.*, 2004; Phillips *et al.*, 2004; Silva Mendez *et al.*, 1999) and, probably, in the control of digestive functions, such as mucosal blood flow, motility and possibly secretion and absorption (Duncan *et al.*, 1997). Estimated nitrite concentrations were in the range of 0.1–1000 μM in different parts of the digestive system, approaching a concentration of 1 mM in saliva following a high-nitrate/nitrite test meal (Duncan *et al.*, 1995; McKnight *et al.*, 1997).

Similarly, nonenzymatic NO formation has been demonstrated in human skin (Weller *et al.*, 1996) and urine (Carlsson *et al.*, 2001; Lundberg *et al.*, 1997), with a suggested physiological role in the inhibition of infection by pathogenic microscopic fungi (Anyim *et al.*, 2005), as well as in the modulation of cutaneous T-cell function, skin blood flow and keratinocyte differentiation (Vallette *et al.*, 1998). Evidently, nitrites formed in the skin can act as important bacteriostatic agents and be directly involved in wound healing (Benjamin *et al.*, 1997). Similarly, nonenzymatically generated NO and related species in the digestive tract can be both toxic for potential pathogens and able to perform regulatory functions such as control of circulation, uptake mechanisms and modulation of muscle contractions in the gut. The maintenance of a very acidic stomach pH favors the reduction of nitrites from various food sources and supposedly would be essential for ecological adaptations in herbivore species. The combination of low pH and high nitrite levels (~1 mM) results in NO levels that can sometimes exceed 4 μM (*i.e.*, more than 10,000 times higher than the levels required for vasodilatation) (Lundberg *et al.*, 2004).

Nonenzymatic NO formation from nitrites also plays an important role in vascular control (Lundberg and Weitzberg, 2005). Large quantities of NO (similar to, or even higher than, those produced by NOS) can be formed in ischemic heart tissues by a mechanism that is not enzyme dependent and not blocked by inhibitors of NOS (Zweier *et al.*, 1995a,b). Zweier *et al.* concluded that this NO formation is a consequence of acidification (pH~5.5), which serves to reduce the large pool of nitrites present within tissue. The mean nitrite concentration of the ischemic myocardium (12 μM) is sufficient to generate the detected amount of NO. It was concluded that enzyme-independent NO formation not only contributes to the process of postischemic injury, but also eliminates the protective effect of NOS inhibitors (Zweier *et al.*, 1995a).

Thus, tissue nitrite can serve as a significant "hypoxic buffer" of NO that can be released during hypoxia when oxygen-dependent NOS activity is suppressed or even eliminated (Gladwin *et al.*, 2006). It is important that the half-life ($t_{1/2}$) of nitrites in biological tissues is significantly longer than that of NO. For example, in blood/plasma under physiological conditions, the $t_{1/2}$ for nitrites can be 1–5 min while that for NO is 1–2 ms. It was also shown that NO generation from nitrites under physiological conditions and induced aortic ring relaxation is further increased by ascorbate (Modin *et al.*, 2001).

Conditions for intracellular nonenzymatic NO formation: Neurons as models

Nonenzymatic NO formation can be substantial inside NOS-containing cells, particularly NOS-containing neurons. First, the intracellular nitrite concentrations can be significantly higher than those estimated from plasma or tissue homogenates. Estimations from homogenates of whole neuronal tissues (Salter *et al.*, 1996) give values of 50–200 µM, which are substantially higher than those in the ischemic heart, but only 1–2% of neurons expressed NOS. Our recent direct single-cell measurements (Moroz *et al.*, 2005) indicate that actual intracellular nitrite levels can be as high as 1–5 mM, more than enough to produce NO nonenzymatically. Furthermore, the distribution of NO_2^- is not uniform, but instead appears to be neuron-specific, with higher concentrations being found in neurons expressing NOS. Yet, some NOS-negative neurons might also have relatively high nitrite levels, possibly due to extensive synaptic inputs from NOS-containing presynaptic neurons (Moroz, 2000a, 2005).

Most importantly, the brain has almost the highest ascorbate level in the human body (Hornig, 1975; Oke *et al.*, 1987). Up to 400–600 µM of L-ascorbate has been detected extracellularly (Miele and Fillenz, 1996; Rice and Nicholson, 1987). Intracellular concentrations are estimated at 2–5 mM in neurons (Grunewald, 1993) and $\geqslant 7$ mM in glial cells (Siushansian and Wilson, 1993). Direct single-cell measurements of intracellular ascorbate concentrations using capillary electrophoresis have confirmed millimolar levels of ascorbate in neurons (Kim *et al.*, 2002; Zhang *et al.*, 2002). Ascorbate accumulation differs among various neuronal groups and can be modified by electrical activity and neurotransmitters (Grunewald, 1993). Thus, we hypothesize that either intracellular ascorbate or ascorbate released from neurons during activity can generate NO from a conventional chemical reduction of intra- and extracellular nitrite ions. It has been suggested that extracellular ascorbate-induced NO production could subtly and accurately match oxygen transport to the local metabolic demands of the nerve cells by vasodilation of cerebral blood vessels under hypoxic conditions (Millar, 1995).

Are physiological variations in pH sufficient to induce nonenzymatic NO release from nitrites in nervous tissues? pH homeostasis in nervous tissue has been studied in some detail; various transporters such as Na^+/H^+, Na^+-dependent Cl^-/HCO_3^- and Ca^{2+}/H^+ exchangers, as well as intra- and extracellular carbonic anhydrase, are involved in pH regulation. Moreover, V-ATPases (vacuolar-type H^+-ATPases) translocate H^+ into vacuoles and synaptic vesicles, providing further avenues for variations in intracellular pH.

Reported physiological pH variations in selected intracellular compartments are very broad, and the range of intracellular pH changes can be
greater that 2 units (Mellman, 1992). The pH values of lysosomes, synaptic
vesicles and some other organelles can be as low as 4.5–5 (Kornfeld and
Mellaman, 1989). It is also known that the release of the synaptic vesicles
results in acidification (to a pH value of ~1) of the synaptic cleft. Based
on model chemical measurements, these pH values, in combination with
the measured nitrite/ascorbate levels, are more than sufficient to generate
high micromolar concentrations of NO non-enzymatically in neurons.

We suggest that a similar situation can also be found in other cell types
(e.g., secretory cells) as well as in many physiological processes such as
development, phagocytosis, etc.

One of the unexpected findings related to non-enzymatic NO synthesis
was the demonstration that "classical" NO inhibitors can produce NO
nonenzymatically (Moroz et al., 1998). While NO is generated enzymatically by NO synthase (NOS) from L-arginine, overproduction of NO
contributes to cell and tissue damage as sequelae of infection and stroke.
Strategies to suppress NO synthesis rely heavily on guanidino-substituted
L-arginine analogs (L-NAME, L-NA, L-NMMA, L-NIO) as competitive
inhibitors of NOS; these are often used in high doses to compete with
millimolar concentrations of intracellular arginine. Surprisingly, these
analogs are also a source for nonenzymatically produced NO. Enzyme-
independent NO release occurs in the presence of NADPH, glutathione,
L-cysteine, dithiothreitol and ascorbate (Moroz et al., 1998). This nonenzymatic synthesis of NO can produce potentially toxic, micromolar
concentrations of NO and can oppose the effects of NOS inhibition. NO
production driven by NOS inhibitors was also demonstrated *ex vivo* in
the central nervous and peripheral tissues of the gastropod molluscs
Aplysia and *Pleurobranchaea* using electron paramagnetic resonance and
spin-trapping techniques (Moroz et al., 1998).

Nonenzymatic NO formation: Nitrite photolysis

Potentially, nitrite photolysis is another nonenzymatic, but biologically
relevant, chemical process associated with NO signaling. Photochemical
generation of NO from nitrites is a notable component of the nitrogen
cycle in the Earth's biosphere. Nitrite absorbs maximally at 356 nm, and
the process can be presented as follows (Zafiriou et al., 1980):

$$NO_2^- + H_2O + hv \ (295 \ nm \leqslant \lambda \leqslant 410 \ nm) \rightarrow NO\bullet + \bullet OH + OH^-$$

This reaction occurs naturally in the surface layers of the world's oceans, predominantly in the central equatorial areas. NO formed by this mechanism may play an important role in marine ecosystems, and must be considered as a NO source to the atmosphere (Zafiriou et al., 1980). The measured partial pressure of NO in the air was less than 8×10^{-12} atm, compared with a P_{NO} calculated for surface seawater of 7×10^{-8} atm (McFarland et al., 1979). The estimated concentrations of NO in the surface film of tropical waters depend strongly on the nitrite distribution; with subnanomolar or nanomolar quantities during the day, dropping to a practically undetectable level after sunset. Nitrites (and, likely, NO) are important endogenous regulators of the biological clocks in planktonic organisms such as the unicellular dinoflagellate Gonyaulax polyedra (Roenneberg and Rehman, 1996). We might therefore speculate that the involvement of NO in the regulation of circadian rhythm, observed in higher animals (Artinian et al., 2001; Ding et al., 1994), might be traced back to the earlier day–night conditions in the ancient oceanic waters.

One practical aspect of nitrite photolysis is UV-induced generation of NO in human skin (reviewed by Suschek et al., 2006). Even a short 3–5 min exposure to sun in the Central European summer leads to significant nonenzymatic NO formation from nitrites ($\sim$10 µM in sweat) or NO-thiols present in normal human skin (Paunel et al., 2005; Suschek et al., 2005). This UV-induced cutaneous NO formation was comparable to or higher than that found in maximally activated human keratinocytes (Bruch-Gerharz et al., 2003). Suschek et al. (2006) have calculated that human skin can be "the largest storage organ for NO derivatives such as nitrites and RSNOs" (S-nitrosothiols) and "non-enzymatic NO generation might represent an initial screening function in human skin." As indicated above, NO can regulate skin pigmentation, growth and differentiation as well as perform antifungal and antibacterial protective functions. Clearly, this type of mechanism of surface pH/UV-dependent NO formation can be widely distributed across animal and plant kingdoms and may be an important factor in various ecological adaptations and development.

In summary, the last decade of research in the field of NO biology has revealed novel physiological functions of abiotic nonenzymatic NO synthesis from nitrites. Furthermore, the biological role of nitrite itself has evolved from an inert by-product of organic nitrate metabolism or NO oxidation into an endogenous signaling molecule and regulator of gene expression (Bryan, 2006; Bryan et al., 2005; Gladwin, 2005; Gladwin et al., 2005). Indeed, recent findings clearly indicate that certain functions of nitrites are unique and cannot be affected by NO scavengers, i.e., they are not mediated by NO (Bryan et al., 2005). Thus, nitrite can act as an independent messenger in the family of inter- and intracellular signal molecules.

A brief overview of the diversity of conventional nitric oxide synthases

In all animal tissues the enzymatic synthesis of NO proceeds according to the following reaction:

$$\text{L-Arginine} + O_2 \rightarrow NO + \text{L-citrulline}$$

Three groups of NOS isoforms catalyze NO synthesis in mammalian tissues (Fig. 2). All of them have been characterized biochemically and cloned from several mammalian species with numerous splicing variants (Griffith and Stuehr, 1995; Roman et al., 2002; Stuehr, 1999). Two isoforms, neuronal NOS (nNOS; also known as NOS I or NOS1) and eNOS (NOS III/NOS3), are Ca^{2+}-dependent and constitutive isoforms (Knowles and Moncada, 1994). Ca^{2+} influxes associated with either ligand-gated or voltage-activated Ca^{2+} channels up-regulate NOS activity and result in transient activation of NO synthesis, release of NO and action on neighboring cells and neuronal terminals. On the other hand, inducible NOS (NOS II/NOS2) is a Ca^{2+}-independent enzyme, which normally cannot be detected in most tissues, but whose expression is dramatically activated after appropriate stimulation (e.g., in the presence of lipopolysaccharides or in response to potentially damaging stimuli), resulting in a high and long-term NO yield. iNOS is primarily involved in defense reactions and cytotoxicity. Structurally, all NOSs consist of two major domains: the oxygenase or catalytic domain [i.e., the N-terminal part of NOS, with L-arginine, tetrahydrobiopterin (BH_4) and heme (Fe) binding sites] and the reductase domain (with FMN, FAD and NADPH binding sites). The formation of dimers comprised of identical subunits is essential for NOS activity (Garcin et al., 2004; Stuehr, 1999; Stuehr et al., 2004).

Figure 3 summarizes the domain organization of major animal NOSs as multi-domain proteins (Ghosh and Salerno, 2003). In contrast, all prokaryotic NOSs are "truncated" and consist of the oxygenase domain alone (Adak et al., 2002a,b; Zemojtel et al., 2003). Thus, the most likely scenario for the origin of eukaryotic NOSs is the fusion of two previously independent genes: one gene representing the oxygenase domain, such as bacterial NOSs, and the second gene encoding the reductase domain, such as is found in cytochrome P-450-type enzymes. This fusion event might have occurred at the time of the origin of the major eukaryotic

24

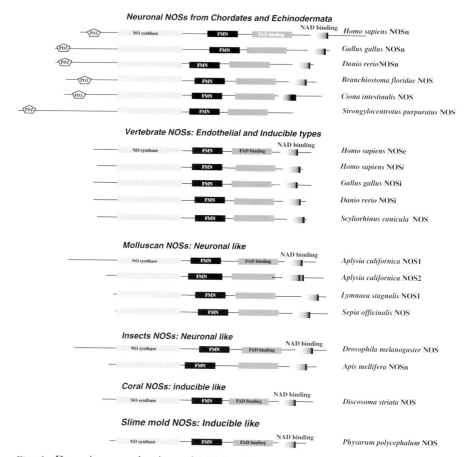

Fig. 3. Domain organization of NOSs. See text and Table 1 for details and accession numbers. Interestingly, only neuronal like NOS from deuterostome lineages (Chordates and echinoderms) have PDZ domain in their N-terminal. *Note*: the length of NOSs in the schematic illustrations correspond to the size of the NOS proteins.

groups as a response to the rise of oxygen concentration in Proterozoic time. It is quite possible that the origin of oxygen-dependent NO synthesis from L-arginine could have been an adaptive response against emerging oxygen toxicity more than 2 Gya. It is not surprising that this is the major molecular function of the precursor of the reductase domain in NOSs, cytochrome P-450 (Gorren and Mayer, 2007).

The first invertebrate NOS enzymes were cloned from insects (Champagne *et al.*, 1995; Nighorn *et al.*, 1998; Regulski and Tully,

1995; Yuda *et al.*, 1996) and molluscs (Korneev *et al.*, 1998; Sadreyev *et al.*, 2000); they all have greater sequence similarity to a constitutive neuronal-like NOS than to other isoforms (iNOS or eNOS) in vertebrates. At first glance, this might suggest that this type of NOS could be evolutionarily close to an "ancestral" NOS prototype. However, the recent discoveries of cnidarian and slime mold NOSs as well as the identification of novel NOS genes from recently sequenced genomes reveal both enormous diversity and rather parallel evolution of different NOS classes.

Cnidarian (*Discosoma*) and slime mold (*Physarum*; Golderer *et al.*, 2001) NOSs lack a distinct structural element that is present as an insertion in the reductase domains of constitutive NOSs but absent in iNOSs of vertebrates. This insert of ~45 amino acids (residues 835–876 in human nNOS) is thought to be an autoinhibitory loop which impedes Ca^{2+}-free calmodulin (CaM) binding and enzymatic activation. Since *Discosoma* NOS is structurally similar to both the only known nonanimal conventional NOS and vertebrate iNOS isoforms, the inducible type of the enzyme can be evolutionarily basal for animal NOSs. Under this scenario, multiple lineages of constitutive-like animal NOSs could be independently derived from iNOS-like proteins, in a process that might include independent events of insertion of the autoinhibitory loop in the NOS reductase domain. This insert provides coupling to Ca^{2+} regulatory mechanisms and reduces potentially toxic NO production following the activation of iNOS.

It is interesting that, in contrast to vertebrate species which have three NOS genes, only one type of NOS isoform has been found in the sequenced genomes from insects, echinoderms and tunicates. However, molluscs have at least two NOS genes, and no NOS genes have been identified in nematodes (*Caenorhabditis elegans*) (Table 1). This situation implies the very interesting possibility that NOS first appeared in basal animal ancestor groups and then was lost in some animal taxa in the course of evolution. On the contrary, in some lineages such as molluscs (with at least two different types of NOS) or chordates (which have two to three NOS genes) these events might have happened more than once, or duplication of the constitutive-type NOS might have occurred. Apparently, duplication events might also have occurred during the evolution of inducible-type NOS, since some fishes have more than one iNOS-like gene (Table 1).

Dendrograms showing the relatedness of the predicted oxygenase domains for NOSs and full-length NOSs are shown in Figs. 4 and 5. As expected, the vertebrate NOSs are all clustered into three distinct subfamilies – iNOS, eNOS and nNOS – with iNOS possibly being the most basal (Figs. 2 and 4). However, the diversity of all nonvertebrate NOSs is

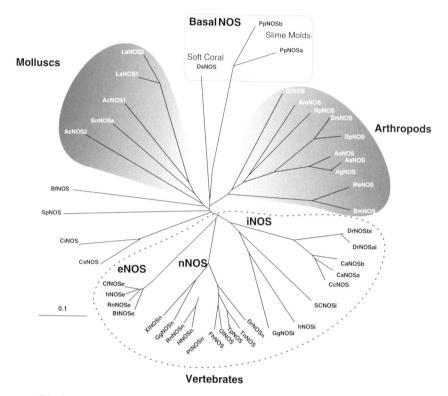

Fig. 4. Phylogeny of animal NOSs: dendrogram of full-length NOS genes. *Note*: all the basal NOSs in corals and slime molds are similar to inducible-like NOS in vertebrate, see text for details. Full-length amino acid sequences were aligned in "ClustalX" version 1.83 using default parameters (Jeanmougin *et al.*, 1998); all gaps were manually removed, to include only the oxygenase domains, in "GeneDoc" (Nicholas *et al.*, 1997). The neighbor-joining method was used to construct the phylogenetic tree using the "ClustalX" program. The scale bar represents 10% divergence. The graphic output was generated using "Treeview" (Page, 1996). See abbreviations for the species listed and gene bank accession numbers for their NOSs in Table 1.

quite significant and cannot be described in the framework of the three mammalian NOS subtypes such as neuronal, endothelial, and inducible (Fig. 2). Molluscs, arthropods and prokaryotes also show appropriate relatedness in their clusters with slime mold NOS as a potential outgroup for all animals. NOSs from cnidaria, echinoderms (the sea urchin) and basal chordates (amphioxus and two ascidians) seem to be the most derived and distinct NOS subtypes. It also appears that only neuronal-type NOSs from deuterostomes (the sea urchin NOS and nNOSs in all chordates) have

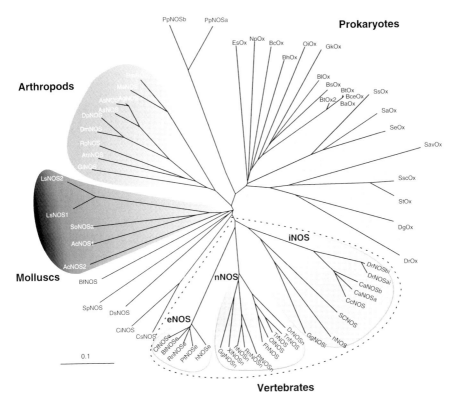

Fig. 5. Phylogeny of all known NOSs: dendrogram of NOS oxygenase domains. Full-length amino acid sequences were aligned in "ClustalX" version 1.83 using default parameters (Jeanmougin *et al.*, 1998); all gaps were manually removed, to include only the oxygenase domains, in "GeneDoc" (Nicholas *et al.*, 1997). The neighbor-joining method was used to construct the phylogenetic tree using the "ClustalX" program. The scale bar represents 10% divergence. The graphic output was generated using "Treeview" (Page, 1996). See abbreviations for the species listed and gene bank accession numbers for their NOSs in Table 1.

developed the PDZ domain (which is probably involved in specific subcellular targeting of NOS and the formation of highly dynamic protein complexes, as shown in postsynaptic regions of mammalian neurons).

Unfortunately, lack of detailed biochemical characterization and pharmacological profiles of such a diverse protein family as invertebrate NOSs is one of the major limitations in the field. Future experiments using purified or expressed NOSs from various phyla are one of the crucial steps in our understanding of the evolution of NO signaling. Such analyses might also provide valuable information for the biomedical industry as part of the development of more specific and efficient NOS inhibitors for different NOS subtypes.

Nonconventional NOSs from plants and animals

NOS-like activity (with the standard pharmacological and biochemical properties of mammalian NOSs) has been described in fungi (Ninnemann and Maier, 1996) and has been reported in plants (Caro and Puntarulo, 1999; Durner and Klessig, 1999; Durner et al., 1998; Ribeiro et al., 1999). It has been proposed that NO synthesis (via the conversion of L-arginine to L-citrulline) in some plant tissues is Ca^{2+}-dependent (Barroso et al., 1999; Cueto et al., 1996; Delledonne et al., 1998), whereas in others it is Ca^{2+}-independent (Cueto et al., 1996), resembling the inducible type of NOS of mammals. Again, in view of the fact that conventional NOS-type enzymes from these species have neither been cloned nor purified, they have not been analyzed in detail. As a result, any discussion of potential homology and similarity between these proteins and NOSs in animals would be highly speculative. Yet again, we stress the interesting situation of claims for the presence of NOS activity, but an absence of recognizable conventional NOS genes, in the genome of *Arabidopsis thaliana*.

This paradox has been partially resolved with the discovery of a non-conventional NOS in plants (Crawford, 2006). An *Arabidopsis* mutant was identified that had impaired NO production, and this was shown to be attributed to a gene which the authors named *AtNOS1* (Guo et al., 2003). Expression of *AtNOS1* on a viral promoter in the mutant plants resulted in overproduction of NO (Guo et al., 2003). The AtNOS1 protein shares greatest identity with the land pulmonate snail *Helix* hypothetical NOS-related protein (Huang et al., 1997) and the bacterial GTPase domain proteins. Expression of this *Helix* protein in bacteria showed a 15-fold increase in NOS-type activity.

To ensure that the novel *AtNOS1* gene encoded an enzyme that had NOS activity, the AtNOS1 protein was expressed in bacteria and showed elevated levels of NOS activity (Guo et al., 2003). These authors further characterized the purified protein and showed NOS activity to be dependent on nicotinamide adenine dinucleotide phosphate (NADP), CaM and Ca^{2+} and inhibited by L-NAME. In other words, the activity of AtNOS1 does resemble the pharmacology of conventional NOSs from mammals.

Recently, putative plant-like NOSs with a conserved GTPase domain have been described in other eukaryotes and many bacteria (Zemojtel et al., 2004). The members of this novel family of NOS-related proteins do not resemble any of the known conventional isoforms of NOSs; they contain neither a heme oxygenase domain nor a flavin reductase domain. Subsequently a mammalian ortholog was cloned from mouse, and named *mAtNOS1* (Zemojtel *et al.*, 2006). These authors constructed a fusion gene and expressed it in cell lines which showed that the fusion protein was localized to the inner mitochondrial compartment. Zemojtel *et al.* (2006) also suggested that the *mAtNOS1* gene has a role in development of neural, hematopoietic and bone organ systems.

We also cloned one of the novel plant-like NOS orthologs from the marine opisthobranch mollusc *Aplysia californica*: *AcAtNOS1* (GenBank

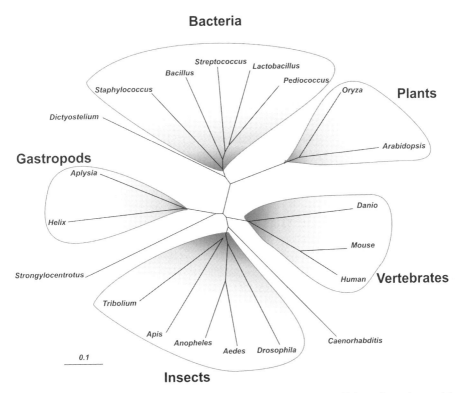

Fig. 6. Phylogeny of nonconventional plant-type NOSs. Full-length amino acid sequences were aligned in "ClustalX" version 1.83 using default parameters (Jeanmougin *et al.*, 1998); all gaps were manually removed, to include only the oxygenase domains, in "GeneDoc" (Nicholas *et al.*, 1997). The neighbor-joining method was used to construct the phylogenetic tree using the "ClustalX" program. The scale bar represents 10% divergence. The graphic output was generated using "Treeview" (Page, 1996). See abbreviations for the species listed and gene bank accession numbers for their NOSs in Table 1.

accession no. EF043280), with a deduced amino acid sequence of 712 residues. The putative AcAtNOS1 protein has a longer N-terminus, as do all the eukaryotic orthologs when compared to their prokaryotic orthologs (Zemojtel *et al.*, 2004). Similar to mammalian mAtNOS1, molluscan AcAtNOS1 contains a mitochondrial-targeting peptide, as predicted by "Target-*P*" with a score of 0.93 (a score of 1 indicates the strongest prediction) (Emanuelsson *et al.*, 2000).

The summary dendrogram showing the relatedness of novel plant-type NOSs with the GTPase domains is shown in Fig. 6. The clades of vertebrates, gastropods, insects, plants and bacteria form five distinct clusters. As predicted, molluscan AtNOSs are most closely related to each other. However, the molecular enzymatic mechanisms of NO synthesis by these hypothetical plant-type NOSs are unknown. It is still possible that this protein may be an important component of a larger complex or chaperone controlling NO production in plants, and possibly all other eukaryotes and even bacteria.

Conclusions

In conclusion, we would like to stress that at least six enzymatic and several nonenzymatic synthetic pathways can be involved in NO formation in living cells and tissues. Many of these pathways are widely distributed across all major prokaryotic and eukaryotic lineages and can be colocalized within the same organism, tissue or cell. Furthermore, enzyme-independent NO production from nitrites can be as important as NOS-dependent NO synthesis. In some cases, nonenzymatic NO formation exceeds the production of NO by specific enzymes.

It also appears that nonenzymatic NO synthesis was a crucial factor in abiotic nitrogen fixation at the dawn of biological evolution, and could even have contributed to the accumulation of "fixed nitrogen" in the Hadean and early Archean eons, providing favorable starting materials for the prebiotic synthesis of complex organic molecules. Thus, NO-related regulatory mechanisms can be traced back to the origins of biological organization. Many of the NO synthetic pathways could have been initially developed under anoxic conditions or in the presence of low oxygen concentrations.

In contrast, the increase in oxygen level in the Precambrian atmosphere and ocean (which started as the "Great Oxidation Event" ~2.4 billion years ago) might have been a factor, contributing to the origin of "conventional" NO synthesis from L-arginine and oxygen by NOS. Moreover, it might also be one of the several protective mechanisms against oxygen toxicity. Thus, trends in the evolution of NO synthesis and signaling may be a reflection of the balance between evolutionary novel NOS/oxygen-dependent and more ancient NOS/oxygen-independent NO syntheses in different lineages. As a result, multiple ecological and even functional adaptations to hypoxic conditions can be described as a metabolic shift between multiple complementary pathways of NO synthesis and its inactivation.

The enormous complexity of redox chemistry of NO and related species further contributes to establishing highly dynamic gradients of NO concentrations within cells and tissues. Consequently, an integrated systemic approach is essential for biomedical studies and clinical applications related to the experimental modulation of NO synthetic and signaling pathways.

Finally, the great diversity of conventional NOSs within the animal kingdom clearly prevents interpretation of the emerging comparative data in terms of the existing tripartite classification of mammalian NOS isoforms (*i.e.*, inducible-like, neuronal-like and endothelial-like NOSs). This classification might reflect the evolution of NOS isoforms within a defined lineage of vertebrates, but it needs to be readjusted as soon as novel cloning data from representatives of the more than 30 other animal phyla become available. The emerging molecular information about NOSs from cnidarians, echinoderms, molluscs and even basal chordates suggests that L-arginine/oxygen NO synthetic pathways might represent examples of parallel evolution of the NOS prototype in different lineages of animals. Recent discoveries of different ways to make NO in plants might also represent novel classes of NO synthetic pathways, and need to be further investigated. Presently, the most important aspect of this field is characterization of the molecular mechanisms of NO synthesis by these plant-type NOSs.

References

Abe, K. and Kimura, H. (1996). The possible role of hydrogen sulfide as an endogenous neuromodulator. *J. Neurosci.* 16,1066–1071.

Adak, S., Aulak, K.S. and Stuehr, D.J. (2002a). Direct evidence for nitric oxide production by a nitric-oxide synthase-like protein from *Bacillus subtilis*. *J. Biol. Chem.* 277,16167–16171.

Adak, S., Bilwes, A.M., Panda, K., Hosfield, D., Aulak, K.S., McDonald, J.F., Tainer, J.A., Getzoff, E.D., Crane, B.R. and Stuehr, D.J. (2002b). Cloning, expression, and characterization of a nitric oxide synthase protein from *Deinococcus radiodurans*. *Proc. Natl. Acad. Sci. USA* 99,107–112.

Allaker, R.P., Silva Mendez, L.S., Hardie, J.M. and Benjamin, N. (2001). Antimicrobial effect of acidified nitrite on periodontal bacteria. *Oral Microbiol. Immunol.* 16,253–256.

Anyim, M., Benjamin, N. and Wilks, M. (2005). Acidified nitrite as a potential antifungal agent. *Int. J. Antimicrob. Agents* 26,85–87.

Archer, S. (1993). Measurement of nitric oxide in biological models. *FASEB J.* 7,349–360.

Artinian, L.R., Ding, J.M. and Gillette, M.U. (2001). Carbon monoxide and nitric oxide: Interacting messengers in muscarinic signaling to the brain's circadian clock. *Exp. Neurol.* 171,293–300.

Baker, P.R., Lin, Y., Schopfer, F.J., Woodcock, S.R., Groeger, A.L., Batthyany, C., Sweeney, S., Long, M.H., Iles, K.E., Baker, L.M., Branchaud, B.P., Chen, Y.E. and Freeman, B.A. (2005). Fatty acid transduction of nitric oxide signaling: Multiple nitrated unsaturated fatty acid derivatives exist in human blood and urine and serve as endogenous peroxisome proliferator-activated receptor ligands. *J. Biol. Chem.* 280,42464–42475.

Baker, P.R., Schopfer, F.J., Sweeney, S. and Freeman, B.A. (2004). Red cell membrane and plasma linoleic acid nitration products: Synthesis, clinical identification, and quantitation. *Proc. Natl. Acad. Sci. USA* 101,11577–11582.

Barroso, J.B., Corpas, F.J., Carreras, A., Sandalio, L.M., Valderrama, R., Palma, J.M., Lupianez, J.A. and del Rio, L.A. (1999). Localization of nitric-oxide synthase in plant peroxisomes. *J. Biol. Chem.* 274,36729–36733.

Bartberger, M.D., Liu, W., Ford, E., Miranda, K.M., Switzer, C., Fukuto, J.M., Farmer, P.J., Wink, D.A. and Houk, K.N. (2002). The reduction potential of nitric oxide (NO) and its importance to NO biochemistry. *Proc. Natl. Acad. Sci. USA* 99,10958–10963.

Benavente, L.M. and Alonso, J.M. (2006). Molecular mechanisms of ethylene signaling in *Arabidopsis. Mol. Biosyst.* 2,165–173.

Benjamin, N., O'Driscoll, F., Dougall, H., Duncan, C., Smith, L., Golden, M. and McKenzie, H. (1994). Stomach NO synthesis. *Nature* 368,502.

Benjamin, N., Pattullo, S., Weller, R., Smith, L. and Ormerod, A. (1997). Wound licking and nitric oxide. *Lancet* 349,1776.

Bhatia, M. (2005). Hydrogen sulfide as a vasodilator. *IUBMB Life* 57,603–606.

Boehning, D. and Snyder, S.H. (2003). Novel neural modulators. *Annu. Rev. Neurosci.* 26,105–131.

Botti, H., Trostchansky, A., Batthyany, C. and Rubbo, H. (2005). Reactivity of peroxynitrite and nitric oxide with LDL. *IUBMB Life* 57,407–412.

Brandes, J.A., Boctor, N.Z., Cody, G.D., Cooper, B.A., Hazen, R.M. and Yoder, H.S., Jr. (1998). Abiotic nitrogen reduction on the early Earth. *Nature* 395,365–367.

Bruch-Gerharz, D., Schnorr, O., Suschek, C., Beck, K.F., Pfeilschifter, J., Ruzicka, T. and Kolb-Bachofen, V. (2003). Arginase 1 overexpression in psoriasis: Limitation of inducible nitric oxide synthase activity as a molecular mechanism for keratinocyte hyperproliferation. *Am. J. Pathol.* 162,203–211.

Brusca, R.C. and and Brusca, G.J. (2003). *Invertebrates*, Sinauer Associates, Inc., Sunderland, MA.

Bryan, N.S. (2006). Nitrite in nitric oxide biology: Cause or consequence? A systems-based review. *Free Radic. Biol. Med.* 41,691–701.

Bryan, N.S., Fernandez, B.O., Bauer, S.M., Garcia-Saura, M.F., Milsom, A.B., Rassaf, T., Maloney, R.E., Bharti, A., Rodriguez, J. and Feelisch, M. (2005). Nitrite is a signaling molecule and regulator of gene expression in mammalian tissues. *Nat. Chem. Biol.* 1,290–297.

Bryan, N.S., Rassaf, T., Maloney, R.E., Rodriguez, C.M., Saijo, F., Rodriguez, J.R. and Feelisch, M. (2004). Cellular targets and mechanisms of nitros(yl)ation: An insight into their nature and kinetics in vivo. *Proc. Natl. Acad. Sci. USA* 101,4308–4313.

Butler, A.R., Flitney, F.W. and Williams, D.L. (1995). NO, nitrosonium ions, nitroxide ions, nitrosothiols and iron-nitrosyls in biology: A chemist's perspective. *Trends Pharmacol. Sci.* 16,18–22.

Carlsson, S., Wiklund, N.P., Engstrand, L., Weitzberg, E. and Lundberg, J.O. (2001). Effects of pH, nitrite, and ascorbic acid on nonenzymatic nitric oxide generation and bacterial growth in urine. *Nitric Oxide* 5,580–586.

Caro, A. and Puntarulo, S. (1999). Nitric oxide generation by soybean embryonic axes. Possible effect on mitochondrial function. *Free Radic. Res.* 31(Suppl.), S205–S212.

Chahl, (2004)Chahl, L. A. (2004). Hydrogen sulphide: an endogenous stimulant of capsaicin-sensitive primary afferent neurons? *Br. J. Pharmacol.* 142,1–2.

Champagne, D.E., Nussenzveig, R.H. and Ribeiro, J.M. (1995). Purification, partial characterization, and cloning of nitric oxide-carrying heme proteins (nitrophorins) from salivary glands of the blood-sucking insect *Rhodnius prolixus*. *J. Biol. Chem.* 270,8691–8695.

Chang, C. and Bleecker, A.B. (2004). Ethylene biology. More than a gas. *Plant Physiol.* 136,2895–2899.

Chang, C. and Shockey, J.A. (1999). The ethylene-response pathway: Signal perception to gene regulation. *Curr. Opin. Plant Biol.* 2,352–358.

Chow, B. and McCourt, P. (2006). Plant hormone receptors: Perception is everything. *Genes Dev.* 20,1998–2008.

Cosby, K., Partovi, K.S., Crawford, J.H., Patel, R.P., Reiter, C.D., Martyr, S., Yang, B.K., Waclawiw, M.A., Zalos, G., Xu, X., Huang, K.T., Shields, H., Kim-Shapiro, D.B., Schechter, A.N., Cannon, R.O., 3rd and Gladwin, M.T. (2003). Nitrite reduction to nitric oxide by deoxyhemoglobin vasodilates the human circulation. *Nat. Med.* 9,1498–1505.

Cox, R.L., Mariano, T., Heck, D.E., Laskin, J.D. and Stegeman, J.J. (2001). Nitric oxide synthase sequences in the marine fish *Stenotomus chrysops* and the sea urchin *Arbacia punctulata*, and phylogenetic analysis of nitric oxide synthase calmodulin-binding domains. *Comp. Biochem. Physiol. B: Biochem. Mol. Biol.* 130,479–491.

Crawford, N.M. (2006). Mechanisms for nitric oxide synthesis in plants. *J. Exp. Bot.* 57,471–478.

Cueto, M., Hernandez-Perera, O., Martin, R., Bentura, M.L., Rodrigo, J., Lamas, S. and Golvano, M.P. (1996). Presence of nitric oxide synthase activity in roots and nodules of *Lupinus albus*. *FEBS Lett.* 398,159–164.

Cutruzzola, F. (1999). Bacterial nitric oxide synthesis. *Biochim. Biophys. Acta* 1411,231–249.

Dahn, H., Loewe, L., Luscher, E. and Menasse, R. (1960). Uber die Oxydation von Ascorbinsaure durch salpetrige Saule. TeilI: Stochiometrie und Kinetische Messtechnik. *Helv. Chim. Acta* 43,287–293.

Delledonne, M., Xia, Y., Dixon, R.A. and Lamb, C. (1998). Nitric oxide functions as a signal in plant disease resistance. *Nature* 394,585–588.

Ding, J.M., Chen, D., Weber, E.T., Faiman, L.E., Rea, M.A. and Gillette, M.U. (1994). Resetting the biological clock: Mediation of nocturnal circadian shifts by glutamate and NO. *Science* 266,1713–1717.

Doel, J.J., Godber, B.L., Eisenthal, R. and Harrison, R. (2001). Reduction of organic nitrates catalysed by xanthine oxidoreductase under anaerobic conditions. *Biochim. Biophys. Acta* 1527,81–87.

Donzelli, S., Switzer, C.H., Thomas, D.D., Ridnour, L.A., Espey, M.G., Isenberg, J.S., Tocchetti, C.G., King, S.B., Lazzarino, G., Miranda, K.M., Roberts, D.D., Feelisch, M. and Wink, D.A. (2006). The activation of metabolites of nitric oxide synthase by metals is both redox and oxygen dependent: A new feature of nitrogen oxide signaling. *Antioxid. Redox Signal.* 8,1363–1371.

Duncan, C., Dougall, H., Johnston, P., Green, S., Brogan, R., Leifert, C., Smith, L., Golden, M. and Benjamin, N. (1995). Chemical generation of nitric oxide in the mouth from the enterosalivary circulation of dietary nitrate. *Nat. Med.* 1,546–551.

Duncan, C., Li, H., Dykhuizen, R., Frazer, R., Johnston, P., MacKnight, G., Smith, L., Lamza, K., McKenzie, H., Batt, L., Kelly, D., Golden, M., Benjamin, N. and Leifert, C. (1997). Protection against oral and gastrointestinal diseases: Importance of dietary nitrate uptake, oral nitrate reduction and enterosalivary nitrate circulation. *Comp. Biochem. Physiol.* 118A,939–948.

Durner, J. and Klessig, D.F. (1999). Nitric oxide as a signal in plants. *Curr. Opin. Plant Biol.* 2,369–374.

Durner, J., Wendehenne, D. and Klessig, D.F. (1998). Defense gene induction in tobacco by nitric oxide, cyclic GMP, and cyclic ADP-ribose. *Proc. Natl. Acad. Sci. USA* 95,10328–10333.

Dykhuizen, R.S., Frazer, R., Duncan, C., Smith, C.C., Golden, M., Benjamin, N. and Leifert, C. (1996). Antimicrobial effect of acidified nitrite on gut pathogens: Importance of dietary nitrate in host defense. *Antimicrob. Agents Chemother.* 40,1422–1425.

Dykhuizen, R.S., Fraser, A., McKenzie, H., Golden, M., Leifert, C. and Benjamin, N. (1998). *Helicobacter pylori* is killed by nitrite under acidic conditions. *Gut* 42,334–337.

Ebrahimkhani, M.R., Mani, A.R. and Moore, K. (2005). Hydrogen sulphide and the hyperdynamic circulation in cirrhosis: A hypothesis. *Gut* 54,1668–1671.

Emanuelsson, O., Nielsen, H., Brunak, S. and von Heijne, G. (2000). Predicting subcellular localization of proteins based on their N-terminal amino acid sequence. *J. Mol. Biol.* 300,1005–1016.

Eto, K., Ogasawara, M., Umemura, K., Nagai, Y. and Kimura, H. (2002). Hydrogen sulfide is produced in response to neuronal excitation. *J. Neurosci.* 22,3386–3391.

Feelisch, M. (1993). Biotransformation to nitric oxide of organic nitrates in comparison to other nitrovasodilators. *Eur. Heart J.* 14(Suppl. I), 123–132.

Feelisch, M. and Kelm, M. (1991). Biotransformation of organic nitrates to nitric oxide by vascular smooth muscle and endothelial cells. *Biochem. Biophys. Res. Commun.* 180,286–293.

Feelisch, M. and Noack, E.A. (1987). Correlation between nitric oxide formation during degradation of organic nitrates and activation of guanylate cyclase. *Eur. J. Pharmacol.* 139,19–30.

Feelisch, M. and Stamler, J.S. (1996). Donors of nitrogen oxides. In *Methods in Nitric Oxide Research* (eds M. Feelisch and J.S. Stamler), pp. 71–115, Wiley, Chichester, NY.

Fiorucci, S., Distrutti, E., Cirino, G. and Wallace, J.L. (2006). The emerging roles of hydrogen sulfide in the gastrointestinal tract and liver. *Gastroenterology* 131,259–271.

Ford, P.C., Wink, D.A. and Stanbury, D.M. (1993). Autoxidation kinetics of aqueous nitric oxide. *FEBS Lett.* 326,1–3.

Fukuto, J.M., Cho, J.Y. and Switzer, C.H. (2000). The chemical properties of nitric oxide and related nitrogen oxides. In *Nitric Oxide: Biology and Pathobiology* (ed. L.J. Ignarro), pp. 23–40, Academic Press, San Diego.

Garcin, E.D., Bruns, C.M., Lloyd, S.J., Hosfield, D.J., Tiso, M., Gachhui, R., Stuehr, D.J., Tainer, J.A. and Getzoff, E.D. (2004). Structural basis for isozyme-specific regulation of electron transfer in nitric-oxide synthase. *J. Biol. Chem.* 279,37918–37927.

Garthwaite, J. (2005). Dynamics of cellular NO-cGMP signaling. *Front. Biosci.* 10,1868–1880.

Gautier, C., van Faassen, E., Mikula, I., Martasek, P. and Slama-Schwok, A. (2006). Endothelial nitric oxide synthase reduces nitrite anions to NO under anoxia. *Biochem. Biophys. Res. Commun.* 341,816–821.

Gelperin, A., Flores, J., Raccuia-Behling, F. and Cooke, I.R. (2000). Nitric oxide and carbon monoxide modulate oscillations of olfactory interneurons in a terrestrial mollusk. *J. Neurophysiol.* 83,116–127.

Ghosh, D.K. and Salerno, J.C. (2003). Nitric oxide synthases: Domain structure and alignment in enzyme function and control. *Front. Biosci.* 8,d193–d209.

Gladwin, M.T. (2005). Nitrite as an intrinsic signaling molecule. *Nat. Chem. Biol.* 1,245–246.

Gladwin, M.T., Raat, N.J., Shiva, S., Dezfulian, C., Hogg, N., Kim-Shapiro, D.B. and Patel, R.P. (2006). Nitrite as a vascular endocrine nitric oxide reservoir that contributes to hypoxic signaling, cytoprotection, and vasodilation. *Am. J. Physiol. Heart Circ. Physiol.* 291,H2026–H2035.

Gladwin, M.T., Schechter, A.N., Kim-Shapiro, D.B., Patel, R.P., Hogg, N., Shiva, S., Cannon, R.O., 3rd, Kelm, M., Wink, D.A., Espey, M.G., Oldfield, E.H., Pluta, R.M., Freeman, B.A., Lancaster, J.R., Jr., Feelisch, M. and Lundberg, J.O. (2005). The emerging biology of the nitrite anion. *Nat. Chem. Biol.* 1,308–314.

Golderer, G., Werner, E.R., Leitner, S., Grobner, P. and Werner-Felmayer, G. (2001). Nitric oxide synthase is induced in sporulation of *Physarum polycephalum*. *Genes Dev.* 15,1299–1309.

Gorren, A.C. and Mayer, B. (2007). Nitric-oxide synthase: A cytochrome P450 family foster child. *Biochim. Biophys. Acta.* 1770,432–445.

Griffith, O.W. and Stuehr, D.J. (1995). Nitric oxide synthases: Properties and catalytic mechanism. *Annu. Rev. Physiol.* 57,707–736.

Grisham, M.B., Johnson, G.G. and Lancaster Jr., J.R. (1996). Quantitation of nitrate and nitrite in extracellular fluids. *Methods Enzymol.* 268,237–246.

Grunewald, R.A. (1993). Ascorbic acid in the brain. *Brain Res. Rev.* 18,123–133.

Guo, F.Q., Okamoto, M. and Crawford, N.M. (2003). Identification of a plant nitric oxide synthase gene involved in hormonal signaling. *Science* 302,100–103.

Hall, A.V., Antoniou, H., Wang, Y., Cheung, A.H., Arbus, A.M., Olson, S.L., Lu, W.C., Kau, C.L. and Marsden, P.A. (1994). Structural organization of the human neuronal nitric oxide synthase gene (NOS1). *J. Biol. Chem.* 269,33082–33090.

Hedrick, M.S., Chen, A.K. and Jessop, K.L. (2005). Nitric oxide changes its role as a modulator of respiratory motor activity during development in the bullfrog (*Rana catesbeiana*). *Comp. Biochem. Physiol. A: Mol. Integr. Physiol.* 142,231–240.

Hess, D.T., Matsumoto, A., Kim, S.O., Marshall, H.E. and Stamler, J.S. (2005). Protein S-nitrosylation: Purview and parameters. *Nat. Rev. Mol. Cell. Biol.* 6,150–166.

Hibino, T., Loza-Coll, M., Messier, C., Majeske, A.J., Cohen, A.H., Terwilliger, D.P., Buckley, K.M., Brockton, V., Nair, S.V., Berney, K., Fugmann, S.D., Anderson, M.K., Pancer, Z., Cameron, R.A., Smith, L.C. and Rast, J.P. (2006). The immune gene repertoire encoded in the purple sea urchin genome. *Dev. Biol.* 300,349–365.

Holmqvist, B., Ellingsen, B., Alm, P., Forsell, J., Oyan, A.M., Goksoyr, A., Fjose, A. and Seo, H.C. (2000). Identification and distribution of nitric oxide synthase in the brain of adult zebrafish. *Neurosci. Lett.* 292,119–122.

Hornig, D. (1975). Distribution of ascorbic acid, metabolites and analogues in man and animals. *Ann. N. Y. Acad. Sci.* 258,103–115.

Hosoki, R., Matsuki, N. and Kimura, H. (1997). The possible role of hydrogen sulfide as an endogenous smooth muscle relaxant in synergy with nitric oxide. *Biochem. Biophys. Res. Commun.* 237,527–531.

Huang, S., Kerschbaum, H.H., Engel, E. and Hermann, A. (1997). Biochemical characterization and histochemical localization of nitric oxide synthase in the nervous system of the snail, *Helix pomatia. J. Neurochem.* 69,2516–2528.

Ignarro, (2000)Ignarro, L. J. (2000). *Nitric Oxide: Biology and Pathobiology*, p. 1003. Academic Press, San Diego.

Ignarro, L.J., Buga, G.M., Wood, K.S., Byrns, R.E. and Chaudhuri, G. (1987). Endothelium-derived relaxing factor produced and released from artery and vein is nitric oxide. *Proc. Natl. Acad. Sci. USA* 84,9265–9269.

Ignarro, L.J., Fukuto, J.M., Griscavage, J.M., Rogers, N.E. and Byrns, R.E. (1993). Oxidation of nitric oxide in aqueous solution to nitrite but not nitrate: Comparison with enzymatically formed nitric oxide from L-arginine. *Proc. Natl. Acad. Sci. USA* 90,8103–8107.

Imamura, M., Yang, J. and Yamakawa, M. (2002). cDNA cloning, characterization and gene expression of nitric oxide synthase from the silkworm, *Bombyx mori. Insect Mol. Biol.* 11,257–265.

Jacklet, J.W. (1997). Nitric oxide signaling in invertebrates. *Invert. Neurosci.* 3,1–14.

Jeanmougin, F., Thompson, J.D., Gouy, M., Higgins, D.G. and Gibson, T.J. (1998). Multiple sequence alignment with Clustal X. *Trends Biochem. Sci.* 23,403–405.

Kalyanaraman, B. (2004). Nitrated lipids: A class of cell-signaling molecules. *Proc. Natl. Acad. Sci. USA* 101,11527–11528.

Kasting, J.F. (1992). Bolide impacts and the oxidation state of carbon in the Earth's early atmosphere. *Orig. Life Evol. Biosph.* 20,199–231.

Kasting, J.F. and Howard, M.T. (2006). Atmospheric composition and climate on the early Earth. *Phil. Trans. R. Soc. Lond. B: Biol. Sci.* 361,1733–1742.

Kasting, J.F. and Ono, S. (2006). Palaeoclimates: The first two billion years. *Phil. Trans. R. Soc. Lond. B: Biol. Sci.* 361,917–929.

Kasting, J.F. and Siefert, J.L. (2001). Biogeochemistry. The nitrogen fix. *Nature* 412,26–27.

Kharitonov, V.G., Sundquist, A.R. and Sharma, V.S. (1994). Kinetics of nitric oxide autoxidation in aqueous solution. *J. Biol. Chem.* 269,5881–5883.

Kharitonov, V.G., Sundquist, A.R. and Sharma, V.S. (1995). Kinetics of nitrosation of thiols by nitric oxide in the presence of oxygen. *J. Biol. Chem.* 270,28158–28164.

Kim, H.P., Ryter, S.W. and Choi, A.M. (2006). CO as a cellular signaling molecule. *Annu. Rev. Pharmacol. Toxicol.* 46,411–449.

Kim, H.W., Batista, L.A., Hoppes, J.L., Lee, K.J. and Mykles, D.L. (2004). A crustacean nitric oxide synthase expressed in nerve ganglia, Y-organ, gill and gonad of the tropical land crab, *Gecarcinus lateralis. J. Exp. Biol.* 207,2845–2857.

Kim, W.S., Dahlgren, R.L., Moroz, L.L. and Sweedler, J.V. (2002). Ascorbic acid assays of individual neurons and neuronal tissues using capillary electrophoresis with laser-induced fluorescence detection. *Anal. Chem.* 74,5614–5620.

Kimura, H. (2002). Hydrogen sulfide as a neuromodulator. *Mol. Neurobiol.* 26,13–19.

Kimura, H., Nagai, Y., Umemura, K. and Kimura, Y. (2005). Physiological roles of hydrogen sulfide: Synaptic modulation, neuroprotection, and smooth muscle relaxation. *Antioxid. Redox Signal.* 7,795–803.

Knowles, R.G. and Moncada, S. (1994). Nitric oxide synthases in mammals. *Biochem. J.* 298,249–258.

Korneev, S. and O'Shea, M. (2002). Evolution of nitric oxide synthase regulatory genes by DNA inversion. *Mol. Biol. Evol.* 19,1228–1233.

Korneev, S.A., Piper, M.R., Picot, J., Phillips, R., Korneeva, E.I. and O'Shea, M. (1998). Molecular characterization of NOS in a mollusc: Expression in a giant modulatory neuron. *J. Neurobiol.* 35,65–76.

Korneev, S.A., Straub, V., Kemenes, I., Korneeva, E.I., Ott, S.R., Benjamin, P.R. and O'Shea, M. (2005). Timed and targeted differential regulation of nitric oxide synthase (NOS) and anti-NOS genes by reward conditioning leading to long-term memory formation. *J. Neurosci.* 25,1188–1192.

Kornfeld, S. and Mellaman, I. (1989). The biogenesis of lysosomes. *Annu. Rev. Cell Biol.* 5,483–525.

Kozlov, A.V., Staniek, K. and Nohl, H. (1999). Nitrite reductase activity is a novel function of mammalian mitochondria. *FEBS Lett.* 454,127–130.

Krasko, A., Schroder, H.C., Perovic, S., Steffen, R., Kruse, M., Reichert, W., Muller, I.M. and Muller, W.E. (1999). Ethylene modulates gene expression in cells of the marine sponge *Suberites domuncula* and reduces the degree of apoptosis. *J. Biol. Chem.* 274,31524–31530.

Lancaster, J.R., Jr. (1994). Simulation of the diffusion and reaction of endogenously produced nitric oxide. *Proc. Natl. Acad. Sci. USA* 91,8137–8141.

Lancaster, J.R., Jr. (1996). Diffusion of free nitric oxide. *Methods Enzymol.* 268,31–50.

Lancaster, J.R., Jr. (1997). A tutorial on the diffusibility and reactivity of free nitric oxide. *Nitric Oxide* 1,18–30.

J.R.Lancaster, Jr. (2000). The physical properties of nitric oxide. In *Nitric Oxide: Biology and Pathobiology* (ed. L.J. Ignarro), pp. 209–224, Academic Press, San Diego.

Leffler, C.W., Parfenova, H., Jaggar, J.H. and Wang, R. (2006). Carbon monoxide and hydrogen sulfide: Gaseous messengers in cerebrovascular circulation. *J. Appl. Physiol.* 100,1065–1076.

Lewis, R.S. and Deen, W.M. (1994). Kinetics of the reaction of nitric oxide with oxygen in aqueous solutions. *Chem. Res. Toxicol.* 7,568–574.

Li, H., Cui, H., Liu, X. and Zweier, J.L. (2005). Xanthine oxidase catalyzes anaerobic transformation of organic nitrates to nitric oxide and nitrosothiols: Characterization

38

of this mechanism and the link between organic nitrate and guanylyl cyclase activation. *J. Biol. Chem.* 280,6594–6600.

Li, H., Samouilov, A., Liu, X. and Zweier, J.L. (2003). Characterization of the magnitude and kinetics of xanthine oxidase-catalyzed nitrate reduction: Evaluation of its role in nitrite and nitric oxide generation in anoxic tissues. *Biochemistry* 42,1150–1159.

Li et al., (2006)Li, L., Bhatia, M. and Moore, P. K. (2006). Hydrogen sulphide – A novel mediator of inflammation? *Curr. Opin. Pharmacol.* 6,125–129.

Lima, E.S., Bonini, M.G., Augusto, O., Barbeiro, H.V., Souza, H.P. and Abdalla, D.S. (2005). Nitrated lipids decompose to nitric oxide and lipid radicals and cause vasorelaxation. *Free Radic. Biol. Med.* 39,532–539.

Lowenstein, C.J. and Snyder, S.H. (1992). Nitric oxide, a novel biologic messenger. *Cell* 70,705–707.

Luckhart, S. and Rosenberg, R. (1999). Gene structure and polymorphism of an invertebrate nitric oxide synthase gene. *Gene* 232,25–34.

Lundberg, J.O., Carlsson, S., Engstrand, L., Morcos, E., Wiklund, N.P. and Weitzberg, E. (1997). Urinary nitrite: More than a marker of infection. *Urology* 50,189–191.

Lundberg, J.O. and Govoni, M. (2004). Inorganic nitrate is a possible source for systemic generation of nitric oxide. *Free Radic. Biol. Med.* 37,395–400.

Lundberg, J.O. and Weitzberg, E. (2005). NO generation from nitrite and its role in vascular control. *Arterioscler. Thromb. Vasc. Biol.* 25,915–922.

Lundberg, J.O., Weitzberg, E., Cole, J.A. and Benjamin, N. (2004). Nitrate, bacteria and human health. *Nat. Rev. Microbiol.* 2,593–602.

Lundberg, J.O., Weitzberg, E., Lundberg, J.M. and Alving, K. (1994). Intragastric nitric oxide production in humans: Measurements in expelled air. *Gut* 35,1543–1546.

Mancardi, D., Ridnour, L.A., Thomas, D.D., Katori, T., Tocchetti, C.G., Espey, M.G., Miranda, K.M., Paolocci, N. and Wink, D.A. (2004). The chemical dynamics of NO and reactive nitrogen oxides: A practical guide. *Curr. Mol. Med.* 4,723–740.

McFarland, M., D., K., Drummond, J.W., Schmeltekopf, A.L. and Winkler, R.M. (1979). Nitric oxide measurements in the equatorial Pacific region. *Geophys. Res. Lett.* 6,605–609.

McKnight, G.M., Smith, L.M., Drummond, R.S., Duncan, C.W., Golden, M. and Benjamin, N. (1997). Chemical synthesis of nitric oxide in the stomach from dietary nitrate in humans. *Gut* 40,211–214.

McNeill, B. and Perry, S.F. (2006). The interactive effects of hypoxia and nitric oxide on catecholamine secretion in rainbow trout (*Oncorhynchus mykiss*). *J. Exp. Biol.* 209,4214–4223.

Mellman, I. (1992). The importance of being acid: The role of acidification in intracellular membrane traffic. *J. Exp. Biol.* 172,39–45.

Miele, M. and Fillenz, M. (1996). In vivo determination of extracellular brain ascorbate. *J. Neurosci. Methods* 70,15–19.

Millar, J. (1995). The nitric oxide/ascorbate cycle: How neurons may control their own oxygen supply. *Med. Hypotheses* 45,21–26.

Millar, T.M., Stevens, C.R., Benjamin, N., Eisenthal, R., Harrison, R. and Blake, D.R. (1998). Xanthine oxidoreductase catalyses the reduction of nitrates and nitrite to nitric oxide under hypoxic conditions. *FEBS Lett.* 427,225–228.

Miranda, K.M., Espey, M.G., Jourd'heuil, D., Grisham, M.B., Fukuto, J.M., Feelisch, M. and Wink, D.A. (2000). The chemical biology of nitric oxide. In *Nitric*

Oxide: Biology and Pathobiology (ed. L.J. Ignarro), pp. 41–55, Academic Press, San Diego.

Miranda, K.M., Katori, T., Torres de Holding, C.L., Thomas, L., Ridnour, L.A., McLendon, W.J., Cologna, S.M., Dutton, A.S., Champion, H.C., Mancardi, D., Tocchetti, C.G., Saavedra, J.E., Keefer, L.K., Houk, K.N., Fukuto, J.M., Kass, D.A., Paolocci, N. and Wink, D.A. (2005). Comparison of the NO and HNO donating properties of diazeniumdiolates: Primary amine adducts release HNO in vivo. *J. Med. Chem.* 48,8220–8228.

Mirvish, S.S., Wallcave, L., Eagen, M. and Shubik, P. (1972). Ascorbate-nitrite reaction: Possible means of blocking the formation of carcinogenic *N*-nitroso compounds. *Science* 177,65–68.

Modin, A., Bjorne, H., Herulf, M., Alving, K., Weitzberg, E. and Lundberg, J.O. (2001). Nitrite-derived nitric oxide: a possible mediator of 'acidic-metabolic' vasodilation. *Acta Physiol. Scand.* 171,9–16.

Moncada, S. and Bolanos, J.P. (2006). Nitric oxide, cell bioenergetics and neurodegeneration. *J. Neurochem.* 97,1676–1689.

Moncada, S., Palmer, R.M. and Higgs, E.A. (1991). Nitric oxide: Physiology, pathophysiology, and pharmacology. *Pharmacol. Rev.* 43,109–142.

Moroz, L.L. (2000a). Giant identified NO-releasing neurons and comparative histochemistry of putative nitrergic systems in gastropod molluscs. *Microsc. Res. Tech.* 49,557–569.

Moroz, L.L. (2000b). On the origin and early evolution of neuronal NO signaling: A comparative analysis. In *Nitric Oxide and Free Radicals in Peripheral Neurotransmission* (ed. S. Kalsner), pp. 1–34, Springer-Verlag, New York.

Moroz, L.L. (2001). Gaseous transmission across time and species. *Am. Zoologist* 41,304–320.

Moroz, L.L., Dahlgren, R.L., Boudko, D., Sweedler, J.V. and Lovell, P. (2005). Direct single cell determination of nitric oxide synthase related metabolites in identified nitrergic neurons. *J. Inorg. Biochem.* 99,929–939.

Moroz, L.L., Norby, S.W., Cruz, L., Sweedler, J.V., Gillette, R. and Clarkson, R.B. (1998). Non-enzymatic production of nitric oxide (NO) from NO synthase inhibitors. *Biochem. Biophys. Res. Commun.* 253,571–576.

Muller, W.E., Ushijima, H., Batel, R., Krasko, A., Borejko, A., Muller, I.M. and Schroder, H.C. (2006). Novel mechanism for the radiation-induced bystander effect: Nitric oxide and ethylene determine the response in sponge cells. *Mutat. Res.* 597,62–72.

Mur, L.A., Carver, T.L. and Prats, E. (2006). NO way to live; the various roles of nitric oxide in plant–pathogen interactions. *J. Exp. Bot.* 57,489–505.

Nagababu, E., Ramasamy, S., Abernethy, D.R. and Rifkind, J.M. (2003). Active nitric oxide produced in the red cell under hypoxic conditions by deoxyhemoglobin-mediated nitrite reduction. *J. Biol. Chem.* 278,46349–46356.

Nagai, Y., Tsugane, M., Oka, J. and Kimura, H. (2004). Hydrogen sulfide induces calcium waves in astrocytes. *FASEB J.* 18,557–559.

Nathan, C. (1992). Nitric oxide as a secretory product of mammalian cells. *FASEB J.* 6,3051–3064.

Nathan, C. (2004). The moving frontier in nitric oxide-dependent signaling. *Sci. STKE* 2004,52.

40

Navarro-Gonzalez, R., McKay, C.P. and Mvondo, D.N. (2001). A possible nitrogen crisis for Archaean life due to reduced nitrogen fixation by lightning. *Nature* 412,61–64.

Nicholas, K., Simpson, K., Wilson, M., Trott, J. and Shaw, D. (1997). The tammar wallaby: A model to study putative autocrine-induced changes in milk composition. *J. Mammary Gland Biol. Neoplasia* 2,299–310.

Nighorn, A., Gibson, N.J., Rivers, D.M., Hildebrand, J.G. and Morton, D.B. (1998). The nitric oxide-cGMP pathway may mediate communication between sensory afferents and projection neurons in the antennal lobe of *Manduca sexta*. *J. Neurosci.* 18,7244–7255.

Ninnemann, H. and Maier, J. (1996). Indications for the occurrence of nitric oxide synthases in fungi and plants and the involvement in photoconidiation of *Neurospora crassa*. *Photochem. Photobiol.* 64,393–398.

Nisoli, E., Falcone, S., Tonello, C., Cozzi, V., Palomba, L., Fiorani, M., Pisconti, A., Brunelli, S., Cardile, A., Francolini, M., Cantoni, O., Carruba, M.O., Moncada, S. and Clementi, E. (2004). Mitochondrial biogenesis by NO yields functionally active mitochondria in mammals. *Proc. Natl. Acad. Sci. USA* 101,16507–16512.

Nisoli, E., Tonello, C., Cardile, A., Cozzi, V., Bracale, R., Tedesco, L., Falcone, S., Valerio, A., Cantoni, O., Clementi, E., Moncada, S. and Carruba, M.O. (2005). Calorie restriction promotes mitochondrial biogenesis by inducing the expression of eNOS. *Science* 310,314–317.

Nna-Mvondo, D., Navarro-Gonzalez, R., McKay, C.P., Coll, P. and Raulin, F. (2001). Production of nitrogen oxides by lightning and coronae discharges in simulated early Earth, Venus and Mars environments. *Adv. Space Res.* 27,217–223.

Nna-Mvondo, D., Navarro-Gonzalez, R., Raulin, F. and Coll, P. (2005). Nitrogen fixation by corona discharge on the early precambrian Earth. *Orig. Life Evol. Biosph.* 35,401–409.

Nohl, H., Staniek, K. and Kozlov, A.V. (2005). The existence and significance of a mitochondrial nitrite reductase. *Redox Rep.* 10,281–286.

Oke, A.F., May, L. and Adams, R.N. (1987). Ascorbic acid distribution patterns in human brain. A comparison with nonhuman mammalian species. *Ann. N. Y. Acad. Sci.* 498,1–12.

Page, R.D. (1996). TreeView: An application to display phylogenetic trees on personal computers. *Comput. Appl. Biosci.* 12,357–358.

Palmer, R.M., Ferrige, A.G. and Moncada, S. (1987). Nitric oxide release accounts for the biological activity of endothelium-derived relaxing factor. *Nature* 327,524–526.

Palumbo, A. (2005). Nitric oxide in marine invertebrates: A comparative perspective. *Comp. Biochem. Physiol. A: Mol. Integr. Physiol.* 142,241–248.

Paunel, A.N., Dejam, A., Thelen, S., Kirsch, M., Horstjann, M., Gharini, P., Murtz, M., Kelm, M., de Groot, H., Kolb-Bachofen, V. and Suschek, C.V. (2005). Enzyme-independent nitric oxide formation during UVA challenge of human skin: Characterization, molecular sources, and mechanisms. *Free Radic. Biol. Med.* 38,606–615.

Perovic, S., Seack, J., Gamulin, V., Muller, W.E. and Schroder, H.C. (2001). Modulation of intracellular calcium and proliferative activity of invertebrate and vertebrate cells by ethylene. *BMC Cell Biol.* 2,7.

Philippot, L. (2002). Denitrifying genes in bacterial and Archaeal genomes. *Biochim. Biophys. Acta* 1577,355–376.

Phillips, R., Kuijper, S., Benjamin, N., Wansbrough-Jones, M., Wilks, M. and Kolk, A.H. (2004). In vitro killing of *Mycobacterium ulcerans* by acidified nitrite. *Antimicrob. Agents Chemother.* 48,3130–3132.

Poon, K.L., Richardson, M., Lam, C.S., Khoo, H.E. and Korzh, V. (2003). Expression pattern of neuronal nitric oxide synthase in embryonic zebrafish. *Gene Expr. Patterns* 3,463–466.

Quintero, M., Colombo, S.L., Godfrey, A. and Moncada, S. (2006). Mitochondria as signaling organelles in the vascular endothelium. *Proc. Natl. Acad. Sci. USA* 103,5379–5384.

Regulski, M. and Tully, T. (1995). Molecular and biochemical characterization of dNOS: A *Drosophila* Ca^{2+}/calmodulin-dependent nitric oxide synthase. *Proc. Natl. Acad. Sci. USA* 92,9072–9076.

Reutov, V.P. (2002). Nitric oxide cycle in mammals and the cyclicity principle. *Biochemistry (Mosc.)* 67,293–311.

Ribeiro, E.A., Jr., Cunha, F.Q., Tamashiro, W.M. and Martins, I.S. (1999). Growth phase-dependent subcellular localization of nitric oxide synthase in maize cells. *FEBS Lett.* 445,283–286.

Rice, M.E. and Nicholson, C. (1987). Interstitial ascorbate in turtle brain is modulated by release and extracellular volume changes. *J. Neurochem.* 49,1096–1104.

Roenneberg, T. and Rehman, J. (1996). Nitrate, a nonphotic signal for the circadian system. *FASEB J.* 10,1443–1447.

Roman, L.J., Martasek, P. and Masters, B.S. (2002). Intrinsic and extrinsic modulation of nitric oxide synthase activity. *Chem. Rev.* 102,1179–1190.

Roy, B. and Garthwaite, J. (2006). Nitric oxide activation of guanylyl cyclase in cells revisited. *Proc. Natl. Acad. Sci. USA* 103,12185–12190.

Ryter, S.W., Alam, J. and Choi, A.M. (2006). Heme oxygenase-1/carbon monoxide: From basic science to therapeutic applications. *Physiol. Rev.* 86,583–650.

Ryter et al., (2004)Ryter, S. W., Morse, D. and Choi, A. M. (2004). Carbon monoxide: To boldly go where NO has gone before. *Sci. STKE 2004*, RE6.

Ryter, S.W. and Otterbein, L.E. (2004). Carbon monoxide in biology and medicine. *Bioessays* 26,270–280.

Sadreyev, R.I., Panchin, Y., Uvarov, P., Belyavski, A., Matz, M. and Moroz, L.L. (2000). Cloning of nitric oxide synthase (NOS) from Aplysia californica. *Soc. Neurosci. Abstr.* 26.

Saeij, J.P., Stet, R.J., Groeneveld, A., Verburg-van Kemenade, L.B., van Muiswinkel, W.B. and Wiegertjes, G.F. (2000). Molecular and functional characterization of a fish inducible-type nitric oxide synthase. *Immunogenetics* 51,339–346.

Salter, M., Duffy, C., Garthwaite, J. and Strigbos, P.J.L.M. (1996). *Ex vivo* measurement of brain tissue nitrite and nitrate accurately reflects nitric oxide synthase activity in vivo. *J. Neurochem.* 66,1683–1690.

Saran, M. and Bors, W. (1994). Pulse radiolysis for investigation of nitric oxide related reactions. *Methods Enzymol.* 233,20–34.

Scheinker, V., Fiore, G., Di Cristo, C., Di Cosmo, A., d'Ischia, M., Enikolopov, G. and Palumbo, A. (2005). Nitric oxide synthase in the nervous system and ink gland of the cuttlefish *Sepia officinalis*: Molecular cloning and expression. *Biochem. Biophys. Res. Commun.* 338,1204–1215.

42

Schopfer, F.J., Baker, P.R., Giles, G., Chumley, P., Batthyany, C., Crawford, J., Patel, R.P., Hogg, N., Branchaud, B.P., Lancaster, J.R., Jr. and Freeman, B.A. (2005). Fatty acid transduction of nitric oxide signaling. Nitrolinoleic acid is a hydrophobically stabilized nitric oxide donor. *J. Biol. Chem.* 280,19289–19297.

Scorza, G. and Minetti, M. (1998). One-electron oxidation pathway of thiols by peroxynitrite in biological fluids: Bicarbonate and ascorbate promote the formation of albumin disulphide dimers in human blood plasma. *Biochem. J.* 329(2), 405–413.

Scorza, G., Pietraforte, D. and Minetti, M. (1997). Role of ascorbate and protein thiols in the release of nitric oxide from *S*-nitroso-albumin and *S*-nitroso-glutathione in human plasma. *Free Radic. Biol. Med.* 22,633–642.

Seack, J., Perovic, S., Gamulin, V., Schroder, H.C., Beutelmann, P., Muller, I.M. and Muller, W.E. (2001). Identification of highly conserved genes: SNZ and SNO in the marine sponge *Suberites domuncula*: Their gene structure and promoter activity in mammalian cells (1). *Biochim. Biophys. Acta* 1520,21–34.

Shaw, A.W. and Vosper, A.J. (1977). Solubility of nitric oxide in aqueous and non-aqueous solvents. *J. Chem. Soc. Faraday Trans. I* 8,1239–1244.

Silva Mendez, L.S., Allaker, R.P., Hardie, J.M. and Benjamin, N. (1999). Antimicrobial effect of acidified nitrite on cariogenic bacteria. *Oral Microbiol. Immunol.* 14,391–392.

Singel, D.J. and Stamler, J.S. (2005). Chemical physiology of blood flow regulation by red blood cells: The role of nitric oxide and *S*-nitrosohemoglobin. *Annu. Rev. Physiol.* 67,99–145.

Siushansian, R. and Wilson, J.X. (1993). Ascorbate transport and intracellular concentrations in cerebral astrocytes. *J. Neurochem.* 65,41–49.

Somerville, C. (2000). The twentieth century trajectory of plant biology. *Cell* 100,13–25.

Spiro, S. (2006). Nitric oxide-sensing mechanisms in *Escherichia coli. Biochem. Soc. Trans.* 34,200–202.

Stamler, J.S., Lamas, S. and Fang, F.C. (2001). Nitrosylation: The prototypic redox-based signaling mechanism. *Cell* 106,675–683.

Stamler, J.S., Singel, D.J. and Loscalzo, J. (1992). Biochemistry of nitric oxide and its redox-activated forms. *Science* 258,1898–1902.

Stasiv, Y., Regulski, M., Kuzin, B., Tully, T. and Enikolopov, G. (2001). The *Drosophila* nitric-oxide synthase gene (dNOS) encodes a family of proteins that can modulate NOS activity by acting as dominant negative regulators. *J. Biol. Chem.* 276,42241–42251.

Stipanuk, M.H., Dominy, J.E., Jr., Lee, J.I. and Coloso, R.M. (2006). Mammalian cysteine metabolism: New insights into regulation of cysteine metabolism. *J. Nutr.* 136,1652S–1659S.

Stuehr, D.J. (1999). Mammalian nitric oxide synthases. *Biochim. Biophys. Acta* 1411,217–230.

Stuehr, D.J., Santolini, J., Wang, Z.Q., Wei, C.C. and Adak, S. (2004). Update on mechanism and catalytic regulation in the NO synthases. *J. Biol. Chem.* 279,36167–36170.

Summers, D.P. and Chang, S. (1993). Prebiotic ammonia from reduction of nitrite by iron (II) on the early Earth. *Nature* 365,630–633.

Suschek, C.V., Paunel, A. and Kolb-Bachofen, V. (2005). Nonenzymatic nitric oxide formation during UVA irradiation of human skin: Experimental setups and ways to measure. *Methods Enzymol.* 396,568–578.

Suschek, C.V., Schewe, T., Sies, H. and Kroncke, K.D. (2006). Nitrite, a naturally occurring precursor of nitric oxide that acts like a 'prodrug'. *Biol. Chem.* 387,499–506.

Tang, C., Li, X. and Du, J. (2006). Hydrogen sulfide as a new endogenous gaseous transmitter in the cardiovascular system. *Curr. Vasc. Pharmacol.* 4,17–22.

Thimann, K.V. (1974). Fifty years of plant hormone research. *Plant Physiol.* 54,450–453.

Thomas, D.D., Liu, X., Kantrow, S.P. and Lancaster, J.R., Jr. (2001). The biological lifetime of nitric oxide: implications for the perivascular dynamics of NO and O_2. *Proc. Natl. Acad. Sci. USA* 98,355–360.

Thomas, D.D., Miranda, K.M., Espey, M.G., Citrin, D., Jourd'heuil, D., Paolocci, N., Hewett, S.J., Colton, C.A., Grisham, M.B., Feelisch, M. and Wink, D.A. (2002). Guide for the use of nitric oxide (NO) donors as probes of the chemistry of NO and related redox species in biological systems. *Methods Enzymol.* 359,84–105.

Tischner, R., Planchet, E. and Kaiser, W.M. (2004). Mitochondrial electron transport as a source for nitric oxide in the unicellular green alga *Chlorella sorokiniana*. *FEBS Lett.* 576,151–155.

Vallette, G., Tenaud, I., Branka, J.E., Jarry, A., Sainte-Marie, I., Dreno, B. and Laboisse, C.L. (1998). Control of growth and differentiation of normal human epithelial cells through the manipulation of reactive nitrogen species. *Biochem. J.* 331(3), 713–717.

van Loon, L.C., Geraats, B.P. and Linthorst, H.J. (2006). Ethylene as a modulator of disease resistance in plants. *Trends Plant Sci.* 11,184–191.

Wang, T., Ward, M., Grabowski, P. and Secombes, C.J. (2001). Molecular cloning, gene organization and expression of rainbow trout (*Oncorhynchus mykiss*) inducible nitric oxide synthase (iNOS) gene. *Biochem. J.* 358,747–755.

Wang, Y., DeSilva, A.W., Goldenbaum, G.C. and Dickerson, R.R. (1998). Nitrogen oxide production by simulated lighting: Dependence on current, energy, and pressure. *J. Geophys. Res.* 103,19149–19160.

Weitzberg, E. and Lundberg, J.O. (1998). Nonenzymatic nitric oxide production in humans. *Nitric Oxide* 2,1–7.

Weller, R., Pattullo, S., Smith, L., Golden, M., Ormerod, A. and Benjamin, N. (1996). Nitric oxide is generated on the skin surface by reduction of sweat nitrate. *J. Invest. Dermatol.* 107,327–331.

Wink, D.A., Grisham, M.B., Miles, A.M., Nims, R.W., Krishna, M.C., Pacelli, R., Teague, D., Poore, C.M., Cook, J.A. and Ford, P.C. (1996a). Determination of selectivity of reactive nitrogen oxide species for various substrates. *Methods Enzymol.* 268,120–130.

Wink, D.A., Grisham, M.B., Mitchell, J.B. and Ford, P.C. (1996b). Direct and indirect effects of nitric oxide in chemical reactions relevant to biology. *Methods Enzymol.* 268,12–31.

Wink, D.A., Hanbauer, I., Grisham, M.B., Laval, F., Nims, R.W., Laval, J., Cook, J., Pacelli, R., Liebmann, J., Krishna, M., Ford, P.C. and Mitchell, J.B. (1996c). Chemical biology of nitric oxide: Regulation and protective and toxic mechanisms. *Curr. Top. Cell. Regul.* 34,159–187.

Wink, D.A. and Mitchell, J.B. (1998). Chemical biology of nitric oxide: Insights into regulatory, cytotoxic, and cytoprotective mechanisms of nitric oxide. *Free Radic. Biol. Med.* 25,434–456.

44

Wood, J. and Garthwaite, J. (1994). Models of the diffusional spread of nitric oxide: implications for neural nitric oxide signalling and its pharmacological properties. *Neuropharmacology* 33,1235–1244.

Wright, M.M., Schopfer, F.J., Baker, P.R., Vidyasagar, V., Powell, P., Chumley, P., Iles, K.E., Freeman, B.A. and Agarwal, A. (2006). Fatty acid transduction of nitric oxide signaling: Nitrolinoleic acid potently activates endothelial heme oxygenase 1 expression. *Proc. Natl. Acad. Sci. USA* 103,4299–4304.

Yamasaki, H. (2000). Nitrite-dependent nitric oxide production pathway: Implications for involvement of active nitrogen species in photoinhibition in vivo. *Philos. Trans. R. Soc. Lond. B: Biol. Sci.* 355,1477–1488.

Yuda, M., Hirai, M., Miura, K., Matsumura, H., Ando, K. and Chinzei, Y. (1996). cDna cloning, expression and characterization of nitric-oxide synthase from the salivary glands of the blood-sucking insect *Rhodnius prolixus*. *Eur. J. Biochem.* 242,807–812.

Zafiriou, O.C., McFarland, M. and Bromund, R.H. (1980). Nitric oxide in sea water. *Science* 207,637–639.

Zemojtel, T., Kolanczyk, M., Kossler, N., Stricker, S., Lurz, R., Mikula, I., Duchniewicz, M., Schuelke, M., Ghafourifar, P., Martasek, P., Vingron, M. and Mundlos, S. (2006). Mammalian mitochondrial nitric oxide synthase: Characterization of a novel candidate. *FEBS Lett.* 580,455–462.

Zemojtel, T., Penzkofer, T., Dandekar, T. and Schultz, J. (2004). A novel conserved family of nitric oxide synthase? *Trends Biochem. Sci.* 29,224–226.

Zemojtel, T., Wade, R.C. and Dandekar, T. (2003). In search of the prototype of nitric oxide synthase. *FEBS Lett.* 554,1–5.

Zhang, X., Kim, W.S., Hatcher, N., Potgieter, K., Moroz, L.L., Gillette, R. and Sweedler, J.V. (2002). Interfering with nitric oxide measurements. 4,5-Diaminofluorescein reacts with dehydroascorbic acid and ascorbic acid. *J. Biol. Chem.* 277,48472–48478.

Zhang, Z., Naughton, D., Winyard, P.G., Benjamin, N., Blake, D.R. and Symons, M.C. (1998). Generation of nitric oxide by a nitrite reductase activity of xanthine oxidase: A potential pathway for nitric oxide formation in the absence of nitric oxide synthase activity. *Biochem. Biophys. Res. Commun.* 249,767–772.

Zumft, W.G. (1993). The biological role of nitric oxide in bacteria. *Arch. Microbiol.* 160,253–264.

Zumft, W.G. (2002). Nitric oxide signaling and NO dependent transcriptional control in bacterial denitrification by members of the FNR-CRP regulator family. *J. Mol. Microbiol. Biotechnol.* 4,277–286.

Zweier, J.L., Samouilov, A. and Kuppusamy, P. (1999). Non-enzymatic nitric oxide synthesis in biological systems. *Biochim. Biophys. Acta* 1411,250–262.

Zweier, J.L., Wang, P., Samouilov, A. and Kuppusamy, P. (1995a). Enzymatic/non-enzymatic formation of nitric oxide. *Nat. Med.* 1,1103–1104.

Zweier, J.L., Wang, P., Samouilov, A. and Kuppusamy, P. (1995b). Enzyme-independent formation of nitric oxide in biological tissues. *Nat. Med.* 1,804–809.

Nitric oxide biogenesis, signalling and roles in molluscs: The *Sepia officinalis* paradigm

Anna Palumbo[1],* and Marco d'Ischia[2]

[1]*Zoological Station Anton Dohrn, Villa Comunale, 80121 Naples, Italy*
[2]*Department of Organic Chemistry and Biochemistry, University of Naples Federico II, Via Cinthia 4, 80126 Naples, Italy*

Abstract. The past decade has witnessed a burst of interest in the biological roles of nitric oxide (NO) and its signalling pathway in molluscs. Several roles of NO have been demonstrated in different functions often related to specific behaviours such as olfaction, feeding, learning, defence, development and movement. The complex roles of NO in the ink gland and nervous system of the cuttlefish *Sepia officinalis* are paradigmatic in this respect. Stimulation of NO production via the *N*-methyl-D-aspartate (NMDA) receptor and cyclic guanosine monophosphate (cGMP) signal transduction pathway induces a series of events that, though apparently unrelated, come together to control overall regulation of the ink defence system. It is the aim of this chapter to provide a brief overview of the biogenesis and roles of NO in molluscs with specials reference to studies carried out in the authors' laboratories on the ink system of *S. officinalis*. A prospective analysis of future advances in the field is also offered.

Keywords: dopa; dopamine; ink gland; invertebrates; melanin; molluscs; nitration; nitric oxide; nitric oxide synthase; *S. officinalis*; α-tubulin; tyrosinase; ink; ink sac; catecholamine; melanosome; NMDA receptor; glutamate; cuttlefish; tyrosine.

Introduction

The discovery of nitric oxide (NO) as a unique endogenous regulator of blood flow, a neurotransmitter in the central and peripheral nervous system, and a pathophysiological mediator of inflammation and host defence in humans and mammals has stimulated during the past decades intense and extensive research aimed at unravelling its origin and functions into nearly every living organism (Bruckdorfer, 2005; Moncada *et al.*, 1991). The enzyme NO synthase (NOS) occurs in mammals in at least three distinct isoforms, neuronal NOS or NOS1/NOS I (nNOS), inducible NOS or NOS2/NOS II (iNOS) and endothelial NOS or NOS3/NOS III (eNOS). All NOS isoforms are homodimers with subunits of 130–160 kDa. All have binding sites for NADPH, FAD and FMN near the carboxyl terminus (the reductase domain), and binding sites for tetrahydrobiopterin

---

Corresponding author: Tel.: 39-81-5833276(293). Fax: 39-81-7641355.
E-mail: palumbo@szn.it (A. Palumbo).

ADVANCES IN EXPERIMENTAL BIOLOGY
VOLUME 01   ISSN 1872-2423
DOI: 10.1016/S1872-2423(07)01002-2

(BH_4) and heme near the amino terminus (the oxygenase domain). The reductase and oxygenase domains are linked by a calmodulin (CaM) binding site, which is believed to promote electron transfer from the cofactors in the reductase domain to heme, leading to NO production (Alderton et al., 2001; Ghosh and Salerno, 2003; Li and Poulos, 2005; Stuehr, 2004).

Parallel to the studies on NOS and NO signalling in humans and mammals, interest in the biogenesis, signalling and roles of NO in invertebrates has grown steadily, following demonstration of the involvement of NO in diverse physiological processes and in a variety of organisms (Bicker, 2001; Davies, 2000; Enikolopov et al., 1999; Moroz, 2000; Moroz and Gillette, 1995; Palumbo, 2005; Torreilles, 2001; Trimmer et al., 2004; Walker et al., 1999). This chapter focuses on NO synthesis, regulation and signalling in molluscs, with special reference to the cuttlefish Sepia officinalis. It aims to update the interested reader on the latest advances in this rapidly unravelling area of NO research, largely as a result of continuing efforts in the authors' laboratory and other research centres. It also seeks to stimulate new studies directed at gaining a more complete understanding of the many roles NO plays in invertebrates as compared to mammals.

Molluscs

An overall view of the roles and functions of NO in molluscs is schematically illustrated in Fig. 1.

Gastropoda

Among the Gastropoda Pulmonata, NO has been reported in Biomphalaria glabrata, Helisoma trivolvis, Achatina fulica, in different species of Helix and Limax, in Lymnaea stagnalis and in Planorbarius corneus.

The snail B. glabrata is intermediate host for the transmission of the trematode parasite Schistosoma mansoni, the causative agent of the human tropical disease schistosomiasis. It has been reported that NO is involved in killing Schistosoma mansoni sporocysts by hemocytes from resistant snails (Bayne et al., 2001; Hahn et al., 2001). In the pond snail H. trivolvis NO regulates early embryonic behaviour, stimulating ciliary beating of embryos within the egg capsule during development (Cole et al., 2002; Doran et al., 2003). It is also a signalling molecule in neuronal development (Trimm and Rehder, 2004; Welshhans and Rehder, 2005). In the snails A. fulica and Helix aspersa NO modulates heart

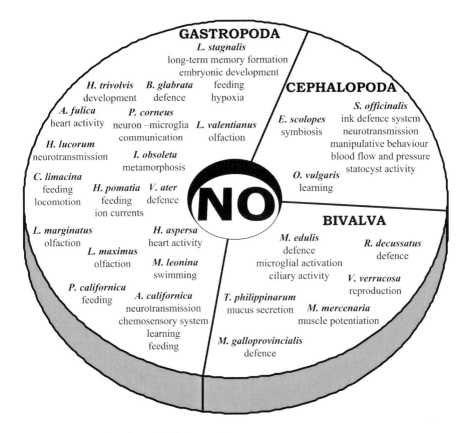

Fig. 1. Biological roles of NO in molluscs.

activity (White *et al.*, 2004). In the common snail *Helix lucorum* NO is involved in neural transmission to intestinal muscles, and its production is blocked during dormancy (Roszer *et al.*, 2004). In the snail *Helix pomatia* NO is involved in food-attraction conditioning (Teyke, 1996). Blocking NOS prior to conditioning significantly affects acquisition of memory, impairing the food-finding ability, whereas memory recall and olfactory orientation are not disturbed, as revealed by the ability of the snail to locate the conditioned food. The effect of NO on ion currents in snail neurons has been studied and it has been found that NO increases excitability by depressing a calcium-activated potassium current (Zsombok *et al.*, 2000).

Interestingly, a new class of NOS enzymes has been identified in *H. pomatia*. From an expression library, a cDNA was isolated which codes for a 60 kDa protein recognised by an antibody to human neuronal NOS (Huang *et al.*, 1997). This protein does not contain consensus binding sites for NADPH, FAD, arginine or CaM and it is similar to GTPase proteins found in bacteria, insects, mammals and plants (Zemojtel *et al.*, 2004).

In the terrestrial slug *Limax marginatus* NO is involved through the cyclic guanosine monophosphate (cGMP) pathway in olfactory processing, modulating electrical oscillations in the procerebral lobes (Fujie *et al.*, 2002, 2005). In the terrestrial mollusc *Limax maximus* NO, together with carbon monoxide, modulates oscillations of olfactory interneurons (Gelperin *et al.*, 1996, 2000). In the terrestrial slug *Limax valentianus* NO plays a crucial role in fine olfactory discrimination (Sakura *et al.*, 2004). In the pond snail *L. stagnalis* NO is involved in long-term memory formation (Kemenes *et al.*, 2002; Korneev *et al.*, 2005), embryonic development (Serfozo and Elekes, 2002), feeding (Elphick *et al.*, 1995; Kobayashi *et al.*, 2000a,b; Korneev *et al.*, 2002; Moroz, 2000), and the cardiorespiratory response to hypoxia (Taylor *et al.*, 2003). Important insights into the mechanisms of NOS regulation have been gained in *L. stagnalis*: in addition to the full-length mRNA encoding for NOS, two pseudo-NOS transcripts, originated by DNA inversion, have been identified. These regulate NOS expression at the translational level by a natural antisense mechanism and at a post-translational level by the formation of nonfunctional heterodimers (Korneev and O'Shea, 2002; Korneev *et al.*, 1998, 1999). Differential regulation of the activity of this group of related genes occurs during long-term memory formation and in the neural network involved in this process (Korneev *et al.*, 2005). Moreover, gene silencing experiments have demonstrated that the expression of NOS is essential for normal feeding behaviour (Korneev *et al.*, 2002). In the freshwater snail *P. corneus* NO acts as a messenger molecule in neuron–microglia communication in the central nervous system (CNS) (Peruzzi *et al.*, 2004).

Among the Gastropoda Prosobranchia, it has been reported that in the marine snail *Ilyanassa obsoleta* NO acts as an endogenous inhibitor of metamorphosis (Bishop and Brandhorst, 2003; Leise *et al.*, 2004), and in *Viviparus ater* NO is used as bactericidal substance in the defence mechanism (Franchini *et al.*, 1995; Ottaviani *et al.*, 1993).

Regarding the Gastropoda Opistobranchia, NO has been reported in *Aplysia californica*, *Clione limacina*, *Melibe leonina* and *Pleurobranchaea californica*. The sea slug *A. californica* is characterised by large identifiable neurons that allow identification of the neurons involved in specific behaviours, and thus related to biochemical events in individual cells and specific functions. In particular, the serotonergic feeding neural circuit, the metacerebral cell, has received a good deal of attention. It has been reported that NO, via cGMP, mediates the membrane responses of metacerebral neurons (Koh and Jacklet, 1999), and that NO, serotonin and histamine all depolarise and increase the excitability of these neurons by diverse mechanisms (Jacklet and Tieman, 2004; Jacklet *et al.*, 2004). NO has also been shown to modulate acetylcholine release in the buccal and the abdominal ganglia (Meulemans *et al.*, 1995) and to be involved in chemosensory processes, as revealed by localisation of putative nitrergic neurons in peripheral chemosensory areas (Moroz, 2006). Learning experiments have revealed that during training in *Aplysia*, NO signalling plays a critical role in the formation of multiple memory processes (Katzoff *et al.*, 2002). Furthermore, NO is involved in the regulation of feeding (Lovell *et al.*, 2000). The NOS from the mollusc *A. californica* has been cloned (NCBI Accession ID AF288780 – Sandreyev *et al.*, 2000) and biochemically characterised (Bodnarova *et al.*, 2005). In the opistobranchia *C. limacina* NO is involved in feeding and locomotion (Moroz *et al.*, 2000). In the nudibranch *M. leonina* NO is used in the CNS to modulate swimming (Newcomb and Watson, III, 2002). In *P. californica* NO is involved in feeding behaviour (Hatcher *et al.*, 2006; Moroz, 2000).

Bivalva

Among the Bivalva, NO has been reported in *Mercenaria mercenaria*, *Mytilus edulis*, *Mytilus galloprovincialis*, *Ruditapes decussatus*, *Tapes philippinarum* and *Venus verrucosa*. In the clam *M. mercenaria* the potentiation of gill muscle is mediated by a NO–cGMP–PKG (protein kinase G) signalling pathway (Gainey and Greenberg, 2003). In the bivalves *M. edulis* and *M. galloprovincialis* NO is involved in defence mechanisms: NO is produced by the hemocytes to kill pathogens (Arumugam *et al.*, 2000; Franchini *et al.*, 1995; Ottaviani *et al.*, 1993). The same mechanism is also used by the carpet shell clam *R. decussatus* (Tafalla *et al.*, 2003). In the mussel *M. edulis* NO also modulates microglial activation and the physiological control of ciliary activity (Cadet, 2004; Stefano *et al.*, 2004). In *T. philippinarum* NO has a regulatory role in mucus secretion (Calabro *et al.*, 2005), and in *V. verrucosa* it is involved in the control of reproduction (Barbin *et al.*, 2003).

Cephalopoda

NO is involved in the complex mechanisms implicated in the initiation and maintenance of symbiont infection of the light organ of the Hawaiian bobtail squid, *Euprymna scolopes*, by the bacterium *Vibrio fischeri* (Davidson *et al.*, 2004). The production of NO begins during light organ embryogenesis, and reaches its highest levels in the light organ of newly hatched animals. Interestingly, it has been suggested that NO regulates the number of bacteria. Indeed, NO is released into the mucus secreted by the light organ, where the symbiotic bacteria aggregate before migrating into the final sites of colonisation. In *Octopus vulgaris* NO is reported to be involved in both visual and tactile learning (Robertson *et al.*, 1994, 1996). Intramuscular injections of the NOS inhibitor N^G-nitro-L-arginine methyl ester (L-NAME) block both types of learning.

These studies underscored the importance of molluscs as model systems for investigating NO biogenesis, signalling and roles in different functions, often underlying specific behaviours. A noticeable example is provided by the cephalopod *Sepia officinalis*, in which NO has been implicated in manipulative behaviour, modulation of statocyst activity, regulation of blood flow and blood pressure, and the complex processes and neural pathways associated with the ink defence system. Very recently, *S. officinalis* NOS has been cloned, allowing the first insight into the structure of a cephalopod NOS. A summary of current knowledge of the NO signalling pathway in *S. officinalis* is provided below.

Sepia officinalis

The first evidence for the presence of NOS activity in a cephalopod came in 1994 from a study by Chichery and Chichery (1994). These authors showed that NADPH-diaphorase activity was selectively localised in the CNS of *S. officinalis*. In particular, only the neuropils of the spines of the peduncle and posterior anterior basal lobes were found to exhibit an intense positive staining, which by contrast was completely absent from the cell bodies. These two regions of the brain are thought to constitute cerebellar analogues (Hobbs and Young, 1973; Messenger, 1967a,b). In light of the structural analogies of the peduncle and anterior basal lobes with the cerebellum, these findings were taken to suggest that NO may be involved as a signal molecule in learned motor skills.

Some years later, our group reported for the first time the occurrence of a Ca^{2+}/CaM-dependent NOS activity in the ink gland of *S. officinalis* (Palumbo *et al.*, 1997) by measuring the formation of radiolabelled citrulline from [^{14}C] L-arginine. Moreover, immunohistochemical analyses showed the presence in the ink gland of NMDAR1 (*N*-methyl-D-aspartate receptor 1) glutamate receptors, raising the possibility that the glutamate–NO neurotransmission pathway played a regulatory role in melanin synthesis and related processes associated with the inking mechanism.

Inking is a characteristic behaviour adopted by nearly all coleoid cephalopods, *i.e., S. officinalis, Loligo vulgaris* and *Octopus vulgaris*, to confuse predators and alert conspecifics to danger while retreating. It represents the final event of the defence behaviour which initially consists of the development of a series of chromatophore patterns involving mantle spots and strips, followed by body blanching and darkening, jet propulsion movements, erratic jetting and finally inking (Hanlon and Messenger, 1996). The ink is a suspension of black melanin granules produced by the ink gland. The ink gland is composed of two distinct zones possessing different histological and biochemical characteristics: the inner glandular epithelium and the outer glandular epithelium (Fig. 2). In the inner glandular epithelium the cells are immature, do not produce melanin and are considered young cells. They gradually mature and migrate towards the external portion where they differentiate,

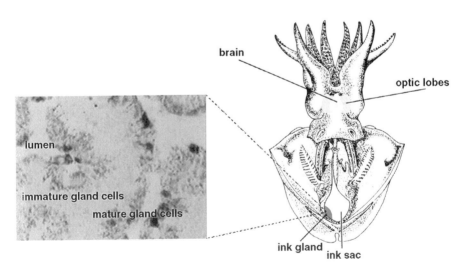

Fig. 2. The cuttlefish *S. officinalis.* The central nervous system, comprising the brain and optic lobes, is shown together with the ink-producing system. Note the ink gland and the ink sac. Inset: microscopic structure of the ink gland. (See Colour Plate Section in this book).

acquire the ability to produce melanin, and form the outer glandular epithelium. Melanin formation within mature cells shares many features in common with the process of melanogenesis in epidermal melanocytes (Hearing, 1999, 2005). The initial step of the process is the tyrosinase-catalysed oxidation of L-tyrosine to an unstable ortho-quinone, which is rapidly converted to melanin within specific organelles called melanosomes. An overview of the biochemical events leading to melanin formation in *S. officinalis* has recently been provided (Palumbo, 2003). In addition to melanin, ink gland cells are able to synthesise 3,4-dihydroxyphenylalanine (DOPA) and dopamine (DA) through the tyrosine hydroxylase and DOPA decarboxylase route (Fiore *et al.*, 2004) (Fig. 3).

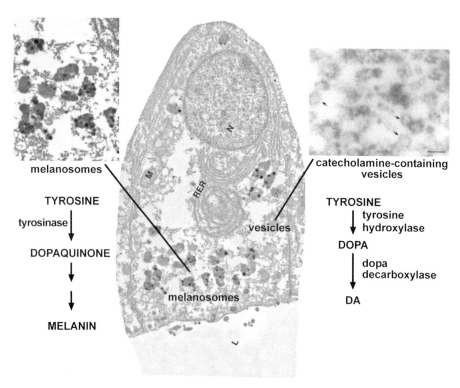

Fig. 3. Tyrosine metabolism in a mature ink gland cell. The ultrastructure of a mature ink gland cell shows the nucleus (N), a highly developed rough endoplasmic reticulum (RER), a mitochondrion (M), several melanosomes and catecholamine-containing vesicles. The two biochemical pathways leading to melanin production and 3,4-dihydroxyphenylalanine (DOPA) and dopamine (DA) biosynthesis are highlighted.

Within mature cells, DA and melanin are present in separate cellular compartments: whereas melanin is confined to the melanosomes, DA appears to be segregated within electron-dense vesicles resembling the catecholamine-containing vesicles reported in other systems. When maturation of the cells is complete, the melanin and other cellular components are shed into the lumen of the gland by a poorly defined mechanism. They are then transferred to the ink sac, which serves as reservoir for the black ink that the animal ejects when frightened (Fig. 2).

The biochemical mechanisms involved in the regulation of tyrosinase activity and ink production are still little understood. The discovery in 1997 of the presence of NOS and NMDAR1 glutamate receptors in the ink gland (Palumbo et al., 1997) raised the possibility that the glutamate–NO transmitter pathway is involved in the specific signalling system that keeps the biochemical machinery for ink production active to ensure that sufficient amounts of melanin are available when necessary. To test this hypothesis, ink glands were incubated in the presence of various compounds with the aim of stimulating or inhibiting the endogenous production of NO, and the effects on the activity of tyrosinase were determined (Palumbo et al., 2000). Glutamate and NMDA were found to cause a dramatic increase in tyrosinase activity, whereas the NOS inhibitor L-NAME suppressed the NMDA-induced stimulation of tyrosinase. The role of NO in glutamate-induced activation of tyrosinase was confirmed by incubation experiments with the NO donor 2-(N,N-diethylamino)-diazenolate-2-oxide (DEA/NO), which elicited an increase in tyrosinase activity. 8-Bromo-cyclic GMP, a permeable and nonhydrolysable analogue of cGMP, resulted in a substantial increase of tyrosinase activity, suggesting that cGMP mediates the stimulating effect of NMDA. This conclusion was corroborated by the finding that stimulation of ink glands with NMDA resulted in a more than six-fold increase of cGMP levels compared to basal levels. Immunohistochemical evidence indicated that enhanced cGMP production is localised largely in the mature part of the ink gland. Tyrosinase activation by NO apparently involved phosphorylation through the action of PKG, without de novo synthesis of the enzyme. NMDA receptor stimulation and exposure to NO in the presence of DOPA, an excellent substrate of tyrosinase, resulted in a marked increase in the melanin content of the ink gland. These results demonstrate that the excitatory neurotransmitter L-glutamate promotes activation of tyrosinase at the appropriate stage of maturation and maintains the enzyme in an active state, acting via NMDA receptors, as schematically outlined in Fig. 4. Activation of the NMDA glutamate receptor causes an influx of calcium, which binds to CaM and activates

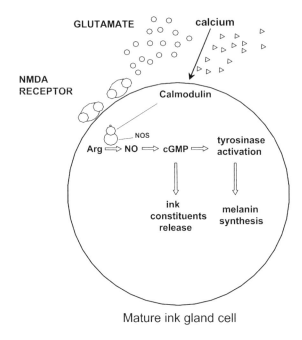

Fig. 4. NMDA–NO–cGMP signalling pathway in mature ink gland cells: tyrosinase activation, melanin synthesis and secretion of ink constituents.

the NOS to produce NO. NO targets guanylyl cyclase to produce higher levels of cGMP, which activates tyrosinase by phosphorylation and promotes melanin formation.

Besides acting on melanin production, NO was found to affect the secretion of ink constituents into the lumen (Fiore *et al.*, 2004). NMDA stimulation or treatment with a NO donor of ink glands exposed to [^{14}C] tyrosine resulted in a marked decrease in the levels of radioactive DOPA and DA along with a consistent loss of DA immunoreactivity in mature cells. This loss was due to release of radiolabelled DOPA and DA in the incubation waters. The involvement of cGMP in the ink constituents' release was shown by incubation experiments with [^{14}C] tyrosine in the presence of 8-bromo-cyclic GMP or a guanylyl cyclase inhibitor. Thus, DOPA and DA appear to be released from the ink glands by processes controlled through the NMDA–NO–cGMP signalling pathway (Fig. 4).

Enzymatic assays have revealed that a Ca^{2+}-dependent NOS is also found in the protein extracts of brain and optic lobes (Palumbo *et al.*, 1999). Moreover, a NOS-like immunoreactivity and operation of the glutamate–NO–cGMP signalling pathway have also been demonstrated in the cuttlefish nervous system, both at the central and peripheral levels. Cephalopods have arguably the largest and most complex nervous systems amongst the invertebrates, but despite the squid giant axon being one of the best-studied nerve cells in neuroscience, and the availability of much information on the morphology of some cephalopod brains, there is surprisingly little known about the operation of the neural networks that underlie the sophisticated range of behaviour these animals display. The brain of *S. officinalis* is formed by the complex of supraoesophageal mass and suboesophageal mass, as distinct from the optic lobes. It is essentially organised hierarchically: motor programmes, usually originating in the optic lobes, are executed via higher motor centres, where specific motor commands are generated and directed to the appropriate sets of motoneurons in the lower centres (Messenger, 1983). Immunohistochemical mapping of *S. officinalis* CNS showed specific localisation of NOS-like and NMDAR2/3-like immunoreactivities in the regions and nervous fibres controlling the inking system, *i.e.,* the latero-ventral palliovisceral lobe, the visceral lobe, the pallial and visceral nerves, as well as in the sphincters and wall of the ink sac (Palumbo *et al.,* 1999). Parallel studies revealed that NOS is also present in other regions of the CNS involved in a variety of functions, including feeding, motor, learning, visual and olfactory systems (Di Cosmo *et al.,* 2000).

In addition to the cGMP-dependent signalling pathway, in *S. officinalis* brain the NO signal can also be transduced through cGMP-independent processes, such as protein nitration (Schopfer *et al.,* 2003). Interestingly, a major target of nitration proved to be α-tubulin (Palumbo *et al.,* 2002). After activation of endogenous NOS by treating optic lobes with NMDA, an approximately 50% decrease of the α-tubulin band at 30 min was observed, with respect to the control, followed by a partial recovery (75%) at 4 h. The 3-nitrotyrosine immunopositivity shows an opposite trend, with an increase after 30 min stimulation followed by a substantial decrease. Overall these results point to the existence of an α-tubulin-mediated signalling in the optic lobes (Fig. 5). After NOS activation, the nitrated α-tubulin is formed, and as soon as the nitration process exceeds a threshold value, a specific mechanism becomes operative which not only efficiently degrades the modified protein but also acts to restore basal levels of functionally active α-tubulin by protein synthesis. On this basis, α-tubulin nitration may represent a natural mechanism of cytoskeletal protein turnover.

Structural insight into cuttlefish NOS has recently been gained by cloning of the enzyme (Scheinker *et al.,* 2005). Two splicing variants, *Sepia* NOSa and NOSb, differing by 18 nucleotides, were found in the

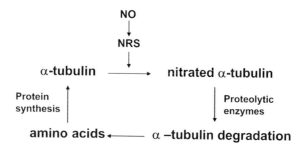

Fig. 5. α-Tubulin nitration and turnover in *S. officinalis* nervous system. NRS, nitrogen-reactive species.

nervous system and in the ink gland. RT-PCR (reverse transcriptase polymerase chain reaction) of *S. officinalis* NOS mRNA showed expression of both forms in the optic lobes and preferential expression of *Sepia* NOSb in the ink gland, suggesting specific biological functions for these isoforms. Phylogenetic analysis of NOS sequences in various species confirms the conserved nature of *S. officinalis* NOS, and particularly conservation of the various cofactor-binding domains: heme, CaM, FMN, FAD and NADPH (Fig. 6). The identity of individual NOSs with that of *S. officinalis* NOSs at the level of the cofactor-binding sites is high. The cuttlefish enzyme, together with the enzyme of another mollusc, *Lymnaea,* has a shorter N-terminal region than mammalian nNOS. *In situ* hybridisation experiments with a *S. officinalis* NOS antisense probe able to recognise both splicing forms of *S. officinalis* NOS revealed that the protein is expressed in the CNS and ink gland. In the CNS, the analysis has been restricted to those regions that are related to the ink defence system, such as the palliovisceral lobe. From this lobe arises the visceral nerve, from which derives a branch, the ink sac nerve, which innervates the ink sac. *S. officinalis* NOS mRNA is expressed in the small inner neurons of both the latero-ventral palliovisceral and central palliovisceral lobes of *S. officinalis.* In the ink gland NOS is expressed both in immature and mature cells.

In addition to being involved in defence and neurotransmission, NO plays other roles in cuttlefish. It regulates blood flow and blood pressure, acting as a vasodilatatory mediator as in mammals (Schipp and Gebauer, 1999). NO donors induce vasodilatation of *S. officinalis* cephalic aorta, although NOS is not present in the endothelium but seems to be located within the nerve tissue supplying the vessel.

NO is also involved in the manipulative behaviour of *S. officinalis* (Halm *et al.*, 2003). Manipulative behaviour requires extensive chemo-tactile

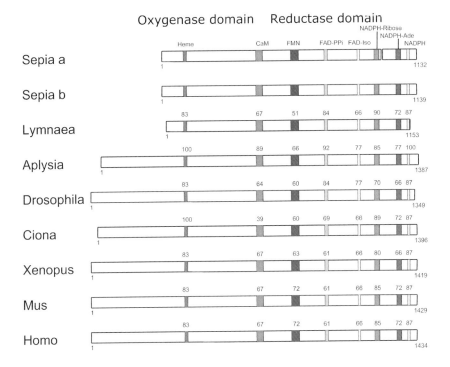

Numbers indicate the identity (percentage) of individual NOSs to Sepia NOSs

Fig. 6. Amino acid alignment of *Sepia* NOSa and *Sepia* NOSb with other known NOS sequences. Numbers indicate the percentage sequence identity of individual NOSs to *Sepia* NOSs at the level of the conserved domains. Mus, *Mus musculus*. NOS from *Mus musculus* and *Homo sapiens* refer to neuronal forms.

sensory processing, fine motor control and probably motor learning processes. Cuttlefish actively explored crabs with their arms and lips, which are NADPH-diaphorase positive, and received information about the toughness of the carapace or the progress of paralysis of the crab. Injection of the NOS inhibitor L-NAME into the vena cava of the cuttlefish resulted in an increase in the latency of prey paralysis, suggesting that NO could play an important role in the transmission of chemical and/or tactile information.

NO is also involved in modulating the resting activity of crista afferent fibres in the statocyst (the equilibrium receptor organ) of *S. officinalis* (Tu and Budelmann, 2000).

58

Conclusions and perspectives

Following publication of the seminal papers on the gastropod *L. stagnalis* (Korneev *et al.*, 1998, 2002, 2005), the presence of NOS has also been demonstrated in other gastropods, bivalves and cephalopods. As a result of intense efforts, molluscs stand today among the most useful model systems in which to investigate NO biogenesis, signalling and roles in different functions, often related to specific behaviours, as highlighted in Fig. 1. These studies on molluscs allowed the demonstration of an association between NO and olfaction, feeding, learning, defence, development and movement, suggesting that the NO signalling pathway has great versatility and has evolved to play a variety of roles in widely varying biological settings and cellular environments. The complex roles of NO in *S. officinalis* ink gland and nervous systems are paradigmatic in this respect. Stimulation of NO production via the NMDA receptor and the cGMP signal transduction pathway has been shown to induce a series of events that, though apparently unrelated, come together to control overall regulation of the ink defence system. Useful insights are coming from investigations of the anatomical distribution and functional significance of NOS in various areas and regions of the CNS of *S. officinalis* as well as other molluscs, mainly in the context of behaviour and neuronal plasticity. Additional advances that appear to be within our grasp will hopefully unravel the complex mechanisms controlling the maturation and differentiation of ink gland cells. Another fascinating area for investigation concerns the role of post-translational protein modification by NO-derived species in the overall regulatory mechanisms mediated by NO. Relevant in this respect is the demonstration of temporary patterns of α-tubulin nitration in *S. officinalis* brain (Palumbo *et al.*, 2002). Thus, what we have discussed so far is probably only the tip of an iceberg, and it is likely that in the near future new regulatory pathways and functional roles will be uncovered.

References

Alderton, W. K., Cooper, C. E. and Knowles, R. G. (2001). Nitric oxide synthases: Structure, function and inhibition. *Biochem. J.* 357,593–615.
Arumugam, M., Romestand, B., Torreilles, J. and Roch, P. (2000). In vitro production of superoxide and nitric oxide (as nitrite and nitrate) by *Mytilus galloprovincialis* haemocytes upon incubation with PMA or laminarin or during yeast phagocytosis. *Eur. J. Cell Biol.* 79,513–519.

Barbin, L., Boarini, I., Borasio, P. G., Barion, P., Fiorini, S., Rossi, R. and Biondi, C. (2003). Nitric oxide-mediated arachidonic acid release from perifused *Venus verrucosa* oocytes. *Gen. Comp. Endocrinol.* 130,215–221.

Bayne, C. J., Hahn, U. K. and Bender, R. C. (2001). Mechanisms of molluscan host resistance and of parasite strategies for survival. *Parasitology* 123,S159–S167.

Bicker, G. (2001). Nitric oxide: An unconventional messenger in the nervous system of an orthopteroid insect. *Arch. Insect Biochem. Physiol.* 48,100–110.

Bishop, C. D. and Brandhorst, B. P. (2003). On nitric oxide signaling, metamorphosis, and the evolution of biphasic life cycles. *Evol. Dev.* 5,542–550.

Bodnarova, M., Martasek, P. and Moroz, L. L. (2005). Calcium/calmodulin-dependent nitric oxide synthase activity in the CNS of *Aplysia californica*: Biochemical characterization and link to cGMP pathways. *J. Inorg. Biochem.* 99,922–928.

Bruckdorfer, R. (2005). The basics about nitric oxide. *Mol. Aspects Med.* 26,3–31.

Cadet, P. (2004). Nitric oxide modulates the physiological control of ciliary activity in the marine mussel *Mytilus edulis* via morphine: Novel mu opiate receptor splice variants. *Neuroendocrinol. Lett.* 25,184–190.

Calabro, C., Albanese, M. P., Martella, S., Licata, P., Lauriano, E. R., Bertuccio, C. and Licata, A. (2005). Glycoconjugate histochemistry and nNOS immunolocalization in the mantle and foot epithelia of *Tapes philippinarum* (bivalve mollusc). *Folia Histochem. Cytobiol.* 43,151–156.

Chichery, R. and Chichery, M. P. (1994). NADPH-diaphorase in a cephalopod brain (Sepia): Presence in an analogue of the cerebellum. *Neuroreport* 5,1273–1276.

Cole, A. G., Mashkournia, A., Parries, S. C. and Goldberg, J. I. (2002). Regulation of early embryonic behavior by nitric oxide in the pond snail *Helisoma trivolvis*. *J. Exp. Biol.* 205,3143–3152.

Davidson, S. K., Koropatnick, T. A., Kossmehl, R., Sycuro, L. and McFall-Ngai, M. J. (2004). NO means 'yes' in the squid-vibrio symbiosis: Nitric oxide (NO) during the initial stages of a beneficial association. *Cell Microbiol.* 6,1139–1151.

Davies, S. (2000). Nitric oxide signalling in insects. *Insect Biochem. Mol. Biol.* 30, 1123–1138.

Di Cosmo, A., Di Cristo, C., Palumbo, A., d'Ischia, M. and Messenger, J. B. (2000). Nitric oxide synthase (NOS) in the brain of the cephalopod *Sepia officinalis*. *J. Comp. Neurol.* 428,411–427.

Doran, S. A., Tran, C. H., Eskicioglu, C., Stachniak, T., Ahn, K. C. and Goldberg, J. I. (2003). Constitutive and permissive roles of nitric oxide activity in embryonic ciliary cells. *Am. J. Physiol. Regul. Integr. Comp. Physiol.* 285,R348–R355.

Elphick, M. R., Kemenes, G., Staras, K. and O'Shea, M. (1995). Behavioral role for nitric oxide in chemosensory activation of feeding in a mollusc. *J. Neurosci.* 15,7653–7664.

Enikolopov, G., Banerji, J. and Kuzin, B. (1999). Nitric oxide and *Drosophila* development. *Cell Death Differ.* 6,956–963.

Fiore, G., Poli, A., Di Cosmo, A., d'Ischia, M. and Palumbo, A. (2004). Dopamine in the ink defence system of *Sepia officinalis*: Biosynthesis, vesicular compartmentation in mature ink gland cells, nitric oxide (NO)/cGMP-induced depletion and fate in secreted ink. *Biochem. J.* 378,785–791.

Franchini, A., Conte, A. and Ottaviani, E. (1995). Nitric oxide: An ancestral immunocyte effector molecule. *Adv. Neuroimmunol.* 5,463–478.

Fujie, S., Aonuma, H., Ito, I., Gelperin, A. and Ito, E. (2002). The nitric oxide/cyclic GMP pathway in the olfactory processing system of the terrestrial slug *Limax marginatus*. *Zoolog. Sci.* 19,15–26.

Fujie, S., Yamamoto, T., Muratami, J., Hatakeyama, D., Shiga, H., Suzuki, N. and Ito, E. (2005). Nitric oxide synthase and soluble guanylyl cyclase underlying the modulation of electrical oscillations in a central olfactory organ. *J. Neurobiol.* 62,14–30.

Gainey, L. F., Jr. and Greenberg, M. J. (2003). Nitric oxide mediates seasonal muscle potentiation in clam gills. *J. Exp. Biol.* 206,3507–3520.

Gelperin, A., Flores, J., Raccuia-Behling, F. and Cooke, I. R. (2000). Nitric oxide and carbon monoxide modulate oscillations of olfactory interneurons in a terrestrial mollusk. *J. Neurophysiol.* 83,116–127.

Gelperin, A., Kleinfeld, D., Denk, W. and Cooke, I. R. (1996). Oscillations and gaseous oxides in invertebrate olfaction. *J. Neurobiol.* 30,110–122.

Ghosh, D. K. and Salerno, J. C. (2003). Nitric oxide synthases: Domain structure and alignment in enzyme function and control. *Front. Biosci.* 8,193–209.

Hahn, U. K., Bender, R. C. and Bayne, C. J. (2001). Involvement of nitric oxide in killing of *Schistosoma mansoni* sporocysts by hemocytes from resistant *Biomphalaria glabrata*. *J. Parasitol.* 87,778–785.

Halm, M. P., Chichery, M. P. and Chichery, R. (2003). Effect of nitric oxide synthase inhibition on the manipulative behaviour of *Sepia officinalis*. *Comp. Biochem. Physiol. C: Toxicol. Pharmacol.* 134,139–146.

Hanlon, R.T. and Messenger, J.B. (1996). Defence. In *Cephalopod Behaviour*, Cambridge University Press, Cambridge, pp. 66–93.

Hatcher, N. G., Sudlow, L. C., Moroz, L. L. and Gillette, R. (2006). Nitric oxide potentiates cAMP-gated cation current in feeding neurons of *Pleurobranchaea californica* independent of cAMP and cGMP signaling pathways. *J. Neurophysiol.* 95,3219–3227.

Hearing, V. J. (1999). Biochemical control of melanogenesis and melanosomal organization. *J. Investig. Dermatol. Symp. Proc.* 4,24–28.

Hearing, V. J. (2005). Biogenesis of pigment granules: A sensitive way to regulate melanocyte function. *J. Dermatol. Sci.* 37,3–14.

Hobbs, M. J. and Young, J. Z. (1973). A cephalopod cerebellum. *Brain Res.* 55,424–430.

Huang, S., Kerschbaum, H. H., Engel, E. and Hermann, A. (1997). Biochemical characterization and histochemical localization of nitric oxide synthase in the nervous system of the snail, *Helix pomatia*. *J. Neurochem.* 69,2516–2528.

Jacklet, J., Grizzaffi, J. and Tieman, D. (2004). Serotonin, nitric oxide and histamine enhance the excitability of neuron MCC by diverse mechanisms. *Acta Biol. Hung.* 55,201–210.

Jacklet, J. W. and Tieman, D. G. (2004). Nitric oxide and histamine induce neuronal excitability by blocking background currents in neuron MCC of *Aplysia*. *J. Neurophysiol.* 91,656–665.

Katzoff, A., Ben-Gedalya, T. and Susswein, A. J. (2002). Nitric oxide is necessary for multiple memory processes after learning that a food is inedible in *Aplysia*. *J. Neurosci.* 22,9581–9594.

Kemenes, I., Kemenes, G., Andrew, R. J., Benjamin, P. R. and O'Shea, M. (2002). Critical time-window for NO-cGMP-dependent long-term memory formation after one-trial appetitive conditioning. *J. Neurosci.* 22,1414–1425.

Kobayashi, S., Ogawa, H., Fujito, Y. and Ito, E. (2000a). Nitric oxide suppresses fictive feeding response in *Lymnaea stagnalis*. *Neurosci. Lett.* 285,209–212.

Kobayashi, S., Sadamoto, H., Ogawa, H., Kitamura, Y., Oka, K., Tanishita, K. and Ito, E. (2000b). Nitric oxide generation around buccal ganglia accompanying feeding behavior in the pond snail, *Lymnaea stagnalis*. *Neurosci. Res.* 38,27–34.

Koh, H. Y. and Jacklet, J. W. (1999). Nitric oxide stimulates cGMP production and mimics synaptic responses in metacerebral neurons of *Aplysia*. *J. Neurosci.* 19,3818–3826.

Korneev, S. A., Kemenes, I., Straub, V., Staras, K., Korneeva, E. I., Kemenes, G., Benjamin, P. R. and O'Shea, M. (2002). Suppression of nitric oxide (NO)-dependent behavior by double-stranded RNA-mediated silencing of a neuronal NO synthase gene. *J. Neurosci.* 22(RC227), 1–5.

Korneev, S. A. and O'Shea, M. (2002). Evolution of nitric oxide synthase regulatory genes by DNA inversion. *Mol. Biol. Evol.* 19,1228–1233.

Korneev, S. A., Park, J. H. and O'Shea, M. (1999). Neuronal expression of neural nitric oxide synthase (nNOS) protein is suppressed by an antisense RNA transcribed from an NOS pseudogene. *J. Neurosci.* 19,7711–7720.

Korneev, S. A., Piper, M. R., Picot, J., Phillips, R., Korneeva, E. I. and O'Shea, M. (1998). Molecular characterization of NOS in a mollusc: Expression in a giant modulatory neuron. *J. Neurobiol.* 35,65–76.

Korneev, S. A., Straub, V., Kemenes, I., Korneeva, E. I., Ott, S. R., Benjamin, P. R. and O'Shea, M. (2005). Timed and targeted differential regulation of nitric oxide synthase (NOS) and anti-NOS genes by reward conditioning leading to long-term memory formation. *J. Neurosci.* 25,1188–1192.

Leise, E. M., Kempf, S. C., Durham, N. R. and Gifondorwa, D. J. (2004). Induction of metamorphosis in the marine gastropod *Ilyanassa obsoleta*: 5HT, NO and programmed cell death. *Acta Biol. Hung.* 55,293–300.

Li, H. and Poulos, T. L. (2005). Structure-function studies on nitric oxide synthases. *J. Inorg. Biochem.* 99,293–305.

Lovell, P. J., Kabotyanski, E. A., Sadreyev, R. I., Boudko, D. Y., Bryne, J. H. and Moroz, L. L. (2000). Nitric oxide activates buccal motor programs in *Aplysia californica*. *Soc. Neurosci. Abstr.* 26,918.

Messenger, J. B. (1967a). The peduncle lobe: A visuo-motor centre in octopus. *Proc. R. Soc. London B: Biol. Sci.* 167,225–251.

Messenger, J. B. (1967b). The effects on locomotion of lesions to the visuo-motor system in octopus. *Proc. R. Soc. London B: Biol. Sci.* 167,252–281.

Messenger, J. B. (1983). Multimodal convergence and the regulation of motor programs in cephalopods. In *Multimodal Convergence in Sensory Systems (Fortschritte der Zoologie 28)* (ed. E. Horn), pp. 79–98, Gustav Fischer, Stuttgart.

Meulemans, A., Mothet, J. P., Schirar, A., Fossier, P., Tauc, L. and Baux, G. (1995). A nitric oxide synthase activity is involved in the modulation of acetylcholine release in *Aplysia* ganglion neurons: A histological, voltammetric and electrophysiological study. *Neuroscience* 69,985–995.

Moncada, S., Palmer, R. M. and Higgs, E. A. (1991). Nitric oxide: Physiology, pathophysiology, and pharmacology. *Pharmacol. Rev.* 43,109–142.

Moroz, L. L. (2000). Giant identified NO-releasing neurons and comparative histochemistry of putative nitrergic systems in gastropod molluscs. *Microsc. Res. Tech.* 49,557–569.

Moroz, L. L. (2006). Localization of putative nitrergic neurons in peripheral chemo-sensory areas and the central nervous system of *Aplysia californica*. *J. Comp. Neurol.* 495,10–20.

Moroz, L. L. and Gillette, R. (1995). From Polyplacophora to Cephalopoda: Comparative analysis of nitric oxide signalling in mollusca. *Acta Biol. Hung.* 46,169–182.

Moroz, L. L., Norekian, T. P., Pirtle, T. J., Robertson, K. J. and Satterlie, R. A. (2000). Distribution of NADPH-diaphorase reactivity and effects of nitric oxide on feeding and locomotory circuitry in the pteropod mollusc, *Clione limacina*. *J. Comp. Neurol.* 427,274–284.

Newcomb, J. M. and Watson, W. H. (2002). Modulation of swimming in the gastropod *Melibe leonina* by nitric oxide. *J. Exp. Biol.* 205,397–403.

Ottaviani, E., Paeman, L. R., Cadet, P. and Stefano, G. B. (1993). Evidence for nitric oxide production and utilization as a bacteriocidal agent by invertebrate immuno-cytes. *Eur. J. Pharmacol.* 248,319–324.

Palumbo, A. (2003). Melanogenesis in the ink gland of *Sepia officinalis*. *Pigment Cell Res.* 16,517–522.

Palumbo, A. (2005). Nitric oxide in marine invertebrates: A comparative perspective. *Comp. Biochem. Physiol. A* 142,241–248.

Palumbo, A., Di Cosmo, A., Gesualdo, I., d'Ischia, M. (1997). A calcium-dependent nitric oxide synthase and NMDA R1 glutamate receptor in the ink gland of *Sepia officinalis*: A hint to a regulatory role of nitric oxide in melanogenesis? *Biochem. Biophys. Res. Commun.* 235, 429–432.

Palumbo, A., Di Cosmo, A., Poli, A., Di Cristo, C. and d'Ischia, M. (1999). A calcium/calmodulin-dependent nitric oxide synthase, NMDAR2/3 receptor subunits, and glutamate in the CNS of the cuttlefish Sepia officinalis: Localization in specific neural pathways controlling the inking system. *J. Neurochem.* 73,1254–1263.

Palumbo, A., Fiore, G., Di Cristo, C., Di Cosmo, A. and d'Ischia, M. (2002). NMDA receptor stimulation induces temporary alpha-tubulin degradation signaled by nitric oxide-mediated tyrosine nitration in the nervous system of Sepia *officinalis*. *Biochem. Biophys. Res. Commun.* 293,1536–1543.

Palumbo, A., Poli, A., Di Cosmo, A. and d'Ischia, M. (2000). *N*-Methyl-d-aspartate receptor stimulation activates tyrosinase and promotes melanin synthesis in the ink gland of the cuttlefish *Sepia officinalis* through the nitric oxide/cGMP signal trans-duction pathway. *J. Biol. Chem.* 275,16885–16890.

Peruzzi, E., Fontana, G. and Sonetti, D. (2004). Presence and role of nitric oxide in the central nervous system of the freshwater snail *Planorbarius corneus*: Possible impli-cation in neuron-microglia communication. *Brain Res.* 1005,9–20.

Robertson, J. D., Bonaventura, J. and Kohm, A. P. (1994). Nitric oxide is required for tactile learning in *Octopus vulgaris*. *Proc. R. Soc. London B: Biol. Sci.* 256, 269–273.

Robertson, J. D., Bonaventura, J., Kohm, A. and Hiscat, M. (1996). Nitric oxide is necessary for visual learning in *Octopus vulgaris*. *Proc. R. Soc. London B: Biol. Sci.* 263,1739–1743.

Roszer, T., Czimmerer, Z., Szentmiklosi, A. J. and Banfalvi, G. (2004). Nitric oxide synthesis is blocked in the enteral nervous system during dormant periods of the snail *Helix lucorum*. *Cell Tissue Res.* 316,255–262.

Sakura, M., Kabetani, M., Watanabe, S. and Kirino, Y. (2004). Impairment of olfactory discrimination by blockade of nitric oxide activity in the terrestrial slug *Limax valentianus*. *Neurosci. Lett.* 370,257–261.

Scheinker, V., Fiore, G., Di Cristo, C., Di Cosmo, A., d'Ischia, M., Enikolopov, G. and Palumbo, A. (2005). Nitric oxide synthase in the nervous system and ink gland of the cuttlefish *Sepia officinalis*: Molecular cloning and expression. *Biochem. Biophys. Res. Commun.* 338,1204–1215.

Schipp, R. and Gebauer, M. (1999). Nitric oxide: A vasodilatatory mediator in the cephalic aorta of *Sepia officinalis* (L) (Cephalopoda). *Invert. Neurosci.* 4,9–15.

Schopfer, F. J., Baker, P. R., Freeman, B. A. (2003). NO-dependent protein nitration: A cell signaling event or an oxidative inflammatory response? *Trends Biochem. Sci.* 28, 646–654.

Serfozo, Z. and Elekes, K. (2002). Nitric oxide level regulates the embryonic development of the pond snail *Lymnaea stagnalis*: Pharmacological, behavioral, and ultrastructural studies. *Cell Tissue Res.* 310,119–130.

Stefano, G. B., Kim, E., Liu, Y., Zhu, W., Casares, F., Mantione, K., Jones, D. A. and Cadet, P. (2004). Nitric oxide modulates microglial activation. *Med. Sci. Monit.* 10,BR17–BR22.

Stuehr, D. J. (2004). Enzymes of the l-arginine to nitric oxide pathway. *J. Nutr.* 134, 2748S–2751S.

Tafalla, C., Gomez-Leon, J., Novoa, B. and Figueras, A. (2003). Nitric oxide production by carpet shell clam (*Ruditapes decussatus*) hemocytes. *Dev. Comp. Immunol.* 27,197–205.

Taylor, B. E., Harris, M. B., Burk, M., Smyth, K., Lukowiak, K. and Remmers, J. E. (2003). Nitric oxide mediates metabolism as well as respiratory and cardiac responses to hypoxia in the snail *lymnaea stagnalis*. *J. Exp. Zoolog. A: Comp. Exp. Biol.* 295,37–46.

Teyke, T. (1996). Nitric oxide, but not serotonin, is involved in acquisition of food-attraction conditioning in the snail *Helix pomatia*. *Neurosci. Lett.* 206,29–32.

Torreilles, J. (2001). Nitric oxide: One of the more conserved and widespread signaling molecules. *Front. Biosci.* 6,D1161–D1172.

Trimm, K. R. and Rehder, V. (2004). Nitric oxide acts as a slow-down and search signal in developing neurites. *Eur. J. Neurosci.* 19,809–818.

Trimmer, B. A., Aprille, J. and Modica-Napolitano, J. (2004). Nitric oxide signalling: Insect brains and photocytes. *Biochem. Soc. Symp.* 71,65–83.

Tu, Y. and Budelmann, B. U. (2000). Effects of nitric oxide donors on the afferent resting activity in the cephalopod statocyst. *Brain Res.* 865,211–220.

Walker, F., Ribeiro, J. M. and Montfort, W. R. (1999). Novel nitric oxide-liberating heme proteins from the saliva of bloodsucking insects. *Met. Ions Biol. Syst.* 36,621–663.

Welshhans, K. and Rehder, V. (2005). Local activation of the nitric oxide/cyclic guanosine monophosphate pathway in growth cones regulates filopodial length via protein kinase G, cyclic ADP ribose and intracellular Ca^{2+} release. *Eur. J. Neurosci.* 22,3006–3016.

White, A. R., Curtis, S. A. and Walker, R. J. (2004). Evidence for a possible role for nitric oxide in the modulation of heart activity in *Achatina fulica* and *Helix aspersa*. *Comp. Biochem. Physiol. C: Toxicol. Pharmacol.* 137,95–108.

64

Zemojtel, T., Penzkofer, T., Dandekar, T., Schultz, J. (2004). A novel conserved family of nitric oxide synthase? *Trends Biochem. Sci.* 29, 224–226.

Zsombok, A., Schrofner, S., Hermann, A. and Kerschbaum, H. H. (2000). Nitric oxide increases excitability by depressing a calcium activated potassium current in snail neurons. *Neurosci. Lett.* 295, 85–88.

Soluble guanylyl cyclases in invertebrates: Targets for NO and O₂

David B. Morton[1,*] and Anke Vermehren[1]

[1]*Department of Integrative Biosciences, Oregon Health & Science University, Portland, OR 97239, USA*

Abstract. The major cellular targets for NO are soluble guanylyl cyclases (sGCs), which are activated upon binding NO and catalyse the synthesis of cyclic guanosine monophosphate (GMP). Invertebrates and possibly vertebrates have two families of sGCs: conventional NO-sensitive sGCs, and atypical sGCs that are insensitive to NO. Recent evidence suggests that the atypical sGCs act as oxygen sensors, mediating behavioral responses to oxygen content in the environment. Here we review the biochemical properties of both families of sGCs and recent evidence supporting the model that atypical sGCs can act as molecular oxygen sensors.

Keywords: cGMP; *Drosophila melanogaster*; heme proteins; hypoxia; signaling.

Mammalian soluble guanylyl cyclases

Soluble guanylyl cyclase (sGC) is probably the most prevalent target for nitric oxide (NO) in almost all species and tissues (Lucas *et al.*, 2000). The native enzyme is usually a heterodimer (see below for exceptions to this) containing a single heme group (Lucas *et al.*, 2000). In mammals, four subunits have been identified: α1, α2, β1 and β2. The major functional enzyme is the α1/β1 heterodimer (Lucas *et al.*, 2000). The α2 subunit is very similar in sequence to the α1 subunit, and the α2/β1 heterodimer has similar properties to the α1/β1 enzyme (Gibb *et al.*, 2003; Russworm *et al.*, 1998). One significant difference is that the α2 subunit interacts with the scaffold protein PSD95 and hence likely has a different subcellular distribution (Russworm *et al.*, 2001).

All four mammalian subunits are homologous proteins with a similar arrangement of functional domains (Fig. 1). The subunits can be divided into a C-terminal catalytic domain that catalyses the conversion of GTP to cyclic guanosine monophosphate (cGMP), and an N-terminal regulatory domain that functions as a heme-binding region and is required for NO activation of the enzyme (Fig. 1). A variety of studies have identified several residues in each of these domains that are required for function

---

*Corresponding author.
E-mail: mortonda@ohsu.edu (D.B. Morton).

ADVANCES IN EXPERIMENTAL BIOLOGY
VOLUME 01   ISSN 1872-2423
DOI: 10.1016/S1872-2423(07)01003-4

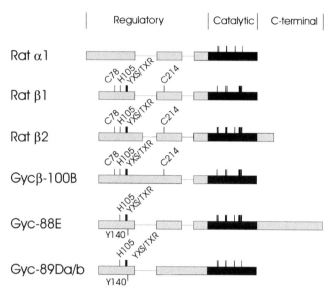

Fig. 1. Schematic representation of the functional domains of select mammalian and *Drosophila* sGCs. All sGCs contain a regulatory and a catalytic domain, whereas only Gyc-88E and the mammalian β2 subunits contain an extended C-terminal domain. In the regulatory domain, *Drosophila* Gycβ-100B and Gyc-89Da/b contain additional sequences at the positions shown. Also shown are the locations of residues necessary for specific biochemical properties. In the regulatory domain these include His105 (H105) and the TyrXSer/ThrXArg (YXS/TXR) motif in all β and β-like subunits, Cys78 and Cys214 (C78 and C214) in the conventional β subunits and mammalian β2, and Tyr140 (Y140) in the *Drosophila* atypical sGCs. In the catalytic domain the residues that interact with the GTP substrate are shown. Note that there is a different set of residues in the conventional α and β subunits – whereas the homodimeric subunits, mammalian β2 and Gyc-88E contain both sets, and Gyc-89Da/b contains the same set as the α subunits.

and these are indicated in Fig. 1. The α1/β1 heterodimer binds a single heme group per heterodimer and the β1 subunit is primarily responsible for this interaction (Lucas *et al.*, 2000). The Fe^{2+} ion in the center of the heme group interacts with His105 of the β1 subunit (Zhao *et al.*, 1998). The heme group interacts with three residues in the β1 subunit via hydrogen bonds that form the YXS/TXR motif (Schmidt *et al.*, 2004). In addition, two cysteine residues (at positions Cys78 and Cys214 in the rat β1 subunit) are necessary for NO activation (Friebe *et al.*, 1997). The relative positions of these residues are shown in Fig. 1.

Modeling of the catalytic domain of sGC, based on the crystal structure of the catalytic domain of mammalian adenylate cyclase, predicts that there are 17 residues that contact the GTP substrate (Liu *et al.*, 1997). The active site is at the interface between the two subunits, so each subunit provides a subset of specific residues (10 for the β subunit and 7 for the α subunit, which are marked on the catalytic domain in Fig. 1). This model predicts that a single GTP molecule will bind per heterodimer (Liu *et al.*, 1997).

The β2 subunit has quite different functional characteristics and appears to be a member of a different subfamily of sGC subunits, which we have termed the atypical sGCs (Morton, 2004a). Although the β2 subunit will form active heterodimers with both the α1 and α2 subunits, it has the unusual property that it is also active as a homodimer (Gibb *et al.*, 2003; Koglin *et al.*, 2001). All of these enzymes are sensitive to NO, but they are all less potently activated by NO than the α1/β1 and α2/β1 heterodimers (Gibb *et al.*, 2003). The regulatory domain of the β2 subunit contains His105, Cys78, Cys214 and the YXS/TXR motif (Fig. 1), which suggests that the α1/β2 and α2/β2 heterodimers bind a single heme group, but it is not known whether the β2 homodimer binds one or two heme groups. The catalytic domain of the β2 subunits contains all 17 of the residues that bind the GTP substrate, providing a structural basis for the formation of an active homodimer (Morton, 2004a). This is similar to the situation with the receptor GCs, which also form active homodimers and contain all 17 residues that interact with GTP. Modeling of the catalytic domain of receptor GCs suggests that two GTP molecules bind per dimer (Liu *et al.*, 1997), a prediction that is supported by the finding that the receptor GC, GC-A, shows kinetics with positive cooperativity (Wong *et al.*, 1995). Members of both conventional and atypical sGC families have also been identified in a variety of invertebrates (Table 1; Morton, 2004a; Vermehren *et al.*, 2006). Phylogenetic and sequence comparisons have also been made across a broad array of prokaryotes and eukaryotes (Schaap, 2005).

Conventional invertebrate sGCs

Structures

Genes that code for sGCs have been identified in both nematodes and insects (Morton, 2004a). Insects have members of both conventional and atypical sGC families, whereas all the sGCs identified in *Caenorhabditis elegans* are members of the atypical sGC family (Table 1; Morton,

Table 1. Summary of the conventional and atypical sGCs in mammals, insects and *C. elegans*. Assignment as conventional or atypical sGCs is based on sequence similarity and biochemical properties (Morton, 2004a). GCs in the same column represent orthologous sequences—the columns within the atypical sGC family are based on the phylogenetic analysis performed by Fitzpatrick *et al.* (2006).

	Conventional sGCs			Atypical sGCs		
Mammals	α1	α2	β1	β2		
Insects						
Manduca	MsGC-α1		MsGC-β1		MsGC-β3	
Drosophila	Gycα-99B		Gycβ-100B		Gyc-88E	Gyc-89Da, Gyc-89Db
C. elegans				GCY-32, GCY-34, GCY-35, GCY-36, GCY-37	GCY-31	GCY-33

2004a). Insects appear to express one conventional α subunit and one conventional β subunit (Vermehren *et al.*, 2006). Both subunits are similar in primary sequence to their mammalian orthologues, with a similar regulatory and catalytic domain (Morton and Hudson, 2002). All of the individual amino acid residues that have been identified as critical to the function of conventional sGCs are conserved between the mammalian and insect orthologues (Fig. 1; Morton and Hudson, 2002). In the regulatory domain of β1, these include His105, which forms the heme axial ligand, and the YXS/TXR motif (Fig. 1). In addition, Cys78 and Cys214, which are necessary for NO activation, are also conserved in the insect conventional sGC β subunits that have been identified (Morton and Hudson, 2002). These sequence comparisons predict that the insect conventional heterodimers bind a single heme group per dimer. Although there is no direct evidence to support this, the sGC inhibitor 1H-[1,2,4] oxadiazolo[4,3,-α]quinoxalin-1-one (ODQ), which acts by oxidizing the heme group, is an effective inhibitor of the insect conventional sGCs, suggesting that they do contain heme (Morton *et al.*, 2005a). One notable difference among the insect β subunit regulatory domains is that although the *Manduca* MsGC-β1 regulatory domain contains the same number of residues as mammalian β subunits (Nighorn *et al.*, 1998), the *Drosophila* Gycβ-100B subunit contains an insertion of 118 amino acids at the center of the regulatory domain (Fig. 1; Shah and Hyde, 1995).

This results in a β subunit of 86 kDa, which is larger than the α subunit, whereas all other β subunits are smaller than the α subunits. Interestingly, the *Anopheles* β subunit also has an insertion of 91 residues in the same position, although there is no sequence conservation of the inserts between the two flies (Caccone *et al.*, 1999). The insect α subunits are very similar in primary sequence to the mammalian α1 subunits. None of the insect α subunit sequences reported contain C-terminal sequences that would predict that they can bind to PDZ domains in a similar fashion to the mammalian α2 subunits (Russworm *et al.*, 2001).

The catalytic domains of both the α and the β subunits are highly conserved across species, with all the residues that interact with GTP from each subunit conserved in the insect orthologue (Fig. 1; Morton and Hudson, 2002). Several studies have described the biochemical properties of insect conventional sGCs by transiently expressing cloned subunits in heterologous cells. These studies have generally shown that the insect enzymes have very similar properties to their mammalian orthologues. Both *Drosophila* and *Manduca* conventional sGCs form obligate heterodimers, which are potently activated by a variety of NO donors and are inhibited by ODQ, which has a similar potency on insect and mammalian sGCs (Morton *et al.*, 2005a; Nighorn *et al.*, 1998; Shah and Hyde, 1995).

Function

Most studies aimed at investigating the physiological functions of the conventional sGCs have focused on the use of sGC inhibitors, such as ODQ, and sGC activators, such as NO donors. One general theme that has emerged from these studies has been the proposed role of the sGCs in a variety of developmental events in the nervous system. Examples include neuronal outgrowth, synapse formation, and neuronal and glial migration in a variety of insect species (Bicker, 2005; Gibbs and Truman, 1998; Gibson *et al.*, 2001; Truman *et al.*, 1996; Wright *et al.*, 1998). Although these studies provide strong evidence to support the role of the NO/cGMP pathway in neuronal development, it should be pointed out that most of the pharmacological manipulations used are not specific for the conventional sGCs. As described below, in addition to the conventional sGCs, insects also express another family of sGCs, the atypical sGCs (Morton, 2004a). In *Drosophila* these enzymes are slightly stimulated by NO donors and are also strongly inhibited by ODQ (Morton *et al.*, 2005a). At the present time the only specific pharmacological agents available for the conventional sGCs are the new family of NO-independent sGC activators such as YC-1 and BAY 41-2272 (Morton et al., 2005a).

Genetic manipulations provide an alternative, nonpharmacological, approach for assessing sGC function. This is potentially complicated by the fact that *Drosophila* that are homozygous for a point mutation within the coding region of the NO synthase (NOS) gene die in the early first instar (Regulski et al., 2004), raising the possibility that null mutations in the conventional sGC subunit genes would also be lethal. It is not known which tissues require the presence of NOS during early development. It is also not known whether any or all of the essential actions of NO require sGCs. However, studies in progress (Vermehren and Morton, unpublished data) show that fly lines that have a transposon within introns of the genes for either the α or the β subunit of the conventional sGC are also homozygous early larval lethal. Because the insertion is also within the intron of a gene on the opposite DNA strand, it is not known whether the lethality is due to disruption of the sGC genes. By contrast, fly lines with point mutations that lead to greatly reduced levels of expression of Gycα-99B were viable, but failed to develop connections between their photoreceptors and the postsynaptic cells (Gibbs et al., 2001). This defect was rescued with exogenous, global expression of Gycα-99B (Gibbs et al., 2001). These results were consistent with the inhibitor-based studies cited above that pharmacologically demonstrated a requirement for sGC activity in photoreceptor development (Gibbs and Truman, 1998).

Atypical sGCS

cDNA cloning and expression studies

The first atypical sGC was demonstrated biochemically in lobster muscle (Prabhakar et al., 1997). In this study, two distinct peaks of GC activity were separated chromatographically from soluble fractions of lobster muscle, one of which was stimulated by NO and the other of which was NO-insensitive (Prabhakar et al., 1997). Whether the NO-insensitive peak represented an sGC orthologue or whether it was similar to MsGC-I (Simpson et al., 1999), a *Manduca* GC more similar to receptor GCs but lacking a transmembrane domain, is not known. The first cloned atypical sGC, MsGC-β3 (Table 1), was identified in the insect, *Manduca sexta* (Nighorn et al., 1999). As described above, conventional sGCs are obligate heterodimers that are potently activated by NO. When MsGC-β3 was transiently expressed in heterologous cells its properties were strikingly different: in cell-free homogenates, it was enzymatically active in the absence of additional subunits and it was insensitive to NO donors (Nighorn et al., 1999). Subsequent analysis showed that it formed active homodimers and could form heterodimers with either of the conventional subunits that showed neither basal nor NO-stimulated activity (Morton and Anderson, 2003).

The mammalian $\beta 2$ subunit also appears to belong to this family of atypical sGC subunits and shares a number of biochemical properties with MsGC-$\beta 3$. The sequence first reported for $\beta 2$ predicted a protein lacking the equivalent residues corresponding to the first 62 residues of the $\beta 1$ subunit (Yuen et al., 1990). When this version of $\beta 2$ was co-expressed with the $\alpha 1$ subunit, the resulting enzyme was substantially less sensitive to NO than the $\alpha 1/\beta 1$ heterodimer (Gupta et al., 1997). When co-expressed with the $\alpha 1/\beta 1$ heterodimer, the $\beta 2$ subunit acted as a dominant negative subunit, reducing the NO sensitivity of the combined subunits (Gupta et al., 1997). We showed that MsGC-$\beta 3$ could act in the same way when co-expressed with both the α and β subunits from Manduca, also reducing the NO activation of the conventional sGC (Morton and Anderson, 2003). The results described by Gupta et al. (1997) were subsequently questioned as others failed to measure enzyme activity when the $\beta 2$ subunit was co-expressed with $\alpha 1$ (Denninger and Marletta, 1999). A more recent report describes the cloning of another cDNA for the $\beta 2$ subunit that contains residues that correspond to the 62 amino acid N-terminus of the $\beta 1$ subunit (Koglin et al., 2001). This variant of the $\beta 2$ subunit is, like MsGC-$\beta 3$, active in the absence of additional subunits, but unlike MsGC-$\beta 3$ is slightly stimulated by NO donors in cell-free homogenates (Koglin et al., 2001). RT-PCR (reverse transcriptase polymerase chain reaction) studies showed that the $\beta 2$ subunit was expressed in the brain, but in situ hybridization experiments failed to reveal specific staining, so the cellular or regional location of the subunit is unknown (Gibb and Garthwaite, 2001). More recent experiments have revealed further subtleties in the properties of the $\beta 2$ subunits (Gibb et al., 2003). The studies by Gupta et al. (1997) and Koglin et al. (2001) used cell homogenates from heterologous cells transiently expressing the subunits to measure enzyme activity. Gibb et al. (2003), by contrast, used intact cells that were transiently transfected with the subunits and then measured the accumulated levels of cGMP. This more recent study showed that the shorter version of the $\beta 2$ subunit when co-expressed with either the $\alpha 1$ or the $\alpha 2$ subunit was active, and although it was stimulated by NO, it showed a much reduced level of activation compared to either the $\alpha 1/\beta 1$ or $\alpha 2/\beta 1$ combinations (Gibb et al., 2003). The $\beta 2$ variant (v$\beta 2$, with the additional 62 amino acids) was active in intact cells in the absence of additional subunits and again was stimulated to a lesser extent than the conventional subunit combinations (Gibb *et al.*, 2003). All the combinations involving β2 or vβ2 subunits showed a marked bell-shaped dose–response curve for NO activation, suggesting rapid desensitization (Gibb *et al.*, 2003). These data indicate that, depending on which splice variant is expressed, β2 can function as either a homodimer or a heterodimer, but in either case it is less sensitive to NO stimulation than the conventional subunits.

Sequencing the genomes of *C. elegans* and *Drosophila melanogaster* revealed a much broader array of atypical guanylyl cyclase subunits than previously imagined (Morton, 2004a). The genome of *C. elegans* contains seven genes (Table 1) that code for guanylyl cyclase subunits (Birnby *et al.*, 2000), and all are predicted to be NO insensitive (Morton *et al.*, 1999). Although their primary sequences are more similar to β subunits than α subunits, they have all been predicted to form various hetero-dimers *in vivo* (Morton, 2004a). Although no biochemical evidence has been obtained to support these predictions, genetic studies confirmed heterodimer formation between two subunits, GCY-35 and GCY-36 (Cheung *et al.*, 2004). The *Drosophila* genome predicts three genes that code for atypical subunits (Table 1): Gyc-88E (an orthologue of MsGC-β3), and two additional genes, Gyc-89Da and Gyc-89Db (Langlais *et al.*, 2004; Morton and Hudson, 2002; Morton, 2004a). Additional genomic sequences have suggested that all insects have an orthologue of MsGC-β3/Gyc-88E, but other insects appear to have a single gene coding for an orthologue of Gyc-89Da/89Db (Vermehren *et al.*, 2006). *Drosophila pseudoobscura* has both Gyc-89Da and Gyc-89Db orthologues, whereas the mosquito *Anopheles gambiae* (also a member of the order Diptera) has a single orthologue, suggesting that there was a relatively recent gene duplication of Gyc-89Da/89Db (Vermehren *et al.*, 2006). A recent phylogenetic analysis of sGCs has grouped two of the *C. elegans* sGCs, GYC-31 and GYC-33, close to the insect atypical sGCs, while the re-maining five *C. elegans* sGCs were grouped closer to the mammalian β2 subunits (Fitzpatrick *et al.*, 2006). The domain structure of the atypical sGCs is similar to that of the conventional sGCs, with both an N-terminal regulatory domain and a more C-terminal catalytic domain (Fig. 1). One notable difference is that Gyc-88E has a large C-terminal domain (Fig. 1). This has no known function and does not contain any known functional protein domains. The *Manduca* and *Anopheles* orthologues to Gyc-88E also have a C-terminal domain that is a similar size (about 300 amino acid residues). There is little sequence similarity between the C-terminal domains between these species, although there is a region of about 20 residues that is highly conserved (Langlais *et al.*, 2004). Interestingly, the β2 subunit and a *C. elegans* orthologue, GCY-31, also contain a C-terminal domain, although there is no sequence conservation in this domain between insects, mammals and *C. elegans*. Since no functional role for this domain has been identified, in MsGC-β3 there is evidence that it may play an auto-inhibitory role (Morton and Anderson, 2003).

All the *Drosophila* subunits have been transiently expressed in heterologous cells and their biochemical properties have been described (Langlais *et al.*, 2004; Morton *et al.*, 2005a). Gyc-88E, like MsGC-β3 and the mammalian vβ2 subunit, was active in the absence of additional subunits (Langlais *et al.*, 2004). In cell-free homogenates, Gyc-88E was similar to vβ2 and was slightly stimulated by some, but not all, NO donors (Langlais *et al.*, 2004; Morton *et al.*, 2005a). Gyc-89Da and Gyc-89Db were inactive on their own, but enhanced the activity of Gyc-88E when they were co-expressed (Langlais *et al.*, 2004; Morton *et al.*, 2005a). These heterodimers were also slightly stimulated by NO and were more sensitive to NO donors than the homodimeric Gyc-88E (Morton *et al.*, 2005a). The NO activation was inhibited by ODQ, a property shared by the conventional sGCs, suggesting that they contain a heme group (Morton *et al.*, 2005a). The heterodimers showed a pronounced bell-shaped dose–response curve to the NO donor DEA-NONOate, which was interpreted as indicating that rapid desensitization occurred, similar to the situation for mammalian β2 subunits (Morton *et al.*, 2005a). *In vivo*, we found that Gyc-88E was frequently co-expressed with either Gyc-89Da or Gyc-89Db, suggesting that the native enzymes were likely to be heterodimers (Langlais *et al.*, 2004; Morton *et al.*, 2005a). Although the maximum stimulation of the atypical heterodimers to NO donors was only two- to fourfold, compared to at least 50-fold stimulation for the conventional heterodimers, the only qualitative difference in biochemical properties revealed by these studies was the response to the NO-independent activator BAY 41-2272. This compound was a potent activator of the conventional sGCs, but had no effect on the atypical subunits (Morton *et al.*, 2005a).

Analysis of the critical residues shown in Fig. 1 is consistent with most of the biochemical properties of the atypical sGCs. They all contain the heme-interacting residues, His105 and the YXS/TXR motif, supporting the prediction that they contain a heme group. As both Gyc-88E and Gyc-89Da/89Db contain these residues it is not clear whether the heterodimers would contain one or two heme groups per heterodimer. All three atypical sGC subunits lack Cys78 and Cys214, required for NO activation in β1. Whereas *Manduca* MsGC-β3 lacks these cysteines and is NO-insensitive, and the mammalian β2 subunit contains these cysteines and is slightly NO-sensitive, the *Drosophila* Gyc-88E, Gyc-89Da and Gyc-89Db are slightly NO-sensitive, suggesting that these residues are not absolutely required for NO activation. In the catalytic domain, Gyc-88E contains all the residues that interact with GTP, consistent with its ability to form homodimers, and Gyc-89Da and Gyc-89Db only contain the residues found in α subunits, consistent with their formation of obligate heterodimers.

Oxygen sensing by the atypical sGCs

The poor responsiveness of the atypical sGCs to NO raises the possibility that this is not the natural ligand. This idea was supported by the finding that in *C. elegans*, a related atypical sGC subunit, GCY-35 (Table 1), was required for oxygen sensitivity (Gray *et al.*, 2004). Conventional sGCs are unusual for heme proteins in that they do not bind oxygen (Lawson *et al.*, 2003). By contrast, the regulatory domain of GCY-35 is capable of binding oxygen via a heme group, suggesting that it could act as a molecular oxygen detector and its activity would be regulated by oxygen concentration (Gray *et al.*, 2004). It has not been possible to test this directly as none of the *C. elegans* sGC subunits exhibit any enzymatic activity when expressed in heterologous cells (Morton, 2004a).

The activity of the *Drosophila* atypical sGCs expressed in heterologous cells, by contrast, was regulated by oxygen. When COS-7 cells, which had been transiently transfected with the *Drosophila* atypical sGC subunits, were incubated in the presence or absence of oxygen, up to 50-fold more cGMP accumulated in the cells incubated in the absence of oxygen compared to normal atmospheric oxygen concentrations, whereas cells transfected with the conventional subunits showed no change (Morton, 2004b; Vermehren *et al.*, 2006). This increase was graded over 0–21% oxygen, a characteristic that would be expected if the enzyme served as a molecular oxygen detector (Morton, 2004b). Incubation in 50% oxygen further inhibited activity, suggesting that at 21% oxygen the enzyme was not fully saturated (Vermehren *et al.*, 2006). Additionally, the activation in the absence of oxygen was blocked by the sGC inhibitor ODQ, suggesting that activation required the heme group (Morton, 2004b). These data suggested that oxygen bound to the heme group in a manner analogous to NO binding to the heme group of conventional guanylyl cyclases, but caused inhibition of the enzyme rather than activation. This raised an apparent paradox: if NO and oxygen bind at the same site, how can one ligand stimulate the enzyme while the other is inhibitory, when they are so chemically similar? Results of an experiment that exposed transfected cells to NO donors in the presence of different concentrations of oxygen suggested a possible explanation (Vermehren et al., 2006). At 21% oxygen, NO was stimulatory – increasing cGMP levels about twofold, whereas NO added to cells incubated at 0% or 10% oxygen was inhibitory, reducing the cGMP levels to about the same levels as measured at 21% oxygen (Vermehren et al., 2006). One interpretation for this result is that although both NO and oxygen bound at the same site and both were inhibitory, NO was less effective as an inhibitor compared to oxygen. Thus, at 21% oxygen, NO displaced some of the oxygen bound to the heme group and relieved some of the oxygen inhibition, causing an apparent stimulation. This is comparable to both NO and CO binding to conventional guanylyl cyclases, but CO being less effective at stimulating enzyme activity compared to NO (Sharma and Magde, 1999). This hypothesis can be tested by examining the absorbance spectra of purified proteins. Soluble GCs, like other heme proteins, have a peak of absorbance at about 430 nm (the Soret peak), which in the presence of a bound gas shifts to a lower wavelength (Stone and Marletta, 1994). Conventional sGCs do not bind oxygen and no shift in the Soret peak is observed when spectra are gathered under aerobic or anaerobic conditions. By contrast, the regulatory domain from a C. elegans atypical sGC (GCY-35) exhibited a shift from 430 nm to 415 nm when exposed to aerobic conditions (Gray et al., 2004). Interestingly, in the presence of NO, the Soret peak had a shoulder, suggesting that the heme group formed two stable nitrosyl complexes (Gray et al., 2004). It is not known whether GCY-35 is activated or inhibited by NO or oxygen, but the difference in the absorption spectra in the presence of NO and oxygen suggests that the enzyme activity could differ in the presence of the two gases.

Modeling and sequence comparison have also indicated the structural basis for the oxygen sensitivity of the atypical sGCs. Although the crystal structure of the heme domain has not been solved for any sGC, the structure of a chemotaxis protein of the obligate anaerobe Thermoanaerobacter tengcongensis, with a GC-like heme domain, has been solved (Nioche et al., 2004; Pellicena et al., 2004). This domain shares primary sequence similarity to sGC subunits and binds both O_2 and NO with high affinity (Karow et al., 2004). Recent studies have identified Tyr140 as a critical residue in determining NO/O_2 selectivity (Boon et al., 2005). These findings allow us to predict which sGC subunits will be able to form oxygen-sensitive GCs. All three of the Drosophila (see Fig. 1) and all seven of the *C. elegans* atypical sGCs have a tyrosine in the equivalent position. We have also analysed additional sequences that have recently been deposited in databases, and these studies suggest that all insects probably have both oxygen-sensitive and oxygen-insensitive (conventional) sGCs (Vermehren *et al.*, 2006).

Behavioral studies indicating roles for atypical sGCs

Responses to hypoxia

The first evidence supporting a behavioral role for atypical sGCs in oxygen sensation was in *C. elegans*. When wild-type *C. elegans* were placed in a gradient of 0–21% oxygen concentrations, they congregated at intermediate concentrations of 5–12%, avoiding both high and low oxygen concentrations (Gray *et al.*, 2004). Animals with null mutations in an atypical sGC, GCY-35, were evenly distributed across this gradient, failing to avoid both normal and anoxic conditions (Gray *et al.*, 2004). Avoidance of high (21%) oxygen concentration required the presence of a cGMP-gated ion channel in the neurons that expressed GCY-35, suggesting that the activity of GCY-35 was required for this behavior (Gray *et al.*, 2004). As described above, the regulatory domain of GCY-35 bound oxygen via a heme group (Gray *et al.*, 2004), suggesting that its activity was directly regulated by oxygen, but no direct evidence for this has been obtained (Morton, 2004a).

The biochemical properties of the *Drosophila* atypical sGCs suggested that they might also function as molecular oxygen detectors (Morton, 2004b). As the activity of the *Drosophila* atypical sGCs is activated by reduced oxygen concentrations, we reasoned that they could mediate responses to hypoxia (Morton *et al.*, 2005b). The *Drosophila* atypical sGCs are expressed in a subset of central and peripheral sensory neurons (Langlais *et al.*, 2004) where they are ideally situated to respond to changing environmental oxygen concentrations and mediate behavioral responses to hypoxia (Morton, 2004b). We have recently begun to test this possibility (Morton *et al.*, 2005b).

Drosophila larvae placed on a small pile of yeast paste, under normal atmospheric oxygen concentrations, will immediately burrow into the yeast and begin feeding with only their posterior spiracles above the surface of the yeast. When the larvae were exposed to low (1–10%) oxygen levels, they rapidly ceased feeding, withdrew from the food and began exploratory behaviors (Wingrove and O'Farrell, 1999). Several lines of evidence indicated that this behavior was mediated by increases in cGMP. Firstly, larvae that differ in the levels of cGMP-dependent protein kinase (G-kinase) due to a polymorphism in the *for* gene (Osborne *et al.*, 1997) showed different response times to hypoxia: those with lower levels of G-kinase responded more slowly (Wingrove and O'Farrell, 1999). In addition, larvae fed NO synthase inhibitors also responded more slowly to hypoxia than control larvae, while larvae over-expressing NO synthase responded faster (Wingrove and O'Farrell, 1999). This suggested that conventional sGCs mediated the behavioral response by utilizing a NO/cGMP/G-kinase pathway (Wingrove and O'Farrell, 1999).

To determine whether the atypical sGCs were involved in this response, we specifically targeted the central and peripheral neurons that express the atypical sGC subunits by generating fly lines that expressed the yeast transcription factor, GAL4, under the control of the predicted promoter regions of the Gyc-89Da and Gyc-89Db genes (Vermehren *et al.*, 2005). Using these fly lines we reduced cGMP levels specifically in these neurons by expressing a cGMP-specific phosphodiesterase (bovine PDE5) using UAS-bPDE5 flies (Broderick *et al.*, 2004). Larvae expressing bPDE5 in either Gyc-89Da or Gyc-89Db neurons showed much slower responses to hypoxia than larvae from either of the parental lines (Morton *et al.*, 2005b).

The results of experiments with the p89Da-bPDE5 animals imply that an increase in cGMP in the Gyc-89Da neurons is necessary for the behavioral response to hypoxia. A major question to resolve is the identity of the GC that is responsible for this increase in cGMP. It is very tempting to assume that it is the Gyc-88E/Gyc-89Da heterodimer within some of the sensory neurons that is directly responding to reduced oxygen levels with an increase in GC activity. Wingrove and O'Farrell (1999), however, suggested that a NO-sensitive conventional sGC is involved. It is not known whether the conventional sGC is co-expressed with any of the atypical subunits. If we assume that they are not co-expressed, it is still possible to reconcile the two sets of data. Wingrove and O'Farrell (1999) showed that larvae that over-expressed NO synthase were more sensitive to reduced oxygen levels than control larvae. As NO can displace oxygen bound to the atypical sGCs, increasing their activity (Vermehren *et al.*, 2006), the increased NO concentration would have a similar effect on the atypical sGCs as decreased oxygen concentration, stimulating the hypoxia escape response.

Chemotaxis to other chemicals

The atypical sGCs are also expressed in the ganglia that innervate the two main chemosensory organs in *Drosophila* larvae (Langlais *et al.*, 2004). The main larval olfactory organ is the dorsal organ, which is inner-vated by the dorsal ganglion (Heimbeck *et al.*, 1999). The main gustatory organ, the terminal organ, is innervated by the terminal ganglion (Heimbeck *et al.*, 1999). *In situ* hybridization experiments showed that all three atypical sGCs are expressed in a few cells in both the terminal and dorsal ganglia (Langlais *et al.*, 2004; Morton *et al.*, 2005a). We used the same promoter:GAL4 and UAS:bPDE5 fly lines described above to determine if there was a role for cGMP in the neurons that express the atypical sGCs in chemotaxis. *Drosophila* larvae are attracted to a wide range of volatile and nonvolatile chemicals, which can easily be quantified by simple behavioral preference tests (Heimbeck *et al.*, 1999). Using the larvae which expressed bPDE5 in neurons that expressed the atypical sGCs, we showed that elevated levels of cGMP were required for chemo-taxis to certain odorants and tastants (Vermehren *et al.*, 2005).

While these data suggest a role for cGMP in the atypical sGC cells in chemotaxis, several questions remain unanswered. As with the hypoxia escape response, the data do not directly show that the atypical sGCs are the source of the cGMP required for the behavioral responses. The cellular basis for the chemotactic deficits is also undefined. In addition to the sensory neurons in the chemosensory organs, the atypical sGCs are also expressed in a large number of neurons in the central nervous system (CNS). It is not known whether the chemotactic deficits were due to reduced cGMP levels in the CNS or in the sensory neurons or both. If the cGMP is required in the sensory neurons, it is also not known whether the cGMP is required in the primary signal transduction process of the cell, or whether it modulates the olfactory and gustatory signal trans-duction. Although many olfactory and gustatory receptors have now been identified in *Drosophila* and other insects (Robertson *et al.*, 2003), recent studies have shown that they are quite distinct in their membrane topology and likely function from olfactory receptors in other species (Benton *et al.*, 2006). The signal transduction that underlies these recep-tors is currently unknown, and hence whether cGMP plays a direct or indirect role in the pathway is also unknown.

These findings also raise the question of why chemotaxis should utilize an oxygen-sensitive sGC. Studies with *C. elegans* have shed some light on this issue. In addition to mediating oxygen sensitivity, GCY-35 also mediates feeding behavior (Cheung *et al.*, 2004). In the presence of bacteria, some strains of *C. elegans* tend to aggregate when feeding, a behavior that requires the presence of GCY-35 (Cheung *et al.*, 2004). This aggregation behavior is not seen when animals are exposed to reduced oxygen concentrations (Gray *et al.*, 2004), and it has been suggested that one of the cues used by *C. elegans* to detect the presence of bacteria is reduced oxygen concentrations (Gray *et al.*, 2004).

Conclusion

We have provided here a brief overview of the properties and functions of sGCs in invertebrates. As in mammals, the conventional sGCs probably act as the major receptor for NO and mediate most of its actions. In addition, invertebrates express another class of sGC: the atypical sGCs, which are oxygen-sensitive. There is also evidence that genes coding for similar enzymes are likely expressed by fish, birds and marsupials (Morton, Vermehren and Langlais, unpublished data). *In vitro*, the *Drosophila* atypical sGCs also respond, albeit weakly, to NO. Whether they are also receptors for NO *in vivo* and how the combined actions of NO and oxygen regulate their functions are important issues to resolve.

Acknowledgements

This work was supported by NIH grant NS29740.

References

Benton, R., Saches, S., Michnick, S. W. and Vosshall, L. B. (2006). Atypical membrane topology and heteromeric function of *Drosophila* odorant receptors in vivo. *PLoS Biol.* 4,240–257.

Bicker, G. (2005). STOP and GO and NO: Nitric oxide as a regulator of cell motility in simple brains. *Bioessays* 27,495–505.

Birnby, D. A., Link, E. M., Vowels, J. J., Tian, H., Colacurcio, P. L. and Thomas, J. H. (2000). A transmembrane guanylyl cyclase (DAF-11) and Hsp90 (DAF-21) regulate a common set of chemosensory behaviors in Caenorhabditis elegans. *Genetics* 155,85–104.

Boon, E. M., Huang, S. H. and Marletta, M. A. (2005). A molecular basis for NO selectivity in soluble guanylate cyclase. *Nat. Chem. Biol.* 1,53–59.

Broderick, K. E., Kean, L., Dow, J. A., Pyne, N. J. and Davies, S. A. (2004). Ectopic expression of bovine type 5 phosphodiesterase confers a renal phenotype in *Drosophila*. *J. Biol. Chem.* 279,8159–8168.

Caccone, A., Garcia, B. A., Mathiopoulos, K. D., Min, G. S., Moriyama, E. N. and Powell, J. R. (1999). Characterization of the soluble guanylyl cyclase beta-subunit gene in the mosquito *Anopheles gambiae*. *Insect Mol. Biol.* 8,23–30.

Cheung, B. H. H., Arellano-Carbajal, F., Rybicki, I. and de Bono, M. (2004). Soluble guanylate cyclases act in neurons exposed to the body fluid to promote *C elegans* aggregation behavior. *Curr. Biol.* 14,1105–1111.

Denninger, J. W. and Marletta, M. A. (1999). Guanylate cyclase and the NO/cGMP signaling pathway. *Biochim. Biophys. Acta* 1411,334–350.

Fitzpatrick, D. A., O'Halleran, D. M. and Burnell, A. M. (2006). Multiple lineage specific expansions within the guanylyl cyclase gene family. *BMC Evol. Biol.* 6,26.

Friebe, A., Wedel, B., Harteneck, C., Foerster, J., Schultz, G. and Koesling, D. (1997). Functions of conserved cysteines of sGCs. *Biochemistry* 36,1194–1198.

Gibb, B. J. and Garthwaite, J. (2001). Subunits of the nitric oxide receptor, soluble guanylyl cyclase, expressed in rat brain. *Eur. J. Neurosci.* 13,539–544.

Gibb, B. J., Wykes, V. and Garthwaite, J. (2003). Properties of NO-activated guanylyl cyclases expressed in cells. *Br. J. Pharmacol.* 139,1032–1040.

Gibbs, S. M., Becker, A., Hardy, R. W. and Truman, J. W. (2001). Soluble guanylate cyclase is required during development for visual system function in *Drosophila*. *J. Neurosci.* 21,7705–7714.

Gibbs, S. M. and Truman, J. W. (1998). Nitric oxide and cyclic GMP regulate retinal patterning in the optic lobe of *Drosophila*. *Neuron* 20,83–93.

Gibson, N. J., Rössler, W., Nighorn, A. J., Oland, L. A., Hildebrand, J. G. and Tolbert, L. P. (2001). Neuron-glial communication via nitric oxide is essential in establishing antennal-lobe structure in *Manduca sexta*. *Dev. Biol.* 240,326–339.

Gray, J. M., Karow, D. S., Lu, H., Chang, A. J., Chang, J. S., Ellis, R. E., Martletta, M. A. and Bargmann, C. I. (2004). Oxygen sensation and social feeding mediated by a *C. elegans* guanylate cyclase homologue. *Nature* 430,317–322.

Gupta, G., Azam, M., Yang, L. and Danziger, R. S. (1997). The β2 subunit inhibits stimulation of the α1/β1 form of soluble guanylyl cyclase by nitric oxide. Potential relevance to regulation of blood pressure. *J. Clin. Invest.* 100,1488–1492.

Heimbeck, G., Bugnon, V., Gendre, N., Habelin, C. and Stocker, R. F. (1999). Smell and taste perception in *Drosophila melanogaster* larva: Toxin expression studies in chemosensory neurons. *J. Neurosci.* 19,6599–6609.

Karow, D. S., Pan, D., Tran, R., Pellicena, P., Presley, A., Mathies, R. A. and Marletta, M. A. (2004). Spectroscopic characterization of the soluble guanylate cyclase-like heme domains from *Vibrio cholerae* and *Thermoanaerobacter tengcongensis*. *Biochemistry* 43,10203–10211.

Koglin, M., Vehse, K., Budaeus, L., Scholz, H. and Behrends, S. (2001). Nitric oxide activates the β2 subunit of soluble guanylyl cyclase in the absence of a second subunit. *J. Biol. Chem.* 276,30737–30743.

Langlais, K. K., Stewart, J. A. and Morton, D. B. (2004). Preliminary characterization of two atypical sGC in the central and peripheral nervous system of *Drosophila melanogaster*. *J. Exp. Biol.* 207,2323–2338.

Lawson, D. M., Stephenson, C. E. M., Andrew, C. R., Gerorge, S. J. and Eady, R. R. (2003). A two-faced molecule offers NO explanation: The proximal binding of nitric oxide to haem. *Biochem. Soc. Trans.* 31,553–557.

Liu, Y., Ruoho, A. E., Rao, V. D. and Hurley, J. H. (1997). Catalytic mechanism of the adenylyl and guanylyl cyclases: Modeling and mutational analysis. *Proc. Natl. Acad. Sci. USA* 94,13414–13419.

Lucas, K. A., Pitari, G. M., Kazerounian, S., Ruiz-Stewart, I., Park, J., Schulz, S., Chepenik, K. P. and Waldman, S. A. (2000). Guanylyl cyclases and signaling by cyclic GMP. *Pharmacol. Rev.* 52,375–413.

Morton, D. B. (2004a). Invertebrates yield a plethora of atypical guanylyl cyclases. *Mol. Neurobiol.* 29,97–115.

Morton, D. B. (2004b). Atypical sGCs in *Drosophila* can function a molecular oxygen sensors. *J. Biol. Chem.* 279,50651–50653.

Morton, D. B. and Anderson, E. (2003). MsGC-β3 forms active homodimers and inactive heterodimers with NO-sensitive sGC subunits. *J. Exp. Biol.* 206,937–947.

Morton, D. B. and Hudson, M. L. (2002). Cyclic GMP regulation and function in insects. *Adv. Insect Physiol.* 29,1–54.

Morton, D. B., Hudson, M. L., Waters, E. and O'Shea, M. (1999). Soluble guanylyl cyclases in *C. elegans* – NO is not the answer. *Curr. Biol.* 9,R546–R547.

Morton, D. B., Langlais, K. K., Steward, J. A. and Vermehren, A. (2005a). Comparison of the properties of the five sGC subunits in *Drosophila melanogaster*. *J. Insect Sci.* 5,12.

Morton, D. B., Langlais, K. K., Steward, J. A. and Vermehren, A., (2005b). Atypical sGCs in *Drosophila* as possible oxygen sensors. In *2005 Abstract Viewer/Itinerary Planner*, Program 613.7. Society for Neuroscience, Washington DC.

Nighorn, A., Byrnes, K. A. and Morton, D. B. (1999). Identification and characterization of a novel beta subunit of sGC that is active in the absence of a second subunit and is relatively insensitive to nitric oxide. *J. Biol. Chem.* 274,2525–2531.

Nighorn, A., Gibson, N. J., Rivers, D. M., Hildebrand, J. G. and Morton, D. B. (1998). The NO/cGMP pathway may mediate communication between sensory afferents and projection neurons in the antennal lobe of *Manduca sexta*. *J. Neurosci.* 18,7244–7255.

Nioche, P., Berka, V., Vipond, J., Minton, N., Tsai, A. L. and Raman, C. S. (2004). Femtomolar sensitivity of a NO sensor from *Clostridium botulinum*. *Science* 306,1550–1553.

Osborne, K., Robichon, A., Burgess, E., Butland, S., Shaw, R., Coulthard, A., Pereira, H., Greenspan, R. and Sokolowski, M. B. (1997). Natural behavior polymorphism due to a cGMP-dependent protein kinase of *Drosophila*. *Science* 277,834–836.

Pellicena, P., Karow, D., Boon, E. M., Marletta, M. A. and Kuriyan, J. (2004). Crystal structure of an oxygen-binding heme domain related to soluble guanylate cyclases. *Proc. Natl Acad. Sci. USA* 101,12854–12859.

Prabhakar, S., Short, D. B., Scholz, N. L. and Goy, M. F. (1997). Identification of nitric oxide-sensitive and -insensitive forms of cytoplasmic guanylate cyclase. *J. Neurochem.* 69,1650–1660.

Regulski, M., Stasiv, Y., Tully, T. and Enikolopov, G. (2004). Essential function of nitric oxide synthase in *Drosophila*. *Curr. Biol.* 14,R881–R882.

Robertson, H. M., Warr, C. G. and Carlson, J. R. (2003). Molecular evolution of the insect chemoreceptor gene molecular family in *Drosophila melanogaster*. *Proc. Natl. Acad. Sci. USA* 100,14537–14542.

Russworm, M., Behrends, S., Harteneck, C. and Koesling, D. (1998). Functional properties of a naturally occurring isoform of soluble guanylyl cyclase. *Biochem. J.* 335, 125–130.

Russworm, M., Wittau, N. and Koesling, D. (2001). Guanylyl cyclase/PSD95 interaction: Targeting of the NO sensitive α2/β1 guanylyl cyclase to synaptic membranes. *J. Biol. Chem.* 276,44647–44652.

Schaap, P. (2005). Guanylyl cyclases across the tree of life. *Front. Biosci.* 10,1485–1498.

Schmidt, P. M., Schramm, M., Schröeder, H., Wunder, F. and Stasch, J. P. (2004). Identification of residues crucially involved in the binding of the heme moiety of sGC. *J. Biol. Chem.* 279,3025–3032.

Shah, S. and Hyde, D. R. (1995). Two *Drosophila* genes that encode the α and β subunits of the brain sGC. *J. Biol. Chem.* 270,15368–15376.

Sharma, V. S. and Magde, D. (1999). Activation of soluble guanylate cyclase by carbon monoxide and nitric oxide: A mechanistic model. *Methods* 19,494–505.

Simpson, P. J., Nighorn, A. and Morton, D. B. (1999). Identification of a novel guanylyl cyclase that is related to receptor guanylyl cyclases, but lacks extracellular and transmembrane domains. *J. Biol. Chem.* 274,4440–4446.

Stone, J. R. and Marletta, M. A. (1994). Soluble guanylate cyclase from bovine lung: Activation of nitric oxide and carbon monoxide and spectral characterization of the ferrous and ferric states. *Biochemistry* 33,5636–5640.

Truman, J. W., De Vente, J. and Ball, E. E. (1996). Nitric oxide-sensitive guanylate cyclase activity is associated with the maturational phase of neuronal development in insects. *Development* 122,3949–3958.

Vermehren, A., Langlais, K. K. and Morton, D. B. (2006). Oxygen-sensitive guanylyl cyclases in insects and their potential roles in oxygen detection and in feeding behaviors. *J. Insect Physiol.* 52,340–348.

Vermehren, A., Langlais, K. K., Steward, J. A. and Morton, D. B. (2005). Atypical sGCs in Drosophila may be involved in taste responses. In *2005 Abstract Viewer/Itinerary Planner.* Program 295.9. Society for Neuroscience, Washington DC.

Wingrove, J. A. and O'Farrell, P. H. (1999). Nitric oxide contributes to behavioral, cellular, and developmental responses to low oxygen in *Drosophila*. *Cell* 98,105–114.

Wong, S. K. F., Ma, C. P., Foster, D. C., Chen, A. Y. and Garbers, D. L. (1995). The guanylyl cyclase-A receptor transduces an atrial natriuretic peptide/ATP activation signal in the absence of other proteins. *J. Biol. Chem.* 270,30818–30822.

Wright, J. W., Schwinof, K. M., Snyder, M. A. and Copenhaver, P. F. (1998). A delayed role for nitric oxide-sensitive guanylate cyclases in a migratory population of embryonic neurons. *Dev. Biol.* 204,15–33.

Yuen, P. S. T., Potter, L. R. and Garbers, D. L. (1990). A new form of guanylyl cyclase is preferentially expressed in rat kidney. *Biochemistry* 29,10872–10878.

Zhao, Y., Schelvis, J., Babcock, G. and Marletta, M. A. (1998). Identification of the histidine 105 in the β1 subunit of soluble guanylate cyclase as the heme proximal ligand. *Biochemistry* 37,4502–4509.

Nitric oxide signalling in insect epithelial transport

Shireen-A. Davies*

Division of Molecular Genetics, Institute of Biomedical and Life Sciences, University of Glasgow, Glasgow G11 6U, UK

Abstract. Nitric oxide (NO) is a key regulator of $3',5'$-cyclic guanosine monophosphate (cGMP) signalling, and has major roles in the function of insect Malpighian (renal) tubules. Insect tubules determine survival of the whole animal, and study of the mechanisms of tubule function *in vivo* have increased our understanding of epithelial function, as well as advancing the development of novel pesticide strategies. NO controls the rate of fluid transport by the Malpighian tubules of the model organism *Drosophila melanogaster*, where the overall physiological effect on the tubule by NO results from interactions of NO with cGMP signalling pathway components, in particular cGMP-hydrolysing phosphodiesterases. NO also modulates fluid transport rates in tubules from medically relevant insect vectors, the mosquito and tsetse fly. Furthermore, the only known family of insect nitridergic neuropeptides – the capa peptides – has also been identified in the mosquito, and shown to stimulate NO/cGMP signalling and fluid transport in tubules from *Anopheles*, *Aedes* and *Glossina* but not *Schistocerca* – suggesting a conservation of capa-induced NO/cGMP signalling across Dipteran species. A newly-discovered role of the *Drosophila* tubule in immune sensing is described. Microbial challenge increases NO and antimicrobial peptide production by the tubule; moreover, transgenic modulation of NO levels *in vivo* in only specific cells of the tubule increases tubule antimicrobial peptide production and confers increased survival of the whole animal upon immune challenge. We show here that *Aedes* tubules can also act as immune sensors, and mount a NO response to immune challenge. Thus, tubules possess a host of functions relevant to insect survival, in which NO plays major roles. The current challenge will be to use the genomic resources available to *Drosophila*, and increasingly to *Anopheles*, to further our understanding of the role of NO signalling in survival of the adult insect.

Keywords: *Drosophila melanogaster*; nitric oxide synthase; cGMP; malpighian tubules; renal; fluid transport; capa; phosphodiesterase; *Diptera*; *Anopheles*; *Aedes*; *Glossina*; *Schistocerca*; immunity; IMD.

Nitric oxide synthase in *D. melanogaster*

Since the initial discovery of nitric oxide (NO) as a signalling molecule, it has been shown that NO function is implicated in many cell types, and modulates neural and immune function in vertebrates (Bogdan, 2001; Friebe and Koesling, 2003; Holscher, 1997) and in invertebrates (Bicker, 2005; Bicker *et al.*, 1996; Trimmer *et al.*, 2004). NO causes its cellular

Corresponding author: Tel.: +44-141-330-2317. Fax: +44-141-330-4878.
E-mail: s.a.davies@bio.gla.ac.uk (S.-A. Davies).

effects mainly via activation of a soluble guanylyl cyclase (sGC) (Zhao
et al., 1999) and thence the 3′,5′-cyclic guanosine monophosphate
(cGMP) pathway (Schmidt *et al.*, 1993), or via *S*-nitrosylation of key
target proteins (Foster *et al.*, 2003). Powerful insights into physiological
roles of the family of nitric oxide synthase (NOS) enzymes which generate NO (Stuehr, 1999) have been obtained by use of genetic models,
specifically mouse knockouts of all three isoforms: NOS1 (nNOS/NOS
I), NOS2 (iNOS/NOS II) and NOS3 (eNOS/NOS III) (Mungrue *et al.*,
2003). In insects, use of the genetic model *Drosophila melanogaster*
has allowed investigations of the role of NO using the sophisticated molecular genetic and transgenic tools available. One such tool is the GAL4/
UAS$_G$ binary expression system (Brand and Perrimon, 1993). Mobilisa-
tion of a mobile DNA element (P-element) containing an expression
construct for the yeast transcription factor GAL4 around the genome
(a P-element screen; see Sozen *et al.*, 1997) results in tissue- or cell-specific
enhancer-mediated expression of GAL4. Each such GAL4 line controls
the expression of transgenes cloned downstream of the upstream activating sequence (UAS) in separate fly lines. The crossing of parental
GAL4 and UAS lines thus allows tissue- or cell-specific expression of any
gene of choice in the progeny. Heat shock and the current availability
of conditional GAL4 drivers, including flippase (FLP) recombinase/FLP
recombinase target (FRT) (McGuire *et al.*, 2004), allows temporally
controlled targetted gene expression to be achieved. The GAL4 system
can also used for targetted gene silencing *in vivo* using RNA interference
technology (Kennerdell and Carthew, 2000).

Only one *Drosophila* NOS gene, *dNOS*, exists (http://flybase.bio.
indiana.edu/.bin/fbidq.html?FBgn0011676). This encodes a full-length
form of calcium/calmodulin-sensitive NOS, DNOS1 (Stasiv *et al.*, 2001),
which is closely related to vertebrate neuronal NOS (NOS1), with extensive similarity both at the DNA and protein level (40–50%) over the
complete sequence, and complete conservation within the catalytic and
cofactor domains (Davies, 2000; Regulski and Tully, 1995). The *dNOS*
gene was originally found to encode two transcripts, *dNOS1* and *dNOS2*
(Regulski and Tully, 1995); however, more recent work has characterised
a further eight transcripts, *dNOS3–dNOS10*, generated by alternative
splicing of the 5′ untranslated region (UTR) of *dNOS* (Stasiv *et al.*, 2001),
encoding proteins ranging in length from 214 amino acids (DNOS3)
to 1,350 amino acids (DNOS1). Most of the short proteins do not
encode functional NOS, although they retain dimerisation domains. Co-
expression and immunoprecipitation experiments identified inhibition of
DNOS activity when DNOS4, DNOS5 and DNOS6 form heterodimers

with DNOS1. Importantly, this suggests a role for the complex regulation of *dNOS* transcription by dominant negative regulation of DNOS1 activity, via heteromeric dimerisation with truncated forms of NOS. This work was further developed using transgenic flies bearing promoter – *dNOS4* constructs (Stasiv *et al.*, 2004). These elegant experiments established that DNOS4 is endogenously expressed in the fly, is co-expressed at the same sites of expression as DNOS1, and forms endogenous DNOS1/DNOS4 heterodimers *in vivo*, which results in inhibition of the anti-proliferative effects of NO. Thus, complex regulation of DNOS occurs *in vivo*, which allows for at least temporal expression of NO throughout development. *dNOS1* is expressed throughout the life cycle of the fly, whereas the truncated NOS isoforms were isolated from larval or embryonic stages, with *dNOS4* being predominantly expressed in the embryo. Generation of a null allele of *dNOS* using chemical mutagenesis, and of the transgenic lines bearing this construct, has allowed the demonstration of an unequivocal role for NOS in fly development. Flies homozygous for the *dNOS* null allele display embryonic and larval lethality (Regulski *et al.*, 2004).

NO modulates fluid transport

While much attention has been focussed on the neuronal and endothelial roles of NO, the widespread expression of NOS isoforms suggested significant roles of NO in other tissues, notably epithelia. NO produced via NOS3 has been documented in lung, kidney, reproductive tissue and gastrointestinal tract (Ortiz and Garvin, 2003). Furthermore, NO produced by the three NOS isoforms, including NOS1 and NOS2, has been mapped in vertebrate kidney and has been shown to contribute to overall kidney function (Herrera and Garvin, 2005). For example, use of a mouse knockout of NOS2 has demonstrated a role for this isoform of NOS in renal sodium and bicarbonate transport (Wang, 2002).

NO has also been shown to directly modulate fluid transport by the *Drosophila* Malpighian tubule (Dow *et al.*, 1994), an organ critically involved with osmoregulation and ion homoeostasis. The tubule is a valuable model for studies of signal transduction and ion transport pathways in an organotypic context, and provides one of the most informative phenotypes available for the study of cell signalling in *Drosophila* (Dow and Davies, 2003). Using the molecular genetic tools available in *Drosophila*, these studies can be applied to genetically identified cell types, thus allowing the analysis of specific pathways in non-excitable, secretory cells *in vivo*.

Tubules express NOS only in the Type I (principal) cells located in the fluid-secreting main segment (Dow and Davies, 2001) (Fig. 1). While many cell types utilise NO in a paracrine manner, the tubule utilises NO as an autocrine signal (Broderick et al., 2003; Davies, 2000). The localisation of DNOS in principal cells (Dow and Davies, 2001) shows that, in tubules, it is these cells which utilise the NO signalling pathway. This is persuasive evidence that NO may be an autocrine signalling molecule in tubules, but does not exclude the possibility that NO produced in the principal cell may be transduced in neighbouring stellate cells. Induction of the *dNOS* transgene via heat shock of transgenic flies results in expression of DNOS in both principal and stellate cells, although only principal cells show increased cGMP content. Thus, stellate cells do not respond to an intracellular NO signal, and are unlikely to respond to a paracrine signal from the principal cell. Also, application to intact tubules of the NO donor sodium nitroprusside, and immunocytochemistry with anti-cGMP antibody, showed that the NO-induced cGMP increase is only observed in principal cells (Broderick, 2002). While this autocrine use of NO may seem unusual, an autocrine role for NO has been proposed for vertebrate macula densa, in which production of NO may inhibit sodium/potassium/chloride co-transport (Welch and Wilcox, 2002).

The capa family of nitridergic neuropeptides (capa-1 and capa-2, endogenous in *Drosophila*) (Kean et al., 2002) and the related peptide CAP_{2b} found in *Manduca sexta* (Davies et al., 1997), comprise the only known insect neuropeptides which activate NO/cGMP signalling in insect tubules. All these peptides increase fluid transport rates by the tubule. DNOS-induced production of NO is increased via activation of the G-protein linked capa receptors (Iversen et al., 2002), with concomitant increases in principal cell intracellular cGMP levels. Use of an sGC inhibitor, methylene blue, inhibits cGMP increases and associated fluid transport induced by both NO donors and capa peptides (Dow et al., 1994; Kean et al., 2002), implicating the existence of NO-sensitive sGC in tubules. This is supported by recent analysis of the tubule transcriptome (Wang et al., 2004), which shows mRNA expression for both sGC subunits, Gycα-99B and Gycβ-100B, in tubules (Table 1), although Gycα-99B is much more highly expressed in other tissues, presumably in the central nervous system.

Although the canonical brain *Drosophila* sGC subunit is a bona fide heterodimeric NO-binding enzyme composed of α and β subunits (Shah and Hyde, 1995), it is possible that the tubule NO-sensitive sGC is different, given the roughly fourfold difference in mRNA levels between

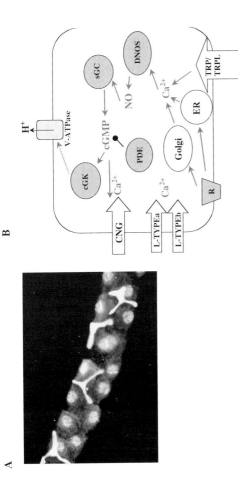

Fig. 1. (A) Main segment of the *Drosophila* Malpighian tubule. Large nuclei of principal cells are stained orange with ethidium bromide, and star-shaped stellate cells are picked out with green fluorescent protein. Photo courtesy of J. A. T. Dow (Rosay *et al.*, 1997). (B) NO/cGMP signalling in the *Drosophila* tubule principal cell. Binding of capa peptides (capa-1, capa-2 and the closely related CAP_{2b}) to their cognate receptor, R, causes elevated intracellular calcium levels (Kean *et al.*, 2002). Calcium influx in the principal cell occurs via L-type calcium channels, Dmca1A and Dmca1D (represented here as a and b), the cyclic nucleotide-gated (CNG) channels (MacPherson *et al.*, 2001) and transient receptor potential (TRP)/transient receptor potential-like (TRPL) channels (MacPherson *et al.*, 2005). Major intracellular calcium stores in the tubule are the Golgi and the endoplasmic reticulum (ER) (Southall *et al.*, 2006), as well as the apical mitochondria (Terhzaz *et al.*, 2006). DNOS, a calcium-activated protein, produces NO, which in turn stimulates a soluble guanylyl cyclase (sGC). Cyclic GMP activates the cGMP-dependent protein kinase(s) (cGK), encoded by *dg1* and *dg2* genes (MacPherson *et al.*, 2004). cGMP can also activate calcium influx via CNG channels. The cGMP signal is efficiently terminated by a family of phosphodiesterases (PDE) (Day *et al.*, 2005). cGMP increases the transepithelial potential of the tubule, suggesting that a target of cGMP (or cGK) may be the apical vacuolar V-ATPase (Davies *et al.*, 1995); however, direct phosphorylation of V-ATPase subunits has yet to be demonstrated. (See Colour Plate Section in this book).

Table 1. Expression of genes encoding soluble guanylyl cyclase subunits in *Drosophila* tubules. Microarray signals (mean ± SEM, $n = 5$) derived as described in Wang *et al.* (2004).

Synonym	Gene name	Chromosomal location	Tubule signal (normalised Affymetrix)	Fly signal (normalised Affymetrix)	Ratio
Gycα-99B	Guanyl cyclase α-subunit at 99B	99B9–99B10	6	45.8	0.2
Gycβ-100B	Guanyl cyclase β-subunit at 1B	100B6–100B8	20.4	26.9	0.8

From http://tubules.freeprohost.com/file/arraysearch.cgi

the α and β subunits. If the major proportion of the tubule sGC was similar to a recently described novel homodimeric form of sGC composed of two β2 subunits, which each bind NO (Koglin *et al.*, 2001), the apparent excess of β subunit mRNA may be explained.

Directed expression of the calcium reporter, aequorin, to specified tubule cell subtypes *in vivo* demonstrated that capa peptides also increase cytosolic calcium in only Type 1 (principal) cells in the tubule's main, fluid-secreting segment (Kean *et al.*, 2002; Rosay *et al.*, 1997). As such, activation of both calcium and cGMP signalling pathways occurs during tubule fluid transport in a cell-specific manner.

The specific localisation of the NO- and capa-induced cGMP signal in principal cells may suggest that stellate cells do not utilise cGMP as a signal at all. However, work using transgenic *Drosophila* which express the guanylyl cyclase GC-A – the receptor for rat atrial natriuretic peptide (ANP) – in stellate cells, shows that stimulation of excised transgenic tubules with ANP results in increased fluid transport, as well as increased cGMP in a dose-dependent manner (Fig. 2). Interestingly, the levels of fluid transport achieved by targetted expression of GC-A in just the stellate cells via the c724 GAL4 driver (Sozen *et al.*, 1997) are similar to those from tubules which express GC-A in just principal cells. Thus, stellate cells contain at least the required downstream elements of the

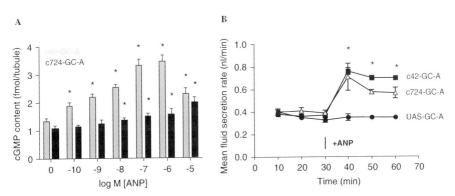

Fig. 2. (A) Dose-dependent atrial natriuretic peptide (ANP)-induced cGMP generation in transgenic c42-GC-A and c724-GC-A tubules. The ANP concentration is expressed as log molar (M). (B) Ectopic expression of GC-A in either principal cells (c42-GC-A) or stellate cells (c724-GC-A) via the GAL4/UAS system results in increased fluid transport in response to ANP (arrowed). Asterisks indicate data significantly different from control (unstimulated) tubules (A) or from the UAS-GC-A parental line (B), $p < 0.05$, determined using the Student's t-test (unpaired samples). Adapted from Kerr *et al.*, 2004.

cGMP signalling pathway in order to transduce a cGMP signal, as well as cGMP-modulated ion transport complexes which are effective at driving fluid transport. Also, the fact that stellate cells do not respond to a NO signal (Broderick *et al.*, 2003), but do generate and respond physiologically to cGMP in response to GC-A activation, suggests that they do not contain sGC but presumably do contain endogenous receptor guanylyl cyclases.

Interactions between NO and signalling components

Although much is known about NOS signalling in tubules, how might variations in NOS levels impact upon the control of fluid transport? Using transgenic flies which over-express a heat shock inducible *dNOS* transgene (Broderick *et al.*, 2003), intriguing results show that the functional effects of increased NO generation are regulated by downstream elements of cGMP signalling. Expression and enzymatic activity of DNOS in tubule principal cells is increased in the transgenic tubules upon heat shock. Surprisingly, however, fluid transport rates do not increase as expected, in spite of elevated cGMP levels. Increased fluid transport is apparent upon induction of the *dNOS* transgene only when a specific inhibitor of cGMP-specific phosphodiesterase (cG-PDE) is applied. Pre-treatment of tubules with such an inhibitor, 'Zaprinast', results in inhibition of cG-PDE activity and in further elevation of the cGMP level, but at the apical membrane of the principal cells. Given that breakdown, and therefore regulation, of the cyclic nucleotide concentration occurs via the phosphodiesterases (PDE), it appears that the physiological response of the tubules to a significant NO load is regulated by a very active cG-PDE(s). Furthermore, the total increased cGMP content of the principal cells is not sufficient to drive the response of the whole tubule. The importance of the cGMP transduction of the NO signal lies in both the concentration and the localisation of the cGMP signal, in this case at the principal cell apical membrane. The genome of *D. melanogaster* encodes six genes for PDEs, including the famous *dunce*, a member of the cAMP-specific PDE4 family, as well as close homologues of vertebrate PDE1C, PDE5/6, PDE8A, PDE9 and PDE11A (Day *et al.*, 2005). Recent biochemical characterisation has shown that the *Drosophila* homologues of PDE1C and PDE11A are indeed dual-specificity enzymes, and that the PDE5/PDE6 homologue, *Dm*PDE6, is a bona fide cG-PDE (Day *et al.*, 2005), which intriguingly also directly modulates cGMP transport in tubule principal cells (Day *et al.*, 2006). Interestingly, tubules contain all six PDEs, confirming the importance of cyclic nucleotide signalling in tubules.

Spatial control of the NO signal has been shown to occur via specific interactions with key proteins (Kone et al., 2003). For example, nNOS (NOS1) associates via with postsynaptic density protein (PSD)-93, PSD-95, alpha-synthrophin and plasma membrane Ca^{2+}/calmodulin-dependent ATPase (PMCA) 4b via a post-synaptic density protein-95, discs large, ZO-1 (PDZ) domain at the N-terminus. The nNOS PDZ domain is also important in mediating dimerisation of the NOS enzyme. In the kidney, PSD-93 is expressed in numerous cell types and regions; in the macula densa, where nNOS is abundant, it anchors a sub-population of NOS to the basolateral membrane, which is presumably critical for transport processes. Given the similarities between DNOS and nNOS, it would be reasonable to assume, then, that the *Drosophila* enzyme is also anchored and compartmentalised in a similar manner. However, examination of the sequence of DNOS shows that PDZ domains are not present in this protein. Also, other possible motifs for protein–protein interactions, *e.g.*, N-terminus glutamine-rich motifs, have been found to be unnecessary for NOS activity and dimerisation (Stasiv et al., 2004), suggesting that compartmentalisation of the NOS signal in insects may occur via a different mechanism from that found in vertebrates. Expression of *dNOS4* has been documented in the tubule (McGettigan and Davies, unpublished), and it is possible that dominant negative regulation of DNOS1 occurs in this tissue.

Function of NO signalling in tubules: Fluid transport

The first demonstration of cGMP modulation of tubule function was described for locust tubules, where stimulation of fluid transport was observed upon application of exogenous cGMP (Morgan and Mordue, 1985). Since then, cGMP has been shown to inhibit fluid transport by *Rhodnius prolixus* (Quinlan et al., 1997) and also *Tenebrio molitor* tubules (Eigenheer et al., 2002). Does this mean that the NO/cGMP-induced stimulation of fluid transport by *Drosophila* tubules is an exception to a general rule? In the work describing antidiuretic effects of cGMP on *Rhodnius* secretion, application of a NO donor had no effect on tubule secretion (Quinlan et al., 1997). Furthermore, the antidiuretic cGMP-mobilising peptide (antidiuretic factor, ADF) in *Tenebrio* acts in a NO-independent manner (Eigenheer et al., 2002). Thus, cGMP – but not the NO-activated cGMP pathway – modulates fluid transport in all these insects.

By contrast, fluid transport in *Drosophila* tubules is stimulated by NO/ cGMP. Application of either the NO-mobilising peptide CAP_{2b}, or cGMP, to tubules increases the transepithelial potential (Davies *et al.*, 1995), indicative of activation of vacuolar H^+-ATPase (V-ATPase), which is localised in the apical membrane of principal cells (Dow, 1999). Thus, a major function of NO/cGMP signalling is to regulate cation transport in this cell type, resulting in increased fluid transport.

Given the antidiuretic effects of cGMP in some insect tubules, is the diuretic role of NO signalling observed in *Drosophila* found in other insects from the same Order? Expression of NOS has been documented in several other closely related fly species, including disease-carrying *Anopheles* mosquito species (Davies, 2000; Pollock *et al.*, 2004). The *Anopheles stephensi* NOS gene has been cloned (Luckhart *et al.*, 1998), and NOS expression has been documented in *Anopheles gambiae* tubules (Dimopoulos *et al.*, 1998). All known insect NOS genes are very similar (Davies, 2000), resulting in virtually identical sequences for the NOS proteins. Thus, conservation of function at the physiological level may be expected, at least within the Diptera. Interestingly, data mining of the *Anopheles gambiae* genome identified capa peptides in this species (Riehle *et al.*, 2002) (Table 2).

Investigation into the expression of endogenous NOS in Dipteran tubules using a universal anti-NOS antibody against a conserved epitope (Pollock *et al.*, 2004) showed that NOS is expressed only in principal cells in mosquitoes (Fig. 3). Interestingly, in the tsetse fly, *Glossina morsitans*, and in the desert locust, *Shistocerca gregaria*, widespread expression of NOS throughout the tubule occurs. Moreover, neither of these insects possesses stellate cells, so the cell-specific expression of NOS seen in higher Dipteran species does not occur in the tsetse and locust.

Does NOS in tubules of these insects have functional significance? Recent work has shown that *Drosophila* capa-1, as well as the *Anopheles*

Table 2. Amino acid sequences for known members of the capa family of neurohormones from Dipteran species.

Origin and peptide name	Amino acid sequence	Reference
Drosophila melanogaster (capa-1)	GANMGLYAFPRVamide	Kean *et al.* (2002)
Drosophila melanogaster (capa-2)	ASGLVAFPRVamide	Kean *et al.* (2002)
Anopheles gambiae (*Ang*CAPA-QGL)	QGLVPFPRVamide	Riehle *et al.* (2002), Pollock *et al.* (2004)
Anopheles gambiae (*Ang*CAPA-GPT)	GPTVGLFAFPRVamide	Riehle *et al.* (2002), Pollock *et al.* (2004)

Adapted from Pollock *et al.* (2004).

A *A. aegypti*

B *A. stephensi*

C *G. morsitans*

D *S. gregaria*

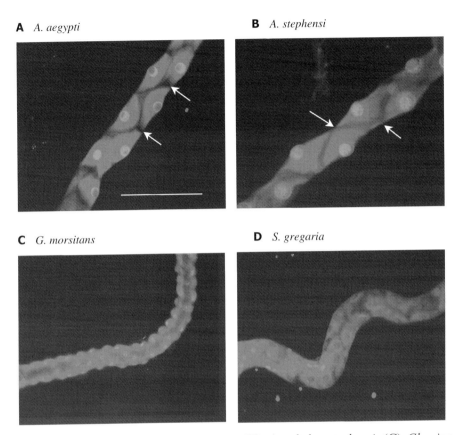

Fig. 3. Tubules from (A) *Aedes aegypti*, (B) *Anopheles stephensi*, (C) *Glossina morsitans* and (D) *Schistocerca gregaria*, based on Pollock *et al.* (2004). Anti-uNOS antibody (Broderick *et al.*, 2003, 2004; Dow and Davies, 2001; Gibbs and Truman, 1998) was used to visualise NOS immunoreactivity (green); cell nuclei were visualised with the nuclear stain 4′,6′-diamidino-2-phenylindole hydrochloride (DAPI; blue) (Broderick *et al.*, 2004). In (A) and (B), no green staining is observed in stellate cells (arrowed). *Glossina* (C) and *Schistocerca* (D) lack this cell type. Scale bar indicates 100 µm in (A) and (B); 200 µm in (C) and (D). (See Colour Plate Section in this book).

gambiae capa peptides *Ang*CAPA-QGL and *Ang*CAPA-GPT, all stimulate NO production in tubules from *D. melanogaster*, *Aedes aegypti*, *Anopheles stephensi* and *G. morsitans*, but not *S. gregaria* (Fig. 4A; Pollock *et al.*, 2004). The NO response is sensitive to a specific inhibitor of NOS, suggesting activation of NOS by capa in these species (Pollock *et al.*, 2004). The cGMP content is also increased by treatment of tubules

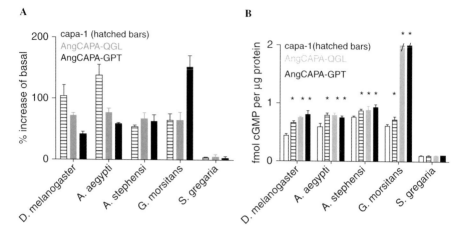

Fig. 4. Capa-1 (hatched bars), *Ang*CAPA-QGL (grey bars) and *Ang*CAPA-GPT (black bars) stimulation of: (A) NOS activity and (B) cGMP content in tubules from the insects shown. For (A), NOS activity data are given as a percentage increase from the basal level in unstimulated tubules; in (B), the cGMP content is normalised across species using the protein concentrations of the tubule preparations. From Pollock *et al.*, 2004.

from the four Diptera (Fig. 4B; Pollock *et al.*, 2004), which demonstrates conservation of capa-induced NO/cGMP signalling in Dipteran tubules. Also, this suggests that sequence and functional conservation of the capa receptors must also exist within the Diptera. Characterisation of the *Drosophila* capa receptor (Iversen *et al.*, 2002; Park *et al.*, 2002) has allowed identification of a putative capa receptor in the *Anopheles* genome (Genbank ID: genbank:XP_312952). Reverse transcription-polymerase chain reaction (RT-PCR) has demonstrated that the *Anopheles* capa receptor homologue is indeed expressed in *Anopheles* tubule,

Fig. 5. Fluid secretion rates in tubules from *D. melanogaster* (unshaded), *Aedes aegypti* (black), *Anopheles stephensi* (red), *G. morsitans* (blue) and *S. gregaria* (green) stimulated by either (A) capa-1, (B) *Ang*CAPA-QGL or (C) *Ang*CAPA-GPT at the concentrations shown (expressed as log molar, M). Basal rates of secretion were measured for 30 min prior to the addition of peptides. Secretion rates were measured for a further 40 min. Secretion rates are expressed as the percentage change of unstimulated tubules for each species, $\pm$SEM ($n = 6-8$). Asterisks denote statistically significant differences from basal values, $P < 0.05$, determined using the Student's *t*-test (unpaired samples). (See Colour Plate Section in this book).

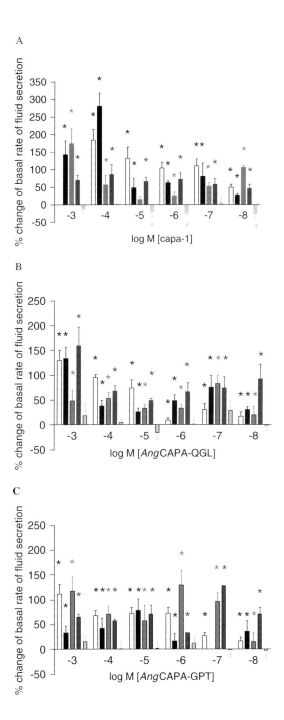

although no such receptor has been identified in *Aedes aegypti* (Pollock *et al.*, 2004).

Stimulation of NO/cGMP signalling in the Diptera via capa peptides results in dose-dependent elevation of fluid transport rates (Fig. 5). However, although NOS genes have been characterized from multiple Orders of insect (Davies, 2000), and locust tubules do contain NOS (Fig. 3D), the capa peptides do not elevate fluid secretion in *S. gregaria*. Capa-1 may even be antidiuretic at some concentrations tested (Fig. 5), although this is not linked to an increase in cGMP content (Fig. 4). Interestingly, *M. sexta* CAP$_{2b}$ does not affect secretion by *Locusta migratoria* tubule, (Coast, 2001; Wegener *et al.*, 2002), supporting the data on *S. gregaria* tubules shown here. The data in Fig. 5 also show measurements of neuropeptide-stimulated secretion rates in *G. morsitans* tubules for the first time, after initial work some 30 years ago on fluid transport by these tubules (Gee, 1976a,b).

In summary, a physiological role for *Anopheles gambiae* capa peptides, *i.e.*, capa-stimulated fluid secretion, is confined to a range of Dipteran insects. The demonstration of conservation of NO-mediated capa signalling in medically important insect vectors suggests new possibilities for novel insecticide targets for pest control.

Function of NO signalling in tubules: Immune function

Work in *Drosophila* has been instrumental in defining the mechanisms of innate immunity (Hoffmann, 2003). Two pathways in *Drosophila* constitute innate immunity: toll and Imd, which involve at least seven different antimicrobial peptides. These peptides are known to be induced either in the fat body, a major site of immune function, or more locally in barrier epithelia (Tzou *et al.*, 2000).

The NO pathway has been shown to modulate immune function in vertebrates (Bogdan, 2001). In addition, NO is also involved in insect immunity; at least two groups have shown that NO activates antimicrobial peptide (diptericin) expression in the fat body of *Drosophila* larvae, as well as in haemocytes (Foley and O'Farrell, 2003; Nappi *et al.*, 2000). The role of epithelia in the production of antimicrobial peptides in both vertebrates (Selsted and Ouellette, 2005) and insects (Tzou *et al.*, 2000) suggests the importance of epithelial tissue in immunity.

Recent novel work from this laboratory has implicated a role for NO signalling in immune responses in the tubule (McGettigan *et al.*, 2005). These data also show that the tubule is a cell-autonomous immune sensor. Molecular evidence shows that tubules express all elements of the

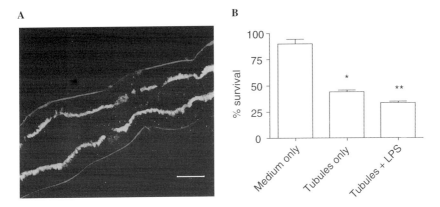

Fig. 6. (A) Lipopolysaccharide (LPS) binds to tubule. Intact tubules from the *vhaSFD* gene-trap line, which express green fluorescent protein (GFP)-tagged *vhaSFD* (Morin *et al.*, 2001), were used in order to delineate the tubule apical membrane. These were incubated with AlexaFluor-LPS (*E. coli*) for up to 10 min. Principal cell nuclei are stained blue using DAPI; tubule apical membrane is defined by green vhaSFD:GFP (Torrie *et al.*, 2004); LPS binding and internalisation is indicated by red staining. Note the presence of red vesicles in the cytosol of the tubule cells. Scale bar = 15 μm. (B) Bacterial killing by excised, intact tubules. Sterile Schneider's medium was either left untreated, incubated with tubules only (6 h), or incubated with LPS-treated tubules (6 h). The medium was subsequently assessed for the effect on *E. coli* populations. Data are expressed as percentage survival of *E. coli* $\pm$ SEM ($n = 4$) from four different biological replicates. Data significantly different from controls are indicated by * and significant differences from tubules only are indicated by ** ($P < 0.05$), Student's *t*-test for unpaired samples. From McGettigan *et al.*, 2005. (See Colour Plate Section in this book).

Imd signalling pathway. Furthermore, incubation of excised tubules in cell culture medium results in killing of *Escherichia coli* populations with only the conditioned medium. Also, conditioned medium from excised tubules challenged with Gram-negative *E. coli*-derived lipopolysaccharide (LPS) – which binds to and is internalised by tubules (Fig. 6A) – results in significantly increased bacterial killing (Fig. 6B). Further investigation revealed that tubules express diptericin in response to LPS but not to peptidoglycan derived from *Staphylococcus aureus*. Diptericin expression is increased twofold in LPS-treated excised tubules, but approximately sevenfold in tubules from LPS-treated flies. Thus, the bacterial killing action of tubules is mediated by the production of antimicrobial peptides. Use of antimicrobial peptide promoter: green fluorescent protein (GFP) fusion transgenic lines (Tzou *et al.*, 2000) showed that immune-challenged tubules displayed increased diptericin expression in principal cells, thus implicating principal cell function in immunity.

Given that NO is implicated in diptericin production in *Drosophila* cells, as well as the important role of NO in tubules, the possibility that NO was involved in the tubule immune response was investigated. Colorimetric analysis and measurement of NO activity upon immune challenge of both excised tubules and tubules from challenged flies showed that NO production is markedly increased in principal cells by either LPS or *E. coli*. As with diptericin production, tubules from infected animals showed a much greater increase in NO production compared to excised, LPS-stimulated tubules, suggesting that while the tubule is capable of autonomous sensing and responding to bacterial infection, communication within the insect may prime these responses *in vivo*.

The consequences of increasing NOS only in the principal cells was investigated by targetting a *dNOS* transgene to only principal cells *in vivo* using a UAS-*dNOS* transgenic line (UAS-dN1-8) together with the c42 GAL4 driver (Broderick *et al.*, 2004; McGettigan *et al.*, 2005; Rosay *et al.*, 1997). This resulted in increased NOS activity, and markedly elevated diptericin expression (McGettigan *et al.*, 2005). LPS treatment of c42/UAS-dN1-8 tubules increased diptericin expression up to nearly 20-fold compared to parental UAS-dN1-8 tubules. Also, resting expression levels of diptericin in c42/UAS-dN1-8 tubules were significantly greater compared to UAS-dN1-8 tubules. Finally, tubules from c42/UAS-dN1-8 flies showed significantly higher levels of diptericin expression compared to parental controls, both in uninfected and LPS-infected animals (Fig. 7A).

If activation of NO signalling in principal cells results in increased diptericin expression in tubules, what is the effect on the whole animal? Data from survival experiments upon immune challenge (Fig. 7B and C) show that targetted over-expression of *dNOS* in only tubule principal cells is sufficient to confer a selective advantage compared to the control parental flies. *E. coli*-infected c42/UAS-dN1-8 flies are significantly fitter than *E. coli*-infected parental c42 or UAS-dN1-8 flies, and behave as mock-infected c42/UAS-dN1-8 flies or mock-infected flies from both parental lines. The survival curves of all three mock-infected lines are indistinguishable upon injury, suggesting that the effects of targetted *dNOS* are not merely an effect of increased tolerance to wounding.

Is the immune role of Malpighian tubules likely to be confined to *Drosophila melanogaster*? In *Anopheles gambiae* it was demonstrated that the largest increase in expression of an immune-responsive gene was that of NOS in the Malpighian tubules (Dimopoulos *et al.*, 1998).

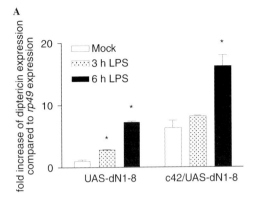

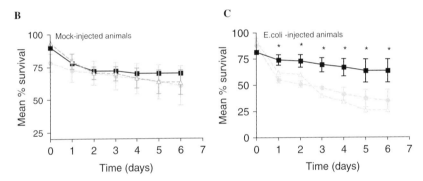

Fig. 7. (A) Expression of a *dNOS* transgene using a principal cell-specific GAL4 driver increases diptericin production in tubules of lipopolysaccharide (LPS)-challenged flies. UAS-dN1-8 and c42/UAS-dN1-8 adult flies were challenged by mock injection (unshaded bars), or by treatment with LPS for 3 h (grey bars) or 6 h (black bars). Diptericin expression was quantified by real-time PCR in tubules of these animals, and normalised against expression of *rp49* mRNA in all samples. Data shown are fold-increases in diptericin mRNA levels compared to those of *rp49* mRNA, $\pm$SEM ($n = 4$). Asterisks indicate data significantly different from the control ($P < 0.05$). (B), (C) c42 (grey circles), UAS-dN1-8 (grey triangles) and c42/UAS-dN1-8 (black squares) flies were either mock challenged (B) or challenged with *E. coli* suspension (C). Survivors were counted 3 h after inoculation (day 0) and at 24 h intervals thereafter (days 1–6). Numbers of survivors for each line after mock or infective treatment were expressed as percentages of the starting number. Data are expressed as % survivors $\pm$SEM ($n = 4$). Data from McGettigan *et al.*, 2005.

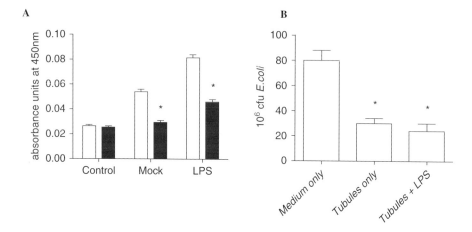

Fig. 8. (A) NO production in tubules excised from immune-challenged *Aedes aegypti* mosquitoes. NO production was quantified by colorimetry as previously described (McGettigan *et al.*, 2005; Pollock *et al.*, 2004), in the absence (unshaded bars) or presence (black bars) of the NOS inhibitor L-NAME (N^G-nitro-L-arginine methyl ester). Mosquitoes were either left untreated (controls), were mock-injected, or were injected with lipopolysaccharide (LPS), as previously described for *Drosophila* (McGettigan *et al.*, 2005), and left to recover prior to removal of tubules for processing (Pollock *et al.*, 2004). Data are expressed as absorbance at $450\,\mathrm{nm} \pm \mathrm{SEM}$ ($n = 8$) for each data set shown. Asterisks indicate significantly different data between L-NAME-treated and untreated tubules from mock- and LPS-treated mosquitoes, $p < 0.05$, Student's *t*-test. (B) *E. coli* killing assay performed with conditioned medium from *Aedes aegypti* tubules, as described for Fig. 6B. Data from different biological replicates are expressed as colony-forming units (cfu) of *E. coli*, $\pm \mathrm{SEM}$ ($n = 4$) for 'Medium only' samples, and $\pm \mathrm{SD}$ ($n = 3$) for 'Tubules only' and 'Tubules + LPS' samples. Individual data points (cfu of *E. coli*) were as follows: 'Medium only', 97, 91, 68, 64; 'Tubules only', 32, 34, 26; 'Tubules + LPS', 24, 18, 30 (J. McGettigan, unpublished data).

Furthermore, given that NO signalling occurs in principal cells in tubules of *Aedes aegypti* and *Anopheles stephensi*, and that the modes of action of functional NO pathways are conserved across Dipteran species (see above and Pollock *et al.*, 2004), it is possible that NO is functionally induced upon immune challenge in mosquitoes. Data shown in Fig. 8A suggest that this is indeed the case, at least for *Aedes aegypti*. Furthermore, like *Drosophila* tubules, excised *Aedes* tubules are capable of killing bacteria (Fig. 8B). It is possible, then, that NO can induce antimicrobial peptide expression across Dipteran species.

In *Anopheles*, increased expression of immune-responsive genes in midgut occurs in response to infection (Dimopoulos *et al.*, 1998). Thus, it is clear – and logical – that epithelia play an immune, as well as a barrier role. Why should the role of the tubule be particularly important? Tubules clear waste metabolites and toxins from the haemolymph, and to achieve this they transport fluid at very high rates in comparison to the haemolymph volume. Given this role, the tubule is one of the first tissues that would be exposed to key molecules, such as LPS, that are associated with immune challenge (Dow and Davies, 2006). Thus, it seems that far from being only osmoregulatory and detoxification organs in Diptera, Malpighian tubules provide a central immune-sensing role, critical for insect survival.

Conclusions

The roles of NO in biology are wide-ranging, and affect almost all known physiological functions. Work on insect tubules, including those of *Drosophila* in particular, has shown that NO has important direct effects on renal transport processes and also in immune-sensing.

The effect of NO on tubule immune function does not seem to be merely incidental to that of other canonical immune tissues, for example the fat body, but rather is critical to the survival of the whole animal under conditions of stress. This important role of tubules may not be confined to insects: recent work in vertebrates has shown that renal tubular epithelial cells (TEC) produce NO in response to pro-inflammatory cytokines (Du *et al.*, 2006). This response is important in determining the susceptibility of TEC to apoptosis, which influences the maintenance of overall kidney function, and therefore is critical for survival. NO is also important in host defence and the function of other types of secretory epithelia, for example airway epithelial cells in the respiratory tract (Bove and van der Vliet, 2006). Thus, while NO is known to have wide-ranging roles in vertebrate kidney function, especially in relation to salt–water balance (Kone, 2004), it is probable that NO also has roles in immune defence in renal tissue. Thus, study of the regulatory mechanisms which govern the function of fluid-secreting epithelia is an important area in biology, and one that greatly benefits from the use of model organisms.

Acknowledgements

Work in the author's laboratory is supported by the Biotechnology and Biological Sciences Research Council, UK. I would like to thank my

102

close collaborator, Professor J. A. T. Dow, for critical reading of this manuscript.

References

Bicker, G. (2005). STOP and GO with NO: Nitric oxide as a regulator of cell motility in simple brains. *Bioessays* 27,495–505.

Bicker, G., Schmachtenberg, O. and De Vente, J. (1996). The nitric oxide/cyclic GMP messenger system in olfactory pathways of the locust brain. *Eur. J. Neurosci.* 8,2635–2643.

Bogdan, C. (2001). Nitric oxide and the immune response. *Nat. Immunol.* 2,907–916.

Bove, P. F. and van der Vliet, A. (2006). Nitric oxide and reactive nitrogen species in airway epithelial signaling and inflammation. *Free Radic. Biol. Med.* 41,515–527.

Brand, A. H. and Perrimon, N. (1993). Targetted gene expression as a means of altering cell fates and generating dominant phenotypes. *Development* 118,401–415.

Broderick K. E. (2002). *Cyclic cGMP-Dependent Signalling in D. Melanogaster Malpighian Tubules* (PhD Thesis). University of Glasgow, Glasgow, UK.

Broderick, K. E., Kean, L., Dow, J. A. T., Pyne, N. J. and Davies, S. A. (2004). Ectopic expression of bovine type 5 phosphodiesterase confers a renal phenotype in Drosophila. *J. Biol. Chem.* 279,8159–8168.

Broderick, K. E., MacPherson, M. R., Regulski, M., Tully, T., Dow, J. A. T. and Davies, S. A. (2003). Interactions between epithelial nitric oxide signaling and phosphodiesterase activity in Drosophila. *Am. J. Physiol. Cell Physiol.* 285,C1207–C1218.

Coast, G. M. (2001). The neuroendocrine regulation of salt and water balance in insects. *Zoology* 103,179–188.

Davies, S.-A. (2000). Nitric oxide signalling in insects. *Insect Biochem. Mol. Biol.* 30, 1123–1138.

Davies, S. A., Huesmann, G. R., Maddrell, S. H. P., O'Donnell, M. J., Skaer, N. J. V., Dow, J. A. T. and Tublitz, N. J. (1995). CAP_{2b}, a cardioacceleratory peptide, is present in *Drosophila* and stimulates tubule fluid secretion via cGMP. *Am. J. Physiol.* 269,R1321–R1326.

Davies, S. A., Stewart, E. J., Huesmann, G. R., Skaer, N. J. V., Maddrell, S. H. P., Tublitz, N. J. and Dow, J. A. T. (1997). Neuropeptide stimulation of the nitric oxide signaling pathway in *Drosophila melanogaster* Malpighian tubules. *Am. J. Physiol.* 42,R823–R827.

Day, J. P., Dow, J. A., Houslay, M. D. and Davies, S. A. (2005). Cyclic nucleotide phosphodiesterases in Drosophila melanogaster. *Biochem. J.* 388,333–342.

Day, J. P., Houslay, M. D. and Davies, S. A. (2006). A novel role for a Drosophila homologue of cGMP-specific phosphodiesterase in the active transport of cGMP. *Biochem. J.* 393,481–488.

Dimopoulos, G., Seeley, D., Wolf, A. and Kafatos, F. C. (1998). Malaria infection of the mosquito *Anopheles gambiae* activates immune-responsive genes during critical transition stages of the parasite life cycle. *EMBO J.* 17,6115–6123.

Dow, J. A. T. (1999). The multifunctional *Drosophila melanogaster* V-ATPase is encoded by a multigene family. *J. Bioenerg. Biomembr.* 31,75–83.

Dow, J. A. T. and Davies, S. A. (2001). The Drosophila Melanogaster Malpighian tubule. *Adv. Insect Physiol.* 28,1–83.

Dow, J. A. T. and Davies, S. A. (2003). Integrative physiology, functional genomics and epithelial function in a genetic model organism. *Physiol. Rev.* 83,687–729.

Dow, J. A. and Davies, S. A. (2006). The Malpighian tubule: Rapid insights from post-genomic biology. *J. Insect Physiol.* 52,365–378.

Dow, J. A. T., Maddrell, S. H., Davies, S. A., Skaer, N. J. and Kaiser, K. (1994). A novel role for the nitric oxide-cGMP signaling pathway: The control of epithelial function in Drosophila. *Am. J. Physiol.* 266,R1716–R1719.

Du, C., Guan, Q., Diao, H., Yin, Z. and Jevnikar, A. M. (2006). Nitric oxide induces apoptosis in renal tubular epithelial cells through activation of caspase-8. *Am. J. Physiol. Renal Physiol.* 290,F1044–F1054.

Eigenheer, R. A., Nicolson, S. W., Schegg, K. M., Hull, J. J. and Schooley, D. A. (2002). Identification of a potent antidiuretic factor acting on beetle Malpighian tubules. *Proc. Natl. Acad. Sci. USA* 99,84–89.

Foley, E. and O'Farrell, P. H. (2003). Nitric oxide contributes to induction of innate immune responses to gram-negative bacteria in Drosophila. *Genes Dev.* 17, 115–125.

Foster, M. W., McMahon, T. J. and Stamler, J. S. (2003). S-nitrosylation in health and disease. *Trends Mol. Med.* 9,160–168.

Friebe, A. and Koesling, D. (2003). Regulation of nitric oxide-sensitive guanylyl cyclase. *Circ. Res.* 93,96–105.

Gee, J. D. (1976a). Active transport of sodium by the malpighian tubules of the tsetse fly Glossian morsitans. *J. Exp. Biol.* 64,357–368.

Gee, J. D. (1976b). Fluid secretion by the malpighian tubules of the tsetse fly Glossina morsitans: The effects of ouabain, ethacrynic acid and amiloride. *J. Exp. Biol.* 65, 323–332.

Gibbs, S. M. and Truman, J. W. (1998). Nitric oxide and cyclic GMP regulate retinal patterning in the optic lobe in *Drosophila*. *Neuron* 20,83–93.

Herrera, M. and Garvin, J. L. (2005). Recent advances in the regulation of nitric oxide in the kidney. *Hypertension* 45,1062–1067.

Hoffmann, J. A. (2003). The immune response of Drosophila. *Nature* 426,33–38.

Holscher, C. (1997). Nitric oxide, the enigmatic neuronal messenger: Its role in synaptic plasticity. *Trends Neurosci.* 20,298–303.

Iversen, A., Cazzamali, G., Williamson, M., Hauser, F. and Grimmelikhuijzen, C. J. (2002). Molecular cloning and functional expression of a Drosophila receptor for the neuropeptides capa-1 and -2. *Biochem. Biophys. Res. Commun.* 299,628–633.

Kean, L., Cazenave, W., Costes, L., Broderick, K. E., Graham, S., Pollock, V. P., Davies, S. A., Veenstra, J. A. and Dow, J. A. T. (2002). Two nitridergic peptides are encoded by the gene capability in Drosophila melanogaster. *Am. J. Physiol. Regul. Integr. Comp. Physiol.* 282,R1297–R1307.

Kennerdell, J. R. and Carthew, R. W. (2000). Heritable gene silencing in Drosophila using double-stranded RNA. *Nat. Biotechnol.* 18,896–898.

Kerr, M., Davies, S. A. and Dow, J. A. (2004). Cell-specific manipulation of second messengers; a toolbox for integrative physiology in Drosophila. *Curr. Biol.* 14, 1468–1474.

Koglin, M., Vehse, K., Budaeus, L., Scholz, H. and Behrends, S. (2001). Nitric oxide activates the beta 2 subunit of soluble guanylyl cyclase in the absence of a second subunit. *J. Biol. Chem.* 276,30737–30743.

Kone, B. C. (2004). Nitric oxide synthesis in the kidney: Isoforms, biosynthesis, and functions in health. *Semin. Nephrol.* 24,299–315.

Kone, B. C., Kuncewicz, T., Zhang, W. and Yu, Z. Y. (2003). Protein interactions with nitric oxide synthases: Controlling the right time, the right place, and the right amount of nitric oxide. *Am. J. Physiol. Renal Physiol.* 285,F178–F190.

Luckhart, S., Vodovotz, Y., Cui, L. and Rosenberg, R. (1998). The mosquito *Anopheles stephensi* limits malaria parasite development with inducible synthesis of nitric oxide. *Proc. Natl. Acad. Sci. USA* 95,5700–5705.

MacPherson, M. R., Lohmann, S. M. and Davies, S. A. (2004). Analysis of Drosophila cGMP-dependent protein kinases and assessment of their in vivo roles by targeted expression in a renal transporting epithelium. *J. Biol. Chem.* 279,40026–40034.

MacPherson, M. R., Pollock, V. P., Broderick, K. B., Kean, L., O'Connell, F. C., Dow, J. A. T. and Davies, S.-A. (2001). Model organisms: New insights into ion channel and transporter function: L-type calcium channels regulate epithelial fluid transport in Drosophila melanogaster. *Am. J. Physiol. Cell Physiol.* 280,C394–C407.

MacPherson, M. R., Pollock, V. P., Kean, L., Southall, T. D., Giannakou, M. E., Broderick, K. E., Dow, J. A., Hardie, R. C. and Davies, S. A. (2005). Transient receptor potential-like channels are essential for calcium signaling and fluid transport in a Drosophila epithelium. *Genetics* 169,1541–1552.

McGettigan, J., McLennan, R. K., Broderick, K. E., Kean, L., Allan, A. K., Cabrero, P., Regulski, M. R., Pollock, V. P., Gould, G. W., Davies, S. A. *et al.* (2005). Insect renal tubules constitute a cell-autonomous immune system that protects the organism against bacterial infection. *Insect Biochem. Mol. Biol.* 35,741–754.

McGuire, S. E., Roman, G. and Davis, R. L. (2004). Gene expression systems in Drosophila: A synthesis of time and space. *Trends Genet.* 20,384–391.

Morgan, P. J. and Mordue, W. (1985). The role of calcium in diuretic hormone action on locust Malpighian tubules. *Mol. Cell. Endocrinol.* 40,221–231.

Morin, X., Daneman, R., Zavortink, M. and Chia, W. (2001). A protein trap strategy to detect GFP-tagged proteins expressed from their endogenous loci in Drosophila. *Proc. Natl. Acad. Sci. USA* 98,15050–15055.

Mungrue, I. N., Bredt, D. S., Stewart, D. J. and Husain, M. (2003). From molecules to mammals: What's NOS got to do with it? *Acta Physiol. Scand.* 179,123–135.

Nappi, A. J., Vass, E., Frey, F. and Carton, Y. (2000). Nitric oxide involvement in Drosophila immunity. *Nitric Oxide* 4,423–430.

Ortiz, P. A. and Garvin, J. L. (2003). Trafficking and activation of eNOS in epithelial cells. *Acta Physiol. Scand.* 179,107–114.

Park, Y., Kim, Y. J. and Adams, M. E. (2002). Identification of G protein-coupled receptors for Drosophila PRXamide peptides, CCAP, corazonin, and AKH supports a theory of ligand-receptor coevolution. *Proc. Natl. Acad. Sci. USA* 99, 11423–11428.

Pollock, V. P., McGettigan, J., Cabrero, P., Maudlin, I. M., Dow, J. A. and Davies, S. A. (2004). Conservation of capa peptide-induced nitric oxide signalling in Diptera. *J. Exp. Biol.* 207,4135–4145.

Quinlan, M. C., Tublitz, N. J. and O'Donnell, M. J. (1997). Anti-diuresis in the blood-feeding insect *Rhodnius prolixus* Stal: The peptide CAP(2b) and cyclic GMP inhibit Malpighian tubule fluid secretion. *J. Exp. Biol.* 200,2363–2367.

Regulski, M., Stasiv, Y., Tully, T. and Enikolopov, G. (2004). Essential function of nitric oxide synthase in Drosophila. *Curr. Biol.* 14,R881–R882.

Regulski, M. and Tully, T. (1995). Molecular and biochemical characterization of dNOS – a *Drosophila* Ca^{2+} calmodulin-dependent nitric-oxide synthase. *Proc. Natl. Acad. Sci. USA* 92,9072–9076.

Riehle, M. A., Garcynski, S. F., Crim, J. W., Hill, C. A. and Brown, M. R. (2002). Neropeptides and peptide hormones in Anopheles gambiae. *Science* 298,172–175.

Rosay, P., Davies, S. A., Yu, Y., Sozen, A., Kaiser, K. and Dow, J. A. T. (1997). Cell-type specific calcium signalling in a *Drosophila* epithelium. *J. Cell. Sci.* 110,1683–1692.

Schmidt, H. H. H. W., Lohmann, S. M. and Walter, U. (1993). The nitric oxide and cGMP signal transduction system: Regulation and mechanism of action. *Biochim. Biophys. Acta* 1178,153–175.

Selsted, M. E. and Ouellette, A. J. (2005). Mammalian defensins in the antimicrobial immune response. *Nat. Immunol.* 6,551–557.

Shah, S. and Hyde, D. R. (1995). Two *Drosophila* genes that encode the alpha and beta subunits of the brain soluble guanylyl cyclase. *J. Biol. Chem.* 270,15368–15376.

Southall, T. D., Terhzaz, S., Cabrero, P., Chintapalli, V. R., Evans, J. M., Dow, J. A. and Davies, S. A. (2006). Novel subcellular locations and functions for secretory pathway Ca^{2+}/Mn^{2+}-ATPases. *Physiol. Genomics* 26,35–45.

Sozen, M. A., Armstrong, J. D., Yang, M. Y., Kaiser, K. and Dow, J. A. T. (1997). Functional domains are specified to single-cell resolution in a *Drosophila* epithelium. *Proc. Natl. Acad. Sci. USA* 94,5207–5212.

Stasiv, Y., Kuzin, B., Regulski, M., Tully, T. and Enikolopov, G. (2004). Regulation of multimers via truncated isoforms: A novel mechanism to control nitric-oxide signaling. *Genes Dev.* 18,1812–1823.

Stasiv, Y., Regulski, M., Kuzin, B., Tully, T. and Enikolopov, G. (2001). The Drosophila nitric-oxide synthase gene (dNOS) encodes a family of proteins that can modulate NOS activity by acting as dominant negative regulators. *J. Biol. Chem.* 276, 42241–42251.

Stuehr, D. J. (1999). Mammalian nitric oxide synthases. *Biochim. Biophys. Acta* 1411, 217–230.

Terhzaz, S., Southall, T. D., Lilley, K. S., Kean, L., Allan, A. K., Davies, S. A. and Dow, J. A. (2006). Differential gel electrophoresis and transgenic mitochondrial calcium reporters demonstrate spatiotemporal filtering in calcium control of mitochondria. *J. Biol. Chem.* 281,18849–18858.

Torrie, L. S., Radford, J. C., Southall, T. D., Kean, L., Dinsmore, A. J., Davies, S. A. and Dow, J. A. (2004). Resolution of the insect ouabain paradox. *Proc. Natl. Acad. Sci. USA* 101,13689–13693.

Trimmer, B. A., Aprille, J. and Modica-Napolitano, J. (2004). Nitric oxide signalling: Insect brains and photocytes. *Biochem. Soc. Symp.* 71,65–83.

Tzou, P., Ohresser, S., Ferrandon, D., Capovilla, M., Reichhart, J. M., Lemaitre, B., Hoffmann, J. A. and Imler, J. L. (2000). Tissue-specific inducible expression of antimicrobial peptide genes in Drosophila surface epithelia. *Immunity* 13,737–748.

Wang, J., Kean, L., Yang, J., Allan, A. K., Davies, S. A., Herzyk, P. and Dow, J. A. (2004). Function-informed transcriptome analysis of Drosophila renal tubule. *Genome Biol.* 5,R69.

Wang, T. (2002). Role of nitric oxide synthase (iNOS and eNOS) in modulating proximal tubule transport and acid-base balance. *Am. J. Physiol. Renal Physiol.* 283, F658–F662.

Wegener, C., Herbert, Z., Eckert, M. and Predel, R. (2002). The periviscerokinin (PVK) peptide family in insects: Evidence for the inclusion of CAP(2b) as a PVK family member. *Peptides* 23,605–611.

Welch, W. J. and Wilcox, C. S. (2002). What is brain nitric oxide synthase doing in the kidney? *Curr. Opin. Nephrol. Hypertens.* 11,109–115.

Zhao, Y., Brandish, P. E., Ballou, D. P. and Marletta, M. A. (1999). A molecular basis for nitric oxide sensing by soluble guanylate cyclase. *Proc. Natl. Acad. Sci. USA* 96,14753–14758.

Nitric oxide/cyclic GMP signaling and insect behavior

Ralf Heinrich[1] and Geoffrey K. Ganter[2,*]

[1]Department of Neurobiology, Institute for Zoology, Berliner Str. 28, 37077 Göttingen, Germany

[2]Department of Biological Sciences, University of New England, 11 Hills Beach Road, Biddeford, ME 04005, USA

Abstract. Behavioral control involves the perception of external information by sensory organs, its integration with proprioceptive information representing an individual's internal state, selection of appropriate actions by central nervous circuits, and their adaptive performance by neuron–muscular and neurosecretory systems. There is accumulating evidence from various species that the gaseous signaling molecule nitric oxide (NO) participates in the control of insect behavior on all these levels. In contrast to the spatially and temporally precise transmission at conventional chemical synapses, NO is formed on demand, freely diffuses through cellular membranes, and may thus coordinate units of neurons without anatomically established synaptic interconnections. As a laterally diffusing messenger, NO may influence populations of sensory afferences, interneurons and efferent cells, contributing to all levels of processing between sensory activation and the activation of neurosecretory and motor functions. After briefly summarizing some general aspects of NO signaling mechanisms and the distribution of their functional components in the insect nervous system, we present a collection of studies demonstrating the direct contribution of NO to the processing of behaviorally relevant sensory information and the selection and coordinated performance of situation-specific behaviors.

Keywords: nitric oxide; nitric oxide synthase; guanylyl cyclase; cyclic GMP; insect behavior; sensory systems; central nervous processing of sensory information; learning and memory; selection of behavior; modulation of secretory and motor systems.

NO in insect nervous systems

The detection of nitric oxide synthase (NOS) and soluble guanylate cyclase (sGC) in the brain of locusts by Elphick and co-workers (Elphick et al., 1993) and the subsequent demonstration of activity- and Ca^{2+}-dependent release of NO in the central nervous system (CNS) by Müller and Bicker (1994) initiated numerous studies on various insect species that implicated NO signaling in the processing of sensory information, mechanisms of habituation, associative learning and memory formation, selection of behavior, and modulation of neurosecretion and synaptic release at the neuromuscular junction (for detailed references see below).

Corresponding author: Tel.: 207-602-2225. Fax: 207-602-5956.
E-mail: gganter@une.edu (G.K. Ganter).

ADVANCES IN EXPERIMENTAL BIOLOGY
VOLUME 01   ISSN 1872-2423
DOI: 10.1016/S1872-2423(07)01005-8

In addition, various functions of NO signaling serving the development, structural organization, and maintenance of nervous neuropils have been demonstrated: cell migration (Bicker, 2005; Gibson et al., 2001; Haase and Bicker, 2003), structuring of neuropils (Gibbs and Truman, 1998), and the formation and maintenance of synapses (Ball and Truman, 1998; Wright et al., 1998). Types of cells that express NOS or display NO-stimulated responses include receptor neurons, interneurons, motoneurons, and neurosecretory cells as well as certain classes of glia.

The function and distribution of NO-producing and NO-responsive neurons in insect nervous systems has previously been reviewed by Bicker (1998, 2001), Müller (1997) and Trimmer et al. (2004).

Components of NO signaling in insect nervous systems: NOS, NO, sGC, cGMP

Nitric oxide synthase (NOS)

NOS converts oxygen and L-arginine into NO and L-citrulline using NADPH as a cofactor (Mayer, 1994). While mammalian tissues express several NOS genes (Bredt and Snyder, 1992; Garthwaite and Boulton, 1995), only a single gene locus has been found in *Drosophila melanogaster* (Regulski and Tully, 1995). This gene nevertheless codes for several transcripts and thus produces a family of NOS-related proteins with largely unknown functions (Enikopolov et al., 1999). Similarly, one NOS has been cloned from the moth *Manduca sexta* (Nighorn et al., 1998) and from the locust *Schistocerca gregaria* (Ogunshola et al., 1995). The majority of insect NOS depend on Ca^{2+}/calmodulin-mediated activation, which has been demonstrated to be coupled to electrical activity in the *M. sexta* CNS (Qazi and Trimmer, 1999) or to synaptic activation by conventional transmitters that mediate increases of cytosolic Ca^{2+} levels: acetylcholine in dissociated locust neurons (Müller and Bicker, 1994), and nicotinic agonists of acetylcholine in *M. sexta* ventral nerve cords (Zayas et al., 2002). However, since 5–10% of the total NOS activity in the nervous systems of *Apis mellifera* and *Drosophila* was found to be independent of Ca^{2+} accumulation in the cytosol (Müller, 1994), and homogenates of locust CNS were found to retain NOS activity in the absence of Ca^{2+} (Müller and Bicker, 1994), the presence of a Ca^{2+}-independent and functionally different isoform of NOS in insects was suggested. Whether specific isoforms of insect NOS assume a particular subcellular organization by association with particular membrane-bound receptors is not known. A PDZ motif (a sequence of approximately 90 amino acids organized in a modular arrangement that interconnects proteins), known to couple mammalian neuronal NOS to NMDA receptors for more effective stimulation by Ca^{2+} influx (Brenman et al., 1996; Denninger and Marletta, 1999), is absent from M. sexta NOS (Nighorn et al., 1998), and other potential contact regions with partner proteins have not been described.

The distribution of NOS in the nervous tissues of various insect species has been described by immunocytochemistry, using a universal anti-NOS antibody raised against the highly conserved NADPH-binding domain of the enzyme, and by NADPH-diaphorase (NADPHd) histochemistry, exploiting the fact that, as in vertebrates (Bredt et al., 1991; Hope et al., 1991), insect NOS activity is largely insensitive to formaldehyde fixation (Elphick et al., 1994; Müller, 1994). Although NADPHd activity may result from both NOS-related and NOS-unrelated diaphorases, close matching of universal NOS immunostaining and NADPHd reaction product localization generally confirmed that NADPHd staining is a useful marker for NOS in insects, enabling the analysis of fine anatomical details of labeled neurons (various studies are summarized in Ott and Elphick, 2002, 2003). Nevertheless, species-specific differences in the formaldehyde sensitivity of NOS and NOS-unrelated diaphorases can generate misleading results (for example see Nighorn et al., 1998; Ott and Burrows, 1999; Ott et al., 2001). Recent studies using a methanol/formalin fixation protocol, which increased both the sensitivity and specificity of NOS-related diaphorase detection, confirmed the general results of earlier studies but revealed a prominent presence of NOS in almost all central nervous neuropils (Kurylas et al., 2005; Ott and Elphick, 2002, 2003), suggesting that the impact of NO signaling on central nervous processing may have been largely underestimated. More than 90% of the NOS activity in most insects studied is located in the brain (Bicker and Hähnlein, 1995), and especially high expression of NOS is consistently seen in the antennal lobes, central complex, and various compartments of mushroom bodies (Müller, 1997). To determine the amount of NOS activation in living tissues, NO-sensitive fluoroprobes (Trimmer et al., 2004) and L-citrulline immunocytochemistry (Cayre et al., 2005) have been successfully applied to insect preparations. These methods are especially suited for the identification of involved central nervous regions and to confirm the functional contribution of NO signaling to natural behaviors.

Nitric oxide

The neural release of this gaseous signaling molecule is not confined to synaptic zones with a specific molecular release machinery. NO is not stored in vesicles but instead is produced on demand, readily diffusing across cellular membranes in potentially all directions and exerting its biological function by directly interacting with various cytosolic targets. NO is not metabolized by specific enzymes but is degraded by spontaneous oxidation or can be inactivated by the addition of scavengers such as 2-phenyl-4,4,5,5-tetramethylimidazoline-3-oxide-1-oxyl (Murata et al., 2004). The effective range and period of the diffusible messenger NO has been estimated to be around 100–300 μm over several seconds (Gonzales-Zulueta, 1997; Kasai and Petersen, 1994; O'Shea et al., 1998; Wood and Garthwaite, 1994), providing it with the potential to affect and coordinate the activity of large populations of neurons. Recent experiments in M. sexta suggested that the actions of NO can be spatially restricted and may therefore not solely depend on its half-life during rapid and uniform diffusion nor on the expression of its target proteins within the range of its unrestricted diffusion (Trimmer et al., 2004).

Soluble guanylate cyclase (sGC)

Soluble guanylate cyclase (sGC) catalyses the synthesis of cyclic GMP (cGMP) in response to NO stimulation and seems to be the principal and most sensitive cellular target of NO in the nervous system (Bredt and Snyder, 1992; Garthwaite, 1991), although other targets have also been identified in both vertebrate and invertebrate nervous systems (Reinking et al., 2005; Stamler et al., 1997; Trimmer et al., 2004). sGC is a heterodimer made up of α- and β-subunits which exist in several isoforms (Mayer, 1994). Genes coding for sGC subunits have been cloned from various insect species including D. melanogaster (Liu et al., 1995; Shah and Hyde, 1995), M. sexta (Nighorn et al., 1998; Simpson et al., 1999), S. gregaria (Ogunshola et al., 1995), and Anopheles gambiae (Caccone et al., 1999). Neurons expressing sGC have been identified in a number of insect species by NO-stimulated accumulation of cGMP, which can be detected by immunocytochemistry using an antibody that was and is generously provided to numerous labs by Jan de Vente from Maastricht University in the Netherlands (de Vente et al., 1987). In both vertebrates (Ko et al., 1994) and insects (Ott et al., 2004), NO-induced production of cGMP can be potentiated by the exogenous allosteric modulator of sGC, YC-1, leading to enhanced cGMP-related immunoreactivity.

cGMP

The intracellular messenger molecule cGMP may directly modulate or gate ion channels, stimulate protein kinase G (PKG) leading to phosphorylation of target proteins, or activate cGMP-dependent phosphodiesterases (reviewed by Müller, 1997). NO-induced cGMP has also been demonstrated to increase cyclic AMP (cAMP) levels and to potentiate cAMP-mediated cellular responses in insects. cGMP directly activates cAMP-dependent protein kinase A (PKA) in *A. mellifera* and probably also in *D. melanogaster* (Müller and Hildebrandt, 2002; Müller and Spatz, 1989), and cGMP-mediated modulation of phosphodiesterase activity has also been demonstrated to impact cAMP levels (Müller, 1997; Wicher *et al.*, 2004). The nature of cellular responses to NO- and/or other signal-stimulated cGMP production is still unknown in most insect preparations studied and the mechanisms are expected to be diverse since they depend on the expression and physiological state of downstream reaction partners, which are subject to modulation by maturational processes and previous activation on a medium to long time scale.

Contribution of NO signaling to insect behavior

In the following section, we present a collection of examples where the role of NO in controlling particular behavioral functions has been revealed in considerable detail. NO contributes to sensory processing, sets behavioral states that promote or inhibit the selection of particular behaviors, and has effects on the performance of behaviors.

NO function in sensory neuropils

Olfaction

The molecular components of NO/cGMP signaling have been detected in chemosensory neuropils of various animal phyla, including primates (Alonso *et al.*, 1998), rodents (Hopkins *et al.*, 1996), molluscs (Gelperin, 1994), and crustaceans (Schachtner *et al.*, 2005). NO may thus be an evolutionarily ancient signal involved in the perception and processing of olfactory information. Consistent with this idea, NOS and NO-responsive neurons have been detected in antennal lobes of various insect species, and physiological experiments have demonstrated a functional contribution of NO to the processing of olfactory information (reviewed by Bicker, 1998; Müller, 1996). As in many other animals, insect olfactory glomeruli are spherically shaped, surrounded by glial cells, and may therefore be ideally suited for signaling via the diffusion of gaseous messengers such as NO (Breer and Sheperd, 1993). The presence of NOS in insect olfactory neuropils is not uniform among species and the precise function of NO signaling may also vary accordingly. With the exception of *S. gregaria* (Bicker *et al.*, 1996; Elphick *et al.*, 1995; Kurylas *et al.*, 2005; Müller and Bicker, 1994), the antennal olfactory receptor neurons of most insects studied appear to contain NOS: *M. sexta* (Collmann *et al.*, 2004; Nighorn *et al.*, 1998), *D. melanogaster* (Müller & Buchner, 1993), crickets, and *A. mellifera* (Müller, 1997). This suggests that NO may act as an anterograde signal. Antennal lobe-intrinsic interneurons contain NOS in locusts and honeybees (Müller & Bicker, 1994; Müller & Hildebrandt, 1995) and may therefore contribute to the accumulation of NO in glomeruli. In contrast, antennal interneurons of *M. sexta* generally do not express NOS (Collmann *et al.*, 2004) and, as yet another peculiarity, *M. sexta*'s olfactory receptors contain sGC and elevate cGMP levels upon NO stimulation (Stengl *et al.*, 2001). Whether and how homologues of vertebrate cGMP-gated channels that have been detected in the antennae of *D. melanogaster* contribute to olfactory signal processing (Baumann *et al.*, 1994) and whether insects may also express types of these channels that are subject to direct NO-mediated modulation (Broillet, 2000; Broillet and Firestein, 1996) remains to be demonstrated. Glomerular interneurons seem to be NO-responsive in all insects, as has been determined by both immunocytochemical detection of sGC and NO-stimulated accumulation of cGMP. In locusts and *M. sexta* (Bicker *et al.*, 1996, Collmann *et al.*, 2004; Seidel and Bicker, 1997) most of these interneurons seem to use GABA as their principal transmitter, but whether and how NO-mediated stimulation of the cGMP pathway affects GABAergic functions in olfactory signal processing is not known. In most insects, portions of projection neurons that relay olfactory information from antennal glomeruli to central brain neuropils such as the mushroom bodies have been demonstrated to accumulate cGMP in response to NO stimulation. Projection neurons that accumulated cGMP in response to a component of the sex pheromone blend were identified in *Bombyx mori* by Seki *et al.* (2005), who suggested that NO/cGMP signaling may regulate the threshold for pheromone-stimulated searching behavior. Similarly, Nighorn *et al.* (1998) detected variable patterns of cGMP-accumulating projection neurons between different individuals of *M. sexta* and suggested that the sensitivity to a particular odorant partly depends on the expression of sGC in projection neurons, which may be altered in response to changing physiological conditions.

Contact chemoreception

Consistent with descriptions of taste receptor cells of rats (Okada *et al.*, 1987), mice (Tonosaki and Funakoshi, 1988), and frogs (Krizhanovsky *et al.*, 2000) NO signaling seems to participate in transduction in particular types of fly (Murata *et al.*, 2004; Wasserman and Itagaki, 2003) and lepidopteran (Stengl *et al.*, 2001) taste receptors. Studies on the blowfly *Phormia regina* revealed that only the sugar receptor cells contained in the hair-shaped taste organs on the labellum contain both NOS and sGC, and that their products NO and cGMP mediate a major portion of the excitatory response to sucrose stimulation (Murata *et al.*, 2004).

Vision

Whether NO signaling contributes to the processing of visual information according to a general scheme in all insects is still an open question since NADPHd staining patterns and total NOS activity in optic lobes vary considerably between different species (Müller, 1996). However, all species investigated, including *Drosophila*, honeybees, crickets, and acridid grasshoppers, contain distinctly labeled NADPHd-positive neurons within their visual lobes. In the locust *S. gregaria*, approximately 40% of monopolar cells, the first-order interneurons in the lamina that receive direct inhibitory synaptic input from the photoreceptors, exhibit NADPHd staining and appear to release NO upon depolarization (Bicker and Schmachtenberg, 1997; Elphick *et al.*, 1996; Ott and Elphick, 2002). NO is believed to act as a retrograde transmitter at the photoreceptor-to-monopolar cell synapse that stimulates the accumulation of cGMP in various cellular compartments including the retinular segments of the photoreceptor cells (Schmachtenberg and Bicker, 1999), which have been shown to contain the α-subunit of sGC (Elphick and Jones, 1998; Jones and Elphick, 1999). Consistent with these cytochemical data, electrophysiological recordings revealed that NO-mediated accumulation of cGMP increases the sensitivity of locust photoreceptor cells (Schmachtenberg and Bicker, 1999). The NO/cGMP system in the locust eye may therefore mediate the adaptational response to dimming of ambient light initiated by reduced release of histamine from photoreceptor cells. Reduction of histamine release leads to relatively more depolarized membrane potentials of monopolar cells, then to increased NO liberation and subsequent stimulation of sGC to finally produce cGMP in photoreceptor cells (reviewed in Bicker, 2001). Biochemical studies by Jones and

Elphick (1999) further demonstrated that NO stimulates the production of cGMP in dark-adapted eyes but not in light-adapted eyes, suggesting that light-induced Ca^{2+} fluxes may inhibit sGC activity in photoreceptor cells. Although fly photoreceptors also accumulate cGMP in response to NO stimulation (Gibbs and Truman, 1998; Hanyu and Franceschini, 1993), a contribution of the signaling pathway to light adaptation is mediated by pigment granule translocation (Hanyu and Franceschini, 1993) rather than by sensitization of the transduction process. Since other neuropils of locust and other insect optic lobes also exhibit distinct NADPHd staining patterns, NO/cGMP signaling likely contributes to further stages of visual signal processing. In addition to visual information from the compound eyes, ocellar pathways may be targets of a similar NO-mediated modulation, since a subpopulation of ocellar interneurons exhibits strong NADPHd staining in locusts (Kurylas et al., 2005), while ocellar photoreceptors and other interneurons within the ocellar tracts contain sGC (Elphick and Jones, 1998).

Mechanical senses

A massive presence of NADPHd-positive fibers within ventral neuropils containing the terminals of mechanosensitive afferents in brains and ventral nerve cords of locusts, crickets, and cockroaches suggested a role for NO/cGMP signaling in the initial stages of mechanosensory information processing (Bullerjahn and Pflüger, 2003; Kurylas et al., 2005; Ott and Burrows, 1998, 1999). Prominent neuropils of this kind are the medial ventral association centers (mVAC) of thoracic ganglia receiving the spatially separated afferents of wing chordotonal organs and auditory receptors (Bullerjahn and Pflüger, 2003; Schürmann et al., 1997) and deutocerebral projection areas of antennal mechanosensory fibers (Kurylas et al., 2005). NADPHd staining in mechanosensory neuropils was generally attributed to interneurons and motoneurons; only locust antennal mechanosensory fibers were also NADPHd positive (Kurylas et al., 2005). Since acetylcholine, the transmitter of insect mechanosensory neurons, has been demonstrated to stimulate the release of NO in neurons of *M. sexta* and *S. gregaria* (Bicker and Kreissl, 1994; Müller and Bicker, 1994; Trimmer and Qazi, 1996; Zayas et al., 2002), NO may be produced in mechanosensory neuropils upon stimulation of mechanosensitive afferents. In contrast, mechanosensory afferents generally seem to contain sGC (Elphick and Jones, 1998; Ott et al., 2000) and may change their functional properties through accumulation of cGMP following retrograde NO signaling. In addition, both interneurons and motoneurons contained in local circuits accumulate cGMP upon NO stimulation (Ott *et al.*, 2000; Zayas *et al.*, 2002) and may provide a substrate for the modulation of sensory-motor circuits underlying reflexes and rhythmic behaviors.

Central nervous mechanisms

Habituation

The contribution of NO/cGMP signaling to mechanisms underlying habituation has been most extensively studied in the honeybee proboscis extension reflex. Repeated stimulation (intervals of 2 s) of one antenna with sucrose solution led to a graded decline of proboscis extension response probability (Braun and Bicker, 1992). Experimental interference with NO/cGMP signaling in the antennal lobe of the stimulated antenna was demonstrated to impact habituation. While NO scavengers and inhibitors of NOS and sGC slow the habituation, NO donors, and cGMP analogues, in contrast, elicit or accelerate its appearance (Müller and Hildebrandt, 1995, 2002). The main circuit mediating the proboscis extension reflex is located in the antennal lobes, suggesting that the NO mediating the habituation of the response may be released by chemosensory afferents and local interneurons, which are known to contain NOS. The NO-induced decrease in response probability during habituation is mediated by a slow and graded increase of PKA activity in the stimulated antennal lobe (Müller and Hildebrandt, 2002). Interneurons that contain both sGC and PKA are therefore the most likely targets for NO that mediate the decline of responsiveness during habituation. In this preparation, NO/cGMP signaling transduces the temporal sequence of repeated stimulations into a cumulative activation of the PKA second messenger pathway that is only transiently activated by single chemosensory stimuli. Whether the NO/cGMP system affects other reflexes and behavioral responses of insects in a similar way remains to be shown.

Learning and memory formation

Studies on the proboscis extension reflex of the honeybee revealed that the NO/cGMP system contributes to both habituation and the formation of associative long-term memory through pairing of an odor stimulus with a sucrose reward, but the underlying mechanisms and responsible networks are clearly different (Müller and Hildebrandt, 2002). Associative learning requires NO release that is limited to specific glomeruli and is even blocked by global release into entire antennal lobes, which in contrast is sufficient to induce habituation of proboscis extension. Multiple-trial conditioning leads to the formation of a long-term memory that lasts for several days, while a single-trial induces a medium-term memory that only lasts for some hours (Hammer and Menzel, 1995). Inhibition of NOS activity during multiple-trial conditioning impairs the usual formation of long-term memory and reduces it to the level of single-trial-induced memory (Müller, 1996, 1997). Formation of long-term memory requires cGMP-mediated activation of PKA (Müller, 2000). Similarly, NO/cGMP signaling connected to PKA activation also seems to mediate the formation of memory of defeat in fights between male crickets (Aonuma et al., 2004). After repeated encounters, the subdominant cricket avoids aggressive behaviors for several hours. Avoidance behavior is abolished when either NOS or sGC activity is inhibited during the initial encounters during which winner and loser were established. The sites within the honeybee and cricket brains where NO mediates long-term memory formation are still unknown, since various neuropils involved in chemosensory processing, including antennal lobes, lateral protocerebral neuropils, and mushroom bodies, contain NOS-expressing neurons that could initiate this process (Müller, 1994). Studies in *Drosophila* revealed that mushroom bodies are essential for the formation of olfactory memory, while simple forms of visual, tactile, and motor learning are not compromised by their inactivation (reviewed in Waddell and Quinn, 2001). Whether NO/cGMP signaling plays a role in the formation of olfactory associative memory in *Drosophila* is not known. Subpopulations of Kenyon cells, the major type of intrinsic interneurons of insect mushroom bodies, have been demonstrated to contain NOS in cockroaches (Ott and Elphick, 2002) and crickets (Cayre et al., 2005; Schürmann, 2000), but not in honeybees (Müller, 1997). In addition, projections from neurons extrinsic to the mushroom bodies (Kurylas et al., 2005; Müller, 1997) and glial cells within the Kenyon cell body regions of the mushroom bodies (Bicker and Hähnlein, 1995) developed NADPHd staining and might release NO upon stimulation. Subsets of Kenyon cells and surrounding glia have been demonstrated to contain sGC in locusts and may potentially serve as targets for NO within the mushroom bodies (Bicker et al., 1996). Studies on adult crickets revealed that neural activity related to olfactory input stimulated NO production and neurogenesis in the mushroom bodies (Cayre et al., 2005). This, together with the earlier observation that suppression of adult neurogenesis impairs the formation of olfactory memories (Scotto-Lomassese et al., 2003), suggested a major contribution of NO-stimulated neurogenesis in mushroom bodies to mechanisms underlying memory formation in crickets. Future functional studies, in which neural activity is selectively compromised in the various brain regions participating in the processing of olfactory information, will identify the sites where NO/cGMP signaling mediates the formation of different types of memories.

Selection of behavior

In addition to adaptive changes of insect behavior mediated by learning, the NO/cGMP signaling system has been demonstrated to influence the selection of situation-specific responses. *Drosophila* larvae rapidly respond to hypoxic conditions by switching from feeding to exploratory behavior (Wingrove and O'Farrell, 1999). Induction of exploratory behavior is mediated by activation of NOS, sGC, and PKG in yet-unidentified peripheral and/or central nervous structures. In addition to this acute response to hypoxic conditions, a general difference in exploratory or foraging behavior has been observed in *Drosophila* larvae carrying different alleles of a gene that encodes one of the two PKG isoforms in *Drosophila* (Osborne *et al.*, 1997), suggesting that the basal level of NO/cGMP signaling may determine an individual's general tendency to perform exploratory behavior.

Similarly, the NO/cGMP signaling system is expressed in the pheromone-processing pathways of the silkmoth *B. mori* and seems to regulate the threshold of female pheromone-induced male searching behavior (Seki *et al.*, 2005; L. Gatellier, personal communication). Projection neurons that are responsive to bombykol, the female pheromone component that elicits male searching behavior (Kanzaki *et al.*, 2003), have been demonstrated to accumulate cGMP upon NO stimulation (Seki *et al.*, 2005). They relay bombykol-related excitation to a lateral protocerebral neuropil regarded as specific for this pheromone component. It is suggested that accumulation of cGMP in these projection neurons may alter their responsiveness to bombykol-related synaptic inputs and thus account for the variations in pheromone-stimulated searching behavior.

A third example suggesting that basal NOS activity may modulate general thresholds to perform particular behaviors is the control of grasshopper acoustic communication mediated by neural circuits in the central body. The central body of locusts and grasshoppers has been demonstrated to contain both NO-generating and NO-responsive neurons in a number of studies (*e.g.,* Kurylas *et al.*, 2005; Ott *et al.*, 2004; Wenzel *et al.*, 2005). Pharmacological studies on restrained but intact grasshoppers (*Chorthippus biguttulus* and other species) revealed that activation of NO/cGMP signaling pathways in the central body suppresses sound production in a reversible and dose-dependent manner (Wenzel *et al.*, 2005). An interaction with the cAMP/PKA signaling cascade, which mediates NO-induced habituation and formation of long-term associative olfactory memory in honeybees (Müller, 1996; Müller and Hildebrandt, 2002) and the formation of social memory in crickets (Aonuma *et al.*, 2004) can be excluded in this preparation since activation of the cAMP/PKA pathway in the central body promotes grasshopper sound production (Heinrich *et al.*, 2001). Recent cytochemical studies in our laboratory have revealed that a subpopulation of NADPHd-positive neurons, recently described by Kurylas and co-workers (2005) in *S. gregaria*, displays intense citrulline immunoreactivity (unpublished results). Since accumulation of citrulline is indicative of considerable NOS activity during the period preceding fixation of the nervous tissue, these neurons might release NO into the central body neuropil to inhibit sound production in unfavorable situations. Whether grasshoppers that have been exposed to situations that favor the performance of mating-related sound production show reduced levels of citrulline accumulation in these neurons is a subject of current investigations. Injections of NOS inhibitors into the hemolymph increased the amount of spontaneous and male song-stimulated sound production in female grasshoppers (unpublished results). In contrast to male grasshoppers, sound production in females depends on the reproductive state, and spontaneous sound production is strictly correlated with high receptivity after prolonged periods without copulation and with periodic oviposition (Loher and Huber, 1964). Since experimental reduction of NOS activity induces sound production under those physiological conditions that normally suppress singing in females, NO/cGMP signaling may couple the reproductive state with neuronal functions that regulate the intrinsic threshold for this mating-related behavior.

The effector level

Muscular systems

Motor patterns related to ecdysis behavior are initiated by a preceding elevation of cGMP in the nervous system in species of various insect orders (Ewer and Truman, 1997; Gammie and Truman, 1997; Shibanaka *et al.*, 1994). In *B. mori* silkworms, the NO/cGMP system constitutes one of the final steps of a signaling cascade by which eclosion hormone triggers ecdysis behavior (Shibanaka *et al.*, 1994). Experimentally applied but also endogenously generated NO can increase spiking activity in motoneurons and interneurons of insect thoracic and abdominal nervous systems (Qazi and Trimmer, 1999), and the release of NO in these regions may be coupled to cholinergic signaling (Zayas et al., 2002). Motoneurons in various species have been demonstrated to contain NOS, and thus might release NO upon appropriate excitation (Schürmann et al., 1997; Zayas et al., 2000), or have been shown to express sGC and respond to NO stimulation with accumulation of cGMP (Ott et al., 2000; Rast, 2001; Wildemann and Bicker, 1999b; Zayas et al., 2002). A direct link between elevation of cGMP levels and neuronal activity has been demonstrated in motoneurons controlling the large longitudinal internal muscles in M. sexta (Trimmer et al., 2004; Zayas and Trimmer, 2001) and in dorsal unpaired median (DUM) neurosecretory neurons of cockroaches (Wicher et al., 2004). However, NO responsiveness of motoneurons seems to be cell-specific, since other types do not change their spike rates, and since some motoneurons are actually inhibited during NO application (Trimmer et al., 1998). Similarly, DUM and ventral unpaired median (VUM) neurons connected to various muscle systems seem to be variable regarding their participation in NO/cGMP signaling. While some of these cells express NOS, as has been found in M. sexta (Zayas et al., 2000) and crickets (Schürmann et al., 1997), studies on locusts indicate that this may not be a general feature of unpaired median neurons, since DUM neurons projecting to the heart muscle but not DUM neurons that innervate abdominal skeletal muscles contain NOS (Bullerjahn and Pflüger, 2003). Studies on Drosophila larvae suggested that NO/cGMP signaling could modulate synaptic release at insect neuromuscular junctions (Wildemann and Bicker, 1999a). NO-stimulated accumulation of cGMP in motoneuron synaptic terminals induced vesicle release. Since the underlying mechanism was found to be independent from Ca^{2+} influx to the presynapse and no potential sources of NO in the vicinity of neuromuscular junctions could be detected by NADPHd staining (Wildemann and Bicker, 1999b), a functional contribution of NO/cGMP signaling to neuromuscular synaptic transmission remains to be confirmed by further studies.

Neurosecretory systems

Neurosecretory cells regulate behavioral states through the release of (neuro-) hormones into the circulation and influence the functions of neural circuits through the release of neuromodulators in particular neuropil regions. Different types of neurosecretory cells in various regions of the nervous system have been demonstrated to either contain NOS or respond to NOS with increased production of cGMP, suggesting that NO/cGMP signaling contributes to the regulation of neurosecretory systems in insects. NADPHd-positive and/or anti-NOS immunoreactive neurosecretory cells have been detected in the pars intercerebralis of locusts (Kurylas *et al.*, 2005; Müller and Bicker, 1994), in median nerves of locust thoracic and abdominal ganglia known to contain neurohemal release sites (Bullerjahn and Pflüger, 2003), and in several neurosecretory cells of *M. sexta*, including various VUM neurons of thoracic and abdominal ganglia and identified neurohemal neurons of abdominal ganglia (Zayas *et al.*, 2000). In addition, octopamine-releasing DUM-type neurosecretory cells of cockroaches have been demonstrated to increase spontaneous firing frequencies following NO-stimulated (and arachidonic acid-stimulated) accumulation of cGMP, eventually leading to modulation of Ca^{2+}-channel function (Wicher *et al.*, 2004). Cytochemical studies revealed that insect hormonal glands like the corpora allata and corpora cardiaca also contain NOS and NO-sensitive sGC in various species – *M. sexta* larvae (Zayas *et al.*, 2000), *Drosophila* larvae (Wildemann and Bicker, 1999b), and grasshoppers (unpublished personal observation) – suggesting that the NO/cGMP contributes to the transduction of neural activity into endocrine signals.

Summary

NO/cGMP signaling impacts the selection and performance of insect behaviors on all functional levels, including sensory activation, central nervous processing, and activation of neurosecretory and motor functions. As a general scheme among all insect orders studied in this respect, NO/cGMP signaling contributes to the initial stages of information processing of olfactory, gustatory, visual, mechanosensory (including auditory), and proprioceptive sensory information. NO-stimulated plasticity of particular neural circuits, including prolonged activation of other intracellular signaling cascades, contributes to mechanisms of habituation and the formation of mid- and long-term memory. In addition, NO-stimulated accumulation of cGMP can regulate thresholds for particular behaviors through modulation of neuronal excitability and synaptic transmission properties in central nervous regions concerned with the selection of behaviors. On the effector level, the presence of both NOS and sGC have been demonstrated in various motoneurons and secretory cells, and direct effects of NO/cGMP-mediated mechanisms on the spiking activity of these cell types have been reported. Thus, NO/cGMP signaling appears to contribute to adjustments of behavioral outputs to an individual's sensory environment by promoting or inhibiting activity in the respective control circuits. The functional status of NOS-, sGC-, and cGMP-initiated mechanisms may therefore reflect different physiological states of an insect. Such a connection has been found between rearing conditions and levels of NOS mRNA expression in crickets (Cayre *et al.*, 2005) and remains to be confirmed for other species and behaviors.

References

Alonso, J. R., Porteros, A., Crespo, C., Arevalo, R., Brinon, J. G., Weruaga, E. and Aijon, J. (1998). Chemical anatomy of the macaque monkey olfactory bulb: NADPH-diaphorase/nitric oxide synthase activity. *J. Comp. Neurol.* 402,419–434.

Aonuma, H., Iwasaki, M. and Niwa, K. (2004). Role of NO signaling in switching mechanisms in the nervous system of insect. *SICE Ann. Conf. 2004* 3,2477–2482.

Ball, E. E. and Truman, J. W. (1998). Developing grasshopper neurons show variable levels of guanylyl cyclase activity on arrival of their targets. *J. Comp. Neurol.* 394,1–13.

Baumann, A., Frings, S., Godde, M., Seifert, R. and Kaupp, U. B. (1994). Primary structure and functional expression of a *Drosophila* cyclic nucleotide-gated channel present in eyes and antennae. *EMBO J.* 13,5040–5050.

Bicker, G. (1998). NO news from insect brains. *Trends Neurosci.* 21,349–355.

Bicker, G. (2001). Sources and targets of nitric oxide signalling in insect nervous systems. *Cell Tissue Res.* 303,137–146.

Bicker, G. (2005). STOP and GO with NO: Nitric oxide as a regulator of cell motility in simple brains. *Bioessays* 27,495–505.

Bicker, G. and Hähnlein, I. (1995). NADPH-diaphorase expression in neurones and glial cells of the locust brain. *Neuroreport* 6,325–328.

Bicker, G. and Kreissl, S. (1994). Calcium imaging reveals nicotinic acetylcholine receptors on cultured mushroom body neurons. *J. Neurophysiol.* 71,808–810.

Bicker, G. and Schmachtenberg, O. (1997). Cytochemical evidence for nitric oxide/cyclic GMP signal transmission in the visual system of the locust. *Eur. J. Neurosci.* 9,189–193.

Bicker, G., Schmachtenberg, O. and DeVente, J. (1996). The nitric oxide/cyclic GMP messenger system in olfactory pathways of the locust brain. *Eur. J. Neurosci.* 8,2635–2643.

Braun, G. and Bicker, G. (1992). Habituation of an appetitive reflex in the honeybee. *J. Neurophysiol.* 67,588–598.

Bredt, D. S., Glatt, C. E., Hwang, P. M., Fotuhi, M., Dawson, T. M. and Snyder, S. H. (1991). Nitric oxide synthase protein and mRNA are discretely localized in neuronal populations of the mammalian CNS together with NADPH diaphorase. *Neuron* 7,615–624.

Bredt, D. S. and Snyder, S. H. (1992). Nitric oxide, a novel neuronal messenger. *Neuron* 8,3–11.

122

Breer, H. and Sheperd, G. M. (1993). Implications of the NO/cGMP system for olfaction. *Trends Neurosci.* 16,5–9.

Brenman, J. E., Chao, D. S., Gee, S. H., McGee, A. W., Craven, S. E., Froehner, S. C. and Bredt, D. S. (1996). Interaction of nitric oxide synthase with the postsynaptic density protein PDS-95 and alpha1-syntrophin mediated by PDZ domains. *Cell* 84,757–767.

Broillet, M. C. (2000). A single intracellular cysteine residue is responsible for the activation of the olfactory cyclic nucleotide-gated channel by NO. *J. Biol. Chem.* 275,15135–15141.

Broillet, M. C. and Firestein, S. (1996). Direct activation of the olfactory cyclic nucleotide gated channel through modification of sulfhydryl groups by NO compounds. *Neuron* 16,377–385.

Bullerjahn, A. and Pflüger, H.-J. (2003). The distribution of putative nitric oxide releasing neurones in the locust abdominal nervous system: A comparison of NADPHd histochemistry and NOS-immunocytochemistry. *Zoology* 106,3–17.

Caccone, A., Garcia, B. A., Mathiopoulos, K. D., Min, G.-S., Moriyama, E. N. and Powell, J. R. (1999). Characterization of the soluble guanylyl cyclase ß-subunit gene in the mosquito *Anopheles gambiae. Insect Mol. Biol.* 8,23–30.

Cayre, M., Malaterre, J., Scotto-Lomassese, S., Hostein, G. R., Martinelli, G. P., Forni, C., Nicolas, S., Aouane, A., Strambi, C. and Strambi, A. (2005). A role for nitric oxide in sensory-induced neurogenesis in an adult insect brain. *Eur. J. Neurosci.* 21,2893–2902.

Collmann, C., Carlsson, M. A., Hansson, B. S. and Nighorn, A. (2004). Odorant-evoked nitric oxide signals in the antennal lobe of *Manduca sexta. J. Neurosci.* 24,6070–6077.

Denninger, J. W. and Marletta, M. A. (1999). Guanylate cyclase and the NI/cGMP signaling pathway. *Biochim. Biophys. Acta* 1411,334–350.

de Vente, J., Steinbusch, H. W. M. and Schipper, J. (1987). A new approach to immunocytochemistry of 3′,5′-cyclic guanosine monophosphate: Preparation, specificity, and initial application of a new antiserum against formaldehyde-fixed 3′,5′-cyclic guanosine monophosphate. *Neuroscience* 22,361–373.

Elphick, M. R., Green, I. C. and O'Shea, M. (1993). Nitric oxide synthesis and action in an invertebrate brain. *Brain Res.* 619,344–346.

Elphick, M. R., Green, I. C. and O'Shea, M. (1994). Nitric oxide signalling in the insect nervous system. In *Insect Neurochemistry and Neurophysiology* (eds A. B. Borkovec and M. J. Loeb), pp. 129–132, CRC Press, Boca Raton, Florida.

Elphick, M. R. and Jones, I. W. (1998). Localization of soluble guanylyl cyclase α-subunit in identified insect neurons. *Brain Res.* 800,174–179.

Elphick, M. R., Rayne, R. C., Riveros-Moreno, V., Moncada, S. and O'Shea, M. (1995). Nitric oxide synthesis and action in locust olfactory interneurones. *J. Exp. Biol.* 198,821–829.

Elphick, M. R., Williams, L. and O'Shea, M. (1996). New features of the locust optic lobe: Evidence of a role for nitric oxide in insect vision. *J. Exp. Biol.* 199,2395–2407.

Enikopolov, G., Banerji, J. and Kuzin, B. (1999). Nitric oxide and Drosophila development. *Cell Death Differ.* 6,956–963.

Ewer, J. and Truman, J. W. (1997). Invariant association of ecdysis with increases in cyclic 3′,5′-guanosine monophosphate immunoreactivity in a small network of peptidergic neurons in the hornworm, *Manduca sexta. J. Comp. Physiol. A* 181,319–330.

Gammie, S. C. and Truman, J. W. (1997). An endogenous elevation of cGMP increases the excitability of identified insect neurosecretory cells. *J. Comp. Physiol. A* 180,329–337.

Garthwaite, J. (1991). Glutamate, nitric oxide and cell-cell signalling in the nervous system. *Trends Neurosci.* 14,60–67.

Garthwaite, J. and Boulton, C. L. (1995). Nitric oxide signalling in the central nervous system. *Annu. Rev. Physiol.* 57,683–706.

Gelperin, A. (1994). Nitric oxide mediates network oscillations of olfactory interneurons in a terrestrial mollusc. *Nature* 369,61–63.

Gibbs, S. M. and Truman, W. T. (1998). Nitric oxide and cyclic GMP regulate retinal patterning in the optic lobe of *Drosophila*. *Neuron* 20,83–93.

Gibson, N. J., Rössler, W., Nighorn, A. J., Oland, L. A., Hildebrand, J. G. and Tolbert, L. P. (2001). Neuron-glia communication via nitric oxide is essential in establishing antennal-lobe structure in *Manduca sexta*. *Dev. Biol.* 240,326–339.

Gonzales-Zulueta, M. (1997). Role of nitric oxide in cell death. *B.I.F. Futura* 12,1–10.

Haase, A. and Bicker, G. (2003). Nitric oxide and cyclic nucleotides are regulators of neuronal migration in an insect embryo. *Development* 130,3977–3987.

Hammer, M. and Menzel, R. (1995). Learning and memory in the honeybee. *J. Neurosci.* 15,1617–1630.

Hanyu, Y. and Franceschini, N. (1993). Pigment granule migration and phototransduction are triggered by separate pathways in fly photoreceptor cells. *Neuroreport* 4,213–215.

Heinrich, R., Wenzel, B. and Elsner, N. (2001). A role for muscarinic excitation: Control of specific singing behavior by activation of the adenylate cyclase pathway in the brain of grasshoppers. *Proc. Natl. Acad. Sci. USA* 98,9919–9923.

Hope, B. T., Michael, G. J., Knigge, K. M. and Vincent, S. R. (1991). Neuronal NADPH diaphorase is a nitric oxide synthase. *Proc. Natl. Acad. Sci. USA* 88,2811–2814.

Hopkins, D. A., Steinbusch, H. W. M., Karkerink-Van Ittensum, M. and de Vente, J. (1996). Nitric oxide synthase, cGMP, and NO-mediated cGMP production in the olfactory bulb of the rat. *J. Comp. Neurol.* 375,641–658.

Jones, I. W. and Elphick, M. R. (1999). Dark-dependent soluble guanylyl cyclase activity in locust photoreceptor cells. *Proc. R. Soc. London* 266,413–419.

Kanzaki, R., Soo, K., Seki, Y. and Wada, S. (2003). Projections to higher olfactory centers from subdivisions of the antennal lobe macroglomerular complex of the male silkmoth. *Chem. Senses* 28,113–130.

Kasai, H. and Petersen, O. H. (1994). Spatial dynamics of second messengers: IP_3 and cAMP as long-range and associative messengers. *Trends Neurosci.* 17,95–101.

Ko, F. N., Wu, C. C., Kuo, S. C., Lee, F. Y. and Teng, C. M. (1994). YC-1, a novel activator of platelet guanylate cyclase. *Blood* 84,4226–4233.

Krizhanovsky, V., Agamy, O. and Naim, M. (2000). Sucrose-stimulated subsecond transient increase in cGMP level in rat intact circumvallate taste bud cells. *Am. J. Physiol. Cell Physiol.* 279,C120–C125.

Kurylas, A. E., Ott, S. R., Schachtner, J., Elphick, M. R., Williams, L. and Homberg, U. (2005). Localization of nitric oxide synthase in the central complex and surrounding midbrain neuropils of the locust *Schistocerca gregaria*. *J. Comp. Neurol.* 484,206–223.

124

Liu, W., Yoon, J., Burg, M., Chen, L. and Pak, W. L. (1995). Molecular characterization of two Drosophila guanylate cyclases expressed in the nervous system. *J. Biol. Chem.* 270,12418–12427.

Loher, W. and Huber, F. (1964). Experimentelle untersuchungen am sexualverhalten des weibchens der heuschrecke *Gomphocerus rufus* L. (Acridinae). *J. Insect Physiol.* 10,13–36.

Mayer, B. (1994). Regulation of nitric oxide synthase and soluble guanylyl cyclase. *Cell Biochem. Funct.* 12,167–177.

Müller, U. (1994). Ca^{2+}/calmodulin-dependent nitric oxide synthase in *Apis mellifera* and *Drosophila melanogaster*. *Eur. J. Neurosci.* 6,1362–1370.

Müller, U. (1996). Inhibition of nitric oxide synthase impairs a distinct form of long-term memory in the honeybee *Apis mellifera*. *Neuron* 16,541–549.

Müller, U. (1997). The nitric oxide system in insects. *Progr. Neurobiol.* 51,363–381.

Müller, U. (2000). Prolonged activation of cAMP-dependent protein kinase during conditioning induces long-term memory in honeybees. *Neuron* 27,159–168.

Müller, U. and Bicker, G. (1994). Calcium activated release of nitric oxide and cellular distribution of nitric oxide synthesizing neurons in the nervous system of the locust. *J. Neurosci.* 14,7521–7528.

Müller, U. and Buchner, E. (1993). Histochemical localization of NADPH-diaphorase in adult Drosophila brain: Is nitric oxide a neuronal messenger also in insects? *Naturwissenschaften* 80,524–526.

Müller, U. and Hildebrandt, H. (1995). The nitric oxide/cGMP system in the antennal lobe of *Apis mellifera* is implicated in integrative processing of chemosensory stimuli. *Eur. J. Neurosci.* 7,2240–2248.

Müller, U. and Hildebrandt, H. (2002). Nitric oxide/cGMP-mediated protein kinase A activation in the antennal lobes plays an important role in appetitive reflex habituation in the honeybee. *J. Neurosci.* 22,8739–8747.

Müller, U. and Spatz, H.-C. (1989). Ca^{2+}-dependent proteolytic modification of the cAMP-dependent protein kinase in Drosophila wilt-type and *dunce* memory mutants. *J. Neurogenet.* 6,95–114.

Murata, Y., Mashiko, M., Ozaki, M., Amakawa, T. and Nakamura, T. (2004). Intrinsic nitric oxide regulates the taste response of the sugar receptor cell in the blowfly, *Phormia regina*. *Chem. Senses* 29,75–81.

Nighorn, A., Gibson, N. J., Rivers, D. M., Hildebrand, J. G. and Morton, D. B. (1998). The nitric oxide-cGMP pathway may mediate communication between sensory afferents and projection neurons in the antennal lobe of *Manduca sexta*. *J. Neurosci.* 18,7244–7255.

Ogunshola, O., Picot, I., Piper, M., Korneev, S. and O'Shea, M. R. (1995). Molecular analysis of the NO-cGMP signalling pathway in insect and molluscan CNS. *Soc. Neurosci. Abstr.* 21,631.

Okada, Y., Miyamoto, T. and Sato, T. (1987). Depolarization induced by injection of cyclic nucleotides into frog taste cells. *Biochim. Biophys. Acta* 904,187–190.

Osborne, K. A., Robichon, A., Burgess, E., Butland, S., Shaw, R. A., Coulthard, A., Pereira, H. S., Greenspan, R. J. and Sokolowski, M. B. (1997). Natural behavior polymorphism due to a cGMP-dependent protein kinase of Drosophila. *Science* 277,834–836.

O'Shea, M., Colbert, R., Williams, L. and Dunn, S. (1998). Nitric oxide compartments in the mushroom bodies of the locust brain. *Neuroreport* 9,333–336.

Ott, S. R. and Burrows, M. (1998). Nitric oxide synthase in the thoracic ganglia of the locust: Distribution in the neuropiles and morphology of neurons. *J. Comp. Neurol.* 395,217–230.

Ott, S. R. and Burrows, M. (1999). NADPH diaphorase histochemistry in the thoracic ganglia of locusts, crickets, and cockroaches: Species differences and the impact of fixation. *J. Comp. Neurol.* 410,387–397.

Ott, S. R., Burrows, M. and Elphick, M. R. (2001). The neuroanatomy of nitric oxide-cyclic GMP signaling in the locust: Functional implications for sensory systems. *Am. Zool.* 41,321–331.

Ott, S. R., Delago, A. and Elphick, M. R. (2004). An evolutionarily conserved mechanism for sensitization of soluble guanylyl cyclase reveals extensive nitric oxide-mediated upregulation of cyclic GMP in insect brain. *Eur. J. Neurosci.* 20,1231–1244.

Ott, S. R. and Elphick, M. R. (2002). Nitric oxide synthase histochemistry in insect nervous systems: Methanol/formalin fixation reveals the neuroarchitecture of formaldehyde-sensitive NADPH diaphorase in the cockroach *Periplaneta americana. J. Comp. Neurol.* 448,165–185.

Ott, S. R. and Elphick, M. R. (2003). New techniques for whole-mount NADPH-diaphorase histochemistry demonstrated in insect ganglia. *J. Histochem. Cytochem.* 51,523–532.

Ott, S. R., Jones, I. W., Burrows, M. and Elphick, M. R. (2000). Sensory afferents and motoneurons as targets for nitric oxide in the locust. *J. Comp. Neurol.* 422,521–532.

Qazi, S. and Trimmer, B. A. (1999). The role of nitric oxide in motoneuron spike activity and muscarinic-evoked changes in cGMP in the CNS of larval *Manduca sexta. J. Comp. Physiol. A* 185,539–550.

Rast, G. F. (2001). Nitric oxide induces centrally generated motor patterns in the locust suboesophageal ganglion. *J. Exp. Biol.* 204,3789–3801.

Regulski, M. and Tully, T. (1995). Molecular and biochemical characterization of *d*NOS: A Drosophila Ca^{2+}/calmodulin dependent nitric oxide synthase. *Proc. Natl. Acad. Sci. USA* 92,9072–9076.

Reinking, J., Lam, M. M., Pardee, K., Sampson, H. M., Liu, S., Yang, P., Williams, S., White, W., Lajoie, G., Edwards, A. and Krause, H. M. (2005). The Drosophila nuclear receptor e75 contains heme and is gas responsive. *Cell* 122,195–207.

Schachtner, J., Schmidt, M. and Homberg, U. (2005). Organization and evolutionary trends of primary olfactory brain centers in Tetraconata (Crustacea + Hexapoda). *Arthropod Struct. Dev.* 34,257–299.

Schmachtenberg, O. and Bicker, G. (1999). Nitric oxide and cyclic GMP modulate photoreceptor cell responses in the visual system of the locust. *J. Exp. Biol.* 202,13–20.

Schürmann, F. W. (2000). Acetylcholine, GABA, glutamate and NO as putative transmitters indicated by immunocytochemistry in the olfactory mushroom body system of the insect brain. *Acta Biol. Hung.* 51,355–362.

Schürmann, F. W., Helle, J., Knierim-Grenzebach, M., Pauls, M. and Spörhase-Eichmann, U. (1997). Identified neuronal cells and sensory neuropiles in the ventral nerve cord of an insect stained by NADPH-diaphorase histochemistry. *Zoology* 100,98–109.

126

Scotto-Lomassese, S., Strambi, C., Strambi, A., Aouane, A., Augier, R., Rougon, G. and Cayre, M. (2003). Suppression of adult neurogenesis impairs olfactory learning and memory in an adult insect. *J. Neurosci.* 23,9289–9296.

Seidel, C. and Bicker, G. (1997). Colocalization of NADPH-diaphorase and GABA-immunoreactivity in the olfactory and visual system of the locust. *Brain Res.* 769,273–280.

Seki, Y., Aonuma, H. and Kanzaki, R. (2005). Pheromone processing center in the protocerebrum of *Bombyx mori* revealed by nitric oxide-induced anti-cGMP immunocytochemistry. *J. Comp. Neurol.* 481,340–351.

Shah, S. and Hyde, D. R. (1995). Two *Drosophila* genes that encode the α- and β-subunits of the brain soluble guanylyl cyclases. *J. Biol. Chem.* 270,15368–15376.

Shibanaka, Y., Hayashi, H., Umemura, I., Fujisawa, Y., Okamoto, M., Takai, M. and Fujita, N. (1994). Eclosion hormone-mediated signal transduction in the silkworm abdominal ganglia: Involvement of a cascade from inositol (1,4,5) trisphosphate to cyclic GMP. *Biochem. Biophys. Res. Commun.* 198(2), 613–618.

Simpson, P. J., Nighorn, A. and Morton, D. B. (1999). Identification of a novel guanylyl cyclase that is related to receptor guanylyl cyclases, but lacks extracellular and trans-membrane domains. *J. Biol. Chem.* 274,4440–4446.

Stamler, J. S., Toone, E. J., Lipton, S. A. and Sucher, N. J. (1997). (S)NO signals: Translocation, regulation, and a consensus motif. *Neuron* 18,691–696.

Stengl, M., Zintl, R., de Vente, J. and Nighorn, A. (2001). Localization of cGMP immunoreactivity and of soluble guanylyl cyclase in antennal sensilla of the hawk-moth, *Manduca sexta. Cell. Tissue Res.* 304,409–421.

Tonosaki, K. and Funakoshi, M. (1988). Cyclic nucleotides may mediate taste trans-duction. *Nature* 331,354–3566.

Trimmer, B. A., Aprille, J. and Modica-Napolitano, J. (2004). Nitric oxide signaling: Insect brains and photocytes. *Biochem. Soc. Symp.* 71,65–83.

Trimmer, B. A. and Qazi, S. (1996). Modulation of second messengers in the nervous system of larval *Manduca sexta* by muscarinic receptors. *J. Neurochem.* 66,1903–1913.

Trimmer, B. A., Qazi, S. and Vermehren, A. (1998). The effects of nitric oxide on the control of abdominal muscles in larval *Manduca sexta. Soc. Neurosci. Abstr.* 24,105.

Waddell, S. and Quinn, W. G. (2001). What can we teach Drosophila? What can they teach us? *Trends Genet.* 17, 719–726.

Wasserman, S. L. and Itagaki, H. (2003). The olfactory responses of the antenna and maxillary palp of the fleshfly, *Neobellieria bullata* (Diptera: Sarcophagidae), and their sensitivity to blockage of nitric oxide synthase. *J. Insect Physiol.* 49,271–280.

Wenzel, B., Kunst, M., Günther, C., Ganter, G. K., Lakes-Harlan, R., Elsner, N. and Heinrich, R. (2005). Nitric oxide/cyclic guanosine monophosphate signaling in the central complex of the grasshopper brain inhibits singing behavior. *J. Comp. Neurol.* 488,129–139.

Wicher, D., Messutat, S., Lavialle, C. and Lapied, B. (2004). A new regulation of non-capacitative calcium entry in insect pacemaker neurosecretory neurons. *J. Biol. Chem.* 279,50410–50419.

Wildemann, B. and Bicker, G. (1999a). Nitric oxide and cyclic GMP induce vesicle release at *Drosophila* neuromuscular junction. *J. Neurobiol.* 39,337–346.

Wildemann, B. and Bicker, G. (1999b). Developmental expression of nitric oxide/cyclic GMP synthesizing cells in the nervous system of *Drosophila melanogaster*. *J. Neurobiol.* 38,1–15.

Wingrove, J. A. and O'Farrell, P. H. (1999). Nitric oxide contributes to behavioral, cellular, and developmental responses to low oxygen in *Drosophila*. *Cell* 98,105–114.

Wood, J. and Garthwaite, J. (1994). Models of the diffusional spread of nitric oxide: Implications for neural nitric oxide signaling and its pharmacological properties. *Neuropharmacology* 33,1235–1244.

Wright, J. W., Schwinhof, K. M., Snyder, M. A. and Copenhaver, F. (1998). A delayed role for nitric oxide-sensitive guanylate cyclases in a migratory population of embryonic neurons. *Dev. Biol.* 204,15–33.

Zayas, R. M., Qazi, S., Morton, D. B. and Trimmer, B. A. (2000). Neurons involved in nitric oxide-mediated cGMP signaling in the tobacco hornworm, *Manduca sexta*. *J. Comp. Neurol.* 419,422–438.

Zayas, R. M., Qazi, S., Morton, D. B. and Trimmer, B. A. (2002). Nicotinic-acetylcholine receptors are functionally coupled to the nitric oxide/cGMP-pathway in insect neurons. *J. Neurochem.* 83,421–431.

Zayas, R. M. and Trimmer, B. A. (2001). Cyclic GMP increases the excitability of identified nitric oxide-sensitive insect neurons. *Soc. Neurosci. Abstr.* 27,13–14.

Impact of nitrative/nitrosative stress in mitochondria: Unraveling targets for malaria chemotherapy

Shirley Luckhart[1], Kazunobu Kato[2] and Cecilia Giulivi[2,*]

[1]*Department of Medical Microbiology and Immunology, School of Medicine, University of California at Davis, Davis, CA 95616, USA*
[2]*Department of Molecular Biosciences, School of Veterinary Medicine, University of California at Davis, Davis, CA 95616, USA*

Abstract. Nitric oxide (NO) has been identified as one of the most important signaling molecules in living organisms. In addition to the direct effects of NO, recent studies have revealed that protein modifications by NO, such as nitration or S-nitrosation, are also of biological significance as regulatory mechanisms that can affect cellular functions. Both protein modifications occur as a consequence of oxidative/nitrative stress. Some of these protein modifications have been found in mitochondria, and in certain cases the corresponding activities and/or pathways affected have been identified. The protein modifications can result as collateral damage during normal or pathological oxidative/nitrative stress, or in more localized, controlled situations that modulate signal transduction pathways. As an example of these complex phenomena, we discuss the interplay among the malaria parasite, mosquito and host, focusing on the role of reactive oxygen and nitrogen species (RONS) and protein modifications in controlling parasite development.

Keywords: mitochondria; nitric oxide; nitric-oxide synthase; mitochondrial nitric-oxide synthase; bioenergetics; protein nitration; tyrosine nitration; nitrosation; nitrosylation; nitrative stress; reactive oxygen species; reactive nitrogen species; malaria; nitrosothiols; mosquito; Anopheles; Plasmodium; thioredoxin; peroxiredoxin; glutathione reductase.

Introduction: Exposure of cells to nitrogen oxides

Since the discovery that cells can generate and release nitric oxide (NO), a molecule that may function as an intracellular and extracellular signal (Moncada *et al.*, 1991), interest in the toxicity of nitrogen oxides has increased rapidly. Recent studies demonstrated that NO exhibits long-term effects through the nitration of proteins, effects that are above and beyond those transient ones mediated through the interaction with cytochrome oxidase or guanylyl cyclase (Beckman and Koppenol, 1996; Brown and Cooper, 1994; Gole *et al.*, 2000; Halliwell *et al.*, 1999; Ischiropoulos *et al.*, 1992; Tatoyan and Giulivi, 1998). C-nitration of proteins has been proposed to constitute a novel post-translational modification or a chemical

Corresponding author: Tel.: 1-530-754-8603. Fax: 1-530-752-4698.
E-mail: cgiulivi@ucdavis.edu (C. Giulivi).

ADVANCES IN EXPERIMENTAL BIOLOGY
VOLUME 01   ISSN 1872-2423
DOI: 10.1016/S1872-2423(07)01006-X

modification that is a consequence of the 'normal' oxidative background (Ischiropoulos et al., 1992; van der Vliet et al., 1995). Tyrosine (Tyr) modification by nitrogen oxides has received much attention because one of the major products formed, 3-nitroTyr, is a stable end product and characteristic of nitrogen oxides.

Nitration reactions introduce a nitro group ($-NO_2$) into an organic molecule, and this chapter focuses mainly on evaluating the roles of nitrated proteins. Nitrosation reactions introduce a nitroso group ($-NO$). Under physiological conditions, nitrosation reactions lead to the formation of three types of X–NO bond: S-nitroso, such as in S-nitrosoCys; N-nitroso, such as in N-nitrosoTrp; and transition metal–nitroso complexes, such as iron-nitrosyl.

Nitration has been associated with various pathological conditions, such as atherosclerosis, Alzheimer's disease, and amyotrophic lateral sclerosis, so it would be desirable to halt this process, especially considering that at least 140 different mammalian proteins have been identified whose activities are dependent upon Tyr residues that may be inactivated by chemically-induced nitration (Nielsen, 1995). Nitration of Tyr residues can also affect Tyr phosphorylation or dephosphorylation, thereby interfering with important signal transduction pathways in cells (Gow et al., 1996). The functional consequences of Tyr nitration for structural proteins can be dramatic because, upon nitration, there is a change in the pK_a of the phenolic group of Tyr of about 2 units, thus altering hydrogen bonding and solubility of that residue (Beckman and Koppenol, 1996).

Protein nitration and S-nitrosation in mitochondria

Nitrated proteins in mitochondria

Given the diverse biochemical characteristics of nitric oxide synthase (NOS), it can be postulated that the production of NO by mitochondria is highly regulated because of the critical role that this molecule has on cellular respiration. Under physiological conditions, the production of NO by mitochondria has important implications for the maintenance of cellular metabolism. For example, NO produced by rat liver mitochondria modulated ATP production sustained by NADPH availability (Giulivi, 1998; Giulivi et al., 1998). This modulation of ATP production is achieved through the reversible inhibition of cytochrome c oxidase by NO. This transient inhibition suits the continuously changing energy and oxygen requirements of the tissue. However, if a sustained inhibition of cytochrome oxidase occurs, or if NO is synthesized at high concentrations, then other deleterious effects may be observed, including inhibition of ATP synthesis (Giulivi, 1998), release of cytochrome c (Ghafourifar et al., 1999), and nitration of critical biomolecules (Elfering et al., 2004). Studies from the Giulivi laboratory have shown that a subset of mitochondrial Tyr-containing proteins acquires a nitro group under conditions of sustained, endogenously-produced NO (Elfering et al., 2004).

Tyr nitration is an irreversible, post-translational modification (acquisition of a nitro moiety) associated with over 50 diseases, including shock, cancer, and stroke (Beckman and Koppenol, 1996; Halliwell et al., 1999; Ischiropoulos, 1998). Multiple nitrating agents have been identified. NO, through its reaction with superoxide anion, forms peroxynitrite, a well-known nitrating agent; alternatively, through the enzymatic oxidation of nitrite, it can form nitrogen dioxide (NO_2), which can also induce nitration (Beckman and Koppenol, 1996).

Few nitrated proteins have been identified, thus restricting our understanding of the role of nitration in pathophysiological situations. Using a proteomic approach, Aulak et al. (2001) identified nitrated proteins that were presumably formed during an inflammatory challenge in vivo; one-third of these were of mitochondrial origin. Although this study served as a platform to examine biological aspects of protein nitration, it did not provide information on the nitration of proteins formed under 'normal' physiological conditions that occur as a consequence of oxidative/nitrative metabolism, or on any correlations of nitration with altered protein activity. Research from the Giulivi laboratory has shown that the endogenous mitochondrial production of NO catalysed by a mitochondrial NOS produces reversible changes without permanently altering mitochondrial function (Giulivi et al., 1998; Haynes et al., 2004; Sarkela et al., 2001). However, under conditions of a sustained production, nitration of proteins ensues (Elfering et al., 2004). Nitrated proteins were frequently located at the membranes and contained a common primary sequence, suggesting that Tyr nitration might be a selective process (Elfering et al., 2004).

Chemical requirements for protein nitration in mitochondria

Previous results from the Giulivi laboratory indicated that mitochondrial protein nitration is not associated with protein abundance, molecular weight, pI, or mean half-life (Elfering et al., 2004) as could had been expected if nitration were a random process, resulting from reaction of a highly-reactive molecule like peroxynitrite with Tyr residues. In contrast, specific nitration (expressed as nmol nitroTyr per gram of mitochondrial protein) was 4.4 times greater in particulate fractions (outer and inner membranes, contact sites) than in soluble fractions (matrix and inter-membrane space). This observation suggested that nitration chemistry in a biological system may require a hydrophobic component, fulfilled by a hydrophobic precursor (which may or may not be NO), and a hydro-phobic target. In support of this hypothesis, a consensus sequence for nitration was identified by aligning several protein segments at which nitration was found. The sequence has a high contribution of hydro-phobic residues and is defined as H-X-[DE]-H-X(2,3)-H(2)-X(2,4)-Y, in which H represents a hydrophobic residue (such as L, M, V, I, P, A, F or W) (Elfering *et al.*, 2004).

Protein nitration has the potential to alter signal transduction path-ways if the target Tyr is involved in phosphorylation/dephosphorylation processes (Gow *et al.*, 1996; Herrero *et al.*, 2001; Kong *et al.*, 1996; Reinehr *et al.*, 2004). Up to now, no mitochondrial protein has been identified for which Tyr phosphorylation is altered by nitration. As such, it will be of biological significance to clarify the effect of nitration on phosphorylation and, more importantly, on protein activity in general (Gow *et al.*, 1996; Herrero *et al.*, 2001; Kong *et al.*, 1996; MacMillan-Crow *et al.*, 2000; Mallozzi *et al.*, 2001; Mondoro *et al.*, 1997; Reinehr *et al.*, 2004).

Based on the chemical character of an introduced nitro group to an organic molecule, it is unlikely that nitration is a reversible process, un-less reduction of the nitro group to an amino group is involved. In this case, aromatic nitro compounds, such as nitroTyr, can be reduced by a wide variety of chemical agents. Among them, a variety of metals (Serbina *et al.*, 2003) usually employed in acidic media, metallic com-pounds such as stannous chloride, anionic sulfur compounds such as sodium sulfides and sodium hydrosulfite, as well as catalytic hydrogen-ation can function to reduce aromatic nitro compounds. However, it is unlikely that a reductant or reducing mechanism from this list would be found among biologically occurring compounds and reactions.

Proteolytic degradation of nitrated proteins is considered the main mechanism for clearance of Tyr-nitrated proteins, and this clearance can result in increased levels of free nitroTyr (Souza *et al.*, 2000). Further-more, nitrated proteins exhibit shortened lifetimes (hours vs. days when compared to native, unmodified proteins), indicating that cellular proteolysis accounts for the high turnover of these proteins (Bota and Davies, 2002; Elfering *et al.*, 2004). Bota and Davies (2002) showed that Lon protease, an ATP-stimulated mitochondrial matrix protein, specifi-cally degraded the oxidized form of mitochondrial aconitase. The preferential proteolysis of oxidized aconitase by Lon protease functions as one of the selective removal mechanisms for oxidatively modified proteins in mitochondria (Bota and Davies, 2002). These clearance mechanisms for nitrated and/or oxidized mitochondrial proteins are probably of significance to defend the organelle against nitrative/oxidative stresses.

Some authors have suggested that nitrated proteins can be denitrated, an assertion that is based on a decrease in cross-reactivity of treated samples relative to controls with antibodies to nitroTyr (Aulak et al., 2004; Irie et al., 2003; Kamisaki et al., 1998; Kuo et al., 1999; Turko and Murad, 2002). For example, Aulak et al. (2004) showed that some mitochondrial proteins were denitrated in response to a decrease in oxygen concentration. The loss of immunoreactivity to nitroTyr occurred rapidly, within 5–20 min after establishing a hypoxic condition. This process was suggested to be promoted by a cellular denitrase. However, given that the nitration process is oxygen-dependent, proteolysis cannot be excluded, no denitrase has been isolated, and no products have been identified from this putative reaction. The potential existence of a 'denitrase' that catalyses denitration or removal of a nitro group from a phenoxyl ring or reduction of the nitro to an amino group warrants further studies.

Impact of nitration on protein activity

Elfering et al. (2004) found that mitochondrial protein nitration increased with age and under conditions of sustained NO production catalysed by mitochondrial NOS. Although these observations were suggestive of cause–consequence, they did not directly correlate protein nitration with effects on protein activity or impact on mitochondrial function. To address this point, the Giulivi lab investigated the effect of β-subunit nitration on ATP synthase activity. From these studies, a reciprocal correlation was identified between β-subunit Tyr nitration and Complex V activity (Kato, Elfering, and Giulivi, in preparation). These results suggested that nitration of the β-subunit may have an impact on the activity of the F_1F_o-ATP synthase and suggested an important pathway by which nitration of a protein modulates the activity of an enzyme by a stable post-translational modification. Changes in enzyme activity will have an impact on cellular bioenergetics and maintenance of ATP levels. These changes are expected to have a significant impact during aging, since mitochondrial protein nitration was observed to increase with age.

The nitration of Tyr is not always linked to protein dysfunction. In this regard, if the Tyr in question is not required for activity or is located in a highly variable region, it is likely that Tyr nitration will not significantly change protein activity (Miller et al., 2006). Thus, clear correlations are required to confirm the effects of Tyr nitration on protein activity. Further, even if changes in activity are detected, increased turnover of the modified protein may overcome a temporary loss of function. In support of this point, Elfering et al. (2004) determined that, in at least three cases, nitrated proteins exhibited a significant increased protein turnover when compared to their native counterparts.

Substantial evidence has indicated that the functions of mitochondrial proteins and other proteins are modulated by nitration. For example, manganese superoxide dismutase (MnSOD) is inactivated by Tyr nitration induced by peroxynitrite (MacMillan-Crow et al., 1998). The nitration of Tyr34 has been implicated in this inactivation (Yamakura et al., 1998). The loss of MnSOD activity has an important biochemical impact on mitochondrial and cellular functions. For example, MnSOD inactivation would be expected to impair reactive oxygen species-mediated signal transduction pathways and to increase oxidative/nitrative stress on biomolecules as a result of increased production of peroxynitrite (Johnson and Giulivi, 2005).

S-nitrosoproteins in Mitochondria

Another important modification for mitochondrial proteins resulting from NO synthesis, although not directly, is the S-nitrosation of thiols. The initial step for S-nitrosation is oxidation of NO, resulting in the formation of a nitrosonium ion (NO^+). Oxidized metals, NAD^+ (under anaerobic conditions), and oxygen have been postulated as electron acceptors during the formation of S-nitrosothiol (RSNO) (Carver et al., 2005; Gow et al., 1997). Following oxidation of NO to NO^+, the reaction between NO^+ and a thiolate will produce an RSNO such as S-nitrosoglutathione (GSNO) or S-nitrosocysteine in proteins (Lancaster and Gaston, 2004; Robinson and Lancaster, 2005). Although the mechanism(s) of S-nitrosation remain largely unknown, the main regulatory mechanisms for S-nitrosation appear to be nonenzymatic (Miersch and Mutus, 2005) and generally reflective of the redox state of the tissue (Bryan et al., 2004).

Recent studies revealed that S-nitrosation could be one of the pathways by which NO initiates changes in mitochondrial physiology (Handy and Loscalzo, 2006; Hess et al., 2005). Several mitochondrial proteins have been identified as molecular targets for S-nitrosation, resulting in alterations of, *e.g.,* apoptosis (Mannick *et al.*, 1999, 2001) and oxygen consumption. In this regard, mitochondrial exposure to a relatively high concentration of NO leads to the inactivation of Complex I via S-nitrosation (Clementi *et al.*, 1998). Studies from the Giulivi laboratory found a significant level of GSNO in mitochondria, and this concentration is modulated by the endogenous production of NO catalysed by a mitochondrial NOS (Steffen *et al.*, 2001). In light of the biochemical importance of GSNO in various physiological and pathological settings (Achuth *et al.*, 2005; Khan *et al.*, 2005; Nozik-Grayck *et al.*, 2006; Zaman *et al.*, 2006), it is reasonable to postulate that mitochondrial GSNO may have specific biological effects. Roles for GSNO may include S-transnitrosation (transfer of NO^+ to other protein thiolates; Perissinotti *et al.*, 2005), delivery of NO to other subcellular compartments, or buffering NO concentrations (Steffen *et al.*, 2001). Further studies are needed to ascertain the function(s) of GSNO in mitochondria.

Although it is not well understood how S-nitrosation is biochemically controlled, close proximity or co-localization of NOS and S-nitrosation targets may be important for specificity (Hess *et al.*, 2005; Miersch and Mutus, 2005). Considering the favorable microenvironment of mitochondria for S-nitrosation, *e.g.,* lower pH, hydrophobic milieu, and availability of reactive oxygen and nitrogen species (RONS) (Carver *et al.*, 2005; Hess *et al.*, 2005), it is not surprising that S-nitrosated proteins are predominantly localized in mitochondria and perimitochondrial compartments in endothelial cells (Yang and Loscalzo, 2005).

Cytoprotective mechanisms against oxidative/nitrative stress

While NO and its metabolites impair/modulate cellular functions through protein modifications such as nitration or S-nitrosation, cells are endowed with protective mechanisms against oxidative/nitrative stress. Non-enzymatic defenses include low molecular weight reducing agents (*e.g.*, glutathione (GSH) and thioredoxin) and a variety of antioxidants (*e.g.*, vitamins E, C, and A and uric acid). Enzymatic defenses include catalase (Rhee *et al.*, 2005), superoxide dismutase (SOD) (Johnson and Giulivi, 2005), peroxidases such as peroxiredoxin or GSH peroxidase, and GSH reductase (Huang and Philbert, 1995; Rhee *et al.*, 2005; Singh *et al.*, 1996).

Among the low molecular weight reducing agents, thioredoxin is a redox-regulating protein that counteracts oxidative stress by direct antioxidant effects or by indirect effects on thioredoxin-interacting proteins (Burke-Gaffney *et al.*, 2005; Nakamura, 2005; Yamawaki and Berk,

2005). Whereas the prototypical thioredoxin 1 is found in cytosol, a second isoform, thioredoxin 2, has been identified in mitochondria (Ejima *et al.*, 1999). Mitochondrial thioredoxin 2 plays a protective role against oxidative stress (Watson *et al.*, 2004) as well as a regulatory role in mitochondrial apoptosis (Tanaka *et al.*, 2002).

Peroxiredoxins, which are thioredoxin-dependent peroxidases, provide enzymatic defenses against oxidative/nitrative stress (Bryk *et al.*, 2000; Dubuisson *et al.*, 2004; Trujillo *et al.*, 2004; Wong *et al.*, 2002). Six isoforms of peroxiredoxin have been identified to date and two of these, peroxiredoxin 3 (Chang *et al.*, 2004; Matsushima *et al.*, 2006) and peroxiredoxin 5 (Declercq *et al.*, 2001; Dubuisson *et al.*, 2004), are localized to the mitochondria. Peroxiredoxin has been identified as an important peroxidase for catabolizing hydrogen peroxide (Chang *et al.*, 2004).

The GSH peroxidase/reductase system has been implicated in the clearance of hydrogen peroxide and lipid hydroperoxides (Huang and Philbert, 1995; Singh *et al.*, 1996). In addition, since increased production of NO in mitochondria can promote the production of GSNO (Steffen *et al.*, 2001), GSH can be considered an important biomolecule in the defense against nitrative stress.

Importantly, these mechanisms are not mutually exclusive, but rather complementary to each other. For example, thioredoxin function overlaps with function of the GSH-dependent systems (Watson *et al.*, 2004). Most of these regulatory mechanisms, however, have not been elucidated for mitochondria, indicating that it will be important to clarify how they cooperate to regulate redox states in mitochondria, the predominant sources of RONS in the cell.

Cytoprotective proteins with critical thiols or essential Tyr residues can be affected by post-translational modifications such as nitrosation and nitration, respectively. GSH reductase is inactivated by peroxynitrite-mediated nitration of Tyr106 and Tyr114 (Savvides *et al.*, 2002). The function of cytosolic thioredoxin 1 is affected by nitration (Jiao *et al.*, 2006) or by S-nitrosation at Cys69 (Haendeler *et al.*, 2002). It has been found that GSNO specifically *S*-nitrosates Cys73 of thioredoxin 1 (Mitchell and Marletta, 2005). S-nitrated thioredoxin 1 inactivates caspase-3, a cytosolic/mitochondrial protease that is involved in apoptosis, by a selective transnitration reaction with Cys163 of caspase-3 (Mitchell and Marletta, 2005). Although most studies have been performed on proteins located in subcellular compartments other than mitochondria, high primary sequence homology among some cytosolic and mitochondrial isoforms suggests that these modifications could also be found in mitochondrial proteins.

These cytoprotective mechanisms are required from mammals to invertebrates to maximize NO-related effects on cellular function while minimizing adverse effects of RONS. In the later half of this chapter, along with these self-defense mechanisms, we will discuss the biological significance of these protein modifications under oxidative/nitrative stress in invertebrates, by exemplifying host defense mechanisms of mosquitoes against malaria infection.

Nitrative and nitrosative stress in a complex biological system: Malaria infection

It has long been recognized that reactive oxygen species and, more recently, reactive nitrogen species function to defend hosts against pathogens (DeGroote and Fang, 1999). While these observations are derived primarily from studies of mammalian hosts, similar observations have been reported for invertebrates. For example, *Anopheles stephensi*, a primary vector of the causative agents for human malaria (*e.g., Plasmodium falciparum, Plasmodium vivax*) in India and the Middle East, limits malaria parasite development with the inducible synthesis of NOS catalysed by *A. stephensi* nitric oxide synthase (AsNOS) (Luckhart and Rosenberg, 1999; Luckhart *et al.*, 1998). In the following section we will discuss various aspects of RONS in light of the complexity of malaria infection and in terms of the role of RONS in the interactions between the vector, the parasite, and the host.

RONS limit parasite development in the mosquito midgut

Malaria parasite development in the mosquito begins with the ingestion of blood containing sexual-stage male (microgametocyte) and female (macrogametocyte) parasites. Parasite fertilization (union of micro- and macrogametes) and development of mobile ookinetes occur in the first 24 h in the blood-filled midgut, a hostile environment of proteases, hemoglobin catabolites, and the immunoresponsive midgut epithelium, which synthesizes RONS (Peterson *et al.*, 2007) and various antimicrobial peptides. While these peptides appear to have modest impact on parasite development (reviewed in Levashina, 2004), other effectors, including RONS, can induce significant parasite losses. Surviving ookinetes invade the midgut epithelium from 24–48 h after a blood meal, to complete their development in the mosquito.

Effects of RONS on the malaria parasite and potential cellular targets

After ingestion of a parasite-infected blood meal, induction of AsNOS expression ensues in the midgut (Luckhart *et al.*, 2003). Provision of L-NAME (N^G-nitro-L-arginine methyl ester) to *A. stephensi* significantly decreased parasite apoptosis in the midgut lumen (Al-Olayan *et al.*, 2002), suggesting that NO synthesis drives the formation of pro-apoptotic RONS (*e.g.*, peroxynitrite or nitroxyl anion; Bai *et al.*, 2001; Szabo, 2003; Szabo and Ohshima, 1997) that inhibit *Plasmodium* development. Minimal toxicity of free NO predicts that inhibition of *Plasmodium* development occurs via the formation of toxic metabolites formed in the mosquito midgut after blood-feeding (Rockett *et al.*, 1991; Peterson *et al.*, 2007).

At lower concentrations, some NO metabolites are cytostatic to *P. falciparum* (Balmer *et al.*, 2000; Taylor-Robinson, 1997). Cytostatic effects could derive from the inhibition of specific cellular targets. For example, chemical donors of NO can dose-dependently inactivate *P. falciparum* falcipain, a member of a papain-like Cys protease family implicated in survival in mammalian erythrocytes and development in the mosquito host (Rosenthal, 2004). Inactivation of falcipains occurs through S-nitrosation of Cys25 and perhaps additional chemical modifications of Tyr (Ascenzi *et al.*, 2004; Bocedi *et al.*, 2004; Venturini *et al.*, 2000). Similar inhibition would be predicted for the Ser repeat antigen proteases (SERAs), another family of malaria parasite Cys proteases (Bzik *et al.*, 1988; Miller *et al.*, 2002). The parasite egress Cys protease-1 (ECP-1), a SERA, is inhibited by S-nitrosation at a Cys residue in the catalytic triad of invariant Cys-His-Asn residues (Aly and Matuschewski, 2005). S-nitrosation of the Cys-rich parasite surface proteins P25 and P21/28 (Hisaeda *et al.*, 2000) and the adhesion protein SOAP (secreted ookinete adhesive protein) (Dessens *et al.*, 2003), which have been implicated in parasite adhesion and invasion of the mosquito midgut epithelium, respectively, could block parasite escape from the midgut.

Other parasite proteins are potential targets for Tyr nitration. *P. falciparum* GSH reductase can be inactivated by nitration of Tyr106 and Tyr114 by peroxynitrite (Savvides *et al.*, 2002). Inactivation of parasite GSH reductase would slow reduction of oxidized GSH and, therefore, could reduce GSH-dependent protection against RONS. *Plasmodium* chitinase is required for parasite escape from the midgut, and chitinase inhibitors and monoclonal antibodies to parasite chitinase effectively block parasite development (Langer *et al.*, 2002; Shahabuddin *et al.*, 1993). One monoclonal antibody epitope is

Tyr-rich (LYDSYAYYGKKYDYVIIMGFTL; Langer *et al.*, 2002), suggesting that Tyr nitration could alter chitinase function. Interestingly, exhaustive searches of the *P. falciparum* genome indicate that, like most other unicellular eukaryotes (Shiu and Li, 2004), malaria parasites do not utilize Tyr kinases (Ward *et al.*, 2004), suggesting that at least one known cell-signaling target of NO inhibition is not present in malaria parasites.

RONS defenses in the mosquito and the parasite

The damaging effects of antiparasite RONS likely extend to host tissues, suggesting that self-protection may be critical to the success of mosquito immunity. Among the variety of cellular antioxidants, peroxiredoxins comprise a family of peroxidases that can protect cells from widely divergent organisms against a variety of nitrosative stress challenges (Bryk *et al.*, 2000; Dubuisson *et al.*, 2004; Trujillo *et al.*, 2004; Wong *et al.*, 2002).

Recent studies from the Luckhart laboratory have demonstrated that an *A. stephensi* 2-Cys peroxiredoxin protects mosquito cells against RONS stress (Peterson and Luckhart, 2006). In this role, the mosquito peroxiredoxin would be expected to undergo 'bystander' oxidation of the peroxidatic Cys to sulfenic acid, which would form a disulfide bond with a Cys residue of a second subunit. Under highly oxidizing conditions, the peroxidatic Cys can be further oxidized to sulfinic acid (Schroder *et al.*, 2000), a moiety that cannot be reduced by thioredoxin, resulting in the inactivation of the peroxiredoxin (Rabilloud *et al.*, 2002; Wood *et al.*, 2003). *A. stephensi* peroxiredoxin, like most eukaryotic 2-Cys peroxiredoxins, contains a GG(V/I/L)G motif and a YF motif, characteristic of those peroxiredoxins that are sensitive to inactivation by hyperoxidation (Wood *et al.*, 2003). In contrast, the *P. falciparum* peroxiredoxin C-terminus motif is modified from YF to YL, suggesting that the parasite peroxiredoxin might be resistant to inactivation by hyperoxidation.

The selective pressure to reverse peroxiredoxin inactivation has likely resulted in the evolution of unique physiological strategies for reactivation. For example, a sulfiredoxin can reverse the inactivation of yeast 2-Cys peroxiredoxin by hyperoxidation (Biteau *et al.*, 2003). Sulfiredoxin orthologs are found only in eukaryotes, consistent with the fact that prokaryotic peroxiredoxin orthologs are insensitive to oxidative inactivation (Biteau *et al.*, 2003). In addition to sulfiredoxins, a family of Cys sulfinyl reductases known as sestrins can increase the recovery of over-oxidized peroxiredoxins (Budanov *et al.*, 2004). Although *in silico* analyses did not identify a sulfiredoxin ortholog in the *Anopheles gambiae* genome, a sequence encoding a sestrin ortholog is apparent (Fig. 1). Interestingly, not only does *P. falciparum* 2-Cys peroxiredoxin lack the signature motif for hyperoxidation, neither sulfiredoxin- nor sestrin-encoding sequences can be identified in the *P. falciparum* genome. Thus, it

First region of similar sequence:

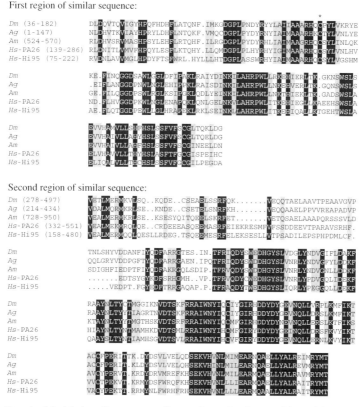

Fig. 1. Multiple alignment of sestrins. Amino acid residues conserved in the aligned sestrin sequences are shaded black for identity and gray for similarity. The conserved catalytic cysteine of this family of cysteine sulfinyl reductases (Budanov *et al.*, 2004) is denoted by an asterisk. Values in parentheses indicate amino acid locations of the aligned proteins. The aligned sestrin sequences are presented as two sections that flank highly divergent sequences. Dm, *Drosophila melanogaster* (Genbank accession no. Q9W1K5). Ag, *Anopheles gambiae* (Genbank accession no. XM_308599). Am, honeybee; *Apis mellifera* (Genbank accession no. XM_394521). Hs-PA26, *Homo sapiens* (Genbank accession no. NP_055269). Hs-Hi95, *Homo sapiens* (Genbank accession no. NP_113647).

is likely that malaria parasite and mosquito peroxiredoxins are regulated differently, with the parasite peroxiredoxin being more like prokaryotic than eukaryotic peroxiredoxins. Given that peroxiredoxins protect both the mosquito (Peterson and Luckhart, 2006) and the parasite (Komaki-Yasuda *et al.*, 2003) from nitrosative stress, these species differences may be exploitable for the development of antiparasitic peroxiredoxin inhibitors to control mosquito-stage malaria parasite infection.

Can we manipulate RONS to prevent human infection?

Parasite invasion and development in the mosquito host may be attenuated if the timing and/or level of AsNOS induction are modified to yield higher, sustained levels of RONS. The generation of RONS may facilitate transport, preserve bioactivity, target specific effectors, and mitigate adverse side reactions of NO synthesis (Stamler *et al.*, 1992). Potential targets of nitration and nitrosation have been identified in malaria parasites, but direct effects of these modifications have not been validated *in vivo*. Once validated, parasite targets that are expressed in the mosquito and that lack orthologs in the mosquito genome would be useful targets in strategies to engineer enhanced antiparasite resistance. Given that these targets may be limited, continued study of the role of RONS-modified proteins in mammalian mitochondrial physiology may suggest novel targets for antiparasite strategies.

The plasmodial mitochondrion has evolved for a parasitic lifestyle in the mammalian host and differs from typical eukaryotic mitochondria. Although the respiratory chain and oxidative phosphorylation are functional in *Plasmodium* (Uyemura *et al.*, 2000, 2004), the presence of malate-quinone oxidoreductase and the alternative NADH-Q oxidoreductase, and the occurrence of a single mitochondrion/cell provide unique targets for drug therapy. Although our understanding of mitochondrial physiology in mosquito-stage parasites has not advanced at the same pace, successful targeting of RONS to differentially expressed mitochondrial enzymes or even to common mitochondrial enzymes with different thresholds for oxidative/nitrative damage could be used in the development of novel antimalarials (Mi-Ichi *et al.*, 2005) and in the development of novel transmission-blocking strategies in the mosquito host. Further, a combined strategy based on inhibiting unique parasite physiologies and enhancing the activity of cytoprotective enzymes may tip the balance in favor of the mosquito, thereby limiting parasite survival and transmission while preserving host fitness.

142

Acknowledgements

This study was supported by National Institutes of Health ES012691 and ES005707 (to CG) and AI50663 and AI60664 (to SL).

References

Achuth, H. N., Moochhala, S. M., Mahendran, R. and Tan, W. T. L. (2005). Nitrosoglutathione triggers collagen deposition in cutaneous wound repair. *Wound Repair Regen* 13,383–389.

Al-Olayan, E. M., Williams, G. T. and Hurd, H. (2002). Apoptosis in the malaria protozoan, *Plasmodium berghei*: A possible mechanism for limiting intensity of infection in the mosquito. *Int. J. Parasitol.* 32,1133–1143.

Aly, A. S. and Matuschewski, K. (2005). A malarial cysteine protease is necessary for Plasmodium sporozoite egress from oocysts. *J. Exp. Med.* 202,225–230.

Ascenzi, P., Bocedi, A., Gentile, M., Visca, P. and Gradoni, L. (2004). Inactivation of parasite cysteine proteinases by the NO-donor 4-(phenylsulfonyl)-3-((2-(dimethylamino)ethyl)thio)-furoxan oxalate. *Biochim. Biophys. Acta* 1703,69–77.

Aulak, K. S., Koeck, T., Crabb, J. W. and Stuehr, D. J. (2004). Dynamics of protein nitration in cells and mitochondria. *Am. J. Physiol. Heart Circ. Physiol.* 286,H30–H38.

Aulak, K. S., Miyagi, M., Yan, L., West, K. A., Massillon, D., Crabb, J. W. and Stuehr, D. J. (2001). Proteomic method identifies proteins nitrated in vivo during inflammatory challenge. *Proc. Natl. Acad. Sci. USA* 98,12056–12061.

Bai, P., Bakondi, E., Szabo, E., Gergely, P., Szabo, C. and Virag, L. (2001). Partial protection by poly(ADP-ribose) polymerase inhibitors from nitroxyl-induced cytotoxity in thymocytes. *Free Radic. Biol. Med.* 31,1616–1623.

Balmer, P., Phillips, H. M., Maestre, A. E., McMonagle, F. A. and Phillips, R. S. (2000). The effect of nitric oxide on the growth of *Plasmodium falciparum, P. chabaudi* and *P. berghei* in vitro. *Parasite Immunol.* 22,97–106.

Beckman, J. S. and Koppenol, W. H. (1996). Nitric oxide, superoxide, and peroxynitrite: The good, the bad, and ugly. *Am. J. Physiol.* 271,C1424–C1437.

Biteau, B., Labarre, J. and Toledano, M. B. (2003). ATP-dependent reduction of cysteine-sulphinic acid by *S. cerevisiae* sulphiredoxin. *Nature* 425,980–984.

Bocedi, A., Gradoni, L., Menegatti, E. and Ascenzi, P. (2004). Kinetics of parasite cysteine proteinase inactivation by NO-donors. *Biochem. Biophys. Res. Commun.* 315,710–718.

Bota, D. A. and Davies, K. J. (2002). Lon protease preferentially degrades oxidized mitochondrial aconitase by an ATP-stimulated mechanism. *Nat. Cell Biol.* 4,674–680.

Brown, G. C. and Cooper, C. E. (1994). Nanomolar concentrations of nitric oxide reversibly inhibit synaptosomal respiration by competing with oxygen at cytochrome oxidase. *FEBS Lett.* 356,295–298.

Bryan, N. S., Rassaf, T., Maloney, R. E., Rodriguez, C. M., Saijo, F., Rodriguez, J. R. and Feelisch, M. (2004). Cellular targets and mechanisms of nitros(yl)ation: An insight into their nature and kinetics in vivo. *Proc. Natl Acad. Sci. USA* 101,4308–4313.

Bryk, R., Griffin, P. and Nathan, C. (2000). Peroxynitrite reductase activity of bacterial peroxiredoxins. *Nature* 407,211–215.

Budanov, A. V., Sablina, A. A., Feinstein, E., Koonin, E. V. and Chumakov, P. M. (2004). Regeneration of peroxiredoxins by p53-regulated sestrins, homologs of bacterial AhpD. *Science* 304,596–600.

Burke-Gaffney, A., Callister, M. E., Nakamura, H. (2005) Thioredoxin: Friend or foe in human disease? *Trends Pharmacol. Sci.* 26,398–404.

Bzik, D. J., Li, W. B., Horii, T. and Inselburg, J. (1988). Amino acid sequence of the serine-repeat antigen (SERA) of *Plasmodium falciparum* determined from cloned cDNA. *Mol. Biochem. Parasitol.* 30,279–288.

Carver, J., Doctor, A., Zaman, K. and Gaston, B. (2005). S-nitrosothiol formation. *Meth. Enzymol.* 396,95–105.

Chang, T. S., Cho, C. S., Park, S., Yu, S., Kang, S. W. and Rhee, S. G. (2004). Peroxiredoxin III, a mitochondrion-specific peroxidase, regulates apoptotic signaling by mitochondria. *J. Biol. Chem.* 279,41975–41984.

Clementi, E., Brown, G. C., Feelisch, M. and Moncada, S. (1998). Persistent inhibition of cell respiration by nitric oxide: crucial role of S-nitrosylation of mitochondrial complex I and protective action of glutathione. *Proc. Natl. Acad. Sci. USA* 95,7631–7636.

Declercq, J. P., Evrard, C., Clippe, A., Stricht, D. V., Bernard, A. and Knoops, B. (2001). Crystal structure of human peroxiredoxin 5, a novel type of mammalian peroxiredoxin at 1.5Å resolution. *J. Mol. Biol.* 311,751–759.

DeGroote, M. A. and Fang, F. C. (1999). Antimicrobial properties of nitric oxide. In *Nitric Oxide and Infection* (ed. F. C. Fang), Kluwer Academic/Plenum.

Dessens, J. T., Siden-Kiamos, I., Mendoza, J., Mahairaki, V., Khater, E., Vlachou, D., Xu, X. J., Kafatos, F. C., Louis, C., Dimopoulos, G. and Sinden, R. E. (2003). SOAP, a novel malaria ookinete protein involved in mosquito midgut invasion and oocyst development. *Mol. Microbiol.* 49,319–329.

Dubuisson, M., Vander Stricht, D., Clippe, A., Etienne, F., Nauser, T., Kissner, R., Koppenol, W. H., Rees, J. F. and Knoops, B. (2004). Human peroxiredoxin 5 is a peroxynitrite reductase. *FEBS Lett.* 571,161–165.

Ejima, K., Nanri, H., Toki, N., Kashimura, M. and Ikeda, M. (1999). Localization of thioredoxin reductase and thioredoxin in normal human placenta and their protective effect against oxidative stress. *Placenta* 20,95–101.

Elfering, S. L., Haynes, V. L., Traaseth, N. J., Ettl, A. and Giulivi, C. (2004). Aspects, mechanism, and biological relevance of mitochondrial protein nitration sustained by mitochondrial nitric oxide synthase. *Am. J. Physiol. Heart Circ. Physiol.* 286,H22–H29.

Ghafourifar, P., Schenk, U., Klein, S. D. and Richter, C. (1999). Mitochondrial nitric-oxide synthase stimulation causes cytochrome *c* release from isolated mitochondria. Evidence for intramitochondrial peroxynitrite formation. *J. Biol. Chem.* 274,31185–31188.

Giulivi, C. (1998). Functional implications of nitric oxide produced by mitochondria in mitochondrial metabolism. *Biochem. J.* 332,673–679.

Giulivi, C., Poderoso, J. J. and Boveris, A. (1998). Production of nitric oxide by mitochondria. *J. Biol. Chem.* 273,11038–11043.

144

Gole, M. D., Souza, J. M., Choi, I., Hertkorn, C., Malcolm, S., Foust, R. F., 3rd., Finkel, B., Lanken, P. N. and Ischiropoulos, H. (2000). Plasma proteins modified by tyrosine nitration in acute respiratory distress syndrome. *Am. J. Physiol. Lung Cell. Mol. Physiol.* 278,L961–L967.

Gow, A. J., Buerk, D. G. and Ischiropoulos, H. (1997). A novel reaction mechanism for the formation of *S*-nitrosothiol in vivo. *J. Biol. Chem.* 272,2841–2845.

Gow, A. J., Duran, D., Malcolm, S. and Ischiropoulos, H. (1996). Effects of peroxynitrite-induced protein modifications on tyrosine phosphorylation and degradation. *FEBS Lett.* 385,63–66.

Haendeler, J., Hoffmann, J., Tischler, V., Berk, B. C., Zeiher, A. M. and Dimmeler, S. (2002). Redox regulatory and anti-apoptotic functions of thioredoxin depend on S-nitrosylation at cysteine 69. *Nat. Cell Biol.* 4,743–749.

Halliwell, B., Zhao, K. and Whiteman, M. (1999). Nitric oxide and peroxynitrite. The ugly, the uglier and the not so good: A personal view of recent controversies. *Free Radic. Res.* 31,651–669.

Handy, D. E. and Loscalzo, J. (2006). Nitric oxide and posttranslational modification of the vascular proteome: S-nitrosation of reactive thiols. *Arterioscler. Thromb. Vasc. Biol.* 26,1207–1214.

Haynes, V., Elfering, S., Traaseth, N. and Giulivi, C. (2004). Mitochondrial nitric-oxide synthase: Enzyme expression, characterization, and regulation. *J. Bioenerg. Biomembr.* 36,341–346.

Herrero, M. B., de Lamirande, E. and Gagnon, C. (2001). Tyrosine nitration in human spermatozoa: A physiological function of peroxynitrite, the reaction product of nitric oxide and superoxide. *Mol. Hum. Reprod.* 7,913–921.

Hess, D. T., Matsumoto, A., Kim, S. O., Marshall, H. E. and Stamler, J. S. (2005). Protein S-nitrosylation: Purview and parameters. *Nat. Rev. Mol. Cell Biol.* 6,150–166.

Hisaeda, H., Stowers, A. W., Tsuboi, T., Collins, W. E., Sattabongkot, J. S., Suwanabun, N., Torii, M. and Kaslow, D. C. (2000). Antibodies to malaria vaccine candidates Pvs25 and Pvs28 completely block the ability of *Plasmodium vivax* to infect mosquitoes. *Infect. Immun.* 68,6618–6623.

Huang, J. and Philbert, M. A. (1995). Distribution of glutathione and glutathione-related enzyme systems in mitochondria and cytosol of cultured cerebellar astrocytes and granule cells. *Brain Res.* 680,16–22.

Irie, Y., Saeki, M., Kamisaki, Y., Martin, E. and Murad, F. (2003). Histone H1.2 is a substrate for denitrase, an activity that reduces nitrotyrosine immunoreactivity in proteins. *Proc. Natl. Acad. Sci. USA* 100,5634–5639.

Ischiropoulos, H. (1998). Biological tyrosine nitration: A pathophysiological function of nitric oxide and reactive oxygen species. *Arch. Biochem. Biophys.* 356,1–11.

Ischiropoulos, H., Zhu, L., Chen, J., Tsai, M., Martin, J. C., Smith, C. D. and Beckman, J. S. (1992). Peroxynitrite-mediated tyrosine nitration catalyzed by superoxide dismutase. *Arch. Biochem. Biophys.* 298,431–437.

Jiao, X., Tao, L., Zhang, H., Christopher, T. A., Lopez, B. L. and Ma, X. L. (2006). In vitro and in vivo evidence that thioredoxin nitration results in its unactivation. *Acad. Emerg. Med.* 13,S133–S134.

Johnson, F. and Giulivi, C. (2005). Superoxide dismutases and their impact upon human health. *Mol. Aspects Med.* 26,340–352.

Kamisaki, Y., Wada, K., Bian, K., Balabanli, B., Davis, K., Martin, E., Behbod, F., Lee, Y. C. and Murad, F. (1998). An activity in rat tissues that modifies nitrotyrosine-containing proteins. *Proc. Natl. Acad. Sci. USA* 95,11584–11589.

Khan, M., Sekhon, B., Giri, S., Jatana, M., Gilg, A. G., Ayasolla, K., Elango, C., Singh, A. K. and Singh, I. (2005). *S*-nitrosoglutathione reduces inflammation and protects brain against focal cerebral ischemia in a rat model of experimental stroke. *J. Cereb. Blood Flow Metab.* 25,177–192.

Komaki-Yasuda, K., Kawazu, S. and Kano, S. (2003). Disruption of the *Plasmodium falciparum* 2-Cys peroxiredoxin gene renders parasites hypersensitive to reactive oxygen and nitrogen species. *FEBS Lett.* 547,140–144.

Kong, S. K., Yim, M. B., Stadtman, E. R. and Chock, P. B. (1996). Peroxynitrite disables the tyrosine phosphorylation regulatory mechanism: Lymphocyte-specific tyrosine kinase fails to phosphorylate nitrated cdc2(6-20)NH2 peptide. *Proc. Natl. Acad. Sci. USA* 93,3377–3382.

Kuo, W. N., Kanadia, R. N., Shanbhag, V. P. and Toro, R. (1999). Denitration of peroxynitrite-treated proteins by 'protein nitratases' from rat brain and heart. *Mol. Cell. Biochem.* 201,11–16.

Lancaster, J. R., Jr. and Gaston, B. (2004). NO and nitrosothiols: spatial confinement and free diffusion. *Am. J. Physiol. Lung Cell. Mol. Physiol.* 287,L465–L466.

Langer, R. C., Li, F., Popov, V., Kurosky, A. and Vinetz, J. M. (2002). Monoclonal antibody against the *Plasmodium falciparum* chitinase, PfCHT1, recognizes a malaria transmission-blocking epitope in *Plasmodium gallinaceum* ookinetes unrelated to the chitinase PgCHT1. *Infect. Immun.* 70,1581–1590.

Levashina, E. A. (2004). Immune responses in *Anopheles gambiae. Insect Biochem. Mol. Biol.* 34,673–678.

Luckhart, S., Crampton, A. L., Zamora, R., Lieber, M. J., Dos Santos, P. C., Peterson, T. M., Emmith, N., Lim, J., Wink, D. A. and Vodovotz, Y. (2003). Mammalian transforming growth factor beta1 activated after ingestion by *Anopheles stephensi* modulates mosquito immunity. *Infect. Immun.* 71,3000–3009.

Luckhart, S. and Rosenberg, R. (1999). Gene structure and polymorphism of an invertebrate nitric oxide synthase gene. *Gene* 232,25–34.

Luckhart, S., Vodovotz, Y., Cui, L. and Rosenberg, R. (1998). The mosquito *Anopheles stephensi* limits malaria parasite development with inducible synthesis of nitric oxide. *Proc. Natl. Acad. Sci. USA* 95,5700–5705.

MacMillan-Crow, L. A., Crow, J. P. and Thompson, J. A. (1998). Peroxynitrite-mediated inactivation of manganese superoxide dismutase involves nitration and oxidation of critical tyrosine residues. *Biochemistry* 37,1613–1622.

MacMillan-Crow, L. A., Greendorfer, J. S., Vickers, S. M. and Thompson, J. A. (2000). Tyrosine nitration of c-SRC tyrosine kinase in human pancreatic ductal adenocarcinoma. *Arch. Biochem. Biophys.* 377,350–356.

Mallozzi, C., Di Stasi, A. M. and Minetti, M. (2001). Nitrotyrosine mimics phosphotyrosine binding to the SH2 domain of the src family tyrosine kinase lyn. *FEBS Lett.* 503,189–195.

146

Mannick, J. B., Hausladen, A., Liu, L., Hess, D. T., Zeng, M., Miao, Q. X., Kane, L. S., Gow, A. J. and Stamler, J. S. (1999). Fas-induced caspase denitrosylation. *Science* 284,651–654.

Mannick, J. B., Schonhoff, C., Papeta, N., Ghafourifar, P., Szibor, M., Fang, K. Z. and Gaston, B. (2001). S-nitrosylation of mitochondrial caspases. *J. Cell Biol.* 154,1111–1116.

Matsushima, S., Ide, T., Yamato, M., Matsusaka, H., Hattori, F., Ikeuchi, M., Kubota, T., Sunagawa, K., Hasegawa, Y., Kurihara, T. *et al.* (2006). Overexpression of mitochondrial peroxiredoxin-3 prevents left ventricular remodeling and failure after myocardial infarction in mice. *Circulation* 113,1779–1786.

Miersch, S. and Mutus, B. (2005). Protein S-nitrosation: Biochemistry and characterization of protein thiol-NO interactions as cellular signals. *Clin. Biochem.* 38,777–791.

Mi-Ichi, F., Miyadera, H., Kobayashi, T., Takamiya, S., Waki, S., Iwata, S., Shibata, S. and Kita, K. (2005). Parasite mitochondria as a target of chemotherapy: Inhibitory effect of licochalcone A on the *Plasmodium falciparum* respiratory chain. *Ann. N Y Acad. Sci.* 1056,46–54.

Miller, S. K., Good, R. T., Drew, D. R., Delorenzi, M., Sanders, P. R., Hodder, A. N., Speed, T. P., Cowman, A. F., de Koning-Ward, T. F. and Crabb, B. S. (2002). A subset of *Plasmodium falciparum* SERA genes are expressed and appear to play an important role in the erythrocytic cycle. *J. Biol. Chem.* 277,47524–47532.

Miller, S., Ross-Inta, C. and Giulivi, C. (2006). Kinetic and proteomic analyses of S-nitrosoglutathione-treated hexokinase A: Consequences for cancer energy metabolism. *Amino Acids.* Epub ahead of print (DOI 10.1007/s00726-006-0424-9).

Mitchell, D. A. and Marletta, M. A. (2005). Thioredoxin catalyzes the S-nitrosation of the caspase-3 active site cysteine. *Nat. Chem. Biol.* 1,154–158.

Moncada, S., Palmer, R. M. and Higgs, E. A. (1991). Nitric oxide: Physiology, pathophysiology, and pharmacology. *Pharmacol. Rev.* 43,109–142.

Mondoro, T. H., Shafer, B. C. and Vostal, J. G. (1997). Peroxynitrite-induced tyrosine nitration and phosphorylation in human platelets. *Free Radic. Biol. Med.* 22,1055–1063.

Nakamura, H. (2005). Thioredoxin and its related molecules: Update 2005. *Antioxid. Redox Signal.* 7,823–828.

Nielsen, A. T. (1995). Introduction and Tetranitromethane. In *Nitrocarbons*, pp. 6–65, VCH Publishers, New York.

Nozik-Grayck, E., Whalen, E. J., Stamler, J. S., McMahon, T. J., Chitano, P. and Piantadosi, C. A. (2006). S-nitrosoglutathione inhibits alpha(1)-adrenergic receptor-mediated vasoconstriction and ligand binding in pulmonary artery. *Am. J. Physiol. Lung Cell. Mol. Physiol.* 290,L136–L143.

Perissinotti, L. L., Turjanski, A. G., Estrin, D. A. and Doctorovich, F. (2005). Transnitrosation of nitrosothiols: Characterization of an elusive intermediate. *J. Am. Chem. Soc.* 127,486–487.

Peterson, T. M. and Luckhart, S. (2006). A mosquito 2-Cys peroxiredoxin protects against nitrosative and oxidative stresses associated with malaria parasite infection. *Free Radic. Biol. Med.* 40,1067–1082.

Peterson, T. M. L., Gow, A. J. and Luckhart, S. (2007). Nitric oxide metabolites induced in *Anopheles stephensi* control malaria parasite infection. *Free Radic. Biol. Med.* 42,132–142.

Rabilloud, T., Heller, M., Gasnier, F., Luche, S., Rey, C., Aebersold, R., Benahmed, M., Louisot, P. and Lunardi, J. (2002). Proteomics analysis of cellular response to oxidative stress. Evidence for in vivo overoxidation of peroxiredoxins at their active site. *J. Biol. Chem.* 277,19396–19401.

Reinehr, R., Gorg, B., Hongen, A. and Haussinger, D. (2004). CD95-tyrosine nitration inhibits hyperosmotic and CD95 ligand-induced CD95 activation in rat hepatocytes. *J. Biol. Chem.* 279,10364–10373.

Rhee, S. G., Yang, K. S., Kang, S. W., Woo, H. A. and Chang, T. S. (2005). Controlled elimination of intracellular H(2)O(2): Regulation of peroxiredoxin, catalase, and glutathione peroxidase via post-translational modification. *Antioxid. Redox Signal.* 7,619–626.

Robinson, J. M. and Lancaster, J. R., Jr. (2005). Hemoglobin-mediated, hypoxia-induced vasodilation via nitric oxide: mechanism(s) and physiologic versus pathophysiologic relevance. *Am. J. Respir. Cell Mol. Biol.* 32,257–261.

Rockett, K. A., Awburn, M. M., Cowden, W. B. and Clark, I. A. (1991). Killing of *Plasmodium falciparum* in vitro by nitric oxide derivatives. *Infect. Immun.* 59,3280–3283.

Rosenthal, P. J. (2004). Cysteine proteases of malaria parasites. *Int. J. Parasitol.* 34,1489–1499.

Sarkela, T. M., Berthiaume, J., Elfering, S., Gybina, A. A. and Giulivi, C. (2001). The modulation of oxygen radical production by nitric oxide in mitochondria. *J. Biol. Chem.* 276,6945–6949.

Savvides, S. N., Scheiwein, M., Bohme, C. C., Arteel, G. E., Karplus, P. A., Becker, K. and Schirmer, R. H. (2002). Crystal structure of the antioxidant enzyme glutathione reductase inactivated by peroxynitrite. *J. Biol. Chem.* 277,2779–2784.

Schroder, E., Littlechild, J. A., Lebedev, A. A., Errington, N., Vagin, A. A. and Isupov, M. N. (2000). Crystal structure of decameric 2-Cys peroxiredoxin from human erythrocytes at 1.7Å resolution. *Structure* 8,605–615.

Serbina, N. V., Salazar-Mather, T. P., Biron, C. A., Kuziel, W. A. and Pamer, E. G. (2003). TNF/iNOS-producing dendritic cells mediate innate immune defense against bacterial infection. *Immunity* 19,59–70.

Shahabuddin, M., Toyoshima, T., Aikawa, M. and Kaslow, D. C. (1993). Transmission-blocking activity of a chitinase inhibitor and activation of malarial parasite chitinase by mosquito protease. *Proc. Natl. Acad. Sci. USA* 90,4266–4270.

Shiu, S. H. and Li, W. H. (2004). Origins, lineage-specific expansions and multiple losses of tyrosine kinases in eukaryotes. *Mol. Biol. Evol.* 21,828–840.

Singh, S. P., Wishnok, J. S., Keshive, M., Deen, W. M. and Tannenbaum, S. R. (1996). The chemistry of the *S*-nitrosoglutathione/glutathione system. *Proc. Natl. Acad. Sci. USA* 93,14428–14433.

Souza, J. M., Choi, I., Chen, Q., Weisse, M., Daikhin, E., Yudkoff, M., Obin, M., Ara, J., Horwitz, J. and Ischiropoulos, H. (2000). Proteolytic degradation of tyrosine nitrated proteins. *Arch. Biochem. Biophys.* 380,360–366.

148

Stamler, J. S., Singel, D. J. and Loscalzo, J. (1992). Biochemistry of nitric oxide and its redox-activated forms. *Science* 258,1898–1902.

Steffen, M., Sarkela, T. M., Gybina, A. A., Steele, T. W., Trasseth, N. J., Kuehl, D. and Giulivi, C. (2001). Metabolism of *S*-nitrosoglutathione in intact mitochondria. *Biochem. J.* 356,395–402.

Szabo, C. (2003). Multiple pathways of peroxynitrite cytotoxicity. *Toxicol. Lett.* 140/141,105–112.

Szabo, C. and Ohshima, H. (1997). DNA damage induced by peroxynitrite: Subsequent biological effects. *Nitric Oxide* 1,373–385.

Tanaka, T., Hosoi, F., Yamaguchi-Iwai, Y., Nakamura, H., Masutani, H., Ueda, S., Nishiyama, A., Takeda, S., Wada, H., Spyrou, G. and Yodoi, J. (2002). Thioredoxin-2 (TRX-2) is an essential gene regulating mitochondria-dependent apoptosis. *EMBO J* 21,1695–1703.

Tatoyan, A. and Giulivi, C. (1998). Purification and characterization of a nitric-oxide synthase from rat liver mitochondria. *J. Biol. Chem.* 273,11044–11048.

Taylor-Robinson, A. W. (1997). Antimalarial activity of nitric oxide: Cytostasis and cytotoxicity towards *Plasmodium falciparum*. *Biochem. Soc. Trans.* 25,262S.

Trujillo, M., Budde, H., Pineyro, M. D., Stehr, M., Robello, C., Flohe, L. and Radi, R. (2004). *Trypanosoma brucei* and *Trypanosoma cruzi* tryparedoxin peroxidases catalytically detoxify peroxynitrite via oxidation of fast reacting thiols. *J. Biol. Chem.* 279,34175–34182.

Turko, I. V. and Murad, F. (2002). Protein nitration in cardiovascular diseases. *Pharmacol. Rev.* 54,619–634.

Uyemura, S. A., Luo, S., Moreno, S. N. and Docampo, R. (2000). Oxidative phosphorylation, Ca2 + transport and fatty acid-induced uncoupling in malaria parasites mitochondria. *J. Biol. Chem.* 275,9709–9715.

Uyemura, S. A., Luo, S., Vieira, M., Moreno, S. N. and Docampo, R. (2004). Oxidative phosphorylation and rotenone-insensitive malate- and NADH-quinone oxidoreductases in *Plasmodium yoelii yoelii* mitochondria in situ. *J. Biol. Chem.* 279,385–393.

van der Vliet, A., Eiserich, J. P., O'Neill, C. A., Halliwell, B. and Cross, C. E. (1995). Tyrosine modification by reactive nitrogen species: A closer look. *Arch. Biochem. Biophys.* 319,341–349.

Venturini, G., Colasanti, M., Salvati, L., Gradoni, L. and Ascenzi, P. (2000). Nitric oxide inhibits falcipain, the *Plasmodium falciparum* trophozoite cysteine protease. *Biochem. Biophys. Res. Commun.* 267,190–193.

Ward, P., Equinet, L., Packer, J. and Doerig, C. (2004). Protein kinases of the human malaria parasite *Plasmodium falciparum*: The kinome of a divergent eukaryote. *BMC Genomics* 5,79.

Watson, W. H., Yang, X., Choi, Y. E., Jones, D. P. and Kehrer, J. P. (2004). Thioredoxin and its role in toxicology. *Toxicol. Sci.* 78,3–14.

Wong, C. M., Zhou, Y., Ng, R. W., Kung Hf, H. F. and Jin, D. Y. (2002). Cooperation of yeast peroxiredoxins Tsa1p and Tsa2p in the cellular defense against oxidative and nitrosative stress. *J. Biol. Chem.* 277,5385–5394.

Wood, Z. A., Poole, L. B. and Karplus, P. A. (2003). Peroxiredoxin evolution and the regulation of hydrogen peroxide signaling. *Science* 300,650–653.

Yamakura, F., Taka, H., Fujimura, T. and Murayama, K. (1998). Inactivation of human manganese-superoxide dismutase by peroxynitrite is caused by exclusive nitration of tyrosine 34 to 3-nitrotyrosine. *J. Biol. Chem.* 273,14085–14089.

Yamawaki, H. and Berk, B. C. (2005). Thioredoxin: A multifunctional antioxidant enzyme in kidney, heart and vessels. *Curr. Opin. Nephrol. Hypertens.* 14,149–153.

Yang, Y. and Loscalzo, J. (2005). *S*-nitrosoprotein formation and localization in endothelial cells. *Proc. Natl. Acad. Sci. USA* 102,117–122.

Zaman, K., Hanigan, M. H., Smith, A., Vaughan, J., Macdonald, T., Jones, D. R., Hunt, J. F. and Gaston, B. (2006). Endogenous *S*-nitrosoglutathione modifies 5-lipoxygenase expression in airway epithelial cells. *Am. J. Respir. Cell Mol. Biol.* 34,387–393.

Effects of S-nitrosation of nitric oxide synthase

Douglas A. Mitchell[1], Thomas Michel[4,5] and Michael A. Marletta[1,2,3,*]

[1]Department of Chemistry, University of California, Berkeley, CA 94720, USA
[2]Department of Molecular and Cell Biology, University of California, Berkeley, CA 94720, USA
[3]Division of Physical Biosciences, Lawrence Berkeley National Laboratory, University of California, Berkeley, CA 94720, USA
[4]Cardiovascular Division, Brigham and Women's Hospital, Harvard Medical School, Boston, MA 02115, USA
[5]Veterans Affairs Boston Healthcare System, West Roxbury, MA 02132, USA

Abstract. Nitric oxide (NO) is a key mammalian signaling molecule that affects numerous physiological processes. Mammals possess three isoforms of nitric oxide synthase (NOS): endothelial (eNOS), neuronal (nNOS), and inducible (iNOS). These isoforms differ in their tissue distribution, cellular location, regulation, and NO output. NO synthesized by eNOS and nNOS acts in a paracrine fashion, whereby NO generated in one cell acts upon an adjacent cell by binding to and activating soluble guanylyl cyclase (sGC). Activation of sGC by NO leads to a several hundredfold enhancement of cyclic guanidine mono-phosphate (cGMP) synthesis. Notable outcomes of elevated cGMP levels are neuro-transmission and smooth muscle relaxation. iNOS is capable of synthesizing much higher steady-state levels of NO, and the toxicity of NO is harnessed as part of the innate immune response. In recent years, it has become increasingly clear that NO targets proteins other than sGC. S-nitrosation is an example of nonclassical NO signaling, defined as sGC/cGMP-independent, and has garnered attention with regard to the regulation of NOS activity both *in vitro* and *in vivo*. The purpose of this review is to cover what is currently known about the S-nitrosation of NOS isoforms.

Keywords: nitric oxide; nitric oxide synthase; S-nitrosation; S-nitrosylation; transnitro-sation; zinc tetrathiolate; NOS; biotin switch; nitrosothiol relay; signal transduction; posttranslational modifications; specificity of nitrosation; inducible NOS; endothelial NOS; neuronal NOS.

NOS enzymology and physiology

Nitric oxide (NO) is a fundamental signaling molecule and plays vital roles in neurotransmission, smooth muscle relaxation, and the immune response (Alderton *et al.*, 2001; Marletta *et al.*, 1998; Snyder and Bredt, 1991). Under signaling conditions, NO is synthesized in mammals by a constitutively expressed isoform of NO synthase (NOS), commonly

Corresponding author: Tel.: (510) 643-9325. Fax: (510) 643-9388.
E-mail: marletta@berkeley.edu (M.A. Marletta).

referred to as endothelial NOS (eNOS; also called NOS3 or NOS III) or neuronal NOS (nNOS; also called NOS1 or NOS I) (EC 1.14.13.39) (Dudzinski *et al.*, 2006; Forstermann *et al.*, 1998). These two isoforms of NOS typically produce low nanomolar steady-state concentrations of NO. eNOS and nNOS dynamically associate with calmodulin (CaM) in response to fluxes in Ca^{2+} concentration. The lipopolysaccharide (LPS) and gamma-interferon (INF-γ) inducible isoform of NOS (iNOS; also called NOS2 or NOS II) are expressed in many cell types, most notably macrophages and other immune cells (Aktan, 2004). Activated 'constitutively' via a very tight association with Ca^{2+} and CaM, iNOS is able to produce a cytotoxic steady-state concentration of NO in the micromolar range. In cells that also produce superoxide (O_2^-), mainly through the actions of NADPH oxidase, NO can react via radical recombination to form the potent oxidant peroxynitrite ($ONOO^-$) (Ellis *et al.*, 2002; Lee *et al.*, 2000; Xia and Zweier, 1997). Peroxynitrite indiscriminately peroxidates lipids and can oxidize and nitrate electron-rich aromatics in nucleic acids (guanines) and proteins (tyrosines) (Burney *et al.*, 1999; Szabo, 2003; Virag *et al.*, 2003). This is at least one path whereby NO exerts its cytotoxic effects.

The physiological roles of each NOS isoform have been studied following the genetic deletion of a specific enzyme isoform, in addition to double and triple 'knockouts' of the NOS genes. Numerous studies have emerged detailing the biological features of $NOS^{-/-}$ mice (Huang, 2000; Mashimo and Goyal, 1999). Homologous recombination was used to delete exon 1 of the nNOS gene in mice. These mice are viable, fertile, and do not exhibit any obvious histological abnormalities in their central nervous system (Huang *et al.*, 1993). In subsequent reports, the effects of exon 1 deletion on penile erection (Burnett *et al.*, 1996), ovulation (Klein *et al.*, 1998), aggressive behavior (Demas *et al.*, 1997), and eNOS compensation (Al-Shabrawey *et al.*, 2003; Kavdia and Popel, 2004; Vallance *et al.*, 2004) were studied. It was found that eNOS is upregulated and likely fulfills the need for NO, as there were no differences in erectile function and mating when compared to control mice (Burnett *et al.*, 1996). However, the results are different when exon 6 (coding for the heme-binding domain) of the nNOS gene is disrupted. This deletion results in hypogonadism and infertility (Gyurko *et al.*, 2002). The explanation for the apparent disparity is that deletion of exon 1 does not fully eliminate nNOS activity due to the presence of two nNOS splice variants, β-nNOS and γ-nNOS. The features of eNOS knockout mice have also been extensively characterized. $eNOS^{-/-}$ mice are hypertensive (Huang *et al.*, 1995) and develop large myocardial infarcts after ischemic events (Fagan *et al.*, 2000; Hannan *et al.*, 2000). There is evidence that while eNOS is capable of compensating for the loss of nNOS in some responses, nNOS cannot compensate for the absence of eNOS as demonstrated in dysfunctional leukocyte homing on endothelial tissue (Sanz *et al.*, 2001). Indeed, eNOS appears to have a broad role in nonendothelial tissues (Lowenstein and Michel, 2006). Recent studies of $eNOS^{-/-}$ mice have revealed an unexpected role of eNOS in mediating the anaphylactic immune response (Cauwels *et al.*, 2006), possibly indicating a role for eNOS in the regulation of mast cell function. The key role of iNOS in immune responses has been confirmed in studies of $iNOS^{-/-}$ mice. As anticipated, $iNOS^{-/-}$ mice are more susceptible to many types of infection (Igietseme *et al.*, 1998; Krahenbuhl and Adams, 2000; Zeidler *et al.*, 2003). The triple NOS knockout was expected to be lethal, since no compensation was possible. However, the triple knockout mice were viable and appeared normal, but had decreased survival and fertility rates (Morishita *et al.*, 2005). Further inspection of the mice found symptoms characteristic of nephrogenic diabetes insipidus, indicating a vital role for NOS in kidney homeostasis. It must be cautioned that the triple $NOS^{-/-}$ mice were produced using the exon 1 deletion of nNOS; therefore, it is possible that residual NOS activity is present due to the splice variation as noted above. It would be of interest to determine the phenotype of the triple NOS knockout mouse created by mating $iNOS^{-/-}/eNOS^{-/-}$ mice with $nNOS^{+/-}_{exon\,6}$ mice.

The diverse functions of NO and the regulation of NOS have been intensely studied since proof was provided that NO was the endothelium-derived relaxation factor (Hutchinson *et al.*, 1987). During the course of the two-step NOS reaction, arginine undergoes an overall five-electron oxidation to citrulline and NO through the hydroxylated intermediate, N-hydroxyarginine (NHA) (Marletta *et al.*, 1998). Synthesis of NO requires 2 mol equivalents of molecular oxygen and 1.5 equivalents of NADPH. Several cofactors are also required for catalytic activity: heme, FMN, FAD, and tetrahydrobiopterin H_4B. Each isoform of NOS is homodimeric (Chen *et al.*, 2002b) and has two domains. The C-terminal reductase domain binds the cosubstrate NADPH, and the cofactors FAD and FMN (Hevel *et al.*, 1991). Ca^{2+}/CaM activates electron transfer from the reductase domain to the N-terminal oxygenase domain of NOS (Cho *et al.*, 1992). The oxygenase domain contains a P450-type heme cofactor, H_4B (Cho *et al.*, 1995), a zinc tetrathiolate cluster (Ludwig and Marletta, 1999), and contains the substrate-binding site for arginine. The individual role of each cofactor has been well established in the first step of the NOS reaction, but the role of H_4B in the second step remains elusive, though it is clear that it also functions in electron transfer (Hurshman and Marletta, 1995; Hurshman *et al.*, 1999).

Regulation of NOS

Regulation of NO synthesis in the cell is different for the specific iso-forms. iNOS is mostly regulated at the transcriptional level. Cells that express iNOS produce cytotoxic concentrations of NO, so the transcriptional control over expression is tightly regulated (De Stefano *et al.*, 2006; Kleinert *et al.*, 1996; Nishiya *et al.*, 2000). The redox-active nuclear factor NF-κB binds to the iNOS promoter and initiates expression of iNOS in response to proinflammatory stimuli such as LPS and INF-γ (Stuehr and Marletta, 1985, 1987). The iNOS promoter can also be bound and activated by interferon regulatory factor-1 (IRF-1) and signal transducer and activator of transcription (STAT-1α) (De Stefano *et al.*, 2006; Morris *et al.*, 2003). In addition to transcriptional control, iNOS activity can also be posttranslationally affected by lowering arginine availability (El-Gayar *et al.*, 2003). Arginase has been implicated in this type of regulation.

Unlike iNOS, which remains persistently activated by its avid binding to CaM even at the low levels of intracellular Ca^{2+} in resting cells, the eNOS and nNOS isoforms only bind CaM tightly at higher Ca^{2+} levels. Therefore, eNOS and nNOS undergo transient activation in response to extracellular signals that promote increases in intracellular Ca^{2+} concentration (Forstermann *et al.*, 1998). Physiological processes that increase intracellular Ca^{2+} concentrations, such as stimulation of bradykinin or angiotensin receptors, can lead to eNOS and nNOS activation (Boulanger *et al.*, 1998; Gryglewski *et al.*, 2002; Saito *et al.*, 1996; Zhao *et al.*, 2005). Despite many similarities, there are important structural and regulatory differences between eNOS and nNOS, above and beyond the striking differences in tissue distribution. nNOS contains an N-terminal PDZ domain that influences subcellular location through interaction with integral membrane proteins, cytoskeletal proteins (α1 syntrophin) (Adams *et al.*, 2001), and carboxy-terminal PDZ ligand of nNOS (CAPON) (Fang *et al.*, 2000; Jaffrey *et al.*, 1998, 2002; Miranda *et al.*, 2006). eNOS does not contain a PDZ domain but does undergo a series of N-terminal posttranslational modifications, which were recently reviewed (Dudzinski *et al.*, 2006). These modifications permit dynamic association with the caveolae membrane (Feron *et al.*, 1996; Michel, 1999). Cotranslational processing of eNOS involves cleavage of Met1 and irreversible *N*-myristoylation of the new N-terminus of eNOS, Gly2

(Busconi and Michel, 1993). After the myristoyl group is successfully appended, eNOS is targeted to caveolae by reversible thiopalmitoylation of Cys15 and Cys26 (Belhassen *et al.*, 1997; Robinson *et al.*, 1995). eNOS responds to agonist stimulation, most notably vascular endothelial growth factor (VEGF) and insulin, by rapid depalmitoylation (Yeh *et al.*, 1999) and translocation to internal membrane structures. Basal activity is achieved after repalmitoylation and retargeting to caveolae. Caveolin-1 is a major negative regulator protein for eNOS (Feron *et al.*, 1996).

In addition to the regulation described above, O-phosphorylation has been shown to affect the function of all three NOSs. nNOS (Ser1412) (Adak *et al.*, 2001) and eNOS (Ser116, Thr495, Ser617, Ser635, and Ser1177) (Corson *et al.*, 1996; Fleming *et al.*, 2001; Fulton *et al.*, 2002; Gallis *et al.*, 1999; Gonzalez *et al.*, 2002; Michel *et al.*, 1993; Robinson *et al.*, 1996) have been shown to be phosphorylated by protein kinase A (PKA) or protein kinase B (PKB/Akt) (Dimmeler *et al.*, 1999; Fulton *et al.*, 1999) with an increase in catalytic activity (Michell *et al.*, 2002). Presumably, phosphorylation induces a conformational change that increases the efficiency of electron transfer to the oxygenase domain. To complicate the issue, phosphorylation can also inhibit NOS activity. nNOS turnover can be attenuated by phosphorylation of Ser847 by CaM-kinase IIα (Komeima *et al.*, 2000; Osuka *et al.*, 2002; Watanabe *et al.*, 2003). Similarly, phosphorylation of eNOS at Thr497 is inhibitory (Harris *et al.*, 2001; Lin *et al.*, 2003). Protein phosphatase 2A restores both eNOS and nNOS activities by dephosphorylating these residues (Greif *et al.*, 2002). A recent report has found that Src can phosphorylate iNOS on Tyr151, resulting in a modest decrease in activity (~30%) and redistribution to the insoluble cellular fraction (Hausel *et al.*, 2006). It is clear that NOS activity is regulated in a complex manner involving several posttranslational modifications. This topic has been the subject of many other reviews (Dudzinski *et al.*, 2006; Fulton *et al.*, 2001; Venema, 2002). The posttranslational regulation of interest to this review involves NO-dependent modification of the zinc tetrathiolate center of all three isoforms of NOS.

Zinc tetrathiolate cluster and reactivity toward NO

Stabilization of the NOS homodimer is influenced by numerous ligands and allosteric modulators including CaM (Cho *et al.*, 1992; Panda *et al.*, 2001; Zemojtel *et al.*, 2003), H_4B and arginine binding (Cho *et al.*, 1995; Klatt *et al.*, 1995, 1996), and the zinc tetrathiolate cluster (Hemmens *et al.*, 2000; Ludwig and Marletta, 1999; Raman *et al.*, 1998). It has been

156

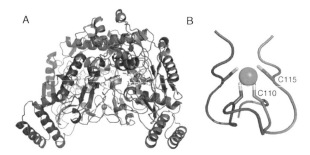

Fig. 1. Structure of the human iNOS oxygenase domain and zinc tetrathiolate center. The oxygenase domain of NOS is homodimeric. (A) The entire oxygenase domain. The heme cofactor is shown as yellow lines, H_4B as light blue lines, and Zn^{2+} as a green sphere with the ligating cysteine thiols (Cys110 and Cys115 in human iNOS) shown as yellow lines. The zinc tetrathiolate and H_4B moieties make numerous interchain contacts to promote dimeric stability. Arginine was omitted for clarity. (B) Closeup view of the zinc tetrathiolate center. Figure was constructed with PyMol (DeLanoScientific LLC, San Francisco, USA) using Protein DataBank entry 1NSI. (See Colour Plate Section in this book).

shown that monomeric NOS is catalytically inactive (Xie *et al.*, 1996). The homodimeric status of all three isoforms has been demonstrated by X-ray crystallography on the heme-containing oxygenase domain (Alderton *et al.*, 2001; Crane *et al.*, 1997, 1998; Flinspach *et al.*, 2004; Garcin *et al.*, 2004; Li *et al.*, 1999; Raman *et al.*, 2001). Figure 1 shows the structure of the human iNOS oxygenase domain (PDB entry ID: 1NSI). The zinc tetrathiolate cluster is comprised of two cysteine residues from each oxygenase monomer (Cys104 and Cys109 in murine iNOS, and Cys110 and Cys115 in human iNOS), which coordinate the redox-inert Zn^{2+} with tetrahedral geometry. Zinc clusters in proteins, including zinc finger transcription factors, are commonly viewed as playing only structural roles. However, when a cysteine thiol is ligated to a potent Lewis acid, such as Zn^{2+}, the thiol becomes significantly more acidic. In fact, several *in vitro* (Chen *et al.*, 2002a; Gergel and Cederbaum, 1996; Rozema and Poulter, 1999) and computational (Dudev and Lim, 2002) studies have confirmed that due to zinc ligation, the cysteine thiol is in fact a thiolate at physiological pH. Reactions of thiolates with electrophiles proceed 2–3 orders of magnitude more rapidly than reactions of the corresponding thiols. The increased nucleophilicity of the zinc tetra-thiolate cluster cysteines, combined with the structural importance of the cluster for dimeric stability and activity, suggests that modification of the cluster will lead to reduced NOS activity. NO is an electrophile, and reaction at the zinc center forms a negative feedback loop for enzymatic turnover. Early reports on NOS activity indicated that prolonged exposure of NOS to NO lowers activity (Abu-Soud et al., 2001; Buga et al., 1993). Also, overexpression of the oxidoreductase thioredoxin prevents NO-dependent loss in NOS activity (Shao et al., 2002; Zhang et al., 1998). Based on these reports and the chemical partnership between zinc and sulfur (Maret, 2004), our laboratories and others discovered that S-nitrosation of the zinc tetrathiolate cluster accounts, at least in part, for the above observations.

S-nitrosation, also known as S-nitrosylation

Besides binding to heme proteins and recombining with other radicals, NO can also react with molecular oxygen (Fig. 2) (McCleverty, 2004). In a termolecular reaction that is second-order-dependent on NO concentration, the electrophile N_2O_3 is produced. N_2O_3 undergoes rapid hydrolysis in water but also reacts with abundant thiol-based nucleophiles, such as glutathione (GSH). GSH is present at low millimolar concentrations in all mammalian cells and serves to keep a reduced intracellular environment. It must be cautioned that N_2O_3 is probably found at very low levels in biological systems, since NO concentrations are quite low (even in activated immune cells) and hydrolysis is facile (Wink et al., 1994). However, if N_2O_3 persists long enough to react with a thiol, a nitrosothiol and a nitrite are formed through formal addition of nitrosonium (^+NO) to the thiol (Fig. 2) (Keshive et al., 1996). This reaction is termed S-nitrosation. This is not to be confused with nitrosylation, which is the formal addition of NO to another group. Nitrosothiols can also be formed by direct addition of NO with a thiol (S-nitrosylation) followed by one-electron oxidation. Oxygen is present in normoxic cells at mid- to high micromolar concentrations and is a suitable candidate for an electron acceptor in this reaction, thus forming superoxide. There is confusion in the literature about which term more accurately describes the mechanism of modification. Chemists tend to use the term S-nitrosation because even if the second mechanism is more relevant in cells, the oxidation state of the thiol changes to S^{-1}. In a true S-nitrosylation reaction, the thiol would remain in the fully reduced S^{-2} state. Nitrosothiols formed by these reactions can be transferred to other proteins or small molecules through transnitrosation reactions or can be eliminated via thiolation, which forms a disulfide and nitroxyl (Fig. 2).

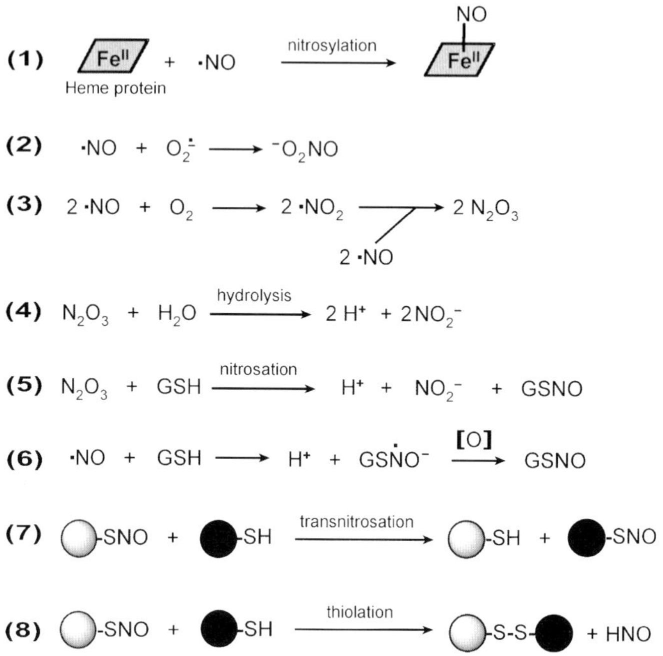

$$\text{(1)} \quad \boxed{Fe^{II}} + \cdot NO \xrightarrow{\text{nitrosylation}} \boxed{Fe^{II}} \overset{NO}{|}$$

Heme protein

$$\text{(2)} \quad \cdot NO + O_2^{\cdot -} \longrightarrow {}^-O_2NO$$

$$\text{(3)} \quad 2 \cdot NO + O_2 \longrightarrow 2 \cdot NO_2 \nearrow 2 N_2O_3 \\ 2 \cdot NO$$

$$\text{(4)} \quad N_2O_3 + H_2O \xrightarrow{\text{hydrolysis}} 2 H^+ + 2 NO_2^-$$

$$\text{(5)} \quad N_2O_3 + GSH \xrightarrow{\text{nitrosation}} H^+ + NO_2^- + GSNO$$

$$\text{(6)} \quad \cdot NO + GSH \longrightarrow H^+ + GS\dot{N}O^- \xrightarrow{[O]} GSNO$$

$$\text{(7)} \quad \bigcirc\!\!-SNO + \bullet\!\!-SH \xrightarrow{\text{transnitrosation}} \bigcirc\!\!-SH + \bullet\!\!-SNO$$

$$\text{(8)} \quad \bigcirc\!\!-SNO + \bullet\!\!-SH \xrightarrow{\text{thiolation}} \bigcirc\!\!-S\text{-}S\!\!-\bullet + HNO$$

Fig. 2. Diversity of biologically relevant reactions of NO. NO can take part in many reactions that are of biological interest. NO nitrosylates heme proteins (1), reacts with superoxide to form peroxynitrite (2), reacts with oxygen to form N_2O_3 (3), which can undergo aqueous hydrolysis (4) and nitrosation reactions (5), both of which release nitrite as a byproduct. Thiols can also be S-nitrosated by nucleophilic attack on NO with subsequent one-electron oxidation (6). Once nitrosothiols (SNO) are formed, they can be transferred to other thiols by transnitrosation reactions (7) or eliminated by disulfide formation and ejection of nitroxyl (8). GSH, glutathione. GSNO, *S*-nitrosoglutathione.

Gaps in our current understanding of S-nitrosation

S-nitrosation is a posttranslational modification in which a cysteine thiol in a protein is converted to a nitrosothiol (Foster *et al.*, 2003; Hess *et al.*, 2005). Ramifications of S-nitrosation are obvious when the modified cysteine participates in catalytic chemistry of an enzyme, such as the cysteine protease caspase-3. In these cases, complete inhibition of activity is an expected outcome. Many have compared S-nitrosation to *O*-phosphorylation, but several key questions remain to be answered to justify this comparison (Lane *et al.*, 2001). In *O*-phosphorylation, families of kinases and phosphatases regulate protein modification, in many cases with exquisite specificity (Aitken *et al.*, 2002; Faux and Scott, 1996; Hofer, 1996). Similar enzymes that affect protein S-nitrosation have yet to be discovered but there is *in vitro* evidence that specific, protein-mediated, transnitrosation reactions can occur (Mitchell and Marletta, 2005). Importantly, the activity of each isoform of NOS has been correlated to nitrosothiol content in a variety of tissues (Gow *et al.*, 2002).

The interpretation of many observations of S-nitrosation is hampered by the fact that many studies treat putative S-nitrosated proteins using an exogenous source of NO at a concentration that is physiologically unattainable. For instance, NO donors or nitrosating agents are often used at concentrations in the high micromolar or even millimolar concentration range, which represent NO concentrations many orders of magnitude higher than activated macrophages can synthesize. In many of the cases in which the number of modified thiol residues was determined, it is not surprising that multiple cysteine residues were found to be S-nitrosated. An example of poly-S-nitrosation is shown in the case of the ryanodine receptor (Xu *et al.*, 1998). This protein has 84 cysteine residues, of which 12 are modified to nitrosothiols on reaction with 1 mM *S*-nitrosoglutathione (GSNO). In this case, such a large amount of GSNO was used that many surface-accessible cysteines were S-nitrosated. Subsequently, it was determined by the same group that specific S-nitrosation of Cys3635 accounts for the NO-dependent regulation of the ryanodine receptor (Sun *et al.*, 2001).

We endorse the proposed criteria of Lancaster and Gaston (2004) for establishing that a specific bioactivity is associated with S-nitrosation or denitrosation of a particular protein. These criteria are important enough to reiterate here:

(1) The activity of the target protein is associated with a change of activity of a NOS isoform.

(2) S-nitrosation of the target protein isolated from cells following NOS activation can be demonstrated by more than one independent assay. In addition, the extent of nitrosation from endogenous NO is sufficient in magnitude to affect the activity of the protein, and nitrosation/denitrosation occurs rapidly enough to account for regulated changes in activity.

(3) Mutation of a specific cysteine residue in the target protein results in the loss of the NOS-responsive bioactivity and the inability to identify the S-nitrosated protein following NOS activation.

(4) In association with termination of the bioactivity, loss of S-nitrosation of the cellular protein should be demonstrated.

(5) Alteration of the function of the purified protein can be demonstrated in association with S-nitrosation under conditions relevant to the protein's cellular environment (*i.e.*, biologically relevant NO concentration, pH, etc.).

(6) Pharmacological experiments demonstrate that cyclic guanidine monophosphate (cGMP) is not exclusively involved in mediating the bioactivity.

(7) Pharmacological experiments suggest that a thiol modification is involved (*i.e.*, the altered bioactivity is blocked by pretreatment with *N*-ethylmaleimide and/or reversed by excess dithiothreitol).

(8) Pharmacological modifications of specific nitrosothiol (SNO) metabolic enzymes relevant to the putative signaling process (such as γ-glutamyl transpeptidase or GSH-dependent formaldehyde dehydrogenase) appropriately alter the bioactivity.

A word of caution is needed regarding item (3). Many assume that serine is the best substitution for cysteine, since it seems to be the most modest perturbation of side chain functionality. This is true if the cysteine of interest is in a hydrophilic region of the protein. However, buried cysteines should not be mutated to serine because this can dramatically decrease the local hydrophobic packing and even weaken the global fold of the protein (Chou and Matthews, 1989; He and Quiocho, 1991; Lu *et al.*, 1992). In these cases, alanine is a more reasonable choice. Indeed, the dipole moment of the buried thiol is not far from that of an alkyl group, and the alcohol group of serine is not capable of 'mimicking grease.'

Another example of poly-S-nitrosation that is physiologically irrelevant has been confirmed in our laboratory using caspase-3 (Mitchell and Marletta, 2005). Caspase-3 has eight cysteines, and on reaction with 250 μM of a NO donor (diethylamine diazeniumdiolate, DEA/NO), we identified four modified cysteines. This is in agreement with earlier work on this enzyme (Zech *et al.*, 1999). Reaction of caspase-3 with 50 μM DEA/NO yields nitrosothiols on the same four cysteines, as demonstrated by the biotin switch assay (Jaffrey and Snyder, 2001) and mass spectrometry, but the overall extent of nitrosothiol formation was lower. At room temperature, the steady-state concentration of NO in solution from decomposition of 50 μM DEA/NO is close to the concentration of NO obtained in an activated immune cell (Davies *et al.*, 2001), so why is poly-S-nitrosation still occurring when it is known that only one cysteine of caspase-3 is S-nitrosated *in vivo* (Mannick *et al.*, 1997, 1999)? There are two interrelated answers to this question: (i) NO exhibits promiscuous solution chemistry, as demonstrated in Fig. 2, and (ii) the mammalian cell has machinery that allows the specific modification of protein substrates that is missing in all *in vitro* experiments to date. Indeed, poly-S-nitrosation has only been described in cells when an exogenous source of NO was applied. We favor the hypothesis that protein-mediated S-nitrosation predominates *in vivo*. Support for this comes from two straightforward observations. First, signaling concentrations of NO are in the nanomolar range, but the concentrations of other species in the cell that react with and destroy NO at extraordinary rates are almost 1 million times higher (all heme proteins, all reductants, oxygen, superoxide, iron–sulfur clusters, etc.). Secondly, poly-S-nitrosation occurs on *in vitro* S-nitrosated proteins, while only mono-S-nitrosation is observed *in vivo*. Without assistance and regulation, accounting for these observations proves difficult. Without assistance, NO cannot even reach its target protein, let alone modify multiple cysteines.

The specificity of S-nitrosation

A key finding by Stamler and colleagues was the report that Fas-induced denitrosation of caspase-3 in human T-cells (Mannick *et al.*, 1999). Procaspase-3 was found to be S-nitrosated exclusively on the catalytic cysteine (Cys163) in unstimulated human cell lines, and denitrosated on activation of the Fas apoptotic pathway. Intracellular caspase activity increased after denitrosation. While this work clearly shows the importance of nitrosothiols in apoptosis, it leaves critical questions unanswered, such as how do nitrosothiols form, what determines the specificity in signaling, and is the process regulated? To further support the hypothesis of protein-catalyzed S-nitrosation, the active site of caspase-3 is effective at excluding the tripeptide GSH. This is exemplified by the facts that cellular concentrations of GSH (low millimolar) do not inhibit enzymatic activity, and large excesses of GSH do not efficiently restore the activity of caspase-3-Cys163-SNO (Mitchell and Marletta, 2005; Zech *et al.*, 1999). This is directly relevant because the Stamler laboratory has also described a formaldehyde dehydrogenase exhibiting GSNO reductase (GSNOR) activity (Liu *et al.*, 2001). Studies by others have found that GSNOR is specific for GSNO (Sanghani *et al.*, 2000, 2002a,b). Thus, protein-SNOs that are not in free exchange with GSH (such as caspase-3) are not reduced by GSNOR. This has ramifications for the current understanding of apoptosis in NO-producing cells. Without an efficient mechanism for reactivating caspases these cells would not be able to apoptose, which they certainly do with great efficiency.

The precise mechanism by which nitrosothiols are formed *in vivo* is currently unknown. If S-nitrosation is a deliberate cellular posttranslational modification, there is an expectation that cellular machinery will regulate the synthesis of nitrosating equivalents, and that the reaction with proteins will be temporally and spatially controlled. It is clear that NO produced at the heme cofactor of NOS eventually S-nitrosates the zinc tetrathiolate cysteines of all three NOS isoforms *in vitro* and in cell culture. eNOS has additionally been demonstrated to be S-nitrosated on these reactive cysteines in mouse aorta (Erwin *et al.*, 2006).

S-nitrosation of eNOS

eNOS was the first isoform of NOS reported to undergo S-nitrosation (Ravi *et al.*, 2004). In this study, the authors found that after treating bovine aortic endothelial cells (BAECs) and purified eNOS with an exogenous source of NO (100 μM spermine diazeniumdiolate), NOS activity was reduced and an increase in monomer formation was observed. It was rationalized that the mechanism accounting for monomer formation was S-nitrosation of the zinc tetrathiolate center with subsequent destabilization of the cluster. After zinc release, the presence of monomeric eNOS was confirmed by gel filtration. The presence of thioredoxin and thioredoxin reductase prevented loss of eNOS activity. The only direct evidence given for the involvement of the zinc tetrathiolate center in the monomerization reaction was through gel filtration data that contrasted the oligomeric structure of wild-type eNOS and the Cys99Ala mutant, and led to the conclusion that the wild-type protein was a more stable dimer.

These results were subsequently extended, and the results indicated that eNOS S-nitrosation probably occurs *in vivo* (Erwin *et al.*, 2005). First, it was shown that formation of nitrosothiol residues on eNOS in BAECs was a feature of the enzyme in living cells and did not require the addition of an exogenous source of NO. Secondly, eNOS agonists (VEGF and insulin) could dynamically control the S-nitrosation state of eNOS in cell culture (Fig. 3A) and intact blood vessels (Erwin *et al.*, 2006). Treatment of BAECs with eNOS agonists led to a marked decrease in Cys96/Cys101-S-nitroso-eNOS (eNOS-SNO) in a time course similar to that of receptor-mediated eNOS activation, along with a concurrent increase in Ser1179 phosphorylation. These observations of a temporal association of receptor-mediated eNOS denitrosation with eNOS activation suggested the hypothesis that eNOS is inhibited by S-nitrosation. In this model, eNOS-SNO concentration is highest in the basal state (*i.e.*, when production of NO is the lowest). The NOS inhibitor *N*-methyl-ʟ-arginine (ʟ-NMA) was also capable of inhibiting the renitrosation of eNOS in BAECs. *In vitro*, the extent of eNOS-SNO (recombinant bovine aortic) was correlated with a decrease in enzymatic activity. Finally, the study strengthened the case for S-nitrosation of the zinc tetrathiolate cysteines by mutation of Cys96 and/or Cys101 to Ser, and showed that the mutation of either zinc tetrathiolate cysteine residue eliminated endogenous eNOS-SNO in transfected COS-7 cells (Fig. 3B). Importantly, the robustly S-nitrosated eNOS was found to be mostly dimeric, as assessed by nondenaturing polyacrylamide gel electrophoresis (PAGE). This observation argues strongly against the model that the eNOS dimer dissociates following enzyme S-nitrosation. This study (Erwin *et al.*, 2005) of eNOS S-nitrosation was the first documentation of receptor-modulated, fully reversible protein S-nitrosation. Mannick *et al.* (1999) published the first study of receptor-modulated S-nitrosation, in which procaspase-3 was shown to undergo Fas-dependent denitrosation. In light of these two reports (eNOS and procaspase-3), it could be argued that receptor-mediated (de)nitrosation provides support for the existence of nitrosases and denitrosases analogous to the diverse protein kinases and phosphoprotein phosphatases that have been extensively studied.

In a follow-up study on the S-nitrosation of eNOS, matrix-assisted laser desorption/ionization (MALDI) mass spectrometry and collision-induced dissociation were employed to conclusively identify the S-nitrosation of Cys101 of bovine aortic eNOS (Erwin *et al.*, 2006). Efforts to locate Cys96 by mass spectrometry, modified or unmodified, were unsuccessful. Using electrospray ionization (ESI) mass spectrometry, another group identified modification of the same residue in human eNOS (Cys98) (Taldone *et al.*, 2005). Apparently, proteolytic digestion of both human and bovine eNOSs yields peptides containing the N-terminal zinc tetrathiolate cysteine that do not ionize well and are not amenable to analysis by either MALDI or ESI. In addition to proper verification of the residue(s) of eNOS involved in nitrosothiol formation, it was found that eNOS targeting the plasma membrane was required for the formation of eNOS-SNO and that subcellular translocation was necessary for de/re-S-nitrosation of eNOS (Erwin *et al.*, 2006). Additionally, eNOS-SNO was detected in mouse aorta. In these intact blood vessels, levels of eNOS-SNO were dynamically regulated, just as in BAECs, on agonist stimulation. This indicates that the process most likely occurs *in vivo* and is yet another regulatory mechanism for controlling arterial tone and relaxation.

164

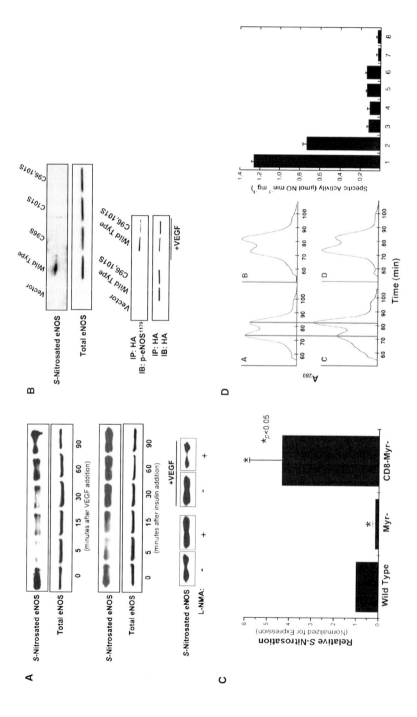

To conclusively show that subcellular targeting is crucial for the state of eNOS S-nitrosation, two mutants were constructed: an acylation-deficient mutant of eNOS (which lacks the ability to be N-myristoylated, Myr^-); and this mutant of eNOS fused to the CD8 transmembrane domain (CD8-Myr^-). Wild-type eNOS dynamically interacts with the caveolae membrane through reversible thiopalmitoylation that is dependent on prior myristoylation. As expected, Myr^- eNOS localized exclusively to the cytosol, while CD8-Myr^- was permanently associated with plasmalemmal caveolae. As a result of differential subcellular localization, CD8-Myr^- eNOS was hyper-S-nitrosated, Myr^- eNOS was

---

Fig. 3. S-Nitrosation of the zinc tetrathiolate of NOS. (A) S-Nitrosation of eNOS is agonist modulated in bovine aortic endothelial cells (BAECs). Shown are Western blots of eNOS immunoprecipitated from lysate and processed using the biotin switch assay. Top: vascular endothelial growth factor (VEGF) treated (10 ng/ml). Middle: insulin treated (1 µM). Bottom: N-methyl-L-arginine (L-NMA) pretreated (5 mM, NOS inhibitor) and VEGF treated (10 ng/ml). L-NMA addition reduced the extent of eNOS renitrosation by almost 60%. (B) Characterization of S-nitrosation and phosphorylation of wild-type eNOS and eNOS mutants at Cys96 and/or Cys101. Top: COS-7 cells were transfected with wild-type eNOS or with eNOS-Cys96S and/or eNOS-Cys101S. eNOS was immunoprecipitated from lysate and processed by the biotin switch method. Biotinylated eNOS was affinity purified with streptavidin–agarose. Immunoblots were probed with anti-eNOS antibody to detect Cys96/Cys101-S-nitroso-eNOS or total eNOS. Bottom: BAEC were transfected with influenza hemagglutinin (HA)-tagged wild-type eNOS and eNOS-Cys96S/101S and later treated with VEGF (10 ng/ml, 5 min). After separation by SDS-PAGE, eNOS was probed with anti-phospho-eNOS-Ser1179 or the HA epitope, as indicated. IP, immunoprecipitate. IB, immunoblot. (C) The extent of eNOS S-nitrosation is dependent on association with the caveolae membrane. After normalizing for total eNOS expression (anti-eNOS Western blot), the relative extent of S-nitrosation was assessed by densitometric analysis of immunoprecipitated eNOS measured by the biotin switch method. Myr^-, an acylation-deficient mutant of eNOS (which lacks the ability to be N-myristoylated). CD8-Myr^-, the Myr^- mutant fused to the CD8 transmembrane domain. (D) iNOS dimer stability and activity is negatively affected by NO. Left: gel filtration of wild-type iNOS – (A) buffer treated; (B) 10 µM diethylamine (DEA)/NO treated; (C) 100 µM DEA/NO treated; (D) 100 µM DEA treated. Increased concentrations of NO lead to increased monomer formation. Right: activity of iNOS as measured by the oxyhemoglobin assay. Odd lanes, DEA-treated (50 µM). Even lanes, DEA/NO-treated (50 µM). Lanes 1 and 2, wild-type; lanes 3 and 4, Cys104Ala; lanes 5 and 6, Cys109Ala; lanes 7 and 8, Cys104Ala/Cys109Ala. Only wild-type protein containing zinc responds to treatment with NO. The cysteine mutants do not bind zinc; therefore, the dimer stability is comprised and lower activity is seen.

not S-nitrosated to any significant extent, and wild-type S-nitrosation was intermediate (Fig. 3C). After agonist stimulation, only wild-type eNOS was denitrosated, suggesting that translocation from the caveolae precedes reactivation of eNOS (Erwin *et al.*, 2006). The denitrosated state of eNOS is short-lived (half-life of ~10 min), as an increase in local NO concentration rapidly leads to renitrosation of eNOS.

S-nitrosation of iNOS

The isoform of NOS that responds to proinflammatory signals is capable of producing cytotoxic levels of NO. After demonstration of S-nitrosation on eNOS, it was not surprising to find that iNOS is also S-nitrosated. At the time of our report on the S-nitrosation of iNOS (Mitchell *et al.*, 2005), it had not yet been conclusively shown that the zinc tetrathiolate region was the moiety involved in monomer formation. Our laboratories carried out experiments to demonstrate that exogenous NO destabilizes dimeric murine iNOS (assessed using gel filtration), with concomitant loss of enzymatic activity measured using the oxyhemoglobin assay (Fig. 3D). A physiologically accessible concentration of NO ($10 \mu M$ DEA/NO yields a maximum concentration of ~3 μM NO at 37°C) significantly increased the dissociation of dimeric iNOS, as assessed by gel filtration at reduced temperature (4°C). This is in contrast to the behavior of the eNOS dimer under similar conditions, although two different methods were used to probe monomer formation. It could be that different NOS isoforms respond differently to S-nitrosation. Zinc loss from dimeric iNOS, as measured by both inductively coupled plasma atomic emission spectroscopy (ICP-AES) and the 4-(pyridyl-2-azo) resorcinol (PAR) colorimetric assay, was correlated to monomer formation and loss of NOS activity. To demonstrate S-nitrosation as the mechanism responsible for these observations, the biotin switch method (Jaffrey and Snyder, 2001) was applied to wild-type murine iNOS and the zinc tetrathiolate mutants Cys104Ala, Cys109Ala, and the double alanine mutant. Mutation of either Cys104 or Cys109 resulted in the inability of iNOS to bind zinc within the detection of our assays and also abolished nitrosothiol formation on iNOS. Proteolytic digestion and mass spectrometry proved that both Cys104 and Cys109 formed nitrosothiols under the conditions employed. Finally, it was found that Cys104/109-*S*-nitroso-iNOS (iNOS-SNO) was detectable in mouse macrophages activated with LPS/INF-γ without the addition of exogenous NO. These data were used to construct the model shown in Fig. 4. This model

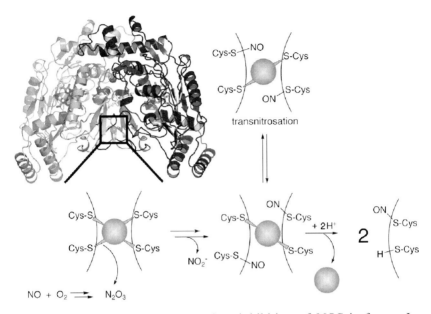

Fig. 4. Proposed model of NO-dependent inhibition of NOS isoforms. In an aerobic environment, NO forms a nitrosothiol on one of the zinc tetrathiolate cysteines of NOS. These cysteines are deprotonated at physiological pH due to ligation to the Zn^{2+} Lewis acid. Modification of one cysteine residue greatly destabilizes the cluster and the zinc ion is released. Intramolecular transnitrosation likely occurs within the cluster before monomers are formed. Monomeric NOS is catalytically inactive.

accurately describes the NO-dependent inhibition of NOS for all three isoforms published to date.

The Snyder laboratory has recently published a report that iNOS binds directly to cyclooxygenase-2 (COX-2) and S-nitrosates COX-2 on Cys526, resulting in a modest increase in prostaglandin formation (Kim *et al.*, 2005). These results bear directly on our understanding of the synergy between reactive nitrogen species and proinflammatory eicosanoids. It was demonstrated by coimmunoprecipitation that the association between iNOS and COX-2 required residues 1–114 of iNOS and residues 484–604 of COX-2. Interestingly, the zinc tetrathiolate cluster is contained in this region (104 and 109) and would be S-nitrosated under their experimental conditions (mouse macrophages, LPS/INF-γ) (Mitchell *et al.*, 2005). COX-2 was not only associated with iNOS in activated macrophages, but was also S-nitrosated with no addition of exogenous NO. The authors were not able to conclusively show that

Cys526 is directly S-nitrosated since this was probed by mutagenesis, and one other cysteine mutation, Cys555Ser, resulted in complete loss of COX-2 catalytic activity. However, the possibly of a direct transnitrosation reaction between these two inflammatory mediators is intriguing. This reaction would theoretically be similar to that shown between the apoptotic mediators thioredoxin and caspase-3 (Mitchell and Marletta, 2005). If the iNOS/COX-2 transnitrosation reaction were demonstrated *in vivo*, it would further support a role for specific transnitrosation reactions being a key component to nonclassical NO signaling.

Also of interest is the demonstration that procaspase-3 and iNOS have an NO-dependent protein–protein interaction (Matsumoto *et al.*, 2003). By utilizing a specially designed yeast-2-hydrid screen with procaspase-3 as bait, it was found that endogenously produced NO leads to an association of procaspase-3 to residues 455–647 of iNOS. This domain of iNOS is conserved between all of the isoforms as it is responsible for CaM affinity (Zhang and Vogel, 1994). The anti-procaspase-3 antibody was then used to successfully coimmunoprecipitate both nNOS and eNOS from human cells (HEK-293 cell line). Is NOS responsible for a direct transnitrosation reaction with procaspase-3? Is thioredoxin then responsible for reactivation of caspase-3 when an apoptotic signal is encountered? Taking into account the potential transnitrosation reaction between NOS and COX-2, could it also be that the original nitrosating agent is formed directly on NOS itself, where the local concentration of NO is the highest and the possibility of side reactions that destroy NO is the lowest? If this turns out to be the case, it would considerably change the current dogma of NO signaling by providing a direct and dual role for NOS in both classical and nonclassical signalings (Fig. 5).

S-nitrosation of nNOS

Modification of nNOS by NO has not been as well studied as eNOS or iNOS. However, nNOS does form nitrosothiols in brain tissue (Jaffrey *et al.*, 2001). The widely used biotin switch assay was first developed and validated using brain lysate from $nNOS^{+/+}$ and $nNOS^{-/-}$ mice. In this original report, the authors found that nNOS was responsible for S-nitrosation of a variety of proteins, including actin, tubulin, glyceraldehyde-3-phosphate dehydrogenase (GAPDH), creatine kinases, and several ion channels. nNOS has also been suspected to affect the localization and function of glucokinase (GK) (Rizzo and Piston, 2003). GK is vital to glucose-stimulated insulin release from pancreatic β-cells. Rizzo and colleagues demonstrated that insulin stimulates the production

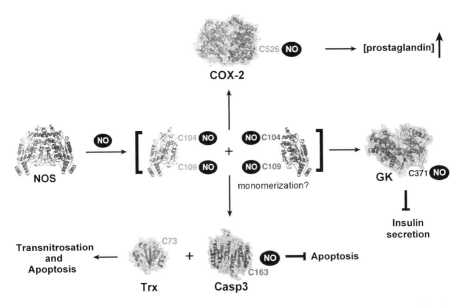

Fig. 5. Potential NOS involvement in cellular nitrosothiol relay. NOS is the source of mammalian NO and undergoes spontaneous S-nitrosation *in vivo.* Through many known protein–protein interactions, it is plausible that NOS directly transfers nitrosothiols to cysteine residues of other proteins with disparate physiological outcomes. Dissociation of the NOS dimer after S-nitrosation *in vivo* requires further study, but *in vitro* reports indicate that it is plausible. This figure was constructed with PyMol (DeLanoScientific LLC, San Francisco, USA) using Protein DataBank entries 1NSI, 1CVU, 1V4S, 1NME, and 1ERU. Casp3, caspase-3. COX-2, cyclooxygenase-2. GK, glucokinase. Trx, thioredoxin-1. (See Colour Plate Section in this book).

of NO in cell culture and leads to the formation of GK-SNO on Cys371 (shown by mutation of all cysteines to serine). The mechanism for GK S-nitrosation could very well be direct transnitrosation with nNOS, because a stable interaction was demonstrated between GK and nNOS by both coimmunoprecipitation and fluorescence resonance energy transfer (FRET).

Implications and conclusions

NO mediates blood vessel relaxation, complex aspects of myocardial function, perfusion and function of all major organs, synaptic plasticity in the brain, platelet aggregation, skin function, and numerous other physiological processes. It is becoming increasingly clear that NO signaling involves targets in addition to soluble guanylyl cyclase (sGC). Reports to date point to S-nitrosation as an alternate explanation for the cGMP-independent (nonclassical) effects of NO. Given the role of NO in human biology, a complete understanding of the molecular details involved in its signaling will have clear application to the understanding and treatment of a broad spectrum of diseases. While previous work on the S-nitrosation of NOS isoforms, in addition to other proteins, shows the importance of nitrosothiols in many physiological functions, essential questions remain to be resolved. The main questions are what determines the specificity in signaling, how are protein nitrosothiols formed, and is the process regulated? Knowledge of the molecular details of *in vivo* S-nitrosation biochemistry could lead to new therapeutic strategies to treat cancer, neurodegeneration, diabetes, autoimmune disorders, and other inflammatory diseases (Foster *et al.*, 2003).

With regard to inflammation, the synergy between iNOS and COX-2 could represent a general mechanism in inflammatory responses. Could anti-inflammatory drugs be designed that block the iNOS/COX-2 protein–protein interaction (Kim *et al.*, 2005)? These drugs could plausibly work in combination with other COX inhibitors. If synergism with existing COX-2-selective drugs is exhibited, then lower doses may be able to be administered, allowing the negative cardiovascular side effects of the COX-2 drugs to be circumvented (Weir *et al.*, 2003).

In general, the goal in the field of S-nitrosation should be to develop a complete molecular-level view of S-nitrosation processes. The extension of this work into understanding their physiological function will provide a rational basis for the understanding and treatment of human disease.

Acknowledgments

This work is supported in part by grants from the NIH to T.M. (GM 36259 and HL46457) and M.A.M. and also the DeBenedictis Fund of UC Berkeley to M.A.M. D.A.M. is supported by an American Heart Association pre-doctoral fellowship.

References

Abu-Soud, H. M., Ichimori, K., Nakazawa, H. and Stuehr, D. J. (2001). Regulation of inducible nitric oxide synthase by self-generated NO. *Biochemistry* 40,6876–6881.
Adak, S., Santolini, J., Tikunova, S., Wang, Q., Johnson, J. D. and Stuehr, D. J. (2001). Neuronal nitric-oxide synthase mutant (Ser-1412 → Asp) demonstrates surprising

connections between heme reduction, NO complex formation, and catalysis. *J. Biol. Chem.* 276,1244–1252.

Adams, M. E., Mueller, H. A. and Froehner, S. C. (2001). In vivo requirement of the alpha-syntrophin PDZ domain for the sarcolemmal localization of nNOS and aquaporin-4. *J. Cell Biol.* 155,113–122.

Aitken, A., Baxter, H., Dubois, T., Clokie, S., Mackie, S., Mitchell, K., Peden, A. and Zemlickova, E. (2002). Specificity of 14-3-3 isoform dimer interactions and phosphorylation. *Biochem. Soc. Trans.* 30,351–360.

Aktan, F. (2004). iNOS-mediated nitric oxide production and its regulation. *Life Sci.* 75,639–653.

Alderton, W. K., Cooper, C. E. and Knowles, R. G. (2001). Nitric oxide synthases: Structure, function and inhibition. *Biochem. J.* 357,593–615.

Al-Shabrawey, M., El-Remessy, A., Gu, X., Brooks, S. S., Hamed, M. S., Huang, P. and Caldwell, R. B. (2003). Normal vascular development in mice deficient in endothelial NO synthase: Possible role of neuronal NO synthase. *Mol. Vis.* 9,549–558.

Belhassen, L., Feron, O., Kaye, D. M., Michel, T. and Kelly, R. A. (1997). Regulation by cAMP of post-translational processing and subcellular targeting of endothelial nitric-oxide synthase (type 3) in cardiac myocytes. *J. Biol. Chem.* 272,11198–11204.

Boulanger, C. M., Heymes, C., Benessiano, J., Geske, R. S., Levy, B. I. and Vanhoutte, P. M. (1998). Neuronal nitric oxide synthase is expressed in rat vascular smooth muscle cells: Activation by angiotensin II in hypertension. *Circ. Res.* 83,1271–1278.

Buga, G. M., Griscavage, J. M., Rogers, N. E. and Ignarro, L. J. (1993). Negative feedback regulation of endothelial cell function by nitric oxide. *Circ. Res.* 73, 808–812.

Burnett, A. L., Nelson, R. J., Calvin, D. C., Liu, J. X., Demas, G. E., Klein, S. L., Kriegsfeld, L. J., Dawson, V. L., Dawson, T. M. and Snyder, S. H. (1996). Nitric oxide-dependent penile erection in mice lacking neuronal nitric oxide synthase. *Mol. Med.* 2,288–296.

Burney, S., Caulfield, J. L., Niles, J. C., Wishnok, J. S. and Tannenbaum, S. R. (1999). The chemistry of DNA damage from nitric oxide and peroxynitrite. *Mutat. Res.* 424,37–49.

Busconi, L. and Michel, T. (1993). Endothelial nitric oxide synthase. N-terminal myristoylation determines subcellular localization. *J. Biol. Chem.* 268,8410–8413.

Cauwels, A., Janssen, B., Buys, E., Sips, P. and Brouckaert, P. (2006). Anaphylactic shock depends on PI3K and eNOS-derived NO. *J. Clin. Invest.* 116,2244–2251.

Chen, Y., Irie, Y., Keung, W. M. and Maret, W. (2002a). *S*-nitrosothiols react preferentially with zinc thiolate clusters of metallothionein III through transnitrosation. *Biochemistry* 41,8360–8367.

Chen, Y., Panda, K. and Stuehr, D. J. (2002b). Control of nitric oxide synthase dimer assembly by a heme-NO-dependent mechanism. *Biochemistry* 41,4618–4625.

Cho, H. J., Martin, E., Xie, Q. W., Sassa, S. and Nathan, C. (1995). Inducible nitric oxide synthase: Identification of amino acid residues essential for dimerization and binding of tetrahydrobiopterin. *Proc. Natl. Acad. Sci. USA* 92,11514–11518.

Cho, H. J., Xie, Q. W., Calaycay, J., Mumford, R. A., Swiderek, K. M., Lee, T. D. and Nathan, C. (1992). Calmodulin is a subunit of nitric oxide synthase from macrophages. *J. Exp. Med.* 176,599–604.

Chou, W. Y. and Matthews, K. S. (1989). Serine to cysteine mutations in trp repressor protein alter tryptophan and operator binding. *J. Biol. Chem.* 264,18314–18319.

Corson, M. A., James, N. L., Latta, S. E., Nerem, R. M., Berk, B. C. and Harrison, D. G. (1996). Phosphorylation of endothelial nitric oxide synthase in response to fluid shear stress. *Circ. Res.* 79,984–991.

Crane, B. R., Arvai, A. S., Gachhui, R., Wu, C., Ghosh, D. K., Getzoff, E. D., Stuehr, D. J. and Tainer, J. A. (1997). The structure of nitric oxide synthase oxygenase domain and inhibitor complexes. *Science* 278,425–431.

Crane, B. R., Arvai, A. S., Ghosh, D. K., Wu, C., Getzoff, E. D., Stuehr, D. J. and Tainer, J. A. (1998). Structure of nitric oxide synthase oxygenase dimer with pterin and substrate. *Science* 279,2121–2126.

Davies, K. M., Wink, D. A., Saavedra, J. E. and Keefer, L. K. (2001). Chemistry of the diazeniumdiolates. 2. Kinetics and mechanism of dissociation to nitric oxide in aqueous solution. *J. Am. Chem. Soc.* 123,5473–5481.

Demas, G. E., Eliasson, M. J., Dawson, T. M., Dawson, V. L., Kriegsfeld, L. J., Nelson, R. J. and Snyder, S. H. (1997). Inhibition of neuronal nitric oxide synthase increases aggressive behavior in mice. *Mol. Med.* 3,610–616.

De Stefano, D., Maiuri, M. C., Iovine, B., Ialenti, A., Bevilacqua, M. A. and Carnuccio, R. (2006). The role of NF-kappaB, IRF-1, and STAT-1alpha transcription factors in the iNOS gene induction by gliadin and IFN-gamma in RAW 264.7 macrophages. *J. Mol. Med.* 84,65–74.

Dimmeler, S., Fleming, I., Fisslthaler, B., Hermann, C., Busse, R. and Zeiher, A. M. (1999). Activation of nitric oxide synthase in endothelial cells by Akt-dependent phosphorylation. *Nature* 399,601–605.

Dudev, T. and Lim, C. (2002). Factors governing the protonation state of cysteines in proteins: An ab initio/CDM study. *J. Am. Chem. Soc.* 124,6759–6766.

Dudzinski, D. M., Igarashi, J., Greif, D. and Michel, T. (2006). The regulation and pharmacology of endothelial nitric oxide synthase. *Annu. Rev. Pharmacol. Toxicol.* 46,235–276.

El-Gayar, S., Thuring-Nahler, H., Pfeilschifter, J., Rollinghoff, M. and Bogdan, C. (2003). Translational control of inducible nitric oxide synthase by IL-13 and arginine availability in inflammatory macrophages. *J. Immunol.* 171,4561–4568.

Ellis, E. A., Guberski, D. L., Hutson, B. and Grant, M. B. (2002). Time course of NADH oxidase, inducible nitric oxide synthase and peroxynitrite in diabetic retinopathy in the BBZ/WOR rat. *Nitric Oxide* 6,295–304.

Erwin, P. A., Lin, A. J., Golan, D. E. and Michel, T. (2005). Receptor-regulated dynamic S-nitrosylation of endothelial nitric-oxide synthase in vascular endothelial cells. *J. Biol. Chem.* 280,19888–19894.

Erwin, P. A., Mitchell, D. A., Sartoretto, J., Marletta, M. A. and Michel, T. (2006). Subcellular targeting and differential S-nitrosylation of endothelial nitric-oxide synthase. *J. Biol. Chem.* 281,151–157.

Fagan, K. A., McMurtry, I. and Rodman, D. M. (2000). Nitric oxide synthase in pulmonary hypertension: Lessons from knockout mice. *Physiol. Res.* 49,539–548.

Fang, M., Jaffrey, S. R., Sawa, A., Ye, K., Luo, X. and Snyder, S. H. (2000). Dexras1: A G protein specifically coupled to neuronal nitric oxide synthase via CAPON. *Neuron* 28,183–193.

Faux, M. C. and Scott, J. D. (1996). More on target with protein phosphorylation: Conferring specificity by location. *Trends Biochem. Sci.* 21,312–315.

Feron, O., Belhassen, L., Kobzik, L., Smith, T. W., Kelly, R. A. and Michel, T. (1996). Endothelial nitric oxide synthase targeting to caveolae. Specific interactions with caveolin isoforms in cardiac myocytes and endothelial cells. *J. Biol. Chem.* 271, 22810–22814.

Fleming, I., Fisslthaler, B., Dimmeler, S., Kemp, B. E. and Busse, R. (2001). Phosphorylation of Thr(495) regulates Ca(2+)/calmodulin-dependent endothelial nitric oxide synthase activity. *Circ. Res.* 88,E68–E75.

Flinspach, M., Li, H., Jamal, J., Yang, W., Huang, H., Silverman, R. B. and Poulos, T. L. (2004). Structures of the neuronal and endothelial nitric oxide synthase heme domain with D-nitroarginine-containing dipeptide inhibitors bound. *Biochemistry* 43, 5181–5187.

Forstermann, U., Boissel, J. P. and Kleinert, H. (1998). Expressional control of the 'constitutive' isoforms of nitric oxide synthase (NOS I and NOS III). *FASEB J.* 12,773–790.

Foster, M. W., McMahon, T. J. and Stamler, J. S. (2003). S-nitrosylation in health and disease. *Trends Mol. Med.* 9,160–168.

Fulton, D., Fontana, J., Sowa, G., Gratton, J. P., Lin, M., Li, K. X., Michell, B., Kemp, B. E., Rodman, D. and Sessa, W. C. (2002). Localization of endothelial nitric-oxide synthase phosphorylated on serine 1179 and nitric oxide in Golgi and plasma membrane defines the existence of two pools of active enzyme. *J. Biol. Chem.* 277,4277–4284.

Fulton, D., Gratton, J. P., McCabe, T. J., Fontana, J., Fujio, Y., Walsh, K., Franke, T. F., Papapetropoulos, A. and Sessa, W. C. (1999). Regulation of endothelium-derived nitric oxide production by the protein kinase Akt. *Nature* 399,597–601.

Fulton, D., Gratton, J. P., Sessa, W. C. (2001). Post-translational control of endothelial nitric oxide synthase: Why isn't calcium/calmodulin enough? *J. Pharmacol. Exp. Ther.* 299, 818–824.

Gallis, B., Corthals, G. L., Goodlett, D. R., Ueba, H., Kim, F., Presnell, S. R., Figeys, D., Harrison, D. G., Berk, B. C., Aebersold, R. and Corson, M. A. (1999). Identification of flow-dependent endothelial nitric-oxide synthase phosphorylation sites by mass spectrometry and regulation of phosphorylation and nitric oxide production by the phosphatidylinositol 3-kinase inhibitor LY294002. *J. Biol. Chem.* 274,30101–30108.

Garcin, E. D., Bruns, C. M., Lloyd, S. J., Hosfield, D. J., Tiso, M., Gachhui, R., Stuehr, D. J., Tainer, J. A. and Getzoff, E. D. (2004). Structural basis for isozyme-specific regulation of electron transfer in nitric-oxide synthase. *J. Biol. Chem.* 279,37918–37927.

Gergel, D. and Cederbaum, A. I. (1996). Inhibition of the catalytic activity of alcohol dehydrogenase by nitric oxide is associated with S-nitrosylation and the release of zinc. *Biochemistry* 35,16186–16194.

Gonzalez, E., Kou, R., Lin, A. J., Golan, D. E. and Michel, T. (2002). Subcellular targeting and agonist-induced site-specific phosphorylation of endothelial nitric-oxide synthase. *J. Biol. Chem.* 277,39554–39560.

Gow, A. J., Chen, Q., Hess, D. T., Day, B. J., Ischiropoulos, H. and Stamler, J. S. (2002). Basal and stimulated protein S-nitrosylation in multiple cell types and tissues. *J. Biol. Chem.* 277,9637–9640.

Greif, D. M., Kou, R. and Michel, T. (2002). Site-specific dephosphorylation of endo-thelial nitric oxide synthase by protein phosphatase 2A: Evidence for crosstalk between phosphorylation sites. *Biochemistry* 41,15845–15853.

Gryglewski, R. J., Uracz, W., Chlopicki, S. and Marcinkiewicz, E. (2002). Bradykinin as a major endogenous regulator of endothelial function. *Pediatr. Pathol. Mol. Med.* 21,279–290.

Gyurko, R., Leupen, S. and Huang, P. L. (2002). Deletion of exon 6 of the neuronal nitric oxide synthase gene in mice results in hypogonadism and infertility. *Endocrinology* 143,2767–2774.

Hannan, R. L., John, M. C., Kouretas, P. C., Hack, B. D., Matherne, G. P. and Laubach, V. E. (2000). Deletion of endothelial nitric oxide synthase exacerbates myocardial stunning in an isolated mouse heart model. *J. Surg. Res.* 93,127–132.

Harris, M. B., Ju, H., Venema, V. J., Liang, H., Zou, R., Michell, B. J., Chen, Z. P., Kemp, B. E. and Venema, R. C. (2001). Reciprocal phosphorylation and regulation of endothelial nitric-oxide synthase in response to bradykinin stimulation. *J. Biol. Chem.* 276,16587–16591.

Hausel, P., Latado, H., Courjault-Gautier, F. and Felley-Bosco, E. (2006). Src-mediated phosphorylation regulates subcellular distribution and activity of human inducible nitric oxide synthase. *Oncogene* 25,198–206.

He, J. J. and Quiocho, F. A. (1991). A nonconservative serine to cysteine mutation in the sulfate-binding protein, a transport receptor. *Science* 251,1479–1481.

Hemmens, B., Goessler, W., Schmidt, K. and Mayer, B. (2000). Role of bound zinc in dimer stabilization but not enzyme activity of neuronal nitric-oxide synthase. *J. Biol. Chem.* 275,35786–35791.

Hess, D. T., Matsumoto, A., Kim, S. O., Marshall, H. E. and Stamler, J. S. (2005). Protein S-nitrosylation: Purview and parameters. *Nat. Rev. Mol. Cell. Biol.* 6,150–166.

Hevel, J. M., White, K. A. and Marletta, M. A. (1991). Purification of the inducible murine macrophage nitric oxide synthase. Identification as a flavoprotein. *J. Biol. Chem.* 266,22789–22791.

Hofer, H. W. (1996). Conservation, evolution, and specificity in cellular control by pro-tein phosphorylation. *Experientia* 52,449–454.

Huang, P. L. (2000). Lessons learned from nitric oxide synthase knockout animals. *Semin. Perinatol.* 24,87–90.

Huang, P. L., Dawson, T. M., Bredt, D. S., Snyder, S. H. and Fishman, M. C. (1993). Targeted disruption of the neuronal nitric oxide synthase gene. *Cell* 75,1273–1286.

Huang, P. L., Huang, Z., Mashimo, H., Bloch, K. D., Moskowitz, M. A., Bevan, J. A. and Fishman, M. C. (1995). Hypertension in mice lacking the gene for endothelial nitric oxide synthase. *Nature* 377,239–242.

Hurshman, A. R., Krebs, C., Edmondson, D. E., Huynh, B. H. and Marletta, M. A. (1999). Formation of a pterin radical in the reaction of the heme domain of inducible nitric oxide synthase with oxygen. *Biochemistry* 38,15689–15696.

Hurshman, A. R. and Marletta, M. A. (1995). Nitric oxide complexes of inducible nitric oxide synthase: Spectral characterization and effect on catalytic activity. *Biochemistry* 34,5627–5634.

Hutchinson, P. J., Palmer, R. M. and Moncada, S. (1987). Comparative pharmacology of EDRF and nitric oxide on vascular strips. *Eur. J. Pharmacol.* 141,445–451.

175

Igietseme, J. U., Perry, L. L., Ananaba, G. A., Uriri, I. M., Ojior, O. O., Kumar, S. N. and Caldwell, H. D. (1998). Chlamydial infection in inducible nitric oxide synthase knockout mice. *Infect. Immun.* 66,1282–1286.

Jaffrey, S. R., Benfenati, F., Snowman, A. M., Czernik, A. J. and Snyder, S. H. (2002). Neuronal nitric-oxide synthase localization mediated by a ternary complex with synapsin and CAPON. *Proc. Natl. Acad. Sci. USA* 99,3199–3204.

Jaffrey, S. R., Erdjument-Bromage, H., Ferris, C. D., Tempst, P. and Snyder, S. H. (2001). Protein S-nitrosylation: A physiological signal for neuronal nitric oxide. *Nat. Cell Biol.* 3,193–197.

Jaffrey, S. R., Snowman, A. M., Eliasson, M. J., Cohen, N. A. and Snyder, S. H. (1998). CAPON: A protein associated with neuronal nitric oxide synthase that regulates its interactions with PSD95. *Neuron* 20,115–124.

Jaffrey, S. R. and Snyder, S. H. (2001). The biotin switch method for the detection of S-nitrosylated proteins. *Sci. STKE* 2001,PL1.

Kavdia, M. and Popel, A. S. (2004). Contribution of nNOS- and eNOS-derived NO to microvascular smooth muscle NO exposure. *J. Appl. Physiol.* 97,293–301.

Keshive, M., Singh, S., Wishnok, J. S., Tannenbaum, S. R. and Deen, W. M. (1996). Kinetics of S-nitrosation of thiols in nitric oxide solutions. *Chem. Res. Toxicol.* 9, 988–993.

Kim, S. F., Huri, D. A. and Snyder, S. H. (2005). Inducible nitric oxide synthase binds, S-nitrosylates, and activates cyclooxygenase-2. *Science* 310,1966–1970.

Klatt, P., Pfeiffer, S., List, B. M., Lehner, D., Glatter, O., Bachinger, H. P., Werner, E. R., Schmidt, K. and Mayer, B. (1996). Characterization of heme-deficient neuronal nitric-oxide synthase reveals a role for heme in subunit dimerization and binding of the amino acid substrate and tetrahydrobiopterin. *J. Biol. Chem.* 271, 7336–7342.

Klatt, P., Schmidt, K., Lehner, D., Glatter, O., Bachinger, H. P. and Mayer, B. (1995). Structural analysis of porcine brain nitric oxide synthase reveals a role for tetrahydrobiopterin and L-arginine in the formation of an SDS-resistant dimer. *EMBO J.* 14,3687–3695.

Klein, S. L., Carnovale, D., Burnett, A. L., Wallach, E. E., Zacur, H. A., Crone, J. K., Dawson, V. L., Nelson, R. J. and Dawson, T. M. (1998). Impaired ovulation in mice with targeted deletion of the neuronal isoform of nitric oxide synthase. *Mol. Med.* 4,658–664.

Kleinert, H., Euchenhofer, C., Ihrig-Biedert, I. and Forstermann, U. (1996). In murine 3T3 fibroblasts, different second messenger pathways resulting in the induction of NO synthase II (iNOS) converge in the activation of transcription factor NF-kappaB. *J. Biol. Chem.* 271,6039–6044.

Komeima, K., Hayashi, Y., Naito, Y. and Watanabe, Y. (2000). Inhibition of neuronal nitric-oxide synthase by calcium/calmodulin-dependent protein kinase IIalpha through Ser847 phosphorylation in NG108-15 neuronal cells. *J. Biol. Chem.* 275, 28139–28143.

Krahenbuhl, J. and Adams, L. B. (2000). Exploitation of gene knockout mice models to study the pathogenesis of leprosy. *Lepr. Rev.* 71(Suppl.), S170–S175.

Lancaster, J. R., Jr. and Gaston, B. (2004). NO and nitrosothiols: Spatial confinement and free diffusion. *Am. J. Physiol. Lung Cell. Mol. Physiol.* 287,L465–L466.

Lane, P., Hao, G. and Gross, S. S. (2001). S-nitrosylation is emerging as a specific and fundamental posttranslational protein modification: Head-to-head comparison with *O*-phosphorylation. *Sci. STKE* 2001,RE1.

Lee, C., Miura, K., Liu, X. and Zweier, J. L. (2000). Biphasic regulation of leukocyte superoxide generation by nitric oxide and peroxynitrite. *J. Biol. Chem.* 275,38965–38972.

Li, H., Raman, C. S., Glaser, C. B., Blasko, E., Young, T. A., Parkinson, J. F., Whitlow, M. and Poulos, T. L. (1999). Crystal structures of zinc-free and -bound heme domain of human inducible nitric-oxide synthase. Implications for dimer stability and comparison with endothelial nitric-oxide synthase. *J. Biol. Chem.* 274,21276–21284.

Lin, M. I., Fulton, D., Babbitt, R., Fleming, I., Busse, R., Pritchard, K. A., Jr. and Sessa, W. C. (2003). Phosphorylation of threonine 497 in endothelial nitric-oxide synthase coordinates the coupling of L-arginine metabolism to efficient nitric oxide production. *J. Biol. Chem.* 278,44719–44726.

Liu, L., Hausladen, A., Zeng, M., Que, L., Heitman, J. and Stamler, J. S. (2001). A metabolic enzyme for *S*-nitrosothiol conserved from bacteria to humans. *Nature* 410,490–494.

Lowenstein, C. J. and Michel, T. (2006). What's in a name? eNOS and anaphylactic shock. *J. Clin. Invest.* 116,2075–2078.

Lu, H. S., Clogston, C. L., Narhi, L. O., Merewether, L. A., Pearl, W. R. and Boone, T. C. (1992). Folding and oxidation of recombinant human granulocyte colony stimulating factor produced in *Escherichia coli*. Characterization of the disulfide-reduced intermediates and cysteine–serine analogs. *J. Biol. Chem.* 267,8770–8777.

Ludwig, M. L. and Marletta, M. A. (1999). A new decoration for nitric oxide synthase – a Zn(Cys)4 site. *Structure* 7,R73–R79.

Mannick, J. B., Hausladen, A., Liu, L., Hess, D. T., Zeng, M., Miao, Q. X., Kane, L. S., Gow, A. J. and Stamler, J. S. (1999). Fas-induced caspase denitrosylation. *Science* 284,651–654.

Mannick, J. B., Miao, X. Q. and Stamler, J. S. (1997). Nitric oxide inhibits Fas-induced apoptosis. *J. Biol. Chem.* 272,24125–24128.

Maret, W. (2004). Zinc and sulfur: A critical biological partnership. *Biochemistry* 43,3301–3309.

Marletta, M. A., Hurshman, A. R. and Rusche, K. M. (1998). Catalysis by nitric oxide synthase. *Curr. Opin. Chem. Biol.* 2,656–663.

Mashimo, H. and Goyal, R. K. (1999). Lessons from genetically engineered animal models. IV. Nitric oxide synthase gene knockout mice. *Am. J. Physiol.* 277,G745–G750.

Matsumoto, A., Comatas, K. E., Liu, L. and Stamler, J. S. (2003). Screening for nitric oxide-dependent protein–protein interactions. *Science* 301,657–661.

McCleverty, J. A. (2004). Chemistry of nitric oxide relevant to biology. *Chem. Rev.* 104,403–418.

Michel, T. (1999). Targeting and translocation of endothelial nitric oxide synthase. *Braz. J. Med. Biol. Res.* 32,1361–1366.

Michel, T., Li, G. K. and Busconi, L. (1993). Phosphorylation and subcellular translocation of endothelial nitric oxide synthase. *Proc. Natl. Acad. Sci. USA* 90,6252–6256.

Michell, B. J., Harris, M. B., Chen, Z. P., Ju, H., Venema, V. J., Blackstone, M. A., Huang, W., Venema, R. C. and Kemp, B. E. (2002). Identification of regulatory sites of phosphorylation of the bovine endothelial nitric-oxide synthase at serine 617 and serine 635. *J. Biol. Chem.* 277,42344–42351.

Miranda, A., Garcia, J., Lopez, C., Gordon, D., Palacio, C., Restrepo, G., Ortiz, J., Montoya, G., Cardeno, C., Calle, J., Lopez, M., Campo, O., Bedoya, G., Ruiz-Linares, A. and Ospina-Duque, J. (2006). Putative association of the carboxy-terminal PDZ ligand of neuronal nitric oxide synthase gene (CAPON) with schizophrenia in a Colombian population. *Schizophr. Res.* 82,283–285.

Mitchell, D. A., Erwin, P. A., Michel, T. and Marletta, M. A. (2005). S-nitrosation and regulation of inducible nitric oxide synthase. *Biochemistry* 44,4636–4647.

Mitchell, D. A. and Marletta, M. A. (2005). Thioredoxin catalyzes the S-nitrosation of the caspase-3 active site cysteine. *Nat. Chem. Biol.* 1,154–158.

Morishita, T., Tsutsui, M., Shimokawa, H., Sabanai, K., Tasaki, H., Suda, O., Nakata, S., Tanimoto, A., Wang, K. Y., Ueta, Y., Sasaguri, Y., Nakashima, Y. and Yanagihara, N. (2005). Nephrogenic diabetes insipidus in mice lacking all nitric oxide synthase isoforms. *Proc. Natl. Acad. Sci. USA* 102,10616–10621.

Morris, K. R., Lutz, R. D., Choi, H. S., Kamitani, T., Chmura, K. and Chan, E. D. (2003). Role of the NF-kappaB signaling pathway and kappaB *cis*-regulatory elements on the IRF-1 and iNOS promoter regions in mycobacterial lipoarabinomannan induction of nitric oxide. *Infect. Immun.* 71,1442–1452.

Nishiya, T., Uehara, T., Kaneko, M. and Nomura, Y. (2000). Involvement of nuclear factor-kappaB (NF-kappaB) signaling in the expression of inducible nitric oxide synthase (iNOS) gene in rat C6 glioma cells. *Biochem. Biophys. Res. Commun.* 275, 268–273.

Osuka, K., Watanabe, Y., Usuda, N., Nakazawa, A., Fukunaga, K., Miyamoto, E., Takayasu, M., Tokuda, M. and Yoshida, J. (2002). Phosphorylation of neuronal nitric oxide synthase at Ser847 by CaM-KII in the hippocampus of rat brain after transient forebrain ischemia. *J. Cereb. Blood Flow Metab.* 22,1098–1106.

Panda, K., Ghosh, S. and Stuehr, D. J. (2001). Calmodulin activates intersubunit electron transfer in the neuronal nitric-oxide synthase dimer. *J. Biol. Chem.* 276,23349–23356.

Raman, C. S., Li, H., Martasek, P., Kral, V., Masters, B. S. and Poulos, T. L. (1998). Crystal structure of constitutive endothelial nitric oxide synthase: A paradigm for pterin function involving a novel metal center. *Cell* 95,939–950.

Raman, C. S., Li, H., Martasek, P., Southan, G., Masters, B. S. and Poulos, T. L. (2001). Crystal structure of nitric oxide synthase bound to nitro indazole reveals a novel inactivation mechanism. *Biochemistry* 40,13448–13455.

Ravi, K., Brennan, L. A., Levic, S., Ross, P. A. and Black, S. M. (2004). S-nitrosylation of endothelial nitric oxide synthase is associated with monomerization and decreased enzyme activity. *Proc. Natl. Acad. Sci. USA* 101,2619–2624.

Rizzo, M. A. and Piston, D. W. (2003). Regulation of beta cell glucokinase by S-nitrosylation and association with nitric oxide synthase. *J. Cell Biol.* 161,243–248.

Robinson, L. J., Busconi, L. and Michel, T. (1995). Agonist-modulated palmitoylation of endothelial nitric oxide synthase. *J. Biol. Chem.* 270,995–998.

Robinson, L. J., Ghanouni, P. and Michel, T. (1996). Posttranslational modifications of endothelial nitric oxide synthase. *Meth. Enzymol.* 268,436–448.

Rozema, D. B. and Poulter, C. D. (1999). Yeast protein farnesyltransferase. pKas of peptide substrates bound as zinc thiolates. *Biochemistry* 38,13138–13146.

178

Saito, S., Hirata, Y., Emori, T., Imai, T. and Marumo, F. (1996). Angiotensin II activates endothelial constitutive nitric oxide synthase via AT1 receptors. *Hypertens. Res.* 19,201–206.

Sanghani, P. C., Bosron, W. F. and Hurley, T. D. (2002a). Human glutathione-dependent formaldehyde dehydrogenase. Structural changes associated with ternary complex formation. *Biochemistry* 41,15189–15194.

Sanghani, P. C., Robinson, H., Bosron, W. F. and Hurley, T. D. (2002b). Human glutathione-dependent formaldehyde dehydrogenase. Structures of apo, binary, and inhibitory ternary complexes. *Biochemistry* 41,10778–10786.

Sanghani, P. C., Stone, C. L., Ray, B. D., Pindel, E. V., Hurley, T. D. and Bosron, W. F. (2000). Kinetic mechanism of human glutathione-dependent formaldehyde dehydrogenase. *Biochemistry* 39,10720–10729.

Sanz, M. J., Hickey, M. J., Johnston, B., McCafferty, D. M., Raharjo, E., Huang, P. L. and Kubes, P. (2001). Neuronal nitric oxide synthase (NOS) regulates leukocyte–endothelial cell interactions in endothelial NOS deficient mice. *Br. J. Pharmacol.* 134, 305–312.

Shao, L. E., Tanaka, T., Gribi, R. and Yu, J. (2002). Thioredoxin-related regulation of NO/NOS activities. *Ann. N. Y. Acad. Sci.* 962,140–150.

Snyder, S. H. and Bredt, D. S. (1991). Nitric oxide as a neuronal messenger. *Trends Pharmacol. Sci.* 12,125–128.

Stuehr, D. J. and Marletta, M. A. (1985). Mammalian nitrate biosynthesis: Mouse macrophages produce nitrite and nitrate in response to *Escherichia coli* lipopolysaccharide. *Proc. Natl. Acad. Sci. USA* 82,7738–7742.

Stuehr, D. J. and Marletta, M. A. (1987). Induction of nitrite/nitrate synthesis in murine macrophages by BCG infection, lymphokines, or interferon-gamma. *J. Immunol.* 139,518–525.

Sun, J., Xin, C., Eu, J. P., Stamler, J. S. and Meissner, G. (2001). Cysteine-3635 is responsible for skeletal muscle ryanodine receptor modulation by NO. *Proc. Natl. Acad. Sci. USA* 98,11158–11162.

Szabo, C. (2003). Multiple pathways of peroxynitrite cytotoxicity. *Toxicol. Lett.* 140–141,105–112.

Taldone, F. S., Tummala, M., Goldstein, E. J., Ryzhov, V., Ravi, K. and Black, S. M. (2005). Studying the S-nitrosylation of model peptides and eNOS protein by mass spectrometry. *Nitric Oxide* 13,176–187.

Vallance, B. A., Dijkstra, G., Qiu, B., van der Waaij, L. A., van Goor, H., Jansen, P. L., Mashimo, H. and Collins, S. M. (2004). Relative contributions of NOS isoforms during experimental colitis: Endothelial-derived NOS maintains mucosal integrity. *Am. J. Physiol. Gastrointest. Liver Physiol.* 287,G865–G874.

Venema, R. C. (2002). Post-translational mechanisms of endothelial nitric oxide synthase regulation by bradykinin. *Int. Immunopharmacol.* 2,1755–1762.

Virag, L., Szabo, E., Gergely, P. and Szabo, C. (2003). Peroxynitrite-induced cytotoxicity: Mechanism and opportunities for intervention. *Toxicol. Lett.* 140–141,113–124.

Watanabe, Y., Song, T., Sugimoto, K., Horii, M., Araki, N., Tokumitsu, H., Tezuka, T., Yamamoto, T. and Tokuda, M. (2003). Post-synaptic density-95 promotes calcium/calmodulin-dependent protein kinase II-mediated Ser847 phosphorylation of neuronal nitric oxide synthase. *Biochem. J.* 372,465–471.

Weir, M. R., Sperling, R. S., Reicin, A. and Gertz, B. J. (2003). Selective COX-2 inhibition and cardiovascular effects: A review of the rofecoxib development program. *Am. Heart J.* 146,591–604.

Wink, D. A., Nims, R. W., Darbyshire, J. F., Christodoulou, D., Hanbauer, I., Cox, G. W., Laval, F., Laval, J., Cook, J. A., Krishna, M. C., DeGraff, W. G. and Mitchell, J. B. (1994). Reaction kinetics for nitrosation of cysteine and glutathione in aerobic nitric oxide solutions at neutral pH. Insights into the fate and physiological effects of intermediates generated in the $NO/O2$ reaction. *Chem. Res. Toxicol.* 7,519–525.

Xie, Q. W., Leung, M., Fuortes, M., Sassa, S. and Nathan, C. (1996). Complementation analysis of mutants of nitric oxide synthase reveals that the active site requires two hemes. *Proc. Natl. Acad. Sci. USA* 93,4891–4896.

Xia, Y. and Zweier, J. L. (1997). Superoxide and peroxynitrite generation from inducible nitric oxide synthase in macrophages. *Proc. Natl. Acad. Sci. USA* 94,6954–6958.

Xu, L., Eu, J. P., Meissner, G. and Stamler, J. S. (1998). Activation of the cardiac calcium release channel (ryanodine receptor) by poly-S-nitrosylation. *Science* 279,234–237.

Yeh, D. C., Duncan, J. A., Yamashita, S. and Michel, T. (1999). Depalmitoylation of endothelial nitric-oxide synthase by acyl-protein thioesterase 1 is potentiated by Ca(2+)-calmodulin. *J. Biol. Chem.* 274,33148–33154.

Zech, B., Wilm, M., van Eldik, R. and Brune, B. (1999). Mass spectrometric analysis of nitric oxide-modified caspase-3. *J. Biol. Chem.* 274,20931–20936.

Zeidler, P. C., Roberts, J. R., Castranova, V., Chen, F., Butterworth, L., Andrew, M. E., Robinson, V. A. and Porter, D. W. (2003). Response of alveolar macrophages from inducible nitric oxide synthase knockout or wild-type mice to an *in vitro* lipopolysaccharide or silica exposure. *J. Toxicol. Environ. Health A* 66,995–1013.

Zemojtel, T., Scheele, J. S., Martasek, P., Masters, B. S., Sharma, V. S. and Magde, D. (2003). Role of the interdomain linker probed by kinetics of CO ligation to an endothelial nitric oxide synthase mutant lacking the calmodulin binding peptide (residues 503–517 in bovine). *Biochemistry* 42,6500–6506.

Zhang, J., Li, Y. D., Patel, J. M. and Block, E. R. (1998). Thioredoxin overexpression prevents NO-induced reduction of NO synthase activity in lung endothelial cells. *Am. J. Physiol.* 275,L288–L293.

Zhang, M. and Vogel, H. J. (1994). Characterization of the calmodulin-binding domain of rat cerebellar nitric oxide synthase. *J. Biol. Chem.* 269,981–985.

Zhao, X., Li, X., Trusa, S. and Olson, S. C. (2005). Angiotensin type 1 receptor is linked to inhibition of nitric oxide production in pulmonary endothelial cells. *Regul. Pept.* 132,113–122.

Regulatory role and evolution of unconventional NOS-related RNAs

Sergei Korneev[1,*] and Michael O'Shea[1]

[1]*Sussex Centre for Neuroscience, University of Sussex, Brighton BN1 9QG, UK*

Abstract. Endogenous nitic oxide (NO) produced by the enzyme NO synthase (NOS) has an important role in a variety of physiological processes. The toxic properties of NO however suggest that its production must be tightly regulated. A particularly exciting and novel aspect of the regulation of NO signalling is the possibility that the expression of NOS genes is controlled by unconventional mechanisms that depend on the presence of natural antisense transcripts (NATs). Here we will discuss the properties of three distinct NATs discovered in the central nervous system (CNS) of the pond snail *Lymnaea stagnalis*. These transcripts possess regions of significant complementarity to NOS-encoding mRNAs. Importantly, our experiments on identified molluscan neurons suggest that NATs regulate the expression of the NOS gene through the formation of duplex molecules with the NOS mRNA. Recent discoveries of NOS-related NATs in mammals support the results of our molluscan studies. Thus, it is quite likely that NAT-mediated regulation of NO signalling is a phenomenon shared by a variety of species.

Keywords: nitric oxide; NOS; antisense RNA; gene regulation; DNA inversion; gene evolution; molluscs; mammals; CNS; pseudogene; long-term memory; CGC; bi-functional RNA; miRNA

Introduction

Throughout the late 1980s and early 1990s a considerable body of evidence accumulated indicating that the free radical gas nitric oxide (NO) is a neuronal signalling molecule in the mammalian brain (for a review, see Garthwaite and Boulton, 1995). NO was shown to be synthesised in neurons that express the neuronal isoform of the calcium–calmodulin-activated enzyme NO synthase (NOS) (Bredt and Snyder, 1990; Bredt *et al.*, 1990). Additional support for the transmitter status of NO was provided by evidence that in the brain it activates the soluble form of the enzyme guanylyl cyclase (sGC), resulting in the synthesis of the second messenger cyclic guanosine monophosphate (cGMP) (Bredt and Snyder, 1989; Kimura *et al.*, 1975; Southam and Garthwaite, 1991, 1993). NO was, however, unlike conventional neurotransmitters because it could not be stored in synaptic vesicles, did not activate membrane-associated

Corresponding author: Tel.: 44-(0)1273-872809. Fax: 44-(0)1273-678937.
E-mail: s.korneev@sussex.ac.uk (S. Korneev).

ADVANCES IN EXPERIMENTAL BIOLOGY
VOLUME 01   ISSN 1872-2423
DOI: 10.1016/S1872-2423(07)01008-3

postsynaptic receptor, and could spread in three dimensions away from its site of synthesis largely unhindered by intervening cellular or membrane structures (Edelman and Gally, 1992; O'Shea *et al.*, 1998; Philippides *et al.*, 2000; Wood and Garthwaite, 1994). Apparently, then, NO was an enigmatic and entirely new type of neuronal signalling molecule, and its discovery was greeted with some incredulity in the neuroscience community. Certainly the possibility of signalling by NO opened new dimensions in our thinking about the way information is transmitted by neurons in the brain.

For those of us working with invertebrate model systems, the discovery of the NO–cGMP signalling pathway in the mammalian brain posed exciting and unexpected challenges. First and foremost, could we show that NO is also a transmitter in invertebrates? And if so, could we use the advantages of identified invertebrate neurons to clarify the functional role of NO and to understand how the cellular and molecular machinery required for NO signalling works? In the early 1990s the first evidence that NO does indeed function as a neuronal signalling molecule in invertebrates was provided (Elphick *et al.*, 1993; Gelperin, 1994). Subsequently, the first behavioural role for NO in an invertebrate was demonstrated when the NO–cGMP signalling pathway was shown to be required for the chemosensory activation of feeding behaviour in the freshwater mollusc *Lymnaea stagnalis* (Elphick *et al.*, 1995). This finding was significant because *Lymnaea* was already well established as an excellent model system for understanding the role of identified neurons and synapses in the neural network underlying feeding (for a review, see Benjamin *et al.*, 2000). We were therefore encouraged to use *Lymnaea* to investigate the role of NO signalling in neuronal functions associated with feeding at behavioural, cellular and molecular levels. In so doing we discovered that the NO–cGMP pathway is involved in the plasticity of the feeding response as activated by chemical sensory stimuli. Specifically, using one-trial reward conditioning of feeding behaviour, Kemenes *et al.* (2002) showed that during memory formation there is a brief but critical time window of 5 h post-training when there is an obligatory requirement for NO–cGMP signalling.

In parallel with the behavioural studies we also investigated molecular mechanisms that underlie NO signalling in *Lymnaea* (Korneev *et al.*, 1998). In this article we will summarise the most interesting outcomes of this work. Particularly, we will focus on some unusual RNA molecules and mechanisms regulating the expression of NOS in the central nervous system (CNS). As with so many other aspects of NO signalling in this and other systems, the molecular biological story that emerges is unconventional and full of surprises – especially with respect to molecular genetic mechanisms associated with neuronal plasticity and memory formation.

NOS mRNA in *Lymnaea*: Organisation and paradoxical expression

In 1998 we reported on the cloning and characterisation of the first molluscan mRNA encoding NOS (Korneev *et al.*, 1998). Hereafter we will refer to this transcript as *Lym*-nNOS1 mRNA. This mRNA is expressed in the CNS and encodes a protein of 1,153 amino acids, which has 50% sequence homology with the neuronal isoform of mammalian NOS (nNOS). Further indications of similarity between *Lym*-nNOS1 and the mammalian enzyme derive from the observations that the molluscan enzyme is calcium activated and contains the same cofactor and cosubstrate binding sites as the mammalian nNOS. Although the general organisation of *Lym*-nNOS1 is typical for an nNOS protein, there is a feature that had not been observed before. This is the presence of 12 tandem repeats of a conserved sequence of seven amino acids located just after the NADPH-ribose binding site. The extra 84 amino acid peptide creates a highly flexible and hydrophilic domain between the two NADPH binding sites. The functional significance of this repetitive region remains unclear, but it certainly does not prevent NOS activity (Korneev *et al.*, 1998).

We have studied the cellular expression of *Lym*-nNOS1 using *in situ* hybridisation and revealed a repeatable and restricted population of stained neurons. Significantly, among these cells are the giant identified serotonergic neurons called the cerebral giant cells (CGCs) (Fig. 1A). This is of functional interest because the CGCs are required for the feeding response. They have widespread synaptic connections with the rest of the feeding circuit, including neurons of the feeding central pattern generator (CPG) and feeding motoneurons (Fig. 1B) (Yeoman *et al.*, 1996). Moreover, the CGC is one of the key neurons involved in the modulation of the conditioned feeding behaviour, for which it has a crucial gating function (Yeoman *et al.*, 1994; Straub and Benjamin, 2001). Later we confirmed the presence of *Lym*-nNOS1 mRNA in the CGCs using single-cell reverse transcriptase polymerase chain reaction (RT-PCR) (Korneev *et al.*, 1999). However, when we used the NADPH-diaphorase histological method for the detection of the NOS enzyme, the CGCs almost always showed no staining (Fig. 2A). These apparently contradictory and puzzling observations suggested to us the

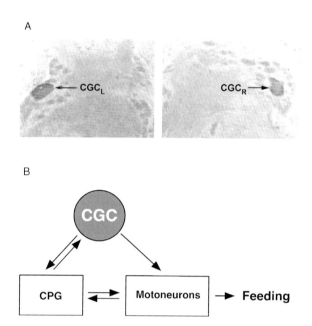

Fig. 1. The CGCs, which are known to be essential for feeding, contain *Lym*-nNOS mRNA. (A) *In situ* hybridisation with probe specifically recognising *Lym*-nNOS mRNA reveals a bilaterally symmetrical pair of large neuronal cell bodies in the left (L) and right (R) cerebral ganglia (Korneev *et al.*, 1998). The cell bodies correspond in size and position to the CGCs. (B) Neural circuit underlying feeding behaviour. The CGCs play an important gating role in feeding behaviour through their modulatory actions on interneurons of the central pattern generator (CPG) and feeding motoneurons.

existence of a mechanism regulating the translation of *Lym*-nNOS1 mRNA in the CGCs that suppresses the synthesis of NOS protein in these cells.

Fig. 2. AntiNOS-1 natural antisense transcript (NAT)-mediated negative regulation of NOS expression in the CGCs. (A) The vast majority of the CGCs shows no NADPH-diaphorase activity (upper image) (Korneev *et al.*, 1999). Single-cell RT-PCR demonstrates the presence in the CGCs of *Lym*-nNOS mRNA as well as antiNOS-1 NAT (lower image). (B) B2 motoneurons localised in the buccal ganglia are strongly NADPH-diaphorase positive (upper image). Single-cell RT-PCR demonstrates the presence in the B2 motoneurons of *Lym*-nNOS mRNA. Note that the antiNOS-1 NAT is not detected (lower image). (C) AntiNOS-1 NAT suppresses translation of *Lym*-nNOS1 mRNA through the formation of RNA–RNA duplex molecules. The antisense region is shown in black (not to scale). (See Colour Plate Section in this book).

185

A

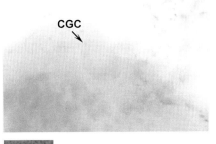

 - *Lym*-nNOS mRNA
- antiNOS-1 NAT

B

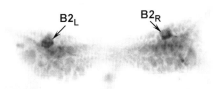

 - *Lym*-nNOS mRNA

C

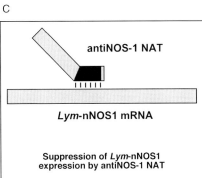

Paradoxical expression explained by an unconventional RNA produced from NOS pseudogene?

Northern blot analysis of RNA extracted from *Lymnaea* CNS has revealed the presence of several different transcripts homologous to NOS. This has been then confirmed by the results of our cloning experiments. Particularly important was the discovery of an unusual non-coding NOS-related polyadenylated RNA of 2,345 nucleotides (nt) transcribed from a NOS pseudogene (Korneev *et al.*, 1999). The major part of the transcript exhibits >80% sequence identity to *Lym*-nNOS1 mRNA but cannot be translated into a protein due to the presence of multiple stop codons in all three reading frames. An unexpected, but functionally important, feature of the transcript is the presence of a region of approximately 150 nt, which shows 80% complementarity to a region close to the middle of the *Lym*-nNOS1 mRNA. We have therefore concluded that this RNA is an example of a natural antisense transcript (NAT) and we have named this NAT antiNOS-1.

Depending on their origin, NATs can be classified into one of two major groups: *cis*-encoded or *trans*-encoded. *Cis*-encoded NATs are produced from the same loci as their sense counterparts, usually represented by protein-encoding mRNAs. Within the antisense region they therefore exhibit 100% complementarity to their sense partners. *Trans*-encoded NATs are transcribed from different loci, and consequently complementarity to their sense partners is not necessarily complete. Importantly, NATs have been proposed to mediate the negative regulation of gene expression through the formation of RNA–RNA duplex molecules in cells in which the sense and antisense transcripts are co-expressed (for a review, see Korneev and O'Shea, 2005). So it was of considerable interest to know whether the antiNOS-1 NAT is involved in the control of NOS expression in the CGCs. We therefore attempted to answer the following questions. Do neurons exist that co-express anti-NOS-1 NAT and *Lym*-nNOS1 mRNA? Can these two transcripts form stable RNA–RNA duplex molecules *in vivo*? Does antiNOS-1 NAT affect the translation of the *Lym*-nNOS1 mRNA? We have succeeded in getting affirmative answers to all these questions (Korneev *et al.*, 1999).

Firstly, we have shown that the antiNOS-1 NAT and *Lym*-nNOS1 mRNA are co-localised in the CGCs. Secondly, using RNase A treatment followed by RT-PCR we have confirmed that the antiNOS-1 NAT and *Lym*-nNOS1 mRNA form a stable RNA–RNA duplex *in vivo*. Thirdly, we have demonstrated that the antiNOS-1 NAT inhibits the synthesis of NOS protein *in vitro*. Finally, we have analysed another uniquely identified *Lym*-nNOS1 mRNA-containing neuron known as the B2 motoneuron, and have found that it does not contain the antiNOS-1 NAT and consistently expresses a functional NOS protein (Fig. 2B).

Taken together these data strongly supported our suggestion that the antiNOS-1 NAT produced by a NOS pseudogene functions as a negative regulator of NOS expression in the CGCs (Fig. 2C). This can explain the paradoxical expression of *Lym*-nNOS1 in the CGCs. We believe that in the CGCs antiNOS-1 NAT mediates post-transcriptional control of *Lym*-nNOS, providing an effective molecular mechanism for achieving rapid changes in NOS protein production in response to internal or external signals. Notice that according to this proposed NAT-mediated regulatory mechanism, a specific function is attributed to a pseudogene.

When we first claimed that a functional pseudogene regulates the expression of a related protein-encoding gene, our proposal could rightly have been regarded as extraordinary. This was because pseudogenes, which are common in all eukaryotic genomes, were regarded as inactive versions of currently functional genes and therefore functionless by definition. Our results clearly contradict this view and suggest that some transcribed pseudogenes might have functions. We should therefore conclude that pseudogenes are not all relics of evolution but are a potential source of a new class of regulatory gene.

Reward conditioning regulates NOS gene and NOS pseudogene expression

Long-term memory (LTM) formation following reward conditioning of feeding behaviour in *Lymnaea* requires the NO-cGMP signalling pathway (Kemenes *et al.*, 2002). This requirement is restricted to a window of time lasting about 5 h following a single pairing of sucrose [the unconditioned stimulus (US)] with amyl acetate [the conditioned stimulus (CS)]. In the light of the proposed role of antiNOS-1 in the regulation of NOS expression in the CNS, it was important therefore to investigate whether this NO-dependent memory formation is associated with specific changes in the expression of the antiNOS-1 NAT.

To address this question we used a single pairing of the US and the CS to condition feeding behaviour (Fig. 3A) and then analysed the temporal dynamics of the post-training expression of the antiNOS-1 NAT in the CGCs by real-time RT-PCR. We found that the antiNOS-1 gene (NOS pseudogene) was significantly down-regulated at 4 h but showed no changes in its activity at 6 h and 24 h after training (Fig. 3B). This dynamic sensitivity of antiNOS-1 to a very weak sensory cue (a single pairing of the US and CS) is remarkable in itself but also provides a unique example of a pseudogene whose expression is regulated by a behavioural stimulus (Korneev et al., 2005). It is interesting to note that the Lym-nNOS1 also exhibited transient training-provoked changes in its expression in the CGCs. Lym-nNOS1 was transiently up-regulated at 6 h, some 2 h after the down-regulation of the antiNOS-1 transcript. The fact that the down-regulation of antiNOS-1 gene precedes the up-regulation of Lym-nNOS1 supports the role of the antiNOS-1 NAT in the regulation of NOS expression. Thus, we can hypothesise that the decrease in the amount of the antiNOS-1 NAT leads to the increase in the amount of Lym-nNOS1 mRNA molecules available for translation. This could result in the induction of NO production, which may have a role in facilitating memory formation.

Evolution of unconventional regulatory NOS-related RNAs

Our present understanding of evolutionary events that led to the creation of the pseudogene from which antiNOS-1 is transcribed began with the discovery of another novel NOS-related transcript which also contains a region of significant antisense homology to the Lym-nNOS1 mRNA (Korneev and O'Shea, 2002). Therefore we referred to it as antiNOS-2

Fig. 3. Training-induced differential regulation of antiNOS-1 NAT and Lym-nNOS1 mRNA in the CGCs. (A) One-trial reward conditioning was performed using a method based on a previously published protocol (Alexander et al., 1984). Experimental (conditioned) animals were exposed to a solution of amyl acetate (CS) and immediately after that to a sucrose solution (US). Unpaired control animals were exposed to the CS and to the US, separated by an interval of 1 h. At different time points (4 h, 6 h and 24 h) after the treatment, individual CGCs were extracted and used for real-time RT-PCR analysis. (B) Real-time RT-PCR analysis of the antiNOS-1 NAT expression in the CGCs. The level of antiNOS-1 NAT in conditioned (C) groups (grey bars) and unpaired control (UC) groups (white bars) normalised to an endogenous reference (β-tubulin) and relative to a calibrator (CAL) was calculated as $2 - DDC_T$, where $DDC_T = DC_T - DC_{T(CAL)}$. DC_T and $DC_{T(CAL)}$ are the differences in threshold cycles for target (antiNOS-1) and reference (β-tubulin) measured in the samples and in the calibrator, respectively. The bars show the mean value of eight samples $\pm$ standard deviation. Note that there is a down-regulation of antiNOS-1 at 4 h after training. The single asterisk indicates a significant difference from controls of $p = 0.05$ (Korneev et al., 2005). (C) Real-time RT-PCR analysis of Lym-nNOS1 mRNA expression in the CGCs. Note that Lym-nNOS1 mRNA is up-regulated at 6 h after training. The double asterisk in the 6 h conditioned group (grey bar) indicates a significant difference from the unpaired control (white bar) of at least $p < 0.005$ (Korneev et al., 2005).

A

Conditioned snails

Unpaired control snails

B

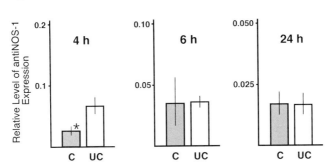

C

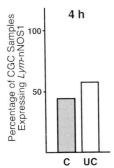

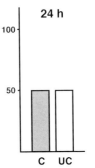

NAT. We have demonstrated that antiNOS-2 NAT is a polyadenylated RNA of about 3,000 nt, which has a functional open reading frame encoding a truncated NOS-homologous protein of 397 amino acids. Notably, in the antiNOS-2 NAT the antisense region is located at the 3' end of the molecule, whereas in antiNOS-1 it is located at the 5' end. Considering the structural organisation of the antiNOS-1 and antiNOS-2 NATs, the positions of the antisense regions, and the localisation of their potential targets in the *Lym*-nNOS1 mRNA, we suggested that the two antiNOS genes were the result of a single internal DNA inversion that occurred in an ancestral gene (Korneev and O'Shea, 2002). Genomic analysis has confirmed this suggestion.

The chain of evolutionary events revealed by our studies was as follows (Fig. 4). Firstly, the ancestral NOS gene duplicated. One copy of the gene retained its original function and organisation and evolved into the present-day *Lym*-nNOS1 gene. Secondly, the function of the other copy was disrupted by a mutation involving an internal DNA inversion. Critically, the inversion also introduced new instructions for the termination and initiation of transcription. This split the duplicated copy, resulting in the creation of two novel and independently expressed transcriptional units, antiNOS-1 and antiNOS-2. We know that the NAT transcribed from the antiNOS-1 gene functions as a negative translational regulator of *Lym*-nNOS1 expression. As for the antiNOS-2 gene, it also produces a NAT but, unlike antiNOS-1, it has an open reading frame encoding a protein with 70% identity to the oxygenase domain of NOS. As the NOS enzyme is active only as a homodimer (Abu-Soud *et al.*, 1995; Crane *et al.*, 1998; Klatt *et al.*, 1996), it is of considerable interest that the molluscan truncated homologue encoded by the antiNOS-2 gene includes the domain required for dimerisation, but lacks other functional regions essential for NO synthesis. We hypothesised that if a dimer were to form between a full-length NOS protein and the truncated version encoded by the antiNOS-2 gene, the resulting complex would lack NO-synthesising activity. Indeed, when heterodimers were formed between the normal nNOS monomer and similarly organised truncated variants of the NOS protein generated *in vitro*, a strong suppressive effect on enzyme activity was observed (Lee *et al.*, 1995). One intriguing possibility therefore is that the antiNOS-2 protein functions as a natural dominant negative regulator of NOS activity through binding to the normal NOS monomer to form a non-functional heterodimer. It is noteworthy that a similar dominant negative regulatory role for a truncated NOS protein has been proposed in *Drosophila* (Stasiv *et al.*, 2001).

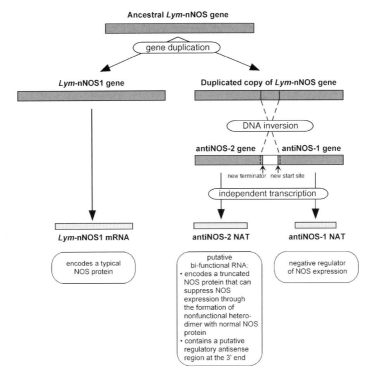

Fig. 4. Evolution of NOS-related regulatory genes by DNA inversion. An ancestral *Lym*-nNOS gene underwent a duplication. The internal DNA inversion in the duplicated copy of the *Lym*-nNOS gene resulted in the generation of a new terminator and new start site which in effect split the gene, resulting in the generation of two new and independently expressed genes, antiNOS-1 (NOS pseudogene) and antiNOS-2. The original functions of the *Lym*-nNOS1 gene were unaffected, but its expression can now be regulated by a NAT transcribed from the antiNOS-1 gene (Korneev *et al.*, 1999). The antiNOS-2 gene produces a message encoding a truncated nNOS homologue. This protein can form a nonfunctional heterodimer and might therefore have a natural dominant negative effect on nNOS enzyme activity (Korneev and O'Shea, 2002).

In summary, the antiNOS-2 NAT can be considered as a putative bi-functional RNA because it can function as a protein-encoding mRNA that also contains a potential regulatory antisense region localised in the 3′ untranslated area. Recently, Cai *et al.* (2004) reported on the identification of a transcript of 3,433 nt which also possessed all typical features of a mRNA and in addition had a sequence that is processed into a mature microRNA (miRNA) called mir-21. Cai *et al.* (2004) showed that the presence of mir-21 precursor within this long primary miRNA (pri-miRNA) only moderately inhibited expression of the linked open reading frame. Consequently they suggested that pri-miRNAs could function both as mRNAs and as sources of miRNAs. MiRNAs are known to be ubiquitous very small non-coding RNAs that appear to regulate gene expression through an antisense mechanism (for a review, see Bartel, 2004).

Our observation that DNA inversion resulted in the creation of two novel genes was unusual. Indeed, DNA inversions are generally known for their devastating effects on the structural and functional integrity of genes in eukaryotic cells (Lupski, 1998). They can lead to the so-called genomic disorders, such as Hunter syndrome and some forms of hae-mophilia (Bondeson et al., 1995; Lakich et al., 1993). Probably the first example suggesting a role for DNA inversions in the creation of trans-encoded NATs was reported by Tosic et al. (1990) and Okano et al. (1991). These authors analysed the level of mRNA encoding myelin basic protein (MBP) in myelin-deficient mice, which are known to be charac-terised by a tandem duplication of the MBP gene (Okano et al., 1988). They demonstrated that the downstream copy of the original MBP gene retained the original function, whereas the upstream copy could not produce mature MBP mRNA due to a large inversion mutation. Remarkably, this mutated copy is transcribed in the brain of myelin-deficient mice as an antisense RNA that is localised in the nuclei and most probably is not polyadenylated. Despite normal transcriptional activity of the MBP gene in the mutant mice, there was a dramatic reduction in the concentration of MBP mRNA in the cytoplasm. This strongly indicated that in mutants either the processing or the transport of the MBP mRNA to the cytoplasm is disrupted through a transNAT-mediated mechanism. This example is of course associated with a mutant phenotype. The results of our Lymnaea studies, however, suggest that NATs generated by DNA inversions can also play an important role in the regulation of gene expression in wild-type brains. Therefore, we believe that DNA inversions should be considered as an essential com-ponent of adaptive gene evolution.

Other NOS-related transcripts in *Lymnaea*

Recently we have cloned a new transcript from *Lymnaea* CNS that exhibits 89% sequence identity to *Lym*-nNOS1 mRNA within its open reading frame (Korneev et al., 2005). This RNA is more than 6,000 nt in length and encodes a typical NOS protein which is about 90% identical to the *Lym*-nNOS1. We referred to it as *Lym*-nNOS2. Although the two proteins are practically identical, the genes encoding them respond very differently to single-trial reward conditioning. By measuring the level of *Lym*-nNOS1 and *Lym*-nNOS2 mRNAs in the CNS at different time points after one-trial conditioning we found that, in contrast to training-induced temporal changes in the expression exhibited by *Lym*-nNOS1, the activity of the *Lym*-nNOS2 gene remained stable at all measured time points. These observations are reminiscent of the situation in mammals, where some NOS genes (nNOS and eNOS) are consistently expressed and one (iNOS) is inducible (for a review, see Stuehr, 1999). Thus, *Lymnaea* appears to be the first example amongst the invertebrates in which the existence of distinct constitutive and inducible NOS genes has been demonstrated. Indeed, in *Drosophila* there is just one gene encoding NOS (Stasiv *et al.*, 2001). However, this gene produces a number of alternatively spliced transcripts encoding proteins with different patterns of expression and function (Stasiv *et al.*, 2004). Molluscs as well as mammals have apparently chosen another way to achieve heterogeneity and flexibility within the system regulating NO signalling. This involves the creation of several differentially expressed homologous NOS genes 'specialised' for different functions.

Completing the complement of NOS and NOS-related transcripts expressed in *Lymnaea* CNS is a novel non-coding RNA, which is an example of a *cis*-encoded NAT. This polyadenylated RNA is about 2,500 nt long and contains a ~250 nt region exhibiting 100% antisense homology to the 5′ untranslated region (UTR) of *Lym*-nNOS1 mRNA (Fig. 5). We

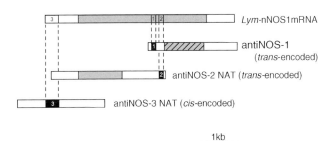

Fig. 5. Three types of NOS-related NATs are present in *Lymnaea* CNS. Grey boxes indicate the protein-encoding regions. The grey hatched box in the antiNOS-1 NAT shows a non-coding region of high similarity to the *Lym*-nNOS1 mRNA. The antisense regions in NATs are indicated by numbered black boxes. Their potential targets within the *Lym*-nNOS1 mRNA are boxed and numbered. A more detailed diagram showing intron–exon boundaries can be found in Korneev and O'Shea, 2002. Scale bar indicates 1,000 base pairs.

will refer to this transcript as antiNOS-3 NAT. Its discovery is probably not unexpected because recent computational analysis of available expressed sequence tags (EST) and genome databases suggested that a large proportion of genes in eukaryotic cells are bi-directionally transcribed (Chen *et al.*, 2004; Yelin *et al.*, 2003). The majority of the overlaps between sense and antisense partners were shown to occur in the 5' and 3' UTRs, which are known to contain elements important for the translation, stability and transport of mRNAs. These observations suggested that *cis*NATs could interact functionally with these important molecular mechanisms of mRNA functioning. Therefore we believe that the anti-NOS-3 *cis*NAT may also possess some important regulatory functions that have not yet been revealed by our experiments.

From molluscs to mammals

Our studies in *Lymnaea* have revealed several NOS-related NATs that are likely to play a role in the regulation of NO signalling. This raises an obvious and important question. Do similar molecules and mechanisms exist in other organisms, including mammals? A paper published by Robb *et al.* (2004) provides an affirmative answer to this question. Specifically, the authors describe the cloning and characterisation of a human *cis*-encoded NAT that is synthesised on the opposite DNA strand to that from which the eNOS mRNA is transcribed. They refer to this NAT as sONE and to its mouse orthologue as NOS3as. Several lines of evidence support the role of the sONE NAT in post-transcriptional regulation of eNOS expression. Firstly, RNA interference-mediated inhibition of sONE expression in vascular smooth muscle cells led to an increase in eNOS expression. Secondly, overexpression of sONE NAT in endothelial cells resulted in a decrease in eNOS protein level. Thirdly, treatment of vascular endothelial cells with trichostatin A, which is known to regulate the expression of eNOS via a post-transcriptional mechanism, increased the level of sONE expression prior to the observed decrease in the amount of eNOS mRNA. From these results it is clear that NATs play a significant role in the regulation of NO signalling not only in molluscs but in mammals as well.

Conclusions

Significant progress has been made in recent years in our understanding of the complexity of the intracellular machinery involved in the regulation of genes. NATs probably represent one of the latest additions to the complex repertoire of molecular mechanisms controlling gene expression. Some of the first indications that NAT-mediated mechanisms can regulate NO signalling came from experiments performed on identified neurons from *Lymnaea* CNS. Complementary studies in mammals indicate that these unconventional regulatory processes are not a unique feature of molluscs but can also take place in other species. Thus, it is now quite clear that the regulation of NO signalling is highly complex, occurs at various levels, and can be mediated by both conventional and unconventional mechanisms.

References

Abu-Soud, H. M., Loftus, M. and Stuehr, D. J. (1995). Subunit dissociation and unfolding of macrophage NO synthase: Relationship between enzyme structure, prosthetic group binding, and catalytic function. *Biochemistry* 34,11167–11175.

Alexander, J., Jr. Audesirk, T. E. and Audesirk, G. J. (1984). One-trial reward learning in the snail *Lymnaea stagnalis*. J. Neurobiol. 15, 67–72.

Bartel, D. P. (2004). MicroRNAs. Genomics, biogenesis, mechanism, and function. *Cell* 116,281–297.

Benjamin, P. R., Staras, K. and Kemenes, G. (2000). A systems approach to the cellular analysis of associative learning in the pond snail *Lymnaea*. *Learn. Mem.* 7,124–131.

Bondeson, M-L., Dahl, N., Malmgren, H., Kleijer, W. J., Tannesen, T., Carlberg, B. M. and Pettersson, U. (1995). Inversion of the IDS gene resulting from recombination with IDS-related sequences is a common cause of the Hunter syndrome. *Hum. Mol. Genet.* 4,615–621.

Bredt, D. S., Hwang, P. M. and Snyder, S. H. (1990). Localization of nitric oxide synthase indicating a neural role for nitric oxide. *Nature* 347,768–770.

Bredt, D. S. and Snyder, S. H. (1989). Nitric oxide mediates glutamate-linked enhancement of cGMP levels in the cerebellum. *Proc. Natl. Acad. Sci. USA* 86,9030–9033.

Bredt, D. S. and Snyder, S. H. (1990). Isolation of nitric oxide synthetase, a calmodulin-requiring enzyme. *Proc. Natl. Acad. Sci. USA* 87,682–685.

Cai, X., Hagedorn, C. H. and Cullen, B. R. (2004). Human microRNAs are processed from capped, polyadenylated transcripts that can also function as mRNAs. *RNA* 10,1957–1966.

Chen, J., Sun, M., Kent, W. J., Huang, X., Xie, H., Wang, W., Zhou, G., Shi, R. Z. and Rowley, J. D. (2004). Over 20% of human transcripts might form sense-antisense pairs. *Nucl. Acids Res.* 32,4812–4820.

Crane, B. R., Arvai, A. S., Ghosh, D. K., Wu, C., Getzoff, E. D., Stuehr, D. J. and Tainer, J. A. (1998). Structure of nitric oxide synthase oxygenase dimer with pterin and substrate. *Science* 279,2122–2126.

Edelman, G. M. and Gally, J. A. (1992). Nitric oxide: Linking space and time in the brain. *Proc. Natl. Acad. Sci. USA* 89,11651–11652.

Elphick, M. R., Green, I. C. and O'Shea, M. (1993). Nitric oxide synthesis and action in an invertebrate brain. *Brain Res.* 619,344–346.

196

Elphick, M. R., Kemenes, G., Staras, K. and O'Shea, M. (1995). Behavioral role for nitric oxide in chemosensory activation of feeding in a mollusc. *J. Neurosci.* 15,7653–7664.
Garthwaite, J. and Boulton, C. L. (1995). Nitric oxide signaling in the central nervous system. *Annu. Rev. Physiol.* 57,683–706.
Gelperin, A. (1994). Nitric oxide mediates network oscillations of olfactory interneurons in a terrestrial mollusc. *Nature* 369,61–63.
Kemenes, I., Kemenes, G., Andrew, R. J., Benjamin, P. R. and O'Shea, M. (2002). Critical time-window for NO-cGMP dependent long-term memory formation after one-trial appetitive conditioning. *J. Neurosci.* 22,1414–1425.
Kimura, H., Mittal, C. K. and Murad, F. (1975). Increases in cyclic GMP levels in brain and liver with sodium azide an activator of guanylate cyclase. *Nature* 257,700–702.
Klatt, P., Pfeiffer, S., List, B. M., Lehner, D., Glatter, O., Bachinger, H. P., Werner, E. R., Schmidt, K. and Mayer, B. (1996). Characterization of heme-deficient neuronal nitric oxide synthase reveals a role for heme in subunit dimerization and binding of the amino acid substrate and tetrahydrobiopterin. *J. Biol. Chem.* 271,7336–7342.
Korneev, S. and O'Shea, M. (2002). Evolution of nitric oxide synthase regulatory genes by DNA inversion. *Mol. Biol. Evol.* 19,1228–1233.
Korneev, S. and O'Shea, M. (2005). Natural antisense RNAs in the nervous system. *Rev. Neurosci.* 16,213–222.
Korneev, S. A., Park, J. H. and O'Shea, M. (1999). Neuronal expression of neural nitric oxide synthase (nNOS) protein is suppressed by an antisense RNA transcribed from an NOS pseudogene. *J. Neurosci.* 19,7711–7720.
Korneev, S. A., Piper, M. R., Picot, J., Phillips, R., Korneeva, E. I. and O'Shea, M. (1998). Molecular characterization of NOS in a mollusc: Expression in a giant modulatory neuron. *J. Neurobiol.* 35,65–76.
Korneev, S. A., Straub, V., Kemenes, I., Korneeva, E. I., Ott, S. R., Benjamin, P. R. and O'Shea, M. (2005). Timed and targeted differential regulation of nitric oxide synthase (NOS) and anti-NOS genes by reward conditioning leading to long-term memory formation. *J. Neurosci.* 25,1188–1192.
Lakich, D., Kazazian, J. H. H., Jr., Antonarakis, S. E. and Gitschier, J. (1993). Inversions disrupting the factor VIII gene are a common cause of severe hemophilia-A. *Nat. Genet.* 5,236–241.
Lee, C. M., Robinson, L. J. and Michel, T. (1995). Oligomerization of endothelial nitric oxide synthase. Evidence for a dominant negative effect of truncation mutants. *J. Biol. Chem.* 270,27403–27406.
Lupski, J. R. (1998). Genomic disorders: Structural features of the genome can lead to DNA rearrangements and human disease traits. *Trends Genet.* 14,417–422.
Okano, H., Aruga, J., Nakagawa, T., Shiota, C. and Mikoshiba, K. (1991). Myelin basic protein gene and the function of antisense RNA in its repression in myelin-deficient mutant mouse. *J. Neurochem.* 56,560–567.
Okano, H., Ikenaka, K. and Mikoshiba, K. (1988) Recombination within the upstream gene of duplicated myelin basic protein genes of myelin deficient shimld mouse results in the production of antisense RNA. *EMBO J.* 7,3407–3412.
O'Shea, M., Colbert, R., Williams, L. and Dunn, S. (1998). Nitric oxide compartments in the mushroom bodies of the locust brain. *Neuroreport* 3,333–336.

Philippides, A., Husbands, P. and O'Shea, M. (2000). Four-dimensional neuronal signaling by nitric oxide: A computational analysis. *J. Neurosci.* 20,1199–1207.

Robb, G. B., Carson, A. R., Tai, S. C., Fish, J. E., Singh, S., Yamada, T., Scherer, S. W., Nakabayashi, K. and Marsden, P. A. (2004). Post-transcriptional regulation of endothelial nitric-oxide synthase by an overlapping antisense mRNA transcript. *J. Biol. Chem.* 279,37982–37996.

Southam, E. and Garthwaite, J. (1991). Comparative effects of some nitric oxide donors on cyclic GMP levels in rat cerebellar slices. *Neurosci. Lett.* 130,107–111.

Southam, E. and Garthwaite, J. (1993). The nitric oxide-cyclic GMP signalling pathway in rat brain. *Neuropharmacology* 32,1267–1277.

Stasiv, Y., Kuzin, B., Regulski, M., Tully, T. and Enikolopov, G. (2004). Regulation of multimers via truncated isoforms: A novel mechanism to control nitric-oxide signaling. *Genes Dev.* 18,1812–1823.

Stasiv, Y., Regulski, M., Kuzin, B., Tully, T. and Enikolopov, G. (2001). The Drosophila nitric-oxide synthase gene (dNOS) encodes a family of proteins that can modulate NOS activity by acting as dominant negative regulators. *J. Biol. Chem.* 276,42241–42251.

Straub, V. A. and Benjamin, P. R. (2001). Extrinsic modulation and motor pattern generation in a feeding network: A cellular study. *J. Neurosci.* 21,1767–1778.

Stuehr, D. J. (1999). Mammalian nitric oxide synthases. *Biochim. Biophys. Acta* 1411,217–230.

Tosic, M., Roach, A., de Rivaz, J. C., Dolivo, M. and Matthieu, J. M. (1990). Post-transcriptional events are responsible for low expression of myelin basic protein in myelin deficient misb: Role of natural antisense RNA. *EMBO J.* 9,401–406.

Wood, J. and Garthwaite, J. (1994). Models of the diffusional spread of nitric oxide: Implications for neural nitric oxide signalling and its pharmacological properties. *Neuropharmacology* 33,1235–1244.

Yelin, R., Dahary, D., Sorek, R., Levanon, E. Y., Goldstein, O., Shoshan, A., Diber, A., Biton, S., Tamir, Y., Khosravi, R., Nemzer, S., Pinner, E., Walach, S., Bernstein, J., Savitsky, K. and Rotman, G. (2003). Widespread occurrence of antisense transcription in the human genome. *Nat. Biotechnol.* 21,379–386.

Yeoman, M. S., Brierley, M. J. and Benjamin, P. R. (1996). Central pattern generator interneurons are targets for the modulatory serotonergic cerebral giant cells in the feeding system of *Lymnaea*. *J. Neurophysiol.* 72,1372–1382.

Yeoman, M. S., Pieneman, A. W., Ferguson, G. P., Ter Maat, A. and Benjamin, P. R. (1994). Modulatory role for the serotonergic cerebral giant cells in the feeding system of the snail, *Lymnaea*. I. Fine wire recording in the intact animal and pharmacology. *J. Neurophysiol.* 72,1357–1371.

The role of blood nitrite in the control of hypoxic vasodilation

Angela Fago[1] and Frank B. Jensen[2],*

[1]*Department of Biological Sciences, University of Aarhus, C.F. Møllers Alle 1131, DK-8000 Aarhus C, Denmark*

[2]*Institute of Biology, University of Southern Denmark, Campusvej 55, DK-5230 Odense M, Denmark*

Abstract. In the past few years circulating nitrite has been increasingly regarded not only as an inert end-product of endogenous nitric oxide (NO), but also as an important physiological compound that participates in the regulation of hypoxic vasodilation. The vasoactivity of nitrite appears to involve its one-electron reduction to NO, whereby nitrite represents a storage pool of NO activity that becomes available during tissue hypoxia. Among the mechanisms for NO formation from nitrite so far identified, the nitrite reductase activity of deoxygenated hemoglobin has been the focus of several recent studies, which have assigned to red blood cells a considerable role in NO generation and local blood flow regulation. This hemoglobin-based mechanism for nitrite vasoactivity involves nitrite transport from plasma into the red blood cells, its reaction with deoxygenated rather than oxygenated hemoglobin to generate NO, the diffusion of NO out of red blood cells, and finally activation of the vasodilatory response. In this review we critically address these steps with the aim of identifying potential control sites along the pathway from plasma nitrite to vasoactive NO, and discuss alternative mechanisms for nitrite-induced blood vessel dilation.

Keywords: anion exchanger; deoxyhemoglobin; hemoglobin oxidation; hypoxic vasodilation; methemoglobin; nitric oxide production from nitrite; nitrite; nitrite detoxification; nitrite membrane permeability; nitrite production; nitrite reactions with hemoglobin; nitrite reduction; nitrite transport in red blood cells; nitrite-induced vasodilation; oxyhemoglobin; red blood cells; *S*-nitrosohemoglobin; *S*-nitrosothiols; vasodilation.

Introduction

Inorganic nitrite has traditionally been regarded as a toxic agent for vertebrates. This concept has become redefined in recent years, as important physiological functions of nitrite are beginning to emerge (Gladwin *et al.*, 2005). There are multiple sources (and, as this review illustrates, fates) of nitrite in animals. Inorganic nitrite is naturally present at low concentrations in animal tissues, because it is the product of the reaction of the physiological signaling molecule nitric oxide (NO)

Corresponding author: Fax: +45-6593-0457.
E-mail: fbj@biology.sdu.dk (F.B. Jensen).

ADVANCES IN EXPERIMENTAL BIOLOGY
VOLUME 01   ISSN 1872-2423
DOI: 10.1016/S1872-2423(07)01009-5

with molecular oxygen dissolved in aqueous solutions. It has been estimated that ~70% of the circulating plasma nitrite (0.15–0.6 μM) in mammals originates from NO produced by the activity of endothelial nitric oxide synthase (eNOS; also called NOS3 or NOS III) (Kleinbongard et al., 2003). Other sources of nitrite are intake via the diet, and reduction of nitrate to nitrite by bacteria in the oral cavity, with subsequent uptake across the intestinal epithelium (Lundberg and Weitzberg, 2005). Some aquatic animals, like freshwater fish, even take up nitrite directly from the environment via active transport across the gills (Jensen, 2003).

Since its discovery in the 1980s, the crucial function of NO as the endothelium-derived relaxing factor (EDRF) produced in the endothelium and relaxing the adjacent smooth muscle has been the focus of extensive investigations. NO from vascular endothelial cells exerts its function by diffusing into underlying vascular smooth muscle, where it reaches and activates soluble guanylyl cyclase, which triggers arterial and arteriolar vasodilation. In being a free radical, NO has an intrinsically high reactivity, particularly for oxygen, superoxide, and ferrous heme, and its physiological half-life is short. Thus, only a fraction of the NO diffusing out of the endothelium will be able to reach and activate soluble guanylyl cyclase in the smooth muscle. Remarkably, in comparison to other tissues, aorta vessels contain much higher basal levels of nitrite and S-nitrosothiols (Bryan et al., 2005), which may be major products of the NO produced in the endothelium. In the blood, NO is mainly inactivated by reacting with plasma O_2 to form nitrite, or by reacting with oxygenated hemoglobin (Hb) inside red blood cells (RBCs) to form nitrate and methemoglobin (metHb; ferric form). MetHb is subsequently converted to the O_2-binding ferrous form by erythrocytic metHb reductase (Jaffé, 1981).

At the concentrations found in vivo, nitrite and nitrate were until recently considered relatively inert end-products of physiological NO oxidation. Nitrite can, however, be converted back to NO by a number of mechanisms, including nonenzymatic acidic reduction and enzymatic reduction via xanthine oxidoreductase or Hb (Cosby et al., 2003; Millar et al., 1998; Zweier et al., 1999). This has led to the suggestion that nitrite may have an important biological role in functioning as a storage pool of available NO activity in blood and other tissues that participates in blood flow regulation (Cosby et al., 2003). Nitrite may also be a signaling molecule on its own and may regulate gene expression (Bryan et al., 2005). Here we discuss the possible mechanisms and the physiological consequences of this recently discovered biological function of nitrite, with focus on the regulation of arterial and arteriolar vasodilation.

The discovery of nitrite as a vasodilator

Even though the first written description of the use of nitrite to treat cardiovascular disorders may date back to China in the 8th century (Gladwin et al., 2005), the first demonstration that nitrite is able to relax aortic strips was made in the mid-20th century (Furchgott and Bradrakom, 1953). However, the high concentrations of nitrite needed to elicit an *in vitro* response seemed to preclude any significant role for nitrite in the physiological control of vasodilation. Along with the advancement in our understanding of the biological chemistry of NO, the role of nitrite in the control of vasodilation has become increasingly evident. Following the breakthroughs in physiological NO research in the 1980s, inorganic nitrite was recognized as a weak vasodilator and stimulator of soluble guanylyl cyclase, particularly at low pH and in the presence of thiols (Ignarro et al., 1981), and its mode of action was associated with the formation of NO (Ignarro, 1989). Gladwin et al. (2000) noticed a significant arterial–venous plasma nitrite difference and an increased nitrite consumption during exercise when eNOS was inhibited. They suggested that nitrite was a stable source of bioavailable NO even at the low normal plasma nitrite concentrations in humans. Infused nitrite was subsequently shown to vasodilate the human forearm circulation at near-physiological blood nitrite concentrations (Cosby et al., 2003), while inhaled nitrite caused vasodilation of the pulmonary circulation and appearance of NO in the expiratory gas of lambs (Hunter et al., 2004). Taken together, these observations point to a common mechanism whereby nitrite is converted to vasoactive NO *in vivo*. It has also been reported that vasodilation of aortic rings requires lower concentrations of nitrite at low than at high oxygen tensions, both in the presence of purified Hb (Cosby et al., 2003; Dalsgaard et al., 2007) and RBCs (Crawford et al., 2006), indicating that hypoxia may facilitate nitrite-mediated vasodilation. This aspect is of crucial importance, given that the activity of eNOS may decrease at low oxygen tensions because of the limited availability of molecular oxygen, which is a co-substrate for NO synthesis from L-arginine. Thus, during *in vivo* hypoxia, NO regeneration from nitrite may overcome the decreased *de novo* NO synthesis via eNOS. The questions therefore arise: where and to which extent does the reduction of nitrite to NO occur, and is it enzymatically catalyzed or not?

The reactions of nitrite with Hb

The reactions of nitrite with oxygenated Hb and deoxygenated Hb differ in their reaction mechanisms, kinetics, and reaction products. The reaction between oxyHb and nitrite is complex and involves a series of intermediate reactions, but its stoichiometry (Kosaka and Tyuma, 1987) reveals that nitrite and oxyHb become oxidized to nitrate and metHb, respectively, and that three of the Hb-bound oxygen molecules become reduced:

$$4Hb(Fe^{2+})O_2 + 4NO_2^- + 4H^+ \rightarrow 4Hb(Fe^{3+}) + 4NO_3^- + O_2 + 2H_2O$$

$$(1)$$

This reaction may be of physiological importance by detoxifying excess nitrite to nontoxic nitrate (Fig. 1) (for details see Jensen, 2003).

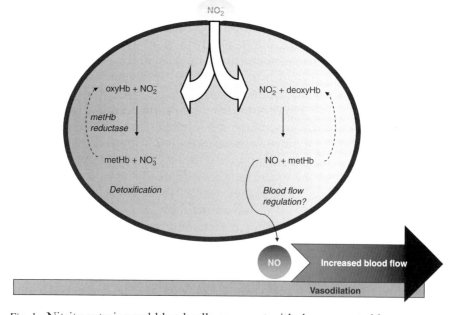

Fig. 1. Nitrite entering red blood cells can react with deoxygenated heme groups to form nitric oxide that may escape from the cells and partake in local blood flow regulation. The reaction of nitrite with oxygenated heme groups can be considered a nitrite detoxification mechanism, because it converts nitrite to nontoxic nitrate, while the oxidized heme groups (metHb) can be reduced to functional hemoglobin (Hb) via metHb reductase. Modified from Jensen (2007).

In the reaction with deoxyhb, nitrite is reduced to NO and Hb is oxidized to methb, whereby deoxyhb functions as a nitrite reductase (Cosby *et al.*, 2003; Doyle *et al.*, 1981; Nagababu *et al.*, 2003):

$$Hb(Fe^{2+}) + NO_2^- + H^+ \rightarrow Hb(Fe^{3+}) + NO + OH^- \qquad (2)$$

The reaction of nitrite with deoxygenated Hb is much slower than the corresponding reaction with oxygenated Hb (Doyle *et al.*, 1981; Kiese, 1974).

The deoxyhb-dependent mechanism of NO generation from nitrite was recently suggested to provide a mechanism by which RBCs and Hb can participate in blood flow regulation. The basic idea is that a decrease in the degree of oxygenation of Hb, as occurs during hypoxia, is linked to an increase in the ability of Hb to generate a vasodilating factor, namely NO, from nitrite, whereby blood flow is locally enhanced in tissues suffering from oxygen deficiency (Fig. 1). The hypothesis involves a series of events: (1) transport of nitrite into the RBCs, (2) reaction of nitrite with Hb, (3) escape of NO from the RBCs, and (4) induction of vaso-dilation by the NO originating from RBCs (Fig. 1). The following section will describe and discuss these individual steps.

Transport of nitrite into RBCs

Early studies on carp showed that nitrite selectively enters deoxygenated RBCs, whereas it hardly enters oxygenated RBCs at physiological pH (Jensen, 1990, 1992). A similar strong oxygenation dependency of nitrite transport was also seen in two other teleost fish (tench and whitefish), suggesting that membrane permeability to nitrite changes drastically with the degree of oxygenation of the RBCs (Jensen, 2003). Such a selective entry of nitrite at low oxygen saturations would seem appropriate for regulating the supply of nitrite to deoxyhb-mediated NO generation. This oxygenation dependency is, however, not ubiquitously present. When added to pig RBC suspensions, nitrite rapidly permeates and equilibrates across the membranes of both oxygenated and deoxygenated RBCs, whereafter nitrite continues to enter the cells (still with little ox-ygenation dependency) as a result of the intracellular removal of nitrite via its reactions with Hb (Jensen, 2005). Similar transport characteristics and a lack of oxygenation dependency of nitrite transport were also observed in rainbow trout RBCs (Jensen and Agnisola, 2005). Thus, there may be substantial species variation in the membrane character-istics that guide nitrite transport.

Several aspects concerning transport kinetics and their oxygenation and pH dependencies are compatible with the idea that nitrite permeates the membrane as the nitrite anion (Jensen, 2003). This transport would require a membrane protein (due to the negative charge, which gives nitrite a low lipid solubility). A likely candidate is the chloride/bicarbonate anion exchanger AE1, because AE1 is the most abundant membrane protein in RBCs, and is capable of mediating large and rapid anion influx and equilibration across the membrane. Erythrocyte nitrite transport is, however, not affected by the AE1 inhibitor DIDS or by the more general inhibitor of facilitated diffusion phloretin (Jensen, 1990, 2005; May et al., 2000). Thus, if AE1 is involved, this should be in a transport mode not affected by DIDS. The latter is possible, as DIDS binds to a different site than the AE1 transport site (Salhany, 2001).

The possibility that nitrite enters as HNO_2 via simple diffusion through the lipid bilayer has also been considered (Jensen, 1990, 2005; May et al., 2000). Nitrite influx increases as pH is lowered (Jensen 1992; May et al., 2000), which could indicate a role for HNO_2 diffusion, because the HNO_2 concentration increases at low pH. However, a lowering of pH also elevates the Donnan-like distribution ratio of permeable ions and thereby the membrane potential, whereby low pH should also increase diffusion of the nitrite anion (Jensen, 1992). Indeed, many aspects of the transport kinetics are compatible with both NO_2^- and HNO_2 diffusion, and therefore cannot be used to distinguish between the transport modes. The low pK_a of nitrous acid (3.3) implies that the concentration of HNO_2 is roughly 12,000 times lower than that of NO_2^- at a pH of 7.4. Furthermore, the extracellular pH remains stable during transport (rather than increasing as a result of NO_2^- combining with H^+), in both the absence and the presence of DIDS (Jensen, 1990, 2005), which argues against HNO_2 diffusion as the major transport mechanism.

Although little evidence is presently available, it has also been suggested that transport could be via gaseous N_2O_3 (May et al., 2000), whose formation from nitrous acid might also occur within membrane lipid bilayers, as both compounds are lipid soluble. Also, a small amount of nitrite may enter human erythrocytes via the sodium-dependent phosphate transporter (May et al., 2000).

Nitrite reduction by hemoglobin

As is evident from the disparate reactions of nitrite with oxyHb and deoxyHb (cf. above), the reduction of nitrite to NO is restricted to deoxygenated heme groups. This potentially links erythrocyte NO formation from nitrite to the saturation of Hb with O_2, and thereby to the allosteric R ('oxy') to T ('deoxy') transition in Hb. The nitrite reductase activity is governed by both heme deoxygenation and heme redox potential. Deoxygenated heme groups are more abundant in the T structure, but their redox potential is higher (*i.e.*, the ability to donate an electron to nitrite is lower) than that for deoxygenated heme groups in the R structure. Studies on solutions of purified adult human Hb suggest that this opposite behavior leads to a maximal nitrite reduction rate when the Hb is ~50% saturated, *i.e.*, at the P_{50} of the Hb (Huang *et al.*, 2005b). This is supported by calculated nitrite reduction rates in intact RBCs and the observation of a higher NO production in RBCs at 50% oxygen saturation than at either 0% or 100% saturation (Crawford *et al.*, 2006).

Rifkind and co-workers (Nagababu *et al.*, 2003) advanced the hypothesis that the reaction between nitrite and deoxyHb would generate vasoactive NO by steady-state formation in the course of the reaction of a ferric (Fe^{3+})-NO-heme intermediate with low affinity for NO. Subsequent studies made at relatively high nitrite concentrations did not identify such a labile intermediate (Huang *et al.*, 2005a). Recent results, on the other hand, have shown that under certain conditions nitrite may produce S-nitrosation of Hb, possibly with (Fe^{3+})-NO heme involved (Angelo *et al.*, 2006; Nagababu *et al.*, 2006). This S-nitrosation process, rather than the generation of NO free radical, could then contribute to the vasodilation effects induced by nitrite and Hb. Clearly, some crucial aspects of the reaction between nitrite and Hb remain to be fully elucidated.

Escape of NO from RBCs

Traditionally, Hb has been considered a scavenger rather than a source of NO, given the extremely rapid, nearly diffusion-limited, reactions of NO with either deoxygenated or oxygenated ferrous heme to form either thermodynamically stable nitrosyl-heme or nitrate, respectively. However, the view that RBCs are exclusively a sink of NO changed in 1996 with the discovery by Stamler and co-workers of small amounts of Hb in the *S*-nitroso form (SNO-Hb) in the blood and with the derived hypothesis that SNO-Hb may have vasoactive functions in being able to liberate NO when deoxygenated (Jia *et al.*, 1996). Although controversial aspects of this theory remain unsolved (an aspect which is beyond the scope of this review), the SNO hypothesis raised for the first time the view that RBCs, in addition to transporting oxygen, could also be involved in the transport of a factor, namely NO, controlling blood flow, and whose delivery was tightly controlled by the oxygenation state of the Hb. However, with the subsequent improvements to the methods for the detection of SNO derivatives *in vivo*, later studies have not found a significant artero-venous gradient of SNO-Hb, and also report much lower amounts of SNO-Hb than those originally measured, at least in humans (Cannon *et al.*, 2001; Gladwin *et al.*, 2002).

One fundamental question is: how are small amounts of NO able to escape erythrocytes without being trapped by Hb present at high concentration ($\sim$20 mM heme) inside the cells? Stamler and co-workers have indicated that this escape is mediated by a series of transnitrosation reactions occurring across the membrane and involving the AE1 anion exchanger of the RBC membrane and low-molecular-weight thiols in the plasma (Pawloski *et al.*, 2001), whereby NO free radical – with a high affinity for heme – would not be generated inside the RBC.

The reduction of nitrite to NO by deoxyHb apparently solves some of the problems associated with the delivery of sufficient NO from RBCs, as nitrite is present at higher concentrations than SNO-thiols, in both plasma and RBCs. Gladwin *et al.* (2004) have advanced the hypothesis that the diffusion of NO out of RBCs would be favored by a localization of catalytic NO formation at the membrane via the local clustering of deoxygenated Hb and carbonic anhydrase on the cytoplasmic domains of the anion exchanger AE1, assuming that AE1 mediates nitrite entry (*cf.* above). The RBC membrane and the unstirred layer of plasma around the RBC have been found to represent an important physical barrier for the diffusion of the NO produced by the endothelium into the red cells (which results in a lower scavenging of endothelial NO than expected from the Hb concentration – *cf.* Kim-Shapiro *et al.*, 2006). However, such a barrier would also limit the diffusion of NO out of the red cells, and thus favor the interaction (which includes scavenging reactions) between Hb and the NO produced within the erythrocyte.

To circumvent the problem associated with the export of NO bioactivity from RBCs, Robinson and Lancaster (2005) recently questioned whether N_2O_3, which can be formed by ferrous heme-catalyzed dehydration of nitrite, might be the species diffusing out of RBCs rather than NO. NO could then be formed outside the RBCs by N_2O_3 homolysis to NO and NO_2. N_2O_3 is also a potent S-nitrosating species, whereby the nitrite-induced vasodilation could, under some conditions, translate into an SNO response, as some recent studies suggest (Angelo *et al.*, 2006; Nagababu *et al.*, 2006).

Even though the mechanism of the export of NO bioactivity from RBCs is still unsolved, many studies support the theory that Hb and

RBCs are able to produce NO at low oxygen saturation and in the presence of nitrite (Cosby *et al.*, 2003; Crawford *et al.*, 2006; Jensen and Agnisola, 2005). However, as discussed below, it remains to be established whether this Hb-mediated NO production is on its own sufficient to produce vasodilation at physiological nitrite levels.

Induction of vasodilation

In vivo infusion of nitrite into the human forearm circulation was shown to increase blood flow by inducing vasodilation at near-physiological nitrite concentrations (Cosby *et al.*, 2003). *In vitro* experiments performed on aortic rings showed that lower concentrations of nitrite were needed to induce vasodilation when the fraction of deoxygenated Hb in RBCs was increased by decreasing the oxygen tension (Crawford *et al.*, 2006). Similarly, a decrease in the oxygen saturation of purified Hb at constant oxygen tension (15 mmHg), due to the presence of the allosteric effector inositol hexaphosphate (IHP), was also able to promote vasodilation at lower nitrite concentrations (Cosby *et al.*, 2003). More recent studies have, however, raised the question whether the vasodilatory effect observed by Cosby *et al.* (2003), when using purified Hb, was due to IHP alone (Dalsgaard *et al.*, 2007; Luchsinger *et al.*, 2005), as IHP has been found to enhance nitrite-mediated vasodilation regardless of the presence of Hb (Dalsgaard *et al.*, 2007).

In the presence of RBCs, the nitrite concentration that stimulated 20% dilation of aortic rings was 5.6 µM at an O_2 tension of 25 mmHg and ~1 µM at an O_2 tension of 15 mmHg (Crawford *et al.*, 2006), which is slightly higher than normal plasma nitrite concentrations in mammals (0.15–0.6 µM; Kleinbongard *et al.*, 2003). At first sight this suggests that the contribution to vasodilation from the nitrite reductase activity of deoxygenated Hb might be small at physiological nitrite concentrations. However, a direct extrapolation from the *in vitro* aortic ring model to *in vivo* conditions may be problematic and should be made with caution, especially when considering that no data are presently available on how nitrite affects vasodilation of the major resistance vessels, the arterioles. Furthermore, due to the strong influence of vessel diameter on resistance to flow, even a minor dilation can translate into a significant effect on blood flow.

An increase in the levels of circulating nitrite in rats did not inhibit platelet aggregation (Bryan *et al.*, 2005), as would be expected if significant levels of NO were generated from nitrite and diffused out of the RBCs and into the plasma. It is possible that a noteworthy part of the *in vivo* vasodilatory effect of nitrite may arise from NO formation within the aortic tissue, rather than from the reductase activity of Hb, as arterial tissue contains higher levels of nitrite ($10\,\mu M$) and S-nitroso compounds (RSNO) ($0.2\,\mu M$) than other tissues. Bryan *et al.* (2005) have shown a direct correlation between tissue RSNO and levels of circulating nitrite in rats. Thus, the generation of a pool of vasoactive S-nitrosothiols from nitrite within the arterial tissue could ensure a local NO supply to activate soluble guanylyl cyclase, especially under reducing and acidic conditions (*i.e.*, tissue hypoxia or ischemia), where S-nitrosothiols are less stable. Also, Modin *et al.* (2001) have shown that acidic conditions approaching those occurring during ischemia or hypoxia in active tissues (pH 6.6) enhance nitrite-induced relaxation of rat aortic rings, and they proposed that nonenzymatic, acid-catalyzed nitrite reduction to NO may contribute significantly to hypoxic vasodilation *in vivo*.

The relative contributions to nitrite-dependent vasodilation from nonenzymatic and enzymatic processes within the vascular tissue and the blood have not yet been fully evaluated. Furthermore, in addition to deoxyHb-mediated reduction of nitrite to NO within the circulation, other heme proteins within the endothelium or vascular/cardiac muscle cells may function as nitrite reductases. The role of 'alternative heme proteins,' including myoglobin, cytochromes, and neuroglobin, in nitrite-induced vasodilation remains to be investigated.

The hypothesis that nitrite is converted into vasoactive NO has primarily been studied in mammalian models. There is limited information on other vertebrates, even though the widespread use of endothelial NO in blood flow regulation among vertebrates implies a potential relevance of nitrite-derived NO as well. Rainbow trout exposed to ambient nitrite were shown to experience acutely increased heart rate well before any significant increases in metHb, extracellular $[K^+]$, or adrenergic stress responses were present (Aggergaard and Jensen, 2001). It was suggested that the mere appearance of nitrite in plasma caused vasodilation (via formation of NO), which was then countered by increased cardiac pumping to reestablish blood pressure (Aggergaard and Jensen, 2001). More recently, the potential role of nitrite was tested in the coronary circulation of isolated trout heart. This study testified to a vasodilatory role for endothelial NO produced under hypoxia, but failed to demonstrate vasodilation from NO in coronary arteries perfused with RBCs and nitrite (Jensen and Agnisola, 2005). Apparently, the nitrite-derived NO was produced in the capillaries beyond the resistance vessels, and the signal was not conducted upstream to the arterioles to induce their dilation (Jensen and Agnisola, 2005).

209

Conclusions and perspectives

The origin of nitrite vasoactivity is not yet fully elucidated, but major advances have recently been made thanks to the stimulating and intense ongoing debate on this issue. The emerging picture is that nitrite reactivity in the blood, like NO reactivity, is governed by a complex interplay between oxygen tension, pH, levels of thiols and reducing equivalents, and the activity of proteins such as Hb and AE1. These factors are linked to the metabolism of RBCs, as well as that of blood vessels and other tissues, and may allow the creation of unique tissue-specific conditions for NO and RSNO generation and availability with functions that extend beyond the control of arterial/arteriolar vasodilation during hypoxia. The discovery that nitrite can exert vasoactivity and other biological functions at its naturally low physiological concentration has opened a new field of research that promises to flourish for years to come.

Acknowledgments

The authors are supported by the Danish Natural Science Research Council and the Novo Nordisk Foundation.

References

Aggergaard, S. and Jensen, F. B. (2001). Cardiovascular changes and physiological response during nitrite exposure in rainbow trout. *J. Fish Biol.* 59,13–27.
Angelo, M., Singel, D. J. and Stamler, J. S. (2006). An *S*-nitrosothiol (SNO) synthase function of hemoglobin that utilizes nitrite as a substrate. *Proc. Natl. Acad. Sci. USA* 103,8366–8371.
Bryan, N. S., Fernandez, B. O., Bauer, S. M., Garcia-Saura, M. F., Milsom, A. B., Rassaf, T., Maloney, R. E., Bharti, A., Rodriguez, J. and Feelisch, M. (2005). Nitrite is a signaling molecule and regulator of gene expression in mammalian tissues. *Nat. Chem. Biol.* 1,290–297.
Cannon, R. O., Schechter, A. N., Panza, J. A., Ognibene, F. P., Pease-Fye, M. E., Waclawiw, M. A., Shelhamer, J. H. and Gladwin, M. T. (2001). Effects of inhaled nitric oxide on regional blood flow are consistent with intravascular nitric oxide delivery. *J. Clin. Invest.* 108,279–287.
Cosby, K., Partovi, K. S., Crawford, J. H., Patel, R. P., Reiter, C. D., Martyr, S., Yang, B. K., Waclawiw, M. A., Zalos, G., Xu, X., Huang, K. T., Shields, H., Kim-Shapiro, D. B., Schechter, A. N., Cannon, R. O. and Gladwin, M. T. (2003). Nitrite reduction to nitric oxide by deoxyhemoglobin vasodilates the human circulation. *Nat. Med.* 9,1498–1505.
Crawford, J. H., Isbell, T. S., Huang, Z., Shiva, S., Chacko, B. K., Schechter, A. N., Darley-Usmar, V. M., Kerby, J. D., Lang, J. D., Kraus, D., Ho, C., Gladwin, M. T.

and Patel, R. P. (2006). Hypoxia, red blood cells, and nitrite regulate NO-dependent hypoxic vasodilation. *Blood* 107,566–574.

Dalsgaard, T., Simonsen, U. and Fago, A. (2007). Nitrite-dependent vasodilation is facilitated by hypoxia and is independent of known NO-generating nitrite-reductase activities. *Am. J. Physiol. Heart Circ. Physiol.* (in press).

Doyle, M. P., Pickering, R. A., DeWeert, T. M., Hoekstra, J. W. and Pater, D. (1981). Kinetics and mechanism of the oxidation of human deoxyhemoglobin by nitrites. *J. Biol. Chem.* 256,12393–12398.

Furchgott, R. F. and Bradrakom, S. (1953). Reactions of strips of rabbit aorta to epinephrine, isopropylarterenol, sodium nitrite and other drugs. *J. Pharmacol. Exp. Ther.* 108,129–143.

Gladwin, M. T., Crawford, J. H. and Patel, R. P. (2004). The biochemistry of nitric oxide, nitrite, and hemoglobin: Role in blood flow regulation. *Free Radic. Biol. Med.* 36,707–717.

Gladwin, M. T., Schechter, A. N., Kim-Shapiro, D. B., Patel, R. P., Hogg, N., Shiva, S., Cannon, R. O., Kelm, M., Wink, D. A., Espey, M. G., Oldfield, E. H., Pluta, R. M., Freeman, B. A., Lancaster, J. R., Feelisch, M. and Lundberg, J. O. (2005). The emerging biology of the nitrite anion. *Nat. Chem. Biol.* 1,308–314.

Gladwin, M. T., Shelhamer, J. H., Schechter, A. N., Pease-Fye, M. E., Waclawiw, M. A., Panza, J. A., Ognibene, F. P. and Cannon, R. O. (2000). Role of circulating nitrite and *S*-nitrosohemoglobin in the regulation of regional blood flow in humans. *Proc. Natl. Acad. Sci. USA* 97,11482–11487.

Gladwin, M. T., Wang, X., Reiter, C. D., Yang, B. K., Vivas, E. X., Bonaventura, C. and Schechter, A. N. (2002). *S*-Nitrosohemoglobin is unstable in the reductive erythrocyte environment and lacks O_2/NO-linked allosteric function. *J. Biol. Chem.* 277,27818–27828.

Huang, K. T., Keszler, A., Patel, N., Patel, R. P., Gladwin, M. T., Kim-Shapiro, D. B. and Hogg, N. (2005a). The reaction between nitrite and deoxyhemoglobin: Reassessment of reaction kinetics and stoichiometry. *J. Biol. Chem.* 280,31126–31131.

Huang, Z., Shiva, S., Kim-Shapiro, D. B., Patel, R. P., Ringwood, L. A., Irby, C. E., Huang, K. T., Ho, C., Hogg, N., Schechter, A. N. and Gladwin, M. T. (2005b). Enzymatic function of hemoglobin as a nitrite reductase that produces NO under allosteric control. *J. Clin. Invest.* 115,2099–2107.

Hunter, C. J., Dejam, A., Blood, A. B., Shields, H., Kim-Shapiro, D. B., Machado, R. F., Tarekegn, S., Mulla, N., Hopper, A. O., Schechter, A. N., Power, G. G. and Gladwin, M. T. (2004). Inhaled nebulized nitrite is a hypoxia-sensitive NO-dependent selective pulmonary vasodilator. *Nat. Med.* 10,1122–1127.

Ignarro, L. J. (1989). Biological actions and properties of endothelium-derived nitric oxide formed and released from artery and vein. *Circ. Res.* 65,1–21.

Ignarro, L. J., Lippton, H., Edwards, J. C., Baricos, W. H., Hyman, A. L., Kadowitz, P. J. and Gruetter, C. A. (1981). Mechanism of vascular smooth-muscle relaxation by organic nitrates, nitrites, nitroprusside and nitric oxide – Evidence for the involvement of *S*-nitrosothiols as active intermediates. *J. Pharmacol. Exp. Ther.* 218,739–749.

Jaffé, E. R. (1981). Methaemoglobinaemia. *Clin. Haematol.* 10,99–122.

Jensen, F. B. (1990). Nitrite and red cell function in carp: Control factors for nitrite entry, membrane potassium ion permeation, oxygen affinity and methaemoglobin formation. *J. Exp. Biol.* 152,149–166.

Jensen, F. B. (1992). Influence of haemoglobin conformation, nitrite and eicosanoids on K^+ transport across the carp red blood cell membrane. *J. Exp. Biol.* 171,349–371.

Jensen, F. B. (2003). Nitrite disrupts multiple physiological functions in aquatic animals. *Comp. Biochem. Physiol.* 135A,9–24.

Jensen, F. B. (2005). Nitrite transport into pig erythrocytes and its potential biological role. *Acta Physiol. Scand.* 184,243–251.

Jensen, F. B. (2007). Physiological effects of nitrite: Balancing the knife's edge between toxic disruption of functions and potential beneficial effects. In *Proceedings of the Ninth International Symposium on Fish Physiology, Toxicology and Water Quality* (eds C. J. Brauner, K. Suvajdzic, G. Nilsson and D. J. Randall), US Environmental Protection Agency, Ecosystems Research Division, pp. 119–132, Athens, GA, USA.

Jensen, F. B. and Agnisola, C. (2005). Perfusion of the isolated trout heart coronary circulation with red blood cells: Effects of oxygen supply and nitrite on coronary flow and myocardial oxygen consumption. *J. Exp. Biol.* 208,3665–3674.

Jia, L., Bonaventura, C., Bonaventura, J. and Stamler, J. S. (1996). *S*-Nitrosohaemoglobin: A dynamic activity of blood involved in vascular control. *Nature* 380,221–226.

Kiese, M. (1974). *Methemoglobinemia: A Comprehensive Treatise*, CRC Press, Cleveland.

Kim-Shapiro, D. B., Schechter, A. N. and Gladwin, M. T. (2006). Unraveling the reactions of nitric oxide, nitrite, and haemoglobin in physiology and therapeutics. *Arterioscler. Thromb. Vasc. Biol.* 26,697–705.

Kleinbongard, P., Dejam, A., Lauer, T., Rassaf, T., Schindler, A., Picker, O., Scheeren, T., Gödecke, A., Schrader, J., Schulz, R., Heusch, G., Schaub, G. A., Bryan, N. S., Feelisch, M. and Kelm, M. (2003). Plasma nitrite reflects constitutive nitric oxide synthase activity in mammals. *Free Radic. Biol. Med.* 35,790–796.

Kosaka, H. and Tyuma, I. (1987). Mechanism of autocatalytic oxidation of oxyhemoglobin by nitrite. *Environ. Health Perspect.* 73,147–151.

Luchsinger, B. P., Ric, E. N., Yan, Y., Williams, E. M., Stamler, J. S. and Singel, D. J. (2005). Assessments of the chemistry and vasodilatory activity of nitrite with haemoglobin under physiologically relevant conditions. *J. Inorg. Biochem.* 99,912–921.

Lundberg, J. O. and Weitzberg, E. (2005). NO generation from nitrite and its role in vascular control. *Arterioscler. Thromb. Vasc. Biol.* 25,915–922.

May, J. M., Qu, Z.-C., Xia, L. and Cobb, C. E. (2000). Nitrite uptake and metabolism and oxidant stress in human erythrocytes. *Am. J. Physiol. Cell Physiol.* 279,C1946–C1954.

Millar, T. M., Stevens, C. R., Benjamin, N., Eisenthal, R., Harrison, R. and Blake, D. R. (1998). Xanthine oxidoreductase catalyses the reduction of nitrates and nitrite to nitric oxide under hypoxic conditions. *FEBS Lett.* 427,225–228.

Modin, A., Björne, H., Herulf, M., Alving, K., Weitzberg, E. and Lundberg, J. O. N. (2001). Nitrite-derived nitric oxide: A possible mediator of 'acidic-metabolic' vasodilation. *Acta Physiol. Scand.* 171,9–16.

Nagababu, E., Ramasamy, S., Abernethy, D. R. and Rifkind, J. M. (2003). Active nitric oxide produced in the red cell under hypoxic conditions by deoxyhemoglobin-mediated nitrite reduction. *J. Biol. Chem.* 278,46349–46356.

Nagababu, E., Ramasamy, S. and Rifkind, J. M. (2006). *S*-Nitrosohemoglobin: A mechanism for its formation in conjunction with nitrite reduction by deoxyhemoglobin. *Nitric Oxide* 15,20–29.

212

Pawloski, J. R., Hess, D. T. and Stamler, J. S. (2001). Export by red blood cells of nitric oxide bioactivity. *Nature* 409,622–626.

Robinson, J.M. and Lancaster Jr., J.R. (2005). Hemoglobin-mediated, hypoxia-induced vasodilation via nitric oxide. *Am. J. Respir. Cell. Mol. Biol.* 32,257–261

Salhany, J. M. (2001). Stilbenedisulfonate binding kinetics to band 3 (AE1): Relationship between transport and stilbenedisulfonate binding sites and role of subunit interactions in transport. *Blood Cell Mol. Dis.* 27,127–134.

Zweier, J. L., Samouilov, A. and Kuppusamy, P. (1999). Non-enzymatic nitric oxide synthesis in biological systems. *Biochim. Biophys. Acta* 1411,250–262.

Nitrite is a vascular store of NO which mediates hypoxic signaling and protects against ischemia/reperfusion injury

Cameron Dezfulian[1,2,3], Mark T. Gladwin[1,2,*] and Sruti Shiva[1]

[1]Vascular Medicine Branch, National Heart Lung Blood Institute, National Institutes of Health, Bethesda, MD 20892, USA
[2]Critical Care Medicine Department, Clinical Center, National Institutes of Health, Bethesda, MD 20892, USA
[3]Division of Pediatric Anesthesia and Critical Care Medicine, Johns Hopkins Hospital, Baltimore, MD 21287, USA

Abstract. The circulating anion nitrite, once thought to be a physiologically inert by-product of nitric oxide (NO) oxidation, has been proposed to be a vascular storage form of bioactive NO. Nitrite is reduced to bioactive NO along a physiological oxygen and pH gradient by its reaction with deoxygenated hemoglobin and other hemoproteins. Through this mechanism, nitrite plays a role in hypoxic vasodilation and is capable of inhibiting mitochondrial respiration during hypoxia. Accumulating data demonstrate that nitrite is a potent mediator of cytoprotection after ischemia/reperfusion (I/R) injury in several organs, including the heart, liver, and brain. However, the mechanisms of nitrite-dependent cytoprotection remain unknown. In this article, we review the role of nitrite as a hypoxic source of NO and discuss the potential mechanisms of nitrite-mediated cytoprotection from I/R injury.

Keywords: nitrite; nitric oxide; ischemia; reperfusion; hypoxia.

Introduction

The field of nitric oxide (NO) biology has grown rapidly since the discovery of the classical NO-vasodilatory pathway. Stimulation of this pathway by shear stress or pharmacologic agents such as acetylcholine or bradykinin results in the production of NO by the endothelial isoform of nitric oxide synthase (NOS) (Vallance, 2000). NO diffuses abluminally into smooth muscle cells, where it activates soluble guanylyl cyclase (sGC), which produces cyclic guanosine monophosphate (cGMP) and triggers a signaling cascade that ultimately leads to smooth muscle relaxation and vasodilation (Ignarro et al., 1987; Palmer et al., 1987). Although this vasodilatory pathway remains the classical NO signaling pathway, we now know that NO is involved in other vascular signaling

*Corresponding author: Tel.: 301-435-2310. Fax: 301-451-7091.
E-mail: mgladwin@mail.nih.gov (M.T. Gladwin).

ADVANCES IN EXPERIMENTAL BIOLOGY   2007 Published by ELSEVIER B.V.
VOLUME 01   ISSN 1872-2423
DOI: 10.1016/S1872-2423(07)01010-1

pathways, including inhibition of platelet activation, smooth muscle proliferation, and hypoxic/ischemic signaling (Bolli, 2001).

Data is accumulating not only about the putative pathways affected by NO signaling but also the means by which this signal is delivered throughout the body both in health and during pathologic circumstances. Traditionally, NO was thought to be a paracrine signaling molecule, mediating effects in close proximity to the site at which it is generated. However, a growing field of data suggests that NO may be stabilized and transported to mediate endocrine signaling distal from its initial site of production (Cannon et al., 2001; Fox-Robichaud et al., 1998; Gladwin et al., 2000a; Ng et al., 2004; Rassaf et al., 2002). Examples of this endocrine activity include the ability of inhaled NO to mediate increases in mesenteric blood flow and decreases in leukocyte adhesion after ischemia/reperfusion (I/R) in cats (Fox-Robichaud et al., 1998; Kubes et al., 1999; Ng et al., 2004), and increases in urinary flow, glomerular filtration rate, and renal blood flow in pigs (Troncy et al., 1997). In human volunteers, inhalation of NO gas increased forearm blood flow in the presence of NOS inhibition (Cannon et al., 2001), and infusion of NO solution in one arm mediated vasodilation in the opposing arm (Rassaf et al., 2002). Since the half-life of NO is less than one second in blood (Fox-Robichaud et al., 1998), an ideal endocrine transporter of NO activity would not only be more stable in vivo but also able to generate NO in the absence of the substrates that NOS requires, such as oxygen. In this context, NOS-independent NO generation may be an integral component of signaling in conditions such as hypoxia or ischemia.

While the concept of the existence of a circulating NO storage form was first proposed by Loscalzo and colleagues and has virtually been accepted (Scharfstein et al., 1994; Stamler et al., 1992), the identity of the NO storage form remains more controversial. Several species have been proposed to be the endocrine mediator of NO, including S-nitrosoalbumin (Ng et al., 2004; Stamler et al., 1992), iron-nitrosyl hemoglobin (Gladwin et al., 2000a), and S-nitrosohemoglobin (Stamler et al., 1997). Most recently, we have proposed that the anion nitrite (NO_2^-) is the largest vascular storage form of NO and may serve this endocrine transport function (Cosby et al., 2003). Here we will review the expanding field of nitrite biology and provide evidence supporting the role for nitrite as a source of NO that is independent of NOS and capable of mediating NO-dependent endocrine effects. We will focus on characterizing the protective effects of nitrite during pathological conditions where NOS activity is limited, particularly I/R.

Nitrite is a vasodilator under normal physiological conditions and during hypoxic or exercise stress

Although nitrite has long been used as an antidote for cyanide poisoning (Pedigo, 1888), this simple anion was traditionally thought to be a physiologically inert oxidation product of NO. This idea was supported by classic *in vitro* vessel tension studies that showed that nitrite could dilate vessels only at supraphysiological $(100\,\mu M{-}1\,mM)$ concentrations (Furchgott and Bhadrakom, 1953). Since circulating concentrations of nitrite are less than $1\,\mu M$ (Dejam *et al.*, 2005; Lauer *et al.*, 2001), these studies appeared to preclude a role for nitrite as a physiological mediator.

More recent studies have shown that nitrite may indeed possess vaso-active properties under physiological relevant conditions and *in vivo* (Cosby *et al.*, 2003; Gladwin *et al.*, 2000b; Hunter *et al.*, 2004). Gladwin and colleagues observed that artery-to-vein gradients of nitrite exist in the forearm of human subjects, and these gradients increase with exercise, suggesting that nitrite is metabolized in the peripheral vasculature and by the tissues it supplies (Gladwin *et al.*, 2000b). In addition, in humans, inhalation of NO resulted in an increase in forearm blood flow in the presence of a NOS inhibitor, and this vasodilation was associated with the formation of nitrite (Cannon *et al.*, 2001).

To test the vasoactivity of nitrite directly, we infused nitrite into the branchial artery of 28 normal volunteers and measured changes in blood flow in the human forearm (Cosby *et al.*, 2003). Infusion of near-physiological concentrations of nitrite increased forearm blood flow at rest in the presence and absence of NOS inhibition. Exercise resulted in a predictable decrease in forearm vein pH and PO_2, and infusion of nitrite during exercise increased blood flow to a greater extent than when subjects were at rest. Interestingly, nitrite-dependent vasodilation was associated with the formation of iron-nitrosyl hemoglobin, and to a lesser extent S-nitrosohemoglobin (Cosby *et al.*, 2003).

Hemoglobin is a nitrite reductase

Several mechanisms are known to exist for the reduction of nitrite to NO, including enzymatic reduction by xanthine oxidoreductase (Millar *et al.*, 1998) and non-enzymatic acidic disproportionation (Zweier *et al.*, 1995). Nohl and colleagues have also described a nitrite reductase activity of mammalian mitochondrion (Kozlov *et al.*, 1999). However, these mechanisms of reduction were unlikely to be responsible for the nitrite-dependent vasodilation observed *in vivo*, since both of the former mechanisms require low pH and hypoxic conditions, and the nitrite reductase activity of mitochondria has only been observed at relatively high nitrite concentrations ($> 50\,\mu M$). Since nitrite-dependent vasodilation was associated with the formation of NO-modified hemoglobin, Huang and colleagues investigated the possibility that nitrite was reduced to bioactive NO by deoxyhemoglobin through the reaction previously characterized by Doyle and colleagues (Doyle *et al.*, 1981; Huang *et al.*, 2005a):

$$\text{Nitritesubind} : + \text{ deoxyhemoglobin } (Fe^{2+}) + H^+ \rightarrow NO$$
$$+ \text{ methemoglobin } (Fe^{3+}) + OH^-$$

By this reaction, deoxyhemoglobin catalyses the conversion of the protonated form of nitrite (*i.e.*, nitrous acid, HNO_2) to NO, and deoxyhemoglobin is oxidized to methemoglobin in the process. Nitric oxide produced by this reaction can then rapidly react with unreacted deoxyhemoglobin to form iron-nitrosyl hemoglobin:

$$NO + \text{ deoxyhemoglobin } \rightarrow \text{ iron-nitrosyl hemoglobin (Hb-NO)}$$

Kinetic analysis of these reactions using purified hemoglobin demonstrated that deoxygenated hemoglobin could indeed reduce nitrite to NO and that the rate of deoxyhemoglobin-dependent nitrite reduction was determined by the allosteric structural transition of hemoglobin (Huang *et al.*, 2005a, b). Interestingly, the maximal rate of nitrite reduction and NO production coincided with the P_{50} of hemoglobin (the point at which hemoglobin is half saturated). This phenomenon is determined by two opposing processes determined by characteristics of the two allosteric conformations of hemoglobin. Since T (deoxy) state hemoglobin is not saturated with oxygen, nitrite can readily bind this species to be reduced. As hemoglobin becomes increasingly saturated with oxygen, the number of available nitrite binding sites decreases, which decelerates the reaction. However, this deceleration is countered by the formation of R (oxy) state hemoglobin, which has a lower reduction potential (making it a better electron donor), which accelerates the rate of nitrite reduction. The optimal balance of these two opposing forces occurs around the P_{50} of hemoglobin, where the maximal rate of nitrite reduction and NO production is observed (Huang *et al.*, 2005b).

Hemoglobin-mediated nitrite reduction is regulated by oxygen and pH

The coupling of NO production to hemoglobin oxygen saturation makes hemoglobin-dependent nitrite reduction an ideal mechanism of hypoxic vasodilation. To determine whether the nitrite-dependent vasodilation observed *in vivo* was dependent on hemoglobin, Patel and colleagues used an *in vitro* aortic ring system to simultaneously measure vessel tone and oxygen concentration in the presence of nitrite and red blood cells (Cosby *et al.*, 2003; Crawford *et al.*, 2006). In the absence of red blood cells, addition of nitrite to pre-constricted aortic rings caused vasodilation only at high ($> 10\,\mu M$) concentrations, consistent with previous data. However, in the presence of red blood cells (0.3% hematocrit), physiological concentrations of nitrite-mediated vasodilation of aortic rings. This effect was potentiated by hypoxia such that, at an oxygen tension of 25 mm Hg, 200 nM of nitrite mediated significant vasodilation of isolated aortic rings (Cosby *et al.*, 2003).

Kinetic experiments in the aortic ring system, in which oxygen concentration was changed over time in the presence of nitrite and hemoglobin, demonstrated that the onset of significant vasodilation began around the P_{50} of hemoglobin, consistent with the kinetics of NO generation observed in the reaction of purified hemoglobin with nitrite (Cosby *et al.*, 2003; Crawford *et al.*, 2006). When mutated forms of hemoglobin with varying P_{50} values were used in this system, the oxygen concentration at the onset of vasodilation correlated with the P_{50} of the hemoglobin used in each case. This series of experiments confirmed that hemoglobin-dependent nitrite reduction could indeed generate bioactive NO, and that this production was maximal under conditions of diminished oxygen tension whereby hemoglobin was approximately 50% saturated (Cosby *et al.*, 2003; Crawford *et al.*, 2006).

These concepts were also tested in a newborn sheep model of hypoxia-induced pulmonary hypertension, in which nebulized nitrite was shown to produce a 50% decrease in pulmonary artery pressure (Hunter *et al.*, 2004). This vasodilation was accompanied by the formation of iron-nitrosyl hemoglobin and an increase in exhaled NO gas consistent with the reduction of nitrite by hemoglobin, and paralleled the effects of 20 ppm inhaled NO. Nebulized nitrite was far less effective at producing vasodilation under conditions of normoxia when pulmonary hypertension was induced by an analog of thromboxane. In this series of experiments, there was also substantially less exhaled NO and iron-nitrosyl hemoglobin formation. This reinforces the ability of nitrite to act as a stable reservoir for NO, with release catalysed by hemoglobin under conditions of diminished oxygen saturation (Hunter *et al.*, 2004).

In addition to oxygen sensor chemistry, the reduction of nitrite by hemoglobin requires a proton, enabling it to be regulated by pH as well (Doyle *et al.*, 1981). Indeed, in kinetic studies of the reaction of nitrite with purified hemoglobin, the reaction rate increased as the pH was decreased (Huang *et al.*, 2005b). In addition, in the *in vivo* sheep model of pulmonary hypertension, the magnitude of decrease in pulmonary artery pressure in response to nitrite was dependent on blood pH, with increased vasodilation as the pH decreased (Hunter *et al.*, 2004). The clear oxygen and pH sensor chemistry of the deoxyhemoglobin–nitrite reaction suggests that this reaction is potentially important in conditions of hypoxia and acidosis, such as those present during ischemia.

Nitrite protects against ischemia/reperfusion (I/R) injury

Ischemia with subsequent reperfusion produces tissue injury characterized by a number of phenomena, including an increase in reactive oxygen species (ROS) leading to oxidative damage (particularly in the mitochondria), inflammation, and activation of apoptotic effectors. The role of NO in I/R injury is complicated, as it is known to affect many pathways involved in the progression of I/R injury (Braunwald and Kloner, 1985; Lefer, 2002; Lefer *et al.*, 1993; McCord *et al.*, 1985). At physiological concentrations, NO is thought to be cytoprotective, while at higher doses NO exacerbates I/R injury (recently reviewed by Bolli, 2001). *In vivo*, production of NO during ischemia must be limited, since oxygen is a requisite substrate for NOS. In these conditions, nitrite may play an integral role as a source of NOS-independent NO generation.

Nitrite has now been shown to protect against I/R-mediated injury in a number of models. In a Langendorff perfusion model, Webb and colleagues have shown that micromolar concentrations ($10 \mu M$) of nitrite protect rat hearts from infarction (Webb *et al.*, 2004). This cytoprotective effect was also seen *in vivo* by Duranski and colleagues, using a murine model of myocardial infarction (Duranski *et al.*, 2005; Lu *et al.*, 2005; Webb *et al.*, 2004). In this model, an increase in plasma nitrite concentration of as low as 200 nM resulted in a 50% decrease in myocardial infarction volume (Duranski *et al.*, 2005). This effect was also translated to a murine model of liver I/R in which a dose of 48 nmol nitrite resulted in a 70% decrease in serum levels of the transaminases alanine aminotransferase (ALT) and aspartate aminotransferase (AST), and a 75% reduction in the number of apoptotic cells, in comparison with levels in mice that underwent hepatic I/R in the absence of nitrite. Although the precise mechanism of nitrite-dependent cytoprotection is not yet known, several possible targets exist.

Potential mechanisms of nitrite-mediated cytoprotection

At present, the precise mechanism(s) of nitrite-induced cytoprotection in I/R injury are unknown. Interestingly, in murine models of both myocardial infarction and hepatic I/R, nitrite has a biphasic protective effect against I/R injury, with low doses conferring dose-dependent protection and doses above 48 nmol demonstrating diminished protection as the dose is increased (Duranski et al., 2005). This biphasic protection is similar to the protective behavior of NO and superoxide dismutase (SOD) mimetics, suggesting that nitrite-dependent cytoprotection may be mediated by similar mechanisms as NO-dependent protection (summarized in Fig. 1). In addition, in all studies to date where nitrite has been used for cytoprotection, the protective effect has been inhibited by the NO scavenger PTIO [2-(4-carboxyphenyl)-4,4,5,5-tetramethylimidazoline-1-oxyl 3-oxide] (Duranski et al., 2005; Lu et al., 2005; Webb et al., 2004).

The mechanism by which nitrite is reduced to NO in ischemic tissue is unknown. Webb and colleagues demonstrated that nitrite-dependent NO production in heart homogenates could be decreased by treatment with allopurinol, an inhibitor of xanthine oxidoreductase (Webb et al., 2004). Other potential mechanisms of tissue-dependent nitrite reduction include cytochrome c oxidase or ubiquinol in the mitochondrion (Castello et al., 2006; Kozlov et al., 1999) and myoglobin in the heart (Huang et al., 2005b). In addition, it has been proposed that nitrite may mediate signaling, including the S-nitrosation of proteins and the regulation of gene expression, independent of NO production (Bryan et al., 2005). It is yet unknown whether NO-independent mechanisms play a role in nitrite-mediated cytoprotection from I/R.

S-nitrosation, caspases and apoptosis

It is known that nitrite can mediate the post-translational modification of metal centers (Huang et al., 2005a, b) and proteins (Bryan et al., 2005; Duranski et al., 2005), and that these types of modification (S-nitrosation and iron-nitrosylation) are associated with nitrite-dependent cytoprotection following I/R (Duranski et al., 2005). It is unclear precisely which proteins are modified by nitrite, but several key proteins may be targets

220

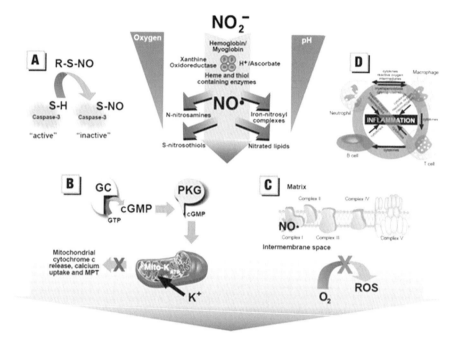

Fig. 1. Potential mechanisms of ischemia/reperfusion (I/R) cytoprotection by nitrite-derived nitric oxide. Nitrite (NO_2^-), in the setting of acidosis and hypoxia present during ischemia, forms nitric oxide ($NO^\bullet$) catalysed by hemoglobin and other pathways. The derived nitric oxide or nitrite itself may modify cellular proteins or lipids, or may directly participate in actions which mitigate the injury occurring at reperfusion. Four potential mechanisms for this cytoprotection are presented: (A) S-nitrosation of critical regulatory thiols on proteins in the apoptotic pathway, such as caspase-3. (B) Stimulation of guanylyl cyclase (GC) to form cGMP, which in turn activates protein kinase G (PKG), resulting in stabilization of mitochondrial K_{ATP} channels. This stabilization is associated with decreased mitochondrial calcium accumulation, prevention of membrane permeability transition (MPT), and release of cytochrome c, actions linked to the development of cellular necrosis and apoptosis. (C) Direct inhibition of complex I of the electron transport chain with resultant decrease in the amount of cytotoxic reactive oxygen species (ROS) formed at the time of reperfusion. (D) Inhibition of acute inflammation, mediated mainly by polymorphonuclear granulocytes. (See Colour Plate Section in this book).

for nitrite-dependent regulation, and these proteins may play a role in the cytoprotective effect. For example, the activation of caspase-3 by upstream caspases 8 and 9 are key pathways by which cells induce apoptosis (Thornberry and Lazebnik, 1998). S-nitrosation of caspase-3 has been linked to its inactivation, which would result in the inhibition of apoptotic cell death (Maejima et al., 2005; Mannick et al., 1999).

Reactive oxygen species (ROS) and cellular injury

S-nitrosation of proteins, particularly in the mitochondrion, may regulate the generation of ROS in the tissues, mediating cytoprotection. ROS have a dual role in I/R injury. Small amounts of ROS generation, particularly in mitochondria, have been determined to be a necessary component of mitochondrial signaling in cytoprotection (Oldenburg et al., 2003, 2004; Pain et al., 2000; Xu et al., 2004). However, the large burst of ROS generated after reperfusion from ischemia is believed to be one of the mechanisms whereby cellular injury and necrosis occurs (Das, 1994; Zweier et al., 1987). S-nitrosation of complex I of the electron transport chain by NO and peroxynitrite is known to inhibit the activity of this complex, which may have protective effects in the tissue by decreasing ROS formed by the mitochondrion itself (Burwell et al., 2006; Dahm et al., 2006).

The role of guanylyl cyclase (GC) and mitochondrial K_{ATP} channels

Mitochondrial K_{ATP} channels are also known targets for NO, and opening of these channels by NO has been shown to be cytoprotective in I/R (Sasaki et al., 2000). In two separate in vitro cardiomyocyte models of NO-mediated protection from I/R injury, NO was shown to activate mitochondrial K_{ATP} channels by a pathway dependent on sGC activation. NO-dependent activation of protein kinase G (PKG) through the sGC/cGMP pathway was shown to open mitochondrial K_{ATP} channels, potentially through phosphorylation of the channels, ultimately conferring cytoprotection.

While it is unclear why opening of the K_{ATP} channels is beneficial, several theories exist. Opening of the mitochondrial K_{ATP} channels appears to result in a small amount of ROS generation as well as a modest dose-dependent depolarization of the mitochondria (Bell et al., 2003). Inhibition of this ROS formation abolishes the benefit in terms of the infarct size seen in isolated heart models. Opening of these channels has also been associated with reduced calcium accumulation in the mitochondria, which is cytoprotective (Rakhit et al., 2001). Korge and colleagues have extended these findings to isolated mitochondria, demonstrating that activation of mitochondrial K_{ATP} channels not only reduced calcium accumulation but also prevented cytochrome c loss from the intermembrane space, thereby preventing apoptosis (Korge et al., 2002).

Although this is an area of active discovery, it is intriguing to consider that nitrite may modulate K_{ATP} channels via a cGMP/PKG-mediated pathway. Opening of these channels may in turn result in several (potentially linked) events associated with cytoprotection, including generation of ROS, prevention of matrix calcium uptake, and prevention of cytochrome c release. The timescale of these events, which is in the order of seconds to minutes, is sufficiently rapid to explain the cytoprotection against I/R injury seen in the recent reports utilizing nitrite (Duranski et al., 2005; Webb et al., 2004). Furthermore, the observation by Duranski et al. (2005) that $1H$-[1,2,4]oxadiazolo[4,3-a]quinoxalin-1-one (ODQ), an inhibitor of GC, abolished nitrite-derived protection against hepatic I/R injury implies that a GC pathway may be of significant importance.

Activation of cAMP-dependent pathways and cardiac contractility

Although NO is typically associated with action through the cGMP pathway (see above), it has been suggested that some of the beneficial cardiac contractile effects of moderate doses of exogenous NO donors are mediated through increases in cyclic adenosine monophosphate (cAMP) concentration (Kojda et al., 1996). Although the authors of this initial report hypothesized that the increases in cAMP levels were due to cGMP inhibition of phosphodiesterase III (PDE III), subsequent research (Vila-Petroff et al., 1999) suggests another mechanism. These authors demonstrated that the NO donor S-nitroso-acetylpenicillamine (SNAP) caused an increase in cAMP level and calcium influx into rat cardiac myocytes, resulting in improved contractility even against a background of tonically inhibited PDE. In this study, SNAP had a direct positive effect on adenylate cyclase activity, resulting in increased cAMP formation. Interestingly, although a low SNAP dose (1 μM) produced positive inotropic effects, higher doses (100 μM) induced negative inotropic effects through a PKG-dependent reduction in myofilament response to calcium. This dose–response curve, wherein doses of NO donors in the 1–10 μM range are beneficial whereas higher doses are harmful, has been observed recurrently in the I/R injury literature (Bell et al., 2003; Jones and Bolli, 2006) and was seen by Duranski and colleagues in the use of nitrite as a cytoprotective agent (Duranski et al., 2005).

Inflammation

The use of inhaled NO (Hataishi *et al.*, 2006) and of NO donors such as SPM-5185 (Lefer *et al.*, 1993) and CAS-754 (Pabla *et al.*, 1996) have been associated with decreased I/R injury in murine and canine myocardial I/R models and isolated rat hearts, respectively. In these studies, NO was given just prior to reperfusion, and resulted in improved heart contractile function. This improvement was correlated with decreased peripheral mononucleocyte (PMN) binding to I/R-damaged endothelium (Lefer *et al.*, 1993) and myocardial PMN infiltration (Hataishi *et al.*, 2006). The fact that the injury only occurred with PMN infusion in the isolated heart model (Pabla *et al.*, 1996) is the strongest evidence that inflammation is a key mediator of I/R injury and NO cytoprotection. However, controversy remains over whether the inflammation noted with I/R and prevented by NO is a cause or effect of necrosis/apoptosis induced by other means.

Conclusions

Although the precise mechanism of nitrite-mediated protection from I/R injury is unknown, several potential mechanisms are currently being explored. However, it is clear that nitrite is not only a physiologically important storage form of NO, but also an important mediator of hypoxic signaling. In addition to its physiological role, the potential use of nitrite as a therapeutic agent to prevent I/R-induced organ injury ensures that the field of nitrite biology will continue to grow.

Disclosure

This work has been funded by the Intramural Research Division of the National Heart Lung and Blood Institute of the National Institutes of Health. Dr. Gladwin is a co-inventor of an NIH provisional patent for the use of nitrite salts for cardiovascular diseases, including ischemia-reperfusion injury.

References

Bell, R. M., Maddock, H. L. and Yellon, D. M. (2003). The cardioprotective and mitochondrial depolarising properties of exogenous nitric oxide in mouse heart. *Cardiovasc. Res.* 57,405–415.

Bolli, R. (2001). Cardioprotective function of inducible nitric oxide synthase and role of nitric oxide in myocardial ischemia and preconditioning: An overview of a decade of research. *J. Mol. Cell. Cardiol.* 33,1897–1918.

Braunwald, E. and Kloner, R. A. (1985). Myocardial reperfusion: A double-edged sword? *J. Clin. Invest.* 76,1713–1719.

Bryan, N. S., Fernandez, B. O., Bauer, S. M., Garcia-Saura, M. F., Milsom, A. B., Rassaf, T., Maloney, R. E., Bharti, A., Rodriguez, J. and Feelisch, M. (2005). Nitrite is a signaling molecule and regulator of gene expression in mammalian tissues. *Nat. Chem. Biol.* 1,290–297.

Burwell, L. S., Nadtochiy, S. M., Tompkins, A. J., Young, S. and Brookes, P. S. (2006). Direct evidence for S-nitrosation of mitochondrial complex I. *Biochem. J.* 394, 627–634.

Cannon, R. O., 3rd., Schechter, A. N., Panza, J. A., Ognibene, F. P., Pease-Fye, M. E., Waclawiw, M. A., Shelhamer, J. H. and Gladwin, M. T. (2001). Effects of inhaled nitric oxide on regional blood flow are consistent with intravascular nitric oxide delivery. *J. Clin. Invest.* 108,279–287.

Castello, P. R., David, P. S., Mcclure, T., Crook, Z. and Poyton, R. O. (2006). Mitochondrial cytochrome oxidase produces nitric oxide under hypoxic conditions: Implications for oxygen sensing and hypoxic signaling in eukaryotes. *Cell. Metab.* 3,277–287.

Cosby, K., Partovi, K. S., Crawford, J. H., Patel, R. P., Reiter, C. D., Martyr, S., Yang, B. K., Waclawiw, M. A., Zalos, G., Xu, X., Huang, K. T., Shields, H., Kim-Shapiro, D. B., Schechter, A. N., Cannon, R. O., 3rd. and Gladwin, M. T. (2003). Nitrite reduction to nitric oxide by deoxyhemoglobin vasodilates the human circulation. *Nat. Med.* 9,1498–1505.

Crawford, J. H., Isbell, T. S., Huang, Z., Shiva, S., Chacko, B. K., Schechter, A. N., Darley-Usmar, V. M., Kerby, J. D., Lang, J. D., Jr., Kraus, D., Ho, C., Gladwin, M. T. and Patel, R. P. (2006). Hypoxia, red blood cells, and nitrite regulate NO-dependent hypoxic vasodilation. *Blood* 107,566–574.

Dahm, C. C., Moore, K. and Murphy, M. P. (2006). Persistent S-nitrosation of complex I and other mitochondrial membrane proteins by *S*-nitrosothiols but not nitric oxide or peroxynitrite: Implications for the interaction of nitric oxide with mitochondria. *J. Biol. Chem.* 281,10056–10065.

Das, D. K. (1994). Cellular, biochemical, and molecular aspects of reperfusion injury. Introduction. *Ann. N.Y. Acad. Sci.* 723,xiii–xvi.

Dejam, A., Hunter, C. J., Pelletier, M. M., Hsu, L. L., Machado, R. F., Shiva, S., Power, G. G., Kelm, M., Gladwin, M. T. and Schechter, A. N. (2005). Erythrocytes are the major intravascular storage sites of nitrite in human blood. *Blood* 106,734–739.

Doyle, M. P., Pickering, R. A., Deweert, T. M., Hoekstra, J. W. and Pater, D. (1981). Kinetics and mechanism of the oxidation of human deoxyhemoglobin by nitrites. *J. Biol. Chem.* 256,12393–12398.

Duranski, M. R., Greer, J. J., Dejam, A., Jaganmohan, S., Hogg, N., Langston, W., Patel, R. P., Yet, S. F., Wang, X., Kevil, C. G., Gladwin, M. T. and Lefer, D. J. (2005). Cytoprotective effects of nitrite during in vivo ischemia-reperfusion of the heart and liver. *J. Clin. Invest.* 115,1232–1240.

225

Fox-Robichaud, A., Payne, D., Hasan, S. U., Ostrovsky, L., Fairhead, T., Reinhardt, P. and Kubes, P. (1998). Inhaled NO as a viable antiadhesive therapy for ischemia/reperfusion injury of distal microvascular beds. *J. Clin. Invest.* 101,2497–2505.

Furchgott, R. F. and Bhadrakom, S. (1953). Reactions of strips of rabbit aorta to epinephrine, isopropylarterenol, sodium nitrite and other drugs. *J. Pharmacol. Exp. Ther.* 108,129–143.

Gladwin, M. T., Ognibene, F. P., Pannell, L. K., Nichols, J. S., Pease-Fye, M. E., Shelhamer, J. H. and Schechter, A. N. (2000a). Relative role of heme nitrosylation and beta-cysteine 93 nitrosation in the transport and metabolism of nitric oxide by hemoglobin in the human circulation. *Proc. Natl. Acad. Sci. USA* 97,9943–9948.

Gladwin, M. T., Shelhamer, J. H., Schechter, A. N., Pease-Fye, M. E., Waclawiw, M. A., Panza, J. A., Ognibene, F. P. and Cannon, R. O., 3rd. (2000b). Role of circulating nitrite and S-nitrosohemoglobin in the regulation of regional blood flow in humans. *Proc. Natl. Acad. Sci. USA* 97,11482–11487.

Hataishi, R., Rodrigues, A. C., Neilan, T. G., Morgan, J. G., Buys, E., Sruti, S., Tambouret, R., Jassal, D. S., Raher, M. J., Furutani, E., Ichinose, F., Gladwin, M. T., Rosenzweig, A., Zapol, W. M., Picard, M. H., Bloch, K. D. and Scherrer-Crosbie, M. (2006). Inhaled nitric oxide decreases infarction size and improves left ventricular function in a murine model of myocardial ischemia-reperfusion injury. *Am. J. Physiol. Heart Circ. Physiol.* 291(1), H379–H384.

Huang, K. T., Keszler, A., Patel, N., Patel, R. P., Gladwin, M. T., Kim-Shapiro, D. B. and Hogg, N. (2005a). The reaction between nitrite and deoxyhemoglobin. Reassessment of reaction kinetics and stoichiometry. *J. Biol. Chem.* 280,31126–31131.

Huang, Z., Shiva, S., Kim-Shapiro, D. B., Patel, R. P., Ringwood, L. A., Irby, C. E., Huang, K. T., Ho, C., Hogg, N., Schechter, A. N. and Gladwin, M. T. (2005b). Enzymatic function of hemoglobin as a nitrite reductase that produces NO under allosteric control. *J. Clin. Invest.* 115,2099–2107.

Hunter, C. J., Dejam, A., Blood, A. B., Shields, H., Kim-Shapiro, D. B., Machado, R. F., Tarekegn, S., Mulla, N., Hopper, A. O., Schechter, A. N., Power, G. G. and Gladwin, M. T. (2004). Inhaled nebulized nitrite is a hypoxia-sensitive NO-dependent selective pulmonary vasodilator. *Nat. Med.* 10,1122–1127.

Ignarro, L. J., Buga, G. M., Wood, K. S., Byrns, R. E. and Chaudhuri, G. (1987). Endothelium-derived relaxing factor produced and released from artery and vein is nitric oxide. *Proc. Natl. Acad. Sci. USA* 84,9265–9269.

Jones, S. P. and Bolli, R. (2006). The ubiquitous role of nitric oxide in cardioprotection. *J. Mol. Cell. Cardiol.* 40,16–23.

Kojda, G., Kottenberg, K., Nix, P., Schluter, K. D., Piper, H. M. and Noack, E. (1996). Low increase in cGMP induced by organic nitrates and nitrovasodilators improves contractile response of rat ventricular myocytes. *Circ. Res.* 78,91–101.

Korge, P., Honda, H. M. and Weiss, J. N. (2002). Protection of cardiac mitochondria by diazoxide and protein kinase C: Implications for ischemic preconditioning. *Proc. Natl. Acad. Sci. USA* 99,3312–3317.

Kozlov, A. V., Staniek, K. and Nohl, H. (1999). Nitrite reductase activity is a novel function of mammalian mitochondria. *FEBS Lett.* 454,127–130.

Kubes, P., Payne, D., Grisham, M. B., Jourd-Heuil, D. and Fox-Robichaud, A. (1999). Inhaled NO impacts vascular but not extravascular compartments in postischemic peripheral organs. *Am. J. Physiol.* 277,H676–H682.

Lauer, T., Preik, M., Rassaf, T., Strauer, B. E., Deussen, A., Feelisch, M. and Kelm, M. (2001). Plasma nitrite rather than nitrate reflects regional endothelial nitric oxide synthase activity but lacks intrinsic vasodilator action. *Proc. Natl. Acad. Sci. USA* 98,12814–12819.

Lefer, D. J. (2002). Do neutrophils contribute to myocardial reperfusion injury? *Basic Res. Cardiol.* 97,263–267.

Lefer, D. J., Nakanishi, K., Johnston, W. E. and Vinten-Johansen, J. (1993). Anti-neutrophil and myocardial protecting actions of a novel nitric oxide donor after acute myocardial ischemia and reperfusion of dogs. *Circulation* 88,2337–2350.

Lu, P., Liu, F., Yao, Z., Wang, C. Y., Chen, D. D., Tian, Y., Zhang, J. H. and Wu, Y. H. (2005). Nitrite-derived nitric oxide by xanthine oxidoreductase protects the liver against ischemia-reperfusion injury. *Hepatobiliary Pancreat. Dis. Int.* 4,350–355.

Maejima, Y., Adachi, S., Morikawa, K., Ito, H. and Isobe, M. (2005). Nitric oxide inhibits myocardial apoptosis by preventing caspase-3 activity via S-nitrosylation. *J. Mol. Cell. Cardiol.* 38,163–174.

Mannick, J. B., Hausladen, A., Liu, L., Hess, D. T., Zeng, M., Miao, Q. X., Kane, L. S., Gow, A. J. and Stamler, J. S. (1999). Fas-induced caspase denitrosylation. *Science* 284,651–654.

McCord, J. M., Roy, R. S. and Schaffer, S. W. (1985). Free radicals and myocardial ischemia. The role of xanthine oxidase. *Adv. Myocardiol.* 5,183–189.

Millar, T. M., Stevens, C. R., Benjamin, N., Eisenthal, R., Harrison, R. and Blake, D. R. (1998). Xanthine oxidoreductase catalyses the reduction of nitrates and nitrite to nitric oxide under hypoxic conditions. *FEBS Lett.* 427,225–228.

Ng, E. S., Jourd'heuil, D., Mccord, J. M., Hernandez, D., Yasui, M., Knight, D. and Kubes, P. (2004). Enhanced S-nitroso-albumin formation from inhaled NO during ischemia/reperfusion. *Circ. Res.* 94,559–565.

Oldenburg, O., Cohen, M. V. and Downey, J. M. (2003). Mitochondrial K(ATP) channels in preconditioning. *J. Mol. Cell. Cardiol.* 35,569–575.

Oldenburg, O., Qin, Q., Krieg, T., Yang, X. M., Philipp, S., Critz, S. D., Cohen, M. V. and Downey, J. M. (2004). Bradykinin induces mitochondrial ROS generation via NO, cGMP, PKG, and mitoKATP channel opening and leads to cardioprotection. *Am. J. Physiol. Heart Circ. Physiol.* 286,H468–H476.

Pabla, R., Buda, A. J., Flynn, D. M., Blesse, S. A., Shin, A. M., Curtis, M. J. and Lefer, D. J. (1996). Nitric oxide attenuates neutrophil-mediated myocardial contractile dysfunction after ischemia and reperfusion. *Circ. Res.* 78,65–72.

Pain, T., Yang, X. M., Critz, S. D., Yue, Y., Nakano, A., Liu, G. S., Heusch, G., Cohen, M. V. and Downey, J. M. (2000). Opening of mitochondrial K(ATP) channels triggers the preconditioned state by generating free radicals. *Circ. Res.* 87,460–466.

Palmer, R. M., Ferrige, A. G. and Moncada, S. (1987). Nitric oxide release accounts for the biological activity of endothelium-derived relaxing factor. *Nature* 327,524–526.

Pedigo, L. (1888). Antagonism between amyl nitrite and prussic acid. *Trans. Med. Soc. Virginia* 19,124–131.

227

Rakhit, R. D., Mojet, M. H., Marber, M. S. and Duchen, M. R. (2001). Mitochondria as targets for nitric oxide-induced protection during simulated ischemia and reoxygenation in isolated neonatal cardiomyocytes. *Circulation* 103,2617–2623.

Rassaf, T., Preik, M., Kleinbongard, P., Lauer, T., Heiss, C., Strauer, B. E., Feelisch, M. and Kelm, M. (2002). Evidence for in vivo transport of bioactive nitric oxide in human plasma. *J. Clin. Invest.* 109,1241–1248.

Sasaki, N., Sato, T., Ohler, A., O'Rourke, B. and Marban, E. (2000). Activation of mitochondrial ATP-dependent potassium channels by nitric oxide. *Circulation* 101,439–445.

Scharfstein, J. S., Keaney, J. F., Jr., Slivka, A., Welch, G. N., Vita, J. A., Stamler, J. S. and Loscalzo, J. (1994). In vivo transfer of nitric oxide between a plasma protein-bound reservoir and low molecular weight thiols. *J. Clin. Invest.* 94,1432–1439.

Stamler, J. S., Jaraki, O., Osborne, J., Simon, D. I., Keaney, J., Vita, J. *et al.* (1992). Nitric oxide circulates in mammalian plasma primarily as an S-nitroso adduct of serum albumin. *Proc. Natl. Acad. Sci. USA* 89,7674–7677.

Stamler, J. S., Jia, L., Eu, J. P., Mcmahon, T. J., Demchenko, I. T., Bonaventura, J., Gernert, K. and Piantadosi, C. A. (1997). Blood flow regulation by S-nitrosohemoglobin in the physiological oxygen gradient. *Science* 276,2034–2037.

Thornberry, N. A. and Lazebnik, Y. (1998). Caspases: Enemies within. *Science* 281,1312–1316.

Troncy, E., Francoeur, M., Salazkin, I., Yang, F., Charbonneau, M., Leclerc, G., Vinay, P. and Blaise, G. (1997). Extra-pulmonary effects of inhaled nitric oxide in swine with and without phenylephrine. *Br. J. Anaesth.* 79,631–640.

Vallance, P. (2000). *Vascular Nitric Oxide in Health and Disease*, Academic Press, San Diego.

Vila-Petroff, M. G., Younes, A., Egan, J., Lakatta, E. G. and Sollott, S. J. (1999). Activation of distinct cAMP-dependent and cGMP-dependent pathways by nitric oxide in cardiac myocytes. *Circ. Res.* 84,1020–1031.

Webb, A., Bond, R., Mclean, P., Uppal, R., Benjamin, N. and Ahluwalia, A. (2004). Reduction of nitrite to nitric oxide during ischemia protects against myocardial ischemia-reperfusion damage. *Proc. Natl. Acad. Sci. USA* 101,13683–13688.

Xu, Z., Ji, X. and Boysen, P. G. (2004). Exogenous nitric oxide generates ROS and induces cardioprotection: Involvement of PKG, mitochondrial KATP channels, and ERK. *Am. J. Physiol. Heart Circ. Physiol.* 286,H1433–H1440.

Zweier, J. L., Flaherty, J. T. and Weisfeldt, M. L. (1987). Direct measurement of free radical generation following reperfusion of ischemic myocardium. *Proc. Natl. Acad. Sci. USA* 84,1404–1407.

Zweier, J. L., Wang, P., Samouilov, A. and Kuppusamy, P. (1995). Enzyme-independent formation of nitric oxide in biological tissues. *Nat. Med.* 1,804–809.

Nitric oxide and the zebrafish (*Danio rerio*): Developmental neurobiology and brain neurogenesis

Bo Holmqvist[1,*], Lars Ebbesson[2] and Per Alm[1]

[1]*Department of Pathology, Lund University, Sölvegatan 25, S-221 85 Lund, Sweden*
[2]*Department of Biology, Bergen University, Thormøhlensgaten 55, N-5020 Bergen, Norway*

Abstract. Nitric oxide synthase (NOS) isoforms produce nitric oxide (NO), which is of vital importance in physiological processes as well as for pathology and recovery in various diseases. NO may also possess important roles in embryonic development and plasticity changes in later life. Knowledge about vertebrate developmental neurobiology has to a large degree come from studies of bony fishes (teleosts) such as the zebrafish (*Danio rerio*, Teleostei), and studies of teleost NO systems have provided insights about the role of NO in these events. We therefore summarize current knowledge about the presence, molecular identity and expression of NOS isoforms in teleosts, *i.e.*, neuronal NOS (NOS I/NOS1), the presumed immune response related inducible NOS (NOS III/NOS2) and the vascular/endothelial NOS (NOS III/NOS3). We describe the spatio-temporal expression of NOS I in relation to the neurotransmitter/hormone differentiation during early development, and present new data about the establishment of NOS I expression in areas of the adult brain with ongoing neurogenesis. We also present further evidence for the influence of NO on early organogenesis, demonstrated by abnormal organ development caused by manipulation of NO systems in embryos. It is concluded that knowledge about the zebrafish NOS/NO systems contribute to the understanding of NO functions in general, and provide an experimental model for studies of functions and cellular mechanisms of NO in vertebrate body morphogenesis and organogenesis. Due to the high capacities for ongoing cell mitosis and neural plasticity throughout life, teleost species may also emerge as important models for studies on retained cell proliferation and neurogenesis in the adult central nervous system (CNS). Future studies need to include more molecular data on identified, but poorly characterized or as-yet unidentified NO-producing NOS isoforms in teleosts, in combination with experimental studies with applications of new investigative tools.

Keywords: zebrafish; vertebrate model; nitric oxide synthase (NOS); NOS-isoforms; ontogeny; morphogenesis; brain; neurogenesis; cardiovascular; developmental pathology.

Introduction

Nitric oxide (NO) is produced by the enzyme nitric oxide synthase (NOS) in biological tissues. Different NOS isoforms are differentially expressed in various tissues and cell types, where they serve individual and well-defined functional roles, sharing NO as a signal molecule. NO has diverse actions that influence cellular and molecular processes, and plays important roles

Corresponding author: Tel.: 46-46-173560. Fax: 46-46-143307.
E-mail: bo.holmqvist@med.lu.se (B. Holmqvist).

ADVANCES IN EXPERIMENTAL BIOLOGY
VOLUME 01   ISSN 1872-2423
DOI: 10.1016/S1872-2423(07)01011-3

in various common physiological processes throughout the vertebrate phylogeny, such as in vascular function, neurotransmission, immune responses and angiogenesis (Guix et al., 2005; Mungrue et al., 2003; Ziche and Morbidelli, 2000, see also contributors in this volume; Peltser, Agnisola and Pellegrino, and Tota et al.). The corresponding NOS systems are present in different vertebrate groups, and new data on NO functions in developmental processes are emerging from different experimental animal species. In lower vertebrates, such as teleosts, recent analyses have aimed at determining the relatively unknown molecular identity of NOS isoforms and the cellular sources of NO-mediated functions. In teleost species like the zebrafish, knowledge about the spatio-temporal embryonic development pattern and the extraordinary capacity of retained growth and plasticity of the central nervous system (CNS) in adults may provide experimental models for investigations of the functional roles, cellular actions and target mechanisms of NO in vertebrate organogenesis and ongoing neurogenesis. The overall aim of this review is therefore to summarize data concerning NO/NOS systems obtained from zebrafish and other teleost model species, and to highlight important issues and remaining questions in order to provide a basis for future studies concerning the role of NO and individual NOS isoforms in developmental processes, with a focus on neurogenesis.

Despite the central role of NO in various cellular mechanisms and biological functions throughout animal phylogeny, molecular and functional analyses have so far focused on the NO/NOS systems in mammals. This is largely due to the involvement of NO in different types of human diseases, particularly in various neurological and cardiovascular disorders, and in different states of inflammation, with actions coupled to both the processes of pathology and recovery (see Duncan and Heales, 2005; Guix et al., 2005; Ming and Song, 2005; Nasseem, 2005; Szabó, 2000). NO is highly diffusible and is readily transported across cell membranes and can thereby affect neighbouring cells. Interactions between NOS isoforms via NO signalling that affect NOS activities provide regulation of total NO production. These NO interactions may be of crucial importance in tissues with spatially closely associated NOS isoforms, NOS-expressing cells or NO target sites. Thus, in addition to the regulation of NOS isoform-specific expression and activity, the spatial localization and temporal activity of NO-producing cells may be important factors determining the functional capacity of NO. During early morphogenesis and adult neurogenesis, the temporal regulation of NOS gene expression and regional actions by NO may in part explain the timing and subsequent diverse roles of NO in these processes.

The three major mammalian NOS isoforms, NOS I (also called NOS1 or nnos), NOS II (also called NOS2 or iNOS) and NOS III (also called NOS3 or eNOS), have been characterized mainly by their distinguishable gene structure, regulation of expression (induced or constitutively expressed) and their cell-type-specific localization: NOS I (brain or neuronal NOS) is constitutively expressed with a prevailing neuronal localization, NOS II (inducible or macrophagial NOS) has a mainly inducible character and, potential expression by various cell types, and NOS III (or endothelial NOS) is constitutively expressed with a prevailing vascular localization (see Ignarro, 2000). In addition, differentiated spatio-temporal expression patterns of the NOS isoforms and variants thereof have been identified in mammals, which may be related to developmental and/or functional processes (Lee *et al.*, 1997; Massman *et al.*, 2000; Matsumoto *et al.*, 1993; Northington *et al.*, 1996, 1997; Ogilvie *et al.*, 1995). The regulatory activation of NOS expression (see Ganster and Geller, 2000; Gratton *et al.*, 2000; Kleinert *et al.*, 2000) may comprise induction, such as in cases of inducible NOS II expression during immune responses, or may involve homeostatic up- or down-regulation in the expression of otherwise constitutively expressed isoforms, such as in the cases of NOS I and NOS III. NO activity depends on regulatory influence by different factors, including NO itself, on NOS expression, at the mRNA and protein level of the individual NOS isoforms.

In teleosts, NOS and NO activities have been detected in various organ structures, and recently more detailed identification and characterization of teleost NOS isoforms have been reported, *i.e.*, genetic and molecular identification and characterization of their mRNAs and protein expression (summarized below). Basically, three NOS isoforms have been shown to be present in teleosts, corresponding to the three major isoforms characterized in mammals. However, the teleost NOS isoforms have been detected with different experimental techniques and in different species, even in phylogenetically separated groups. Furthermore, the NOS isoforms have not been identified at all levels in the same single species, not even in experimental model species. Despite the recent detection of NOS mRNA expression *in situ*, analyses of NOS protein expression in teleosts are still limited to the use of antisera made against mammalian NOS isoforms. As a basis for improvement of future investigations of teleost NOS systems, we summarize current data obtained from different teleost species and evaluate the different investigative tools used in a comparative manner, with focus on the zebrafish (see section below).

NO has recently been proposed to play a key role in processes of development, in early body and organ morphogenesis, and ongoing neurogenesis and neural plasticity throughout life. The main NO-mediated actions in developmental processes have been shown to be via regulation of cell proliferation, which is inhibited or terminated and thereby increases cell differentiation (Cheng et al., 2003; Kuzin et al., 1996; Phung et al., 1999; Tanaka et al., 1994; Wingrove and O'Farrell, 1999). In developmental processes NO has been shown to regulate cell proliferation and differentiation, and in neuronal systems to participate in plasticity as part of neuron and fibre replacement in functional circuits, thereby possessing major influence on embryogenesis, including central roles in development of the CNS (Peunova et al., 2001), bone and limbs (Aguirre et al., 2001; Diket et al., 1994; Fantal et al., 1997; Hefler et al., 2001; Pierce et al., 1995), lung (Han and Stewart, 2006) and the cardiovascular systems (Bloch et al., 1999; Feng et al., 2002; Lee et al., 2000; Peltser et al., 2005). Similar to the situation described for mammals (Currie et al., 2006), in early developing medaka (a teleost) the primary NO target guanylyl cyclase (GC) has a widespread distribution in different body organs, and early NO actions have been shown to have a similar crucial influence on normal organ development by NO stimulation of cGMP production (Yamamoto et al., 2003). Correspondingly, recent studies in developing zebrafish show an early expression of NOS I in different body organs that follows the pattern of early neurogenesis (Holmqvist et al., 2004; Poon et al., 2003). The NOS I cell differentiation may include as-yet unidentified formations of neuronal NOS I target sites in versatile body organs that can influence organogenesis. Here, we describe the early ontogeny of NOS I, and the coinciding differentiation of NOS I and neurotransmitter/hormone-producing cells during brain neurogenesis in zebrafish. In addition, we present data from our ongoing studies providing further evidence of the influence of NO on early organogenesis (Holmqvist, Forsell and Alm, ongoing investigations). These data include the role of NOS on heart gross development, since physiological effects by NO actions on developing cardiovascular and osmoregulatory systems are among the few that are well documented in teleosts, and since a teleost NOS III isoform may be involved in these processes (Eddy, 2005; Pelster et al., 2005; Tota et al., 2005, see also Peltser in this volume).

In mammals, the main ongoing neurogenesis in the adult brain (i.e., the presence of stem cells and retained ongoing cell mitosis, differentiation and migration) is restricted to a few brain regions: the subventricular zone (SVZ), the rostral migratory stream (RMS) and parts of the hippocampus (see Alvarez-Buylla et al., 2001; Gage, 2000; Ming and Song, 2005; Moreno-López et al., 2000; Wendland et al., 1994). In rodents, the highest cell proliferation occurs in the SVZ, from which the cells migrate along the route of the RMS into the olfactory bulb, differentiating along the way and maturating at their final destination. NO is proposed to play a vital role in ongoing neurogenesis, acting as a negative regulator of precursor cell proliferation and as an important promoter of neurogenesis in various brain diseases (see Estrada and Murillo-Carretero, 2005). The main source of the NO-mediated actions in neurogenesis is generally considered to be NOS I of nearby neurons (Chen *et al.*, 2004; Cheng *et al.*, 2003; Gibbs, 2003; Matarredona *et al.*, 2005; Zhang *et al.*, 2001). However, recent studies stress that NOS III participates in adult neurogenesis (Reif *et al.*, 2004), with a possible cellular source located in the SVZ (Alm *et al.*, 2002). In teleosts, extensive NOS I expression has correspondingly been noted in regions of putative cell proliferation and neuronal differentiation, in both developing and adult brains (Holmqvist *et al.*, 1994, 2000, 2004), thus having the ability to induce NO actions during neurogenesis. In contrast to the case for mammals, in teleosts ongoing cell proliferation and neural plasticity is extensive throughout life and widespread in the whole brain (Kirsche, 1967). The basis for the continuously growing teleost brain, the proliferation zones, has recently been defined in detail in the stickleback (Ekström *et al.*, 2001) and zebrafish (Wullimann and Knipp, 2000; Zupanc, 2001). This extraordinary capacity for ongoing neurogenesis in teleosts has received little attention. However, recent documentation on the mechanisms of adult neurogenesis in the zebrafish further stresses the correlating processes in the vertebrate phylogeny, with the presence of stem cells as the origin of active cell proliferation and differentiation, and the incorporation of specific newborn neurotransmitter neurons into brain nuclei throughout life (see Grandel *et al.*, 2006). We here present new data on the detailed distribution of NOS systems in relation to the recently defined regions of neurogenesis of the zebrafish brain, via correlations of NOS mRNA-expressing populations to immunocytochemically labelled brain markers for active cell proliferation, neuronal differentiation, tissue plasticity and cell migration (Holmqvist and Alm, ongoing studies).

Identities of NOS isoforms

All NOS isoforms share binding sites for cofactors such as nicotinamide adenine dinucleotide phosphate (NADPH), flavine adenine dinucleotide (FAD), flavine mononucleotide (FMN) and calmodulin (CaM) (see, *e.g.*, Guix *et al.*, 2005; Mungrue *et al.*, 2003). These sites are also recognized in teleosts (Øyan *et al.*, 2000; Wang *et al.*, 2001), but to date there have been few molecular characterizations of different NOS isoforms. In addition, the detection of NOS has been from different tissues of different teleost species, and has involved the use of different techniques, such as reverse transcriptase polymerase chain reaction (RT-PCR) and molecular cloning techniques, Northern blots, mRNA *in situ* hybridization (ISH), Western blots, immunocytochemistry and nicotinamide adenine dinucleotide phosphate diaphorase (NADPHd) histochemistry. Here we summarize some of the data in order to assess the potential presence of the different NOS isoforms in teleosts, with focus on data from the zebrafish (Tables 1 and 2).

NOS genes and mRNA expression

The complete NOS I gene sequence has been identified in zebrafish (Gen-Bank accession no. AY211528; Poon *et al.*, 2003). The molecular identity of a teleost NOS II has been identified in a few species, the partial NOS II sequence in zebrafish (accession no. AY324390) and the complete sequence in rainbow trout (accession no. AJ295231; Wang *et al.*, 2001) and goldfish (accession no. AY904362 for NOS II form A; AY904363 for NOS II form B). So far there is no molecular evidence for the presence of a teleost NOS III gene. Despite serious attempts in our laboratory, we have not succeeded in identifying NOS gene sequences from either zebrafish or salmon that are structurally similar to NOS III in mammals.

NOS I mRNA gene expression *in situ* has been shown in zebrafish, the brain of adults, and in the brain, eye and different body organs of developing embryos and larvae (see below and Holmqvist *et al.*, 2000, 2004; Poon *et al.*, 2003). In Atlantic salmon, Northern blot data have demonstrated the expression of three different NOS I mRNAs in the brain (Øyan *et al.*, 2000). Collective data from different species show that NOS I is the most conserved isoform among vertebrates, with similar molecular structure and corresponding functional character.

In teleosts, PCR amplification techniques have demonstrated NOS II expression in the brain, gill, kidney and macrophages (Laing *et al.*, 1999; McNeill and Perry, 2005; Neumann *et al.*, 1995; Wang *et al.*, 2001); however, its mRNA expression has not been demonstrated *in situ*. Endotoxins like lipopolysaccharide (LPS) specifically and strongly induce NOS II expression in mammals (see Holmqvist *et al.*, 2006). Correspondingly, in teleosts NOS II is induced by LPS stimulation of macrophages (Laing *et al.*, 1999; Neumann *et al.*, 1995) and viral infection of trout fibroblasts (Wang *et al.*, 2001). Temporal differences in induction are recorded between the applications (between 5 h and 24 h). However, intraperitoneal LPS injections in zebrafish cause a clear (almost twofold) increase in the expression of NOS I mRNA relative to NOS II mRNA in body organs

Table 1. Evidence for nitric oxide synthase isoforms in teleosts.

	NOS I (nNOS)	NOS II (iNOS)	NOS III (eNOS)
Gene sequences (cDNA from brain and body)	*Complete*[a]	Complete[b,c]	–
Expression in			
CNS	*ISH*[d], *ICC*[d], WB[t], NB[f]	PCR[e]	
Peripheral nerves	ICC[k]		WB[p]
Pituitary	ICC[h], WB[j]	ICC[h], WB[j]	ICC[h]
Heart	NB[f]	ICC[q,u], WB[u]	*ICC*[g,q,r]
Gill	ICC[i], WB[o]	ICC[i], PCR[n]	ICC[i]
Kidney	ICC[l]	PCR[e,n], WB[s], ICC[s]	
Retina	*ICC*[m]	*ICC*[m]	
Liver		WB[s], ICC[s]	

Note: Italic indicates data from zebrafish; all other data from other teleosts.
Expression determined using protein immunocytochemistry (ICC), mRNA in situ hybridization (ISH), Northern blot (NB), polymerase chain reaction (PCR) or Western blot (WB).
[a]Poon *et al.* (2003).
[b]Partial zebrafish sequence: GenBank accession no. AY324390.
[c]Complete sequence in rainbow trout: Wang *et al.* (2001).
[d]Holmqvist *et al.* (2000, 2004) and Poon *et al.* (2003).
[e]McNeill and Perry (2005).
[f]Øyan *et al.* (2000).
[g]Fritsche *et al.* (2000).
[h]Ebbesson *et al.* (2004) and Jadhao and Malz (2003).
[i]Ebbesson *et al.* (2005).
[j]Uretsky *et al.* (2003).
[k]Funakoshi *et al.* (1999).
[l]Jimenez *et al.* (2001).
[m]Shin *et al.* (2000).
[n]Campos-Perez *et al.* (2000).
[o]Zaccone *et al.* (2003a).
[p]Cioni *et al.* (2002).
[q]Pellegrino *et al.* (2004).
[r]Tota *et al.* (2005).
[s]Barroso *et al.* (2000).
[t]Virgili *et al.* (2001).
[u]Amelio *et al.* (2006).

and brain, investigated by analysing total RNA isolated from animals during 5–15 h after injection, and detected by means of RT-PCR using primers specific for zebrafish NOS I and NOS II (Holmqvist, Falk Olsson and Alm, unpublished data). Thus, teleost NOS II may have a different

Table 2. Examples of NOS antibodies used in fish to identify NOS-like proteins with Western blots and immunohistochemistry.

NOS isoform	Species of origin	Epitope	Host	Supplier	Species	Tissue	Western blot protein size	Protein localization	Reference
NOS I	Human	1422–1433	rb	Cayman	Goldfish	Pituitary	2 bands 120–160 kDa cytoskeleton + heavy organelles		Uretsky et al. (2003)
NOS I	Human		rb	K-20 Santa Cruz Biochemicals	Tilapia	Spinal cord	150 kDa supernatant 150 kDa pellet	Neurosecretory neurons	Cioni et al. (2002)
NOS I					Catfish	Gill	150 kDa + 2 lower MW bands	Neuroepithelial cells	Zaccone et al. (2003a)
NOS I			rb	TEMA	Goldfish	Brain	150 kDa		Virgili et al. (2001)
NOS I			rb	TEMA	Trout	Brain	150 kDa		Virgili et al. (2001)
NOS III	Human	C-terminus	rb	Transduction Lab	Tilapia	Spinal cord	140 kDa pellet		Cioni et al. (2002)
NOS II	Murine	1131–1144	rb	Uttenthal et al. (1998)	Trout	Kidney	130 kDa	Cells	Barroso et al. (2000)
NOS II	Murine	1131–1144	rb	Uttenthal et al. (1998)	Trout	Liver	130 kDa	Cells	Barroso et al. (2000)
NOS II	Murine	1131–1144	rb	Calbiochem	Goldfish	Pituitary	120 kDa cytosolic + plasma membrane	Somatotropes + gonadotropes	Uretsky et al. (2003)
NOS II	Human		rb		Icefish	Heart	130 kDa	Ventricular myocytes	Amelio et al. (2006)
NOS II	Human	C-terminus	rb		Icefish	Heart	135 kDa	Endocardial endothelial cells	Amelio et al. (2006)
NOS I	Rat	Whole	sh	Alm et al. (1993)	Salmon	Brain		Neurons	Holmqvist et al. (1994)
NOS I	Rat		rb	Gift: P. Emson	Zebrafish	Brain		Neurons	Holmqvist et al. (2000)
NOS I			rb	Aff. Res. Prod.	Catfish	Brain		Neurons	Jadhao and Malz (2003)
NOS I	Rat		rb	Gift: Fahrenkrug	Salmon	Gill		Chloride cells	Ebbesson et al. (2005)

NOS	Animal	Region	Host	Source	Species	Tissue	Localization	Reference
NOS I			rb	Transduction Laboratories	Catfish	Gill vasculature	Nerve fibres in walls of arteries	Zaccone et al. (2003b)
NOS I	Rat	Recombinant whole	rb	Gift: V. Riveros-Moreno	Trout	Kidney	Fibres along blood vessels	Jimenez et al. (2001)
NOS I	Human		rb	K-20 Santa Cruz Biochemicals	Salmon	Pituitary	Cells, fibres	Ebbesson et al. (2004)
NOS I	Rat		rb	Gift: Fahrenkrug	Salmon	Pituitary	Cells	Ebbesson et al. (2004)
NOS I			rb	Aff. Res. Prod.	Catfish	Pituitary	Rostral pars distalis	Jadhao and Malz (2003)
NOS I	Rat	251–270	rb	RBI	Goldfish	Pituitary	Somatotropes + gonadotropes	Uretsky and Chang (2000)
NOS I	Rat	C-terminus	rb	Alm et al. (1993)	Salmon	Retina	Neurons	Östholm et al. (1994)
NOS I	Human		rb	K-20 Santa Cruz Biochemicals	Tilapia	Urophysis	Adjacent blood capillaries	Cioni et al. (2002)
NOS II	Mouse		rb	Affinity Bioreag.	Salmon	Gill	Unidentified cells	Ebbesson et al. (2005)
NOS II	Mouse		rb	Affinity Bioreag.	Salmon	Pituitary	Cells	Ebbesson et al. (2004)
NOS III	Mouse		rb	Affinity Bioreag.	Salmon	Gill	Chloride cells	Ebbesson et al. (2005)
NOS III					Catfish	Gill	Neuroepithelial cells	Zaccone et al. (2003a)
NOS III	Mouse		rb	Affinity Bioreag.	Salmon	Pituitary	Cells	Ebbesson et al. (2004)

Note: MW, molecular weight. rb, rabbit. sh, sheep. RBI, Research Biochemicals International. TEMA, Tema Ricerca, Bologna, Italy.

regulatory character compared to the mammalian isoform, and other iso-forms may participate in inflammatory responses.

To the best of our knowledge, NOS III-like gene sequences have only been identified in different mammalian species. In mammals, NOS III mRNA expression prevails in vascular endothelial cells, and correspond-ingly in teleosts NADPHd histochemistry readily labels vascular struc-tures (Holmqvist et al., 2000; Pellegrino et al., 2004). The failure to identify a teleost NOS III gene suggests that, if present, the gene sequence differs significantly from that in mammals. The difficulties of identifying a teleost NOS III-like protein further support this (see below).

NOS protein expression

In addition to the identification of NOS gene and mRNA nucleotide se-quences, other techniques have been employed to provide indirect evidence of NOS-like proteins and NOS enzyme activity. Most studies to date have detected and partly characterized NOS proteins and their localization in teleosts, by the use of antibodies targeting mammalian proteins with Western blot and immunocytochemical techniques, and with NADPHd histochemistry (Tables 1 and 2).

An important step in the characterization of antibody specificity and protein identification is achieved through Western blots. In mammals, NOS isoforms are separated by size: NOS I (160 kDa), NOS II (120 kDa) and NOS III (135 kDa). Furthermore, NOS I shows four different variants, α, β, γ and μ (ranging in size from 125 kDa to 165 kDa), generated by mRNA splicing (see Guix et al., 2005). In teleosts, Western blots using different NOS antibodies have revealed different-sized bands in a variety of tissues, providing further evidence for the presence in teleosts of NOS I and NOS II protein(s) corresponding to the mammalian isoforms. Still, in the case of NOS I and NOS II, Western blots have demonstrated single and multiple proteins with varying molecular weights depending on the antibody used (Table 2). The presence of different NOS I variants in the goldfish pituitary has been demonstrated by Western blot analysis (Uretsky et al., 2003), supported by Northern blot analyses in the brain of Atlantic salmon (Øyan et al., 2000). Because of the variation in sizes and unknown protein sequences from the mature protein, it is difficult to conclude whether or not the bands are variants, degradation products, or represent antibody cross-reactivity between NOS isoforms or NOS-like proteins. There is only one study that has shown a teleost NOS III-like protein as a single band at 140 kDa (Cioni et al., 2002). To the best of our knowledge, no Western blot data of zebrafish NOS-like proteins have been reported.

In teleost tissues, NOS-immunoreactive proteins and NOS activity (NADPHd activity) have been detected within numerous organs, including brain (Brüning et al., 1995, 1996; Holmqvist and Ekström, 1997; Holmqvist et al., 2000, 1994; Øyan et al., 2000), retina (Östholm et al., 1994), peripheral nervous system (Green and Campbell, 1994; Li and Furness, 1993; Mauceri et al., 1999; Olsson and Holmgren, 1996), caudal neurosecretory system (Cioni et al., 2002), gut (Olsson and Holmgren, 1997), kidney (Jimenez et al., 2001), head kidney (Gallo and Civinini, 2001), heart (Tota et al., 2005) and gill (Ebbesson et al., 2005; Zaccone et al., 2006). The combined use of antibodies has demonstrated differentiated patterns of NOS distribution, thus further emphasizing the presence of different NOS proteins.

Teleost NOS I immunoreactivity has been demonstrated in different species by the use of antibodies against mammalian NOS I, in brain, peripheral nerves, pituitary, retina, heart, kidney and gill (Table 1). In brain tissue sections of adult zebrafish, the identification of specific NOS I protein immunoreactivity has been verified by its corresponding NOS I mRNA expression (Holmqvist et al., 2000), whereas immunohistochemical detection in developing zebrafish has not yet been reported.

Immunocytochemical evidence for teleost NOS II and NOS III is relatively sparse, and stems from the use of heterologous antibodies directed towards mammalian isoforms (see Tables 1 and 2). The presence of a NOS III-like protein in teleosts has been indicated from a Western blot study (Cioni et al., 2002), a few immunohistochemical studies of a vascular-associated NOS III immunoreactivity (see Tota et al., 2005), and NADPHd histochemistry that labels vascular structures in teleosts (Holmqvist et al., 2000; Pellegrino et al., 2004). Functional evidence for NO actions shown to originate from NOS III include regulation of heart performance (for reviews see Tota et al., 2005, and Tota et al., and Agnisola and Pellegrino in this volume) and coronary resistance (Agnisola, 2005). At the same time, immunocytochemistry in trout has suggested the presence of NOS I in fibres surrounding kidney arteries, and in fibres innervating the walls of efferent filament arteries, and in the proximity of the lamellar arteries (Jimenez et al., 2001). This is supported by NADPHd histochemistry staining in catfish (Zaccone et al., 2003b), and further by the demonstration of co-localized NOS I and NOS III in neuroepithelial cells, and NOS I and endothelin in neuroepithelial cells located adjacent to efferent filament arteries (Zaccone et al., 2003a). Thus, the demonstrated central role for NO in, for instance, cardiovascular physiology of fish and cardiovascular development in teleosts (see below) direct towards the importance of future molecular identification of the presumed NOS III, together with more evidence concerning the functional role of NOS I in these events.

Information about the differentiated NOS distributions in tissues, achieved by the use of different antibodies, has indicated the presence of different NOS proteins with spatially separated expression in tissues. Furthermore, discrepancies between neuronal NOS immunoreactivity and NADPHd activity have indicated the presence of NOS I-like isoforms in teleosts, such as in the pineal organ, retina and gill (Brüning et al., 1995, 1996; Ebbesson et al., 2005; Holmqvist et al., 1994, 2000). Recently, we revealed the presence of two separate patterns of NOS immunoreactivity in salmon pituitary using four different NOS antibodies: two different NOS I antibodies (one made against rat NOS I and one made against human NOS I), and NOS II and NOS III antibodies (both made against mouse NOS II and NOS III, respectively). Antibodies made against rat NOS I, mouse NOS II and mouse NOS III produced the same labelling pattern, while the human NOS I antibody showed a different pattern (Ebbesson et al., 2004). In another study of the salmon gill, polyclonal antibodies against rat NOS I and mouse NOS III labelled chloride cells at the base and along the secondary lamellae, whereas antibodies against mouse NOS II labelled small cells deep in the primary filament (Ebbesson et al., 2005).

In conclusion, data from different teleost species, including the zebrafish, provide molecular evidence for the presence and expression of at least two NOS isoforms in teleosts: teleost NOS I and NOS II. Immunocytochemical and Western blot data demonstrate that NOS I has the widest tissue distribution. Furthermore, NOS I is the only NOS isoform mRNA so far demonstrated to be expressed in situ, in both embryonic and adult zebrafish. Protein detection of NOS isoforms in teleosts has so far only been made with antibodies against mammalian isoforms, which may cross-react with at least two different NOS isoforms. The structural characterization of NOS isoforms in teleosts is an important step to improve understanding of the nature of the protein in specific tissues and species, and for the production of specific antibodies as tools to investigate their localization of expression. In future studies, protein sequencing following Western blot analysis will reveal more about unknown teleost-specific NOS isoform protein structures, as is needed for further molecular characterization and further analyses of their cellular distribution and putative target sites. The prevailing molecular characterization of NOS I in teleosts and other vertebrates and its apparent common role in vertebrate organogenesis encouraged us to make further experimental analyses via manipulations of NO systems during embryogenesis (see below).

NOS I ontogeny

NOS I and embryonic brain neurogenesis

In developing zebrafish, the detailed ontogeny of the NOS I system has been described by NOS I mRNA ISH studies (Holmqvist *et al.*, 2004; Poon *et al.*, 2003). NOS I is initially expressed in the brain, with subsequent expression in different body organs. Correspondingly in zebrafish and rodents (Holmqvist *et al.*, 2001; Terada *et al.*, 2001), NOS I expression is first detected in the presumptive preoptic-hypothalamic area. In the developing zebrafish brain, NOS I expression accompanies the pattern of neurogenesis and the ontogeny of neurotransmitter and neurohormone systems, starting in the primary cell clusters of the forebrain and spreading into the whole CNS in a distinct pattern. In teleosts and mammals the corresponding NOS I expression pattern continues into the rhombencephalon, telencephalon, thalamus, collicular/tectal areas, cerebellar regions and medulla.

The spatial expression of NOS I in cells in the developing brain of the zebrafish accompanies the differentiation of cells expressing the transmitter signalling molecules acetylcholine (cholinergic), GABA (GABAergic), serotonin (indolaminergic), dopamine, adrenaline/epinephrine and noradrenaline/norepinephrine (catecholaminergic systems) (Arenzana *et al.*, 2005; Doldan *et al.*, 1999; Ellingsen, 1999; Guo *et al.*, 1999; Hanneman and Westerfield, 1989; Higashijima *et al.*, 2004; MacDonald *et al.*, 1994). All of these early expressed transmitters have also been claimed to participate in developmental processes and later life plasticity changes, such as in precursor cell development and differentiation, and in neuronal plasticity guidance or maintenance of projections (see Buznikov *et al.*, 1996; Lauder, 1988, 1993; Lauder *et al.*, 1998). The ontogeny of the brain neurohormone pituitary controlling systems (the hypophysiotrophic systems) in teleosts is poorly known, although central NO actions on pituitary functions and subsequent hormone-mediated processes are well known in vertebrates (see Rivier, 2002). The neurohormones vasotocin (AVT) and isotocin (IST) in teleosts and the corresponding oxytocin and vasopressin in mammals have been shown to possess important roles in developmental processes and plasticity coupled to changes in hypophysiotrophic functions of teleosts (Ebbesson *et al.*, 2003). In zebrafish, the early ontogeny of IST mRNA expression has been reported (Unger and Glasgow, 2003), and here we add data on the ontogeny of IST protein and the emergence of IST fibres in the preoptic-hypophyseal tract and the pituitary (see next page).

In zebrafish, the first differentiated cells expressing molecules for neurotransmission (acetylcholine and GABA) appear $\sim$14–16 h post-fertilization (hpf) (Doldan et al., 1999; Hanneman and Westerfield, 1989). The neuronal scaffold starts to form $\sim$17 hpf (see Chitnis and Kuwada, 1990). The first NOS I mRNA expression is reported at 16 hpf and is visualized at 19 hpf, in a small cell cluster in the ventral forebrain, the presumptive preoptic-hypothalamic region (the ventro-rostral cell cluster (vrc); see Figs. 1 and 2). This is considered to be the first differentiated neuronal cluster of the primary axogenesis (Chitnis and Kuwada, 1990; Kimmel et al., 1995).

Between 19 hpf and 24 hpf, NOS I is expressed together with GABA, serotonin and catecholamines in the vrc of the presumptive preoptic-hypothalamic area. Here, a small population of serotonergic cells have protrusions into the ventricular space, i.e., the so-called cerebrospinal fluid (CSF)-contacting cells (Figs. 3 and 4), previously noted for the early presence of GABAergic cells. The serotonergic cells of hypothalamic brain regions retain their CSF-contacting properties in later life stages (Fig. 3). The coinciding expression of NOS I and transmitters continues into the ventro-caudal cell cluster, the presumptive brainstem area (putative cholinergic locus coeruleus and indolaminergic raphe nuclei), and the direct comparison (Poon et al., 2003) demonstrates a close spatial association of the early NOS I- and catecholamine-expressing cells (Figs. 1 and 3). Early transmitter-expressing cells present at 24 hpf that do not express NOS I comprise the serotonergic cell clusters in the pineal organ and the catecholaminergic cell clusters of the "arch-associated" cell cluster (neurosecretory cells that correspond to carotid body glomus cells of other vertebrates, cf. Guo et al., 1999). In our studies we also revealed a transient expression of NOS I in the skin, between 20 hpf and time of hatching.

After 24 hpf, the differentiation of NOS I follows in the midbrain (in the ventro-caudal cell cluster), the presumptive telencephalon (in the dorso-rostral cell cluster, up to 30 hpf) and the rhombencephalon (or hind brain cell cluster, $\sim$40 hpf). The relative increase in expression rate $\sim$24 hpf and the subsequent spread of differentiation of NOS I cells (at 30–40 hpf) is followed by a major increase in differentiation prior to hatching (at $\sim$55 hpf). Thus, at 40–55 hpf, NOS I-, serotonin- and catecholamine-expressing cells coincide in cell clusters of the spinal cord and retina.

From 55 hpf, NOS I-expressing cells start to appear also in the peri-ventricular region of the midbrain and spinal cord, and in the ependyme of the telencephalon, and in association with central proliferation zones (see below). NOS I, serotonergic and catecholaminergic expressions are

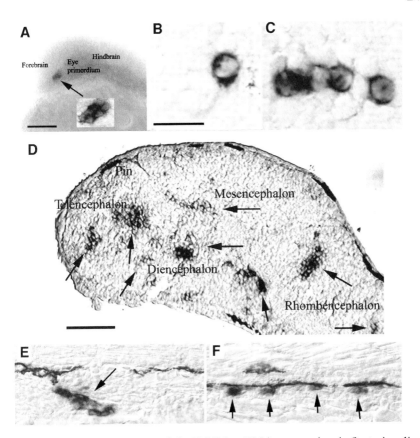

Fig. 1. NOS I ontogeny in zebrafish. NOS I mRNA expression is first visualized in the brain ventro-rostral cell cluster at (A) 19 hpf (see arrow; see also magnified inset at 23 hpf) from whole-mount preparations and (B) 24 hpf (bluish labelling) from a cryosection. (C) NOS I mRNA expression in the ventrocaudal cell cluster at 34 hpf. (D) NOS I mRNA expression at 55 hpf, demonstrating the widespread distribution of NOS I populations in all major brain areas (arrows; "Pin" indicates the pigmented pineal organ, which lacks NOS I expression). (E and F) NOS I mRNA expression in the body at 72 hpf, in a putative enteric ganglia (E) and along the nephritic and alimentary tracts (F). Scale bars represent 100 μm in (A) and (D), 20 μm in (B) and (C), and 10 μm in (E) and (F). (See Colour Plate Section in this book).

retained in cells of the periventricular layers. IST mRNA expression is reported from ~40 hpf in the presumptive preoptic area, demonstrating continuous expression at this location throughout the embryonic stage (Unger and Glasgow, 2003). Correspondingly, IST-immunoreactive cell

244

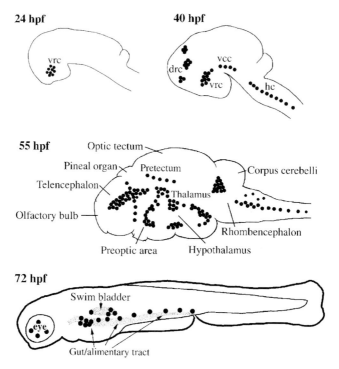

Fig. 2. Schematic representation of NOS I ontogeny in zebrafish, demonstrating the spatio-temporal expression of NOS I mRNA in the CNS and body during embryonic development (from Holmqvist *et al.*, 2004, with permission from The Company of Biologists). drc, dorsoventral cell cluster. hc, hindbrain cell cluster. vcc, ventrocaudal cell cluster. vrc, ventrorostral cell cluster.

bodies are present in the same location, in the preoptic area originating from vrc, but the protein is first detected from ~50 hpf (Fig. 5).

At the time of hatching, large NOS I populations are established in close association with regions anatomically corresponding to the proliferation zones (see below). At this time, NOS I expression also starts in the retina, and in various body organs. In the body, NOS I expression is noted in enteric ganglia and along the presumptive alimentary tract, nephritic system and swim bladder. At this time, NOS I and transmitter cell expressions coincide in various peripheral organs, not documented here. In adult teleosts, NOS activity (NADPHd histochemistry) and/or NOS immunostaining have been reported in these regions (see Villani *et al.*, 2001), including in cranial parasympathetic ganglia and the enteric nervous system (Brüning *et al.*, 1996; Gibbins *et al.*, 1995; Olsson and Holmgren, 1996, 1997).

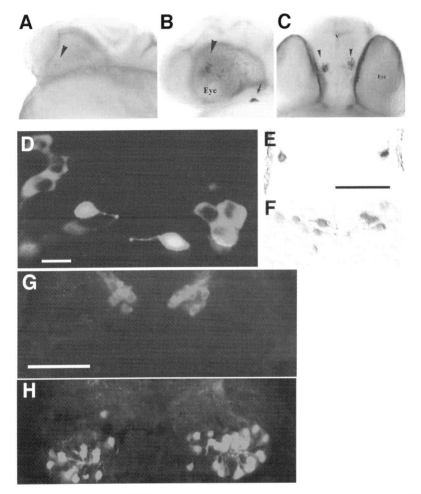

Fig. 3. Brain serotonin (5HT) and catecholamine (tyrosine hydroxylase; TH) ontogeny coincide with that of NOS I in the zebrafish. Images demonstrate the first immunoreactive 5HT-expressing cells (A) and TH-expressing cells (B and C) in whole-mount preparations at 24 hpf and in (D) cryosection (5HT is green and TH is red) at 24 hpf, and in chromogen-immunolabelled cryosections at 40 hpf (E shows serotonin; F shows TH). Note the TH population in the arch-associated cell cluster (arrow in B). (G and H) Double labelling of the same section – (G) shows TH labelled in red and (H) shows 5HT labelled in green – from embryos at 55 hpf, with TH and 5HT cells located in the posterior tubercular region and hypothalamus, respectively, which also contain NOS I at this stage and in adults. Note the processes indicating cerebrospinal fluid-contacting properties of 5HT cells in (D) and (F). Scale bars represent 10 μm in (D), 100 μm in (E) and (F), and 100 μm in (G) and (H). (See Colour Plate Section in this book).

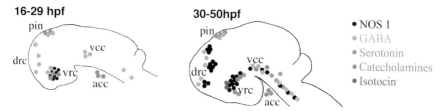

Fig. 4. Schematic representation of the spatial relation between the early differentiating NOS I- and transmitter-expressing cell populations in the zebrafish brain, represented by coloured dots: NOS I (black), GABA (light blue), serotonin (green), catecholamines (tyrosine hydroxylase; red) and isotocin (dark blue). NOS I- and transmitter-expressing cell differentiations coincide in embryonic cell clusters, and follow the general pattern of embryonic brain neurogenesis. Images of embryos represent NOS I- and transmitter-expressing cells that are differentiated during two developmental periods (16–29 hpf and 30–50 hpf). drc, dorsoventral cell cluster. hc, hindbrain cell cluster. pin, pineal organ. vcc, ventrocaudal cell cluster. vrc, ventrorostral cell cluster. (See Colour Plate Section in this book).

In zebrafish, neuronal projections from the early brain transmitter cells start to appear ~22–24 hpf, spreading into various areas of the brain between 55 hpf and hatching, and reaching the spinal cord by 96 hpf (McLean and Fetcho, 2004). At this time, both the NOS I neuronal populations and the other neurotransmitter systems have reached their main mature morphology. At this stage, the NOS I axonal projections and their target sites in peripheral organs of embryos have not yet been identified. IST fibres start to appear at 55 hpf, and are then detected coursing in the preoptic-hypophyseal tract and establishing fibre nets in the presumptive ventral hypothalamus and pituitary at 72 hpf (Fig. 5). More extensive neuronal fibre innervation into the pituitary is first detected at 120 hpf, and the mature status of pronounced nona-peptidergic [IST and AVT] expression in the preoptic area of juveniles develops during the larval stages (Fig. 5). Thus, NOS I does not appear to be expressed by pituitary cells during development, and the relatively late differentiation of hypophysiotrophic projections indicates no actions/interactions by NO on hypophyseal functions until after hatching. In the pituitary, NOS I (and possibly different NOS isoforms) is expressed by pituitary cells of adult and juvenile teleosts only (Ando *et al.*, 2004; Jadhao and Malz, 2003; Uretsky *et al.*, 2003). In juvenile salmon (Ebbesson *et al.*, 2004), NOS I immunoreactivity is located in the pituitary, associated with lactotropes, somatotropes, corticotropes and gonadotropes. NOS I labelling surrounding blood capillaries in the pars distalis and neurointermediate lobe coincides with AVT terminals.

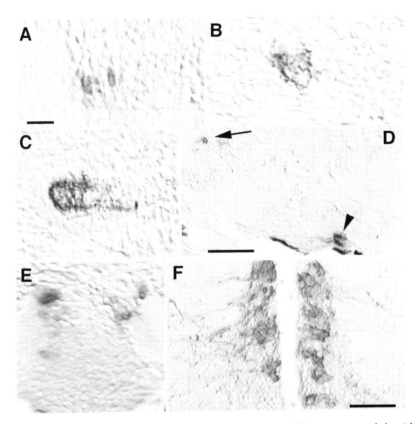

Fig. 5. Isotocin ontogeny in the zebrafish brain. Isotocin immunoreactivity (dark labelling) starts in the ventrorostral cell cluster (A: presumptive preoptic area; 60 hpf, sagittal view), at the same spatial location as the first expression of NOS I, serotonin (5HT) and catecholamine. Projections to the pituitary emerge during embryonic development (B and C at 72 hpf; arrowhead in D at 120 hpf; sagittal views). The number of IST cell bodies present at hatching (D: sagittal view (arrow); E: frontal view) increase during larval development (F: juvenile zebrafish; frontal view). Scale bar represents 40 µm in (A), (B), (C) and (E), 200 µm in (D) and 100 µm in (F).

This vascular association of NOS has also been noted in other tissues of different teleost species (Cioni *et al.*, 2002; Jimenez *et al.*, 2001; Zaccone *et al.*, 2003a). Future studies on the role of NO in pituitary functions in zebrafish should engage identification of NOS isoforms in brain hypophysiotrophic systems, their ontogeny, and the evaluation of putative different NOS isoforms expressed by specific pituitary cells.

During early development, the spatial relations of the primary NOS I and neurohormone/transmitter-expressing cells indicate their capacities for early interactions. Furthermore, during embryogenesis, NOS I populations are noted to be established in close association with regions displaying continued cell proliferation. The coincidence of NOS I expression with the formation of new serotonergic and catecholaminergic neurons (Grandel et al., 2006) incorporated into mature neurotransmitter circuits (see Kaslin and Panula, 2001; McLean and Fetcho, 2004) may also continue into adulthood neurogenesis (see below). NOS immunohistochemistry, using specific antibodies against teleost NOS isoforms, will be a crucial step to further analyse the cellular relation of NOS and neurotransmitter/ hormone systems, and to identify early NOS projections and target sites. Possibly, NOS I neuronal projections from early differentiated NOS I neurons in the brainstem, spinal cord and peripheral organs may have diverse target organs that participate in developmental processes (see below).

NOS and organogenesis

Altered NO-mediated functions during development have been shown to disturb morphogenesis, resulting in altered development of cardiovascular systems (Bloch et al., 1999; Feng et al., 2002; Lee et al., 2000), CNS (Peunova et al., 2001), bone and limbs (Aguirre et al., 2001; Diket et al., 1994; Fantal et al., 1997; Hefler et al., 2001; Pierce et al., 1995) and lung (Han and Stewart, 2006). Direct physiological influence by NOS/NO in teleosts, including effects during development, has been established in cardiovascular systems (Agnisola, 2005; Eddy, 2005; Eddy and Tibbs, 2003; Fritsche et al., 2000; Pelster et al., 2005; Tota et al., 2005). In developing embryos and larvae of zebrafish and trout, recent studies have established that NO influences vascular tone (vasodilator) and heart rate (induces bradycardia), and controls the redirection of blood flow and timing of the formation of certain blood vessels (Eddy, 2005; Eddy and Tibbs, 2003; Fritsche et al., 2000; Pelster et al., 2005). Still, the effects by NO on morphogenesis, or the identity of the NOS isoforms mediating early NO functions in teleosts, are not clear (Pelster et al., 2005). As discussed above, studies in both teleosts and mammals indicate that many NO-mediated actions in CNS development are conveyed specifically by NOS I, but that NOS III may also be an important NO source for processes of general organ development. The widespread distribution of NOS I in development that coincides with central stages of neuronal differentiation in CNS and peripheral nervous system and transmitter differentiation, together with the demonstrated GC/cGMP target sites in various body organs and tissues (Yamamoto *et al.*, 2003), suggests that NOS I may also participate in developmental processes of different organs.

To further investigate the putative role for NOS – NOS I in particular – in early organogenesis we studied the effects on zebrafish organ development, by the exposure of embryonic life stages to NOS inhibitors and a NO donor followed by analyses of heartbeat frequency (HBF) and histopathology in the larval stage (Holmqvist, Forsell and Alm, ongoing studies). Briefly, de-chorionated embryos (performed at 18–19 hpf) were reared at 28.5°C in embryo medium. Embryos were incubated in embryo medium, and were exposed to 1-(2-trifluoromethylphenyl)imidazol (TRIM), a potent and relatively selective NOS I inhibitor in mammals (Handy *et al.*, 1995; Towler *et al.*, 1998). For comparison, embryos were exposed to the most commonly used NOS inhibitor N^G-nitro-L-arginine methyl ester (L-NAME) which affects all isoforms but is relatively selective for NOS I and NOS III (Moore and Handy, 1997), or to *S*-nitroso-*N*-acetyl-penicillamine (SNAP), which is a commonly used NO donor (Ferrero *et al.*, 1999; Gross *et al.*, 1994). The optimal concentrations for the treatments (1 mM for TRIM, 2 mM for SNAP and 100 mM for L-NAME) were set at a 80–90% survival rate, determined by tests at three developmental periods (20–30 hpf, 48–58 hpf and 72–74 hpf). HBF was used to assess the optimal duration of exposure, which showed a clear decrease in HBF up to 2 h followed by a plateau (data not shown). The NO-mediated effects by TRIM and SNAP on GC/cGMP systems were verified via analyses of cGMP concentrations measured in 25 hpf-treated embryos and compared to untreated control animals, showing a decrease caused by TRIM and an increase caused by SNAP (Fig. 6A). In addition, long-term effects by TRIM and SNAP on cGMP levels, analysed 25 h after exposure, revealed that both TRIM and SNAP exposures caused a decrease in cGMP production (Fig. 6A). The relative decrease of cGMP production became stronger with age, which correlates with the increased differentiation of NOS I systems during this period (see Holmqvist *et al.*, 2004). Subsequently, embryos were exposed to the drugs for 2 h at 28–30 hpf, 48–50 hpf or 72–74 hpf, after which animals were transferred into fresh embryo medium and HBF was recorded. Animals were then finally analysed and sampled at 96–100 hpf, *i.e.*, in their larval stage. Following documentation of HBF, and body and organ morphology, animals were sacrificed (in MS 222), and were immersion fixed (4% paraformaldehyde). Paraffin sections were stained with haematoxylin-eosin and analysed in a light microscope equipped with interference Nomarski optics.

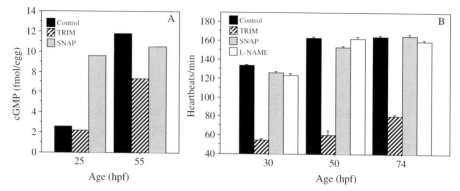

Fig. 6. (A) The effects of the mammalian NOS I inhibitor 1-(2-trifluoromethyl-phenyl)imidazol (TRIM) and the NO donor S-nitroso-N-acetyl-penicillamine (SNAP) on cGMP production in zebrafish embryos at age 25 hpf, after 5 h of treatment, and at age 55 hpf, 25 h after treatment for 5 h. (B) The effects of TRIM, SNAP and the NOS inhibitor L-NAME (N^G-nitro-L-arginine methyl ester) on heartbeat frequency in larvae that had been exposed for 2 h as embryos at ages of 28–30 hpf, 48–50 hpf or 72–74 hpf.

Two-hour treatments of embryonic stages with TRIM, L-NAME or SNAP all affected HBF (Fig. 6B). TRIM showed the strongest effects, significantly reducing HBF by ~50–60% in all developmental stages, whereas L-NAME reduced HBF by 4–8%. The NO donor SNAP reduced HBF by 6% compared to controls. In the 28–30 hpf-treated animals, HBFs were significantly decreased with either TRIM, L-NAME or SNAP compared to controls. In 48–50 hpf-treated animals, TRIM strongly decreased HBF, whereas L-NAME showed no significant effect on HBF at this stage. SNAP-treated animals had a slight decrease in HBF. In 72–74 hpf-treated larvae, TRIM again caused a strongly reduced HBF, whereas there were only minor effects by L-NAME and SNAP (Fig. 6B).

All 28–30 hpf-treated larval groups were less developed compared to the controls (Fig. 7). Abnormal organ morphology and histology were obvious in larvae that had been exposed to TRIM and SNAP as embryos (Figs. 7–9). The most obvious abnormalities noted were effects by TRIM on general body morphology, heart gross morphology and histology, and on cartilage and CNS morphologies of larvae. The vast majority possessed a bent body form, different-sized eyes that tilted ventrally on a high skull, and heart pathologies. The brain appeared relatively undifferentiated, and deformed corresponding to the external morphology of the head (Figs. 7 and 8). In TRIM-treated animals heart pathologies included a thin tube-like heart and a large pericardiac sac (a small bulbus arteri and a thin heart

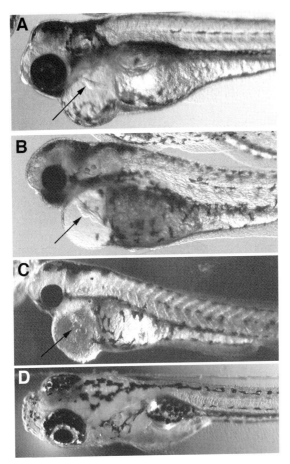

Fig. 7. Body abnormalities in zebrafish larvae, caused by exposure of embryonic life stages to TRIM (A and B) or SNAP (C), in comparison to a non-treated control animal (D). Abnormal development of both head (tilted) and heart (tube-like; arrows) is demonstrated. (See Colour Plate Section in this book).

wall, Figs. 7 and 9). In severely affected animals, the cartilage of the jaw (such as the trabecula, ethmoid plate and ceratohyal) and the pharyngeal arches were deformed, or could not be defined (not shown). We also noted that in some larvae with no obvious deformities, organs/organ systems such as the liver, renal and digestive systems and swim bladder appeared to be less differentiated compared to control larvae. Many SNAP-treated larvae also had abnormal heads and possessed severe abnormalities of the heart (Fig. 7) L-NAME-treated animals did not reveal any obvious

252

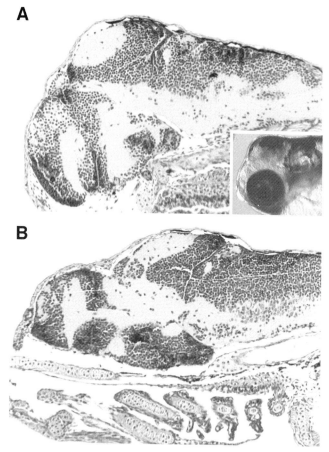

Fig. 8. Brain histopathology in zebrafish larvae caused by exposure of embryonic life stages to TRIM (A; inset shows head), compared to a non-treated control animal (B). In animals exposed to TRIM the brain morphology is disturbed and the differentiation of brain areas, *i.e.*, neuroanatomy, is less clear compared to control animals.

abnormalities. Animals with the most severe structural pathologies were also noted to show abnormal behaviour, in body movement, swimming activity and/or touch response.

This preliminary study shows that manipulations of NO systems for 2 h during embryonic life stages strongly influence organ development in zebrafish. The noted effects of NO on organogenesis are indicated to be due to the disruption of NOS activity and the subsequent actions of NO on cGMP systems, by the inhibition or depletion of NO activity that

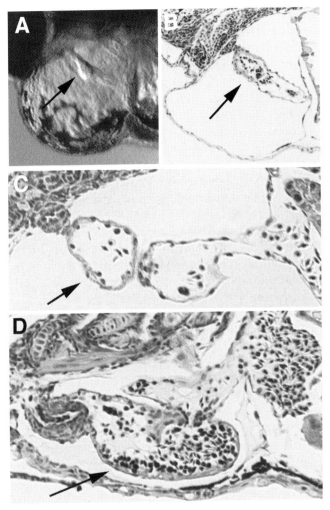

Fig. 9. Heart histopathology in zebrafish larvae caused by exposure of embryonic life stages to TRIM (A–C). Animals exposed to TRIM possess an abnormal, tube-like heart, compared to the normal heart (D). (See Colour Plate Section in this book).

subsequently affects NO-mediated GC/cGMP target activities, as shown previously in the CNS of medaka and mammals (see Krumenacker and Murad, 2006). In zebrafish, the generally impaired organ development combined with decreased NO–cGMP production supports this. In line with the effects seen in previous studies of other vertebrate species, the altered NO systems during zebrafish development resulted in disturbed development of the CNS, cartilage (bone and limbs) and cardiovascular system. The severe effects on heart development may correspond to the abnormal aortal valve formation and development of congenital septal effects seen in mammals, and reported to be due to NOS III deficiencies (Feng et al., 2002; Lee et al., 2000). Reasonably, an optimal concentration of NO is critical for normal developmental processes, as demonstrated by the fact that either an excess or a depletion of NO inhibits normal embryonic development (Chen et al., 2001; Tranguch et al., 2003). Correspondingly, the similar effects on zebrafish organogenesis by NOS inhibitors and a NO donor show that disruption of NO systems in any direction (i.e., NO decrease or increase) can cause similar deficiencies in the similar NO-mediated mechanisms of developmental processes. In the zebrafish, the disrupted deficiency in NOS/NO/cGMP activity caused in embryonic life is shown to be retained in later life stages.

The dominant effects on zebrafish organogenesis by TRIM, the preferential NOS I inhibitor in mammals, indicate inhibition of the early differentiated and widespread NOS I systems. Thus, NOS inhibitors and donors in teleosts are useful tools to determine the functional roles of NO in general, but not of NOS isoforms selectively, since no completely NOS-isoform-specific inhibitors are available. In general, NOS isoform inhibitors have a relatively higher affinity for one isoform, and the affinities vary between species (see Cayman Chemical Company Catalogue, 2004, and references therein). Thus, although NOS I is likely to participate in NO-mediated processes during development, we need more data about the presence and activity of the other teleost NOS isoforms, NOS II and the putative NOS III, to determine any individual role for NOS isoforms in developmental processes. Still, the demonstrated important role for NOS I in body organ development, and specifically in vertebrate brain neurogenesis (see below), underlines the importance of being able to target the putative neuronal NOS I systems in future experimental studies.

NOS I and adult neurogenesis

NO may possess a critical role in adult brain neurogenesis, proposed to promote neurogenesis by modulating ongoing cell proliferation and the fate of neuronal progenitor cells (Chen et al., 2004; Cheng et al., 2003; Gibbs, 2003; Matarredona et al., 2005; Zhang et al., 2001). NOS I is suggested to be the main source of these NO-mediated processes. In the zebrafish brain, large NOS I-expressing cell populations are localized in close association with the proliferation zones at the time of hatching (Holmqvist et al., 2004), and the NOS I expression associated with proliferation zones is indicated to continue into adulthood (Holmqvist *et al.*, 2000), which is described here in detail in relation to sites with active cell mitosis, migration and differentiation.

In the adult zebrafish brain, NOS I cell populations are present in all telencephalic, diencephalic, mesencephalic, cerebellar and rhombencephalic proliferation zones. NOS I mRNA-expressing cells (labelled by ISH; Holmqvist *et al.*, 2000, 2004) are located within or in close vicinity to all anatomically defined proliferation zones (Figs. 10 and 11) that contain newly divided, proliferation nuclear antigen (PCNA)-immunoreactive cells (Figs. 12 and 13; immunohistochemical labelling of PCNA is described by Ekström *et al.*, 2001; Grandel *et al.*, 2006; Wullimann and Knipp, 2000). Here, PCNA-positive cells are preferentially located in the ependymal cell layers. The largest NOS I populations are situated in the inner periventricular cell layers of proliferation zones surrounding the ventricles of telencephalic and diencephalic brain regions. Relatively few NOS I cells are located in the outer ependymal cell layers in the telencephalic dorsal and habenular proliferation zones. In addition, close to proliferation zones, NOS I populations are located in central brain areas, dorsal thalamic, posterior and tubercular nuclei, and the facial and vagal lobes. Thus, NOS I is highly expressed in areas of ongoing cell proliferation, in the same cell layers as dividing cells of restricted brain regions, whereas most populations are located in the more periventricular regions of proliferation zones and do not co-localize with actively dividing cells, thus are rather located in postmitotic regions with putative active cell differentiation and/or migration.

Recently, in a more detailed analysis of the differentiation regions in the adult zebrafish brain, Grandel *et al.* (2006) also report that catecholaminergic and indolaminergic neurons that originate from the proliferation zones may be continuously added into specific brain nuclei. Interestingly, NOS I cell populations are also present in these brain nuclei, previously reported in the adult brain of both zebrafish and salmon (see Holmqvist *et al.*, 2000). To further investigate this we have analysed the relation of NOS I to regions of cell mitosis and neuronal differentiation in adult zebrafish brain (Holmqvist and Alm, ongoing studies), by the detection of newly divided cells expressing histone 3 (H3), disclosure of regions of differentiation and plasticity changes expressing highly polysialylated neural cell adhesion molecule (PSA-NCAM; see Kiss and Rougon, 1997; Marx *et al.*, 2001; Seki and Arai, 1993) and detection of differentiating and mature neurons expressing HU/elav (see Grandel *et al.*, 2006). NADPHd histochemistry was employed to visualize NOS activity (Holmqvist *et al.*, 1994, 2000).

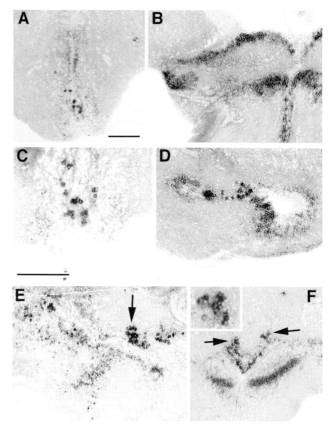

Fig. 10. NOS I expression in anatomically defined proliferation zones in the brain of the adult zebrafish. NOS I mRNA-expressing cell populations are shown in periventricular cell layers of proliferation zones in the ventral telencephalon (A), dorsal and caudal telencephalon (B), rostral preoptic area (C), hypothalamus (D and E: sagittal views; F: frontal view) and the central brain area, such as in the nucleus of the posterior tuberculum (inset and arrows in E and F). Scale bar represents 100 μm in (A), (B), (D)–(F), and 100 μm in (C) and in the inset to (F).

In the adult zebrafish brain, PSA-NCAM immunoreactivity is present in various regions, and is distributed both within and between proliferation zones (Figs. 13–15) in agreement with regions of plasticity and migration routes in the brains of rodents and birds, *i.e.*, hippocampus, SVZ and RMS into the olfactory bulb (Doetsch and Scharff, 2001; Doetsch *et al.*, 1997; Seki and Arai, 1993). Double labelling of NOS activity and PSA (in the same sections) revealed closely coinciding NOS

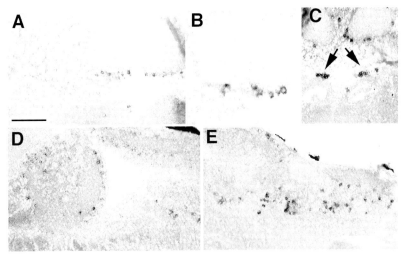

Fig. 11. NOS I expression in the anatomically defined proliferation zones in the brain of the adult zebrafish. NOS I mRNA-expressing cell populations are demonstrated in proliferation zones of the cerebellum (valvula cerebelli; A and B), the area of the griseum centrale/locus coeruleus (C; arrows), and the vagal (D) and facial (E) lobes. Frontal views are shown in (A)–(C) and sagittal views in (E) and (F). Scale bar represents 100 µm in (A), (C)–(E), and 50 µm in (B).

activity and PSA-NCAM expression in most proliferation zones (Fig. 13). NOS activity is also present in the inner periventricular layers, possibly originating from both cell bodies and fibres, thus partly different to the preferential detection of NOS I mRNA-expressing cell bodies in outer periventricular cell layers. Whether NOS I cells stretch through the periventricular layers to reach contact with the CSF, similar to the serotonergic cells in this nuclei, is not clear.

The preferential expression of NOS I cell bodies in regions of cell differentiation and/or migration routes in zebrafish brain is further supported by the relation of labelling of PSA-NCAM and PCNA, H3 or HU/elav (Figs. 14 and 15). The restricted localization of PCNA and H3 expression in cells in relation to the pattern of PSA-NCAM and HU/elav shows that active cell mitosis occurs only at restricted regions of proliferation zones, preferentially in the ependymal cell layers. In addition, the regions of neural differentiations are indicated by the pattern of HU/elav immunoreactivity: lacking in cells in cell layers expressing H3, scarce in regions with PCNA-positive cells, but numerous weakly HU/elav-labelled cells in more periventricular regions that express PSA-NCAM. Together with the direct comparisons of NOS activity and PSA-NCAM, these data further support the view that NOS

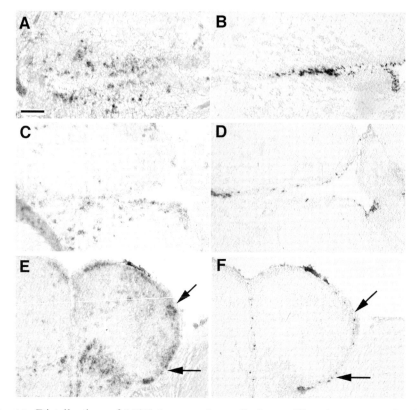

Fig. 12. Distribution of NOS I-expressing cells in proliferation zones of adult zebrafish. NOS I mRNA expression (A, C and E) and proliferation cell nuclear antigen (PCNA) immunoreactivity (B, D and F) in adjacent sections of the proliferation zones of the ventral, caudal and dorsal telencephalon (A–D) and dorsal telencephalon, thalamus and habenular regions (E and F). Note that the NOS I mRNA and PCNA expressions coinciding in proliferation zones are located in different cell populations of periventricular and ependymal cell layers in (A)–(D), but in the same cell layer of dorsal telencephalon and habenula (E and F; arrows). Sagittal views are shown in (A)–(D), and frontal views in (E) and (F). Scale bar represents 100 µm.

I-expressing cells distributed in the periventricular cell layers together with PSA-NCAM- and HU/elav-positive cells (Fig. 15) are located in postmitotic regions with ongoing cell differentiation and cell/neuronal plasticity. Correspondingly in the SVZ of mammals, cell mitosis occurs in restricted regions of the lateral ventricular areas, whereas the cells are differentiating during their transport in the PSA-NCAM-positive extracellular matrix of the periventricular cell layers and along the RMS into the olfactory bulb, in which they finally mature (see Doetsch *et al.*, 1997). In contrast, in the dorsal telencephalic proliferation zones of zebrafish (corresponding to the dentate gyrus of mammals), NOS I expression in the same cellular layer coincides with H3, PCNA and PSA-NCAM, and with relatively strong HU-positive (neuron differentiated) immunoreactivity (Fig. 15), indicating a mixture of mature and immature cells remaining together unlike the situation in most other neurogenic brain regions. Thus, in zebrafish, as in other studied teleosts, the large NOS I populations located in differentiation regions are clearly separate from the NOS I neuronal populations located within mature brain nuclei (Holmqvist and Ekström, 1997; Holmqvist *et al.*, 1994, 2000), of which the NOS I populations in differentiation regions are the most prominent. In differentiation regions, the NOS I-expressing cells may be immature cells, and/or putative progenitor cells with transient NOS I expression. However, simultaneous visualization of NOS I and markers for cell differentiation status has to be performed to determine the nature of maturity for NOS I populations in these regions.

In mature brain nuclei with NOS I neuronal populations (NOS I mRNA- and protein-expressing and NADPHd-positive; Holmqvist *et al.*, 1994, 2000, 2004; Poon *et al.*, 2003), newborn catecholaminergic and indolaminergic neurons are reported to be incorporated (Grandel *et al.*, 2006). The largest number of newborn neurotransmitter cells are reported to be serotonergic, and incorporated into hypothalamic nuclei that also possess the largest NOS I cell populations and highest NOS activity (see Figs. 3, 10 and 13). The newborn catecholaminergic cells are reported to be incorporated into different areas of the brain – the olfactory bulbs, preoptic area, pretectum, posterior tuberculum and the hypothalamus – which all hold differentiated NOS I neuronal populations. The coinciding differentiation of NOS I and catecholaminergic and serotonergic cells that start together in early ontogeny (see above) continues into adulthood, in defined regions with ongoing neurogenesis and with the incorporation of newborn neurotransmitter neurons into specific brain nuclei. Future studies are needed to resolve the cellular relation between NOS I and newly formed cells in regions of ongoing neurogenesis, including simultaneous visualization of NOS I together with PCNA and/or incorporated bromodeoxyuridine (BrDU). Furthermore, the presence of additional NOS isoforms in neurogenic brain regions should be evaluated.

It is concluded that in adult zebrafish, several NOS I cell populations in the brain have the ability to influence NO-mediated processes of ongoing neurogenesis, as previously indicated in mammals. The majority of NOS I-expressing cells in periventricular cell layers are immature cells located

260

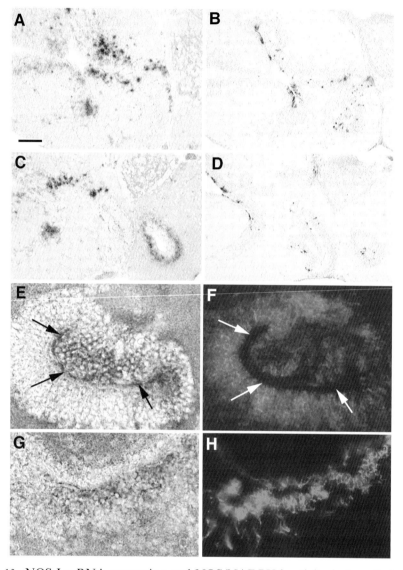

Fig. 13. NOS I mRNA expression and NOS/NADPHd activity in hypothalamic proliferation zones of adult zebrafish. NOS I mRNA expression (A and C) and proliferation nuclear antigen (PCNA) immunoreactivity (B and D) in adjacent sections of the hypothalamic proliferation zones. Note the coincidence of NOS I and PCNA located in different cell populations of periventricular and ependymal cell layers, respectively. NOS/NADPHd activity (E and G) and polysialylated neural cell adhesion molecule (PSA-NCAM) immunofluorescence (green in F and H) double labelling indicate different degrees of cellular co-expression of NOS and PSA-NCAM, and that NOS activity is also present in the inner

261

in regions with ongoing differentiation, and mature NOS I neurons are located in brain nuclei that obtain newborn transmitter neurons. Whether NOS I can be expressed by mature neurons, and/or by immature, putative progenitor cells, cannot yet be concluded. Future experimental studies of the highly active neurogenesis throughout the brain of teleosts may provide further data on these issues, and may shed more light on the individual roles of NOS I, and/or NOS III, in vertebrate neurogenesis.

Conclusions

Data from zebrafish and other teleost species show that NOS I is the best characterized teleost NOS isoform. NOS I and NOS II genes and their expression in different tissues have been documented in teleosts, whereas only NOS I mRNA expression has been identified *in situ* so far. The constitutively expressed NOS I gene is present throughout the teleost phylogeny and it has high structural similarities to NOS I of other vertebrate species. In the zebrafish, NOS I expression is well documented, whereas the character of the NOS I protein is less clear, and the NOS I target sites (neuronal and other) still remain to be identified. The presence and mRNA expression of a NOS II gene has been identified in zebrafish and a few other teleost species, whereas more data are needed on its character of expression (constitutive and/or inducible), sites of expression and functional properties. In teleosts there is so far no molecular evidence for the presence of a NOS isoform corresponding to NOS III in mammals, although there are data on the detection of a NOS III-like protein and NOS activity in vascular systems and cardiovascular NOS III-like functions. The evidence for a NOS III isoform in teleosts comes from immunocytochemistry and Western blot studies using antibodies raised against mammalian NOS proteins or peptides, indicating the presence of a NOS III-like protein. The presence of different NOS isoforms in teleosts has been indicated with various different techniques: Northern blots, Western blots, immunocytochemistry and NADPHd histochemistry. Functional/pharmacology data support NOS III-like actions, but also indicate that NOS I- and NOS II-like proteins can possess similar functional

Fig. 13. Continued.
periventricular and ependymal cell layers (arrows in E and F). Scale bar in (A) represents 100 μm in (A)–(D), and 50 μm in (E)–(H). PSA-NCAM antiserum used was a kind gift from Prof. Urs Rutishauser. (See Colour Plate Section in this book).

262

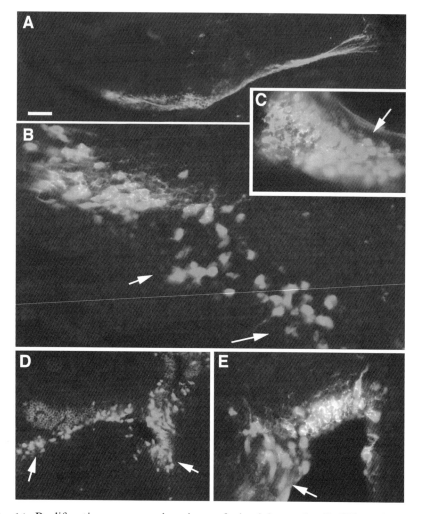

Fig. 14. Proliferation zones and regions of plasticity and cell differentiation in the brain of adult zebrafish. Polysialylated neural cell adhesion molecule (PSA-NCAM; green) immunofluorescence labels the extracellular matrix within and between proliferation zones throughout the brain, demonstrated in the ventral telencephalon (A). Immunofluorescence double labelling of PCNA (red) and PSA-NCAM (green) demonstrates the cellular relation between cell proliferation activity and the putative migration routes of proliferation zones in the ventral telencephalon (B and C), hypothalamus (D) and dorsal thalamus/habenula (E). Note that the labelling of cells ranges from single PCNA expression (arrows), which prevails close to the areas of active cell mitosis of the ependymal and subependymal cell layers (see also Fig. 15), to areas of cellular co-expression of PCNA and PSA-NCAM (green around red cells) and single PSA-NCAM expression, which prevails in periventricular cell layers, *i.e.*, in regions of

roles in teleosts. Antisera against mammalian NOS I have been demonstrated to label teleost NOS I in tissues, whereas few antisera to other mammalian NOS isoforms have been demonstrated to specifically recognize their teleost counterparts. However, it should also be considered that the vast majority of studies in teleosts so far have focused on NOS I. Future studies in zebrafish should therefore include analyses of the molecular identity of the putative NOS II and NOS III isoforms, and determinations of their mRNA and protein expression *in situ*. Antisera against characterized teleost NOS isoforms may provide a helpful tool for various future studies of NOS systems in teleosts.

In the zebrafish, NOS I expression is well documented, in the whole body during early development and in the brain of adults. NOS I-expressing cells are differentiated throughout the brain during embryonic development, accompanying the pattern of neuronal differentiation and coinciding with that of differentiating neurotransmitter systems. A spatial coincidence of NOS I, serotonin and catecholamine expression starts in the same embryonic cell cluster and continues in regions with ongoing neurogenesis in the adult brain. Brain hypophysiotrophic neurohormone systems originate from the same embryonic cell cluster, but appear to differentiate relatively late. In zebrafish, the presence of putative NOS I pituitary projections or NOS expression by pituitary cells is still unclear. Manipulations of NO-dependent processes during zebrafish embryogenesis caused abnormal development of the CNS, cartilage and cardiovascular systems, which is in agreement with the findings of previous studies of vertebrates. Recent data in zebrafish further support previous findings in vertebrates that optimal NO production during embryogenesis is critical for normal organ development, and the activation of cGMP by NO is a central developmental mechanism. The NOS inhibitor TRIM, with high affinity to NOS I in mammals, causes severe developmental abnormalities, but since we do not know the specific NOS I target sites and are lacking evidence on the presence of other NOS isoforms in embryonic life stages, the individual effects of only NOS I cannot be determined. Since available NOS inhibitors are poorly selective for only one NOS isoform in vertebrates, new tools are needed to target the activity of specific NOS isoforms in teleosts, at the gene, mRNA or protein expression levels. Thus, more data and investigative tools are needed to

Fig. 14. Continued.
putative cell migration and differentiation (see Fig. 15). All images show sagittal views. Scale bar represents 100 μm in (A) and (D), and 50 μm in (B), (C) and (E). PSA-NCAM antiserum used was a kind gift from Prof. Urs Rutishauser. (See Colour Plate Section in this book).

264

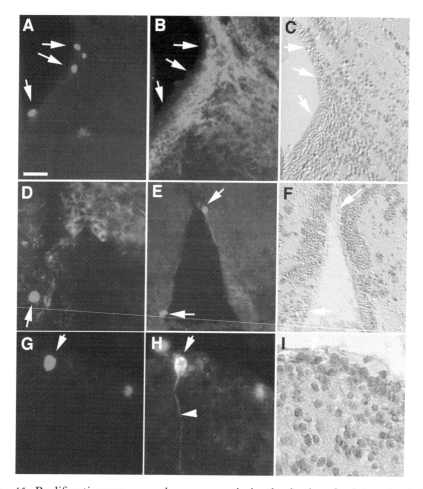

Fig. 15. Proliferation zones and neurogenesis in the brain of adult zebrafish. Distribution of active cell mitosis, regions of plasticity and neuronal differentiation in proliferation zones of adult zebrafish, depicted in the hypothalamus (A–C) and rostral preoptic area (D–F) by immunohistochemical labelling of histon 3 (red), polysialylated neural cell adhesion molecule (PSA-NCAM; green) and HU/elav (weak to strong brownish). Note in (A)–(F) the histon 3-expressing cells in the ependymal cell layers, in relation to prevailing PSA-NCAM and HU labelling in periventricular and central brain regions. Note also the successive HU labelling intensity, ranging from weak in the inner periventricular layers to strong in the outer periventricular layers, in which the majority of NOS I cells are distributed. In contrast, at the same location as NOS I-expressing cells in the dorsal telencephalon (G), newly divided cells with processes into the brain (arrowheads) are co-expressing PSA-NCAM (H) and are located in the same cell layer as HU-positive cells (I), with varying labelling intensities. Scale bar represents 100 μm in (A)–(F), and 50 μm in (G)–(I). PSA-NCAM antiserum used was a kind gift from Prof. Urs Rutishauser. (See Colour Plate Section in this book).

further elucidate the individual roles for the poorly characterized and as-yet unidentified NO-producing NOS isoforms in teleosts.

At the time of hatching, large NOS I-expressing cellular populations are established in association with brain regions with active neurogenesis that lasts into adulthood, distributed mainly in regions with ongoing differentiation. The distribution of NOS I in areas of ongoing neurogenesis in teleosts further supports previous suggestions on the key role for NOS I in NO-mediated processes of neurogenesis, although these regions are more restricted in higher vertebrates. The cell character of these NOS I-expressing cells as mature or immature cells and/or neurons, or putative progenitor cells (with transient expression of NOS I), remains to be characterized.

The current documentation of data on NOS/NO systems in teleosts demonstrates great potential for the use of experimental model species like the zebrafish for studies of versatile functions and cellular mechanisms of NO in vertebrates, in developmental biology concerning processes of body morphogenesis and organogenesis, and in vertebrate brain neurogenesis in both embryos and adults. In conjunction with the high capacity for ongoing neurogenesis in the CNS of teleosts and a plausible common role for NOS I in these processes, experimental species such as the zebrafish are of special interest for comparative studies on mechanisms of NOS I and its NO-mediated actions in adult neurogenesis and the neural plasticity of vertebrates.

Acknowledgements

We thank Lillemor Thuresson, Berit Ellingsen and Christina Falk Olsson for excellent technical help with immunohistochemical and RT-PCR analyses. We thank Dr. Johan Forsell (Novo Nordisk, Denmark) for help with the studies of NOS inhibitors, and Professor Irina Zhdanova (Boston University, USA) for measurements of cGMP. We also thank the support given by ImaGene-iT (BH) and the Norwegian Research Council (LE) during cell imaging analyses and preparation of the manuscript.

References

Agnisola, C. (2005). Role of nitric oxide in the control of coronary resistance in teleosts. *Comp. Biochem. Physiol. A: Mol. Integr. Physiol.* 142,178–187.

Aguirre, J., Buttery, L., O'Shaughnessy, M., Afsai, F., de Marticorena, I. F., Hukkanen, M., Huang, P., MacIntyre, I. and Polak, J. (2001). Endothelial nitric oxide synthase gene-deficient mice demonstrate marked retardation in postnatal bone formation, reduced bone volume, and defects in osteoblast maturation and activity. *Am. J. Pathol.* 158,247–257.

Alm, P., Holmqvist, B., Lundberg, L. M., Falk-Olsson, C., Larsson, B. and Skagerberg, G. (2002). Neuronal and endothelial nitric oxide synthase coincide with polysialic acid in forebrain areas with ongoing neurogenesis. In *32nd Annual Meeting of the Society for Neuroscience,* Abstracts 229.17, Orlando, FL, USA.

Alm, P., Larsson, B., Ekblad, E., Sundler, F. and Andersson, K. E. (1993). Immunohistochemical localization of peripheral nitric oxide synthase-containing nerves using antibodies raised against synthesized C- and N-terminal fragments of a cloned enzyme from rat brain. *Acta Physiol. Scand.* 148(4), 421–429.

Alvarez-Buylla, A., García-Verdugo, J. M. and Tramontin, A. D. (2001). A unified hypothesis on the lineage of neural stem cells. *Nat. Rev. Neurosci.* 2,287–293.

Amelio, D., Garofalo, F., Pellegrino, D., Giordano, F., Tota, B. and Cerra, M. C. (2006). Cardiac expression and distribution of nitric oxide synthases in the ventricle of the cold-adapted Antarctic teleosts, the hemoglobinless *Chionodraco hamatus* and the red-blooded *Trematomus bernacchii*. *Nitric Oxide* 15(3), 190–198.

Ando, H., Shi, Q., Kusakabe, T., Ohya, T., Suzuki, N. and Urano, A. (2004). Localization of mRNAs encoding alpha and beta subunits of soluble guanylyl cyclase in the brain of rainbow trout: Comparison with the distribution of neuronal nitric oxide synthase. *Brain Res.* 1013,13–29.

Arenzana, F. J., Clemente, D., Sanchez-Gonzalez, R., Porteros, A., Aijon, J. and Arevalo, R. (2005). Development of the cholinergic system in the brain and retina of the zebrafish. *Brain Res. Bull.* 66,421–425.

Barroso, J. B., Carreras, A., Esteban, F. J., Peinado, M. A., Martinez-Lara, E., Valderrama, R., Jimenez, A., Rodrigo, J. and Lupianez, J. A. (2000). Molecular and kinetic characterization and cell type location of inducible nitric oxide synthase in fish. *Am. J. Physiol. Regul. Integr. Comp. Physiol.* 279,R650–R656.

Bloch, W., Fleischmann, B. K., Lorke, D. E., Andressen, C., Hops, B., Hescheler, J. and Addicks, K. (1999). Nitric oxide synthase expression and role during cardiogenesis. *Cardiovasc. Res.* 43,675–684.

Brüning, G., Hattwig, K. and Mayer, B. (1996). Nitric oxide synthase in the peripheral nervous system of the goldfish, *Carassius auratus*. *Cell Tissue Res.* 284,87–98.

Brüning, G., Katzbach, R. and Mayer, B. (1995). Histochemical and immunocytochemical localization of nitric oxide synthase in the central nervous system of the goldfish, *Carassius auratus*. *J. Comp. Neurol.* 358,353–382.

Buznikov, G. A., Shmukler, Y. B., Lauder, J. M. (1996). From oocyte to neuron: Do neurotransmitters function in the same way throughout development? *Cell. Mol. Neurobiol.* 16,537–559.

Campos-Perez, J. J., Ward, M., Grabowski, P. S., Ellis, A. E. and Secombes, C. J. (2000). The gills are an important site of iNOS expression in rainbow trout *Oncorhynchus mykiss* after challenge with the Gram-positive pathogen *Renibacterium salmoninarum*. *Immunology* 99,153–161.

Cayman Chemical Company Catalogue (2004). *NOS Inhibitors*, p. 220, http://www.caymanchem.com/images/scientificIllustrations/full/cayman2056.jpg

Chen, H. W., Jiang, W. S. and Tzeng, C. R. (2001). Nitric oxide as a regulator in preimplantation embryo development and apoptosis. *Fertil. Steril.* 75,1163–1171.

Chen, J., Tu, Y., Moon, C., Matarazzo, V., Palmer, A. M. and Ronnett, G. V. (2004). The localization of neuronal nitric oxide synthase may influence its role in neuronal precursor proliferation and synaptic maintenance. *Dev. Biol.* 269,165–182.

Cheng, A., Wang, S., Cai, J., Rao, M. S. and Mattson, M. P. (2003). Nitric oxide acts in a positive feed back loop with BDNF to regulate neural progenitor cell proliferation and differentiation in the mammalian brain. *Dev. Biol.* 258,319–333.

Chitnis, A. B. and Kuwada, J. Y. (1990). Axonogenesis in the brain of zebrafish embryos. *J. Neurosci.* 10,1892–1905.

Cioni, C., Bordieri, L. and De Vito, L. (2002). Nitric oxide and neuromodulation in the caudal neurosecretory system of teleosts. *Comp. Biochem. Physiol.* B 132,57–68.

Currie, D. A., de Vente, J. and Moody, W. J. (2006). Developmental appearance of cyclic guanosine monophosphate (cGMP) production and nitric oxide responsiveness in embryonic mouse cortex and striatum. *Dev. Dyn.* 235,1668–1677.

Diket, A. L., Pierce, M. R., Munshi, U. K., Voelöker, C. A., Eloby-Childress, S., Greenberg, S. S., Zhang, X.-J., Clark, D. A. and Miller, M. J. S. (1994). Nitric oxide inhibition causes intrauterine growth retardation and hind-limb disruption in rats. *Am. J. Obstet. Gynecol.* 171,1243–1250.

Doetsch, F., Garcia-Verdugo, G. and Alvarez-Buylla, A. (1997). Cellular composition and three-dimensional organization of the subventricular zone in the adult mammalian brain. *J. Neurosci.* 17,5046–5061.

Doetsch, F. and Scharff, C. (2001). Challenges for brain repair: Insights from adult neurogenesis in birds and mammals. *Brain Behav. Evol.* 58,306–322.

Doldan, M. J., Prego, B., Holmqvist, B. I. and de Miguel, E. (1999). Distribution of GABA-immunolabeling in the early zebrafish (*Danio rerio*) brain. *Eur. J. Morphol.* 37, 126–129.

Duncan, A. J. and Heales, S. J. R. (2005). Nitric oxide and neurological disorders. *Mol. Aspects Med.* 26,67–96.

Ebbesson, L. O. E., Ekstrom, P., Ebbesson, S. O. E., Stefansson, S. O. and Holmqvist, B. (2003). Neural circuits and their structural and chemical reorganization in the light–brain–pituitary axis during parr–smolt transformation in salmon. *Aquaculture* 222,59–70.

Ebbesson, L. O., Stefansson, S. O., Holmqvist, B. (2004). Distribution of nitric oxide synthase in the salmon pituitary. In *5th Int. Symp. Fish. Endo.* Castellon, Spain. P34.

Ebbesson, L. O. E., Tipsmark, C. K., Holmqvist, B., Nilsen, T., Andersson, E., Stefansson, S. O. and Madsen, S. S. (2005). Nitric oxide synthase in the gill of Atlantic salmon: Colocalization with and inhibition of Na^+, K^+-ATPase. *J. Exp. Biol.* 208,1011–1017.

Eddy, F. B. (2005). Role of nitric oxide in larval and juvenile fish. *Comp. Biochem. Physiol. A: Mol. Integr. Physiol.* 142,221–230.

Eddy, F. B. and Tibbs, P. (2003). Effects of nitric oxide synthase inhibitors and a substrate, L-arginine, on the cardiac function of juvenile salmonid fish. *Comp. Biochem. Physiol. C: Toxicol. Pharmacol.* 135,137–144.

Ekström, P., Johnsson, C. M. and Ohlin, L. M. (2001). Ventricular proliferation zones in the brain of an adult teleost fish and their relation to neuromeres and migration (secondary matrix) zones. *J. Comp. Neurol.* 436,92–110.

268

Ellingsen, B. K. N. (1999). *Neuronal Differentiation and Transmitter Expression in the Brain of Embryonic Zebrafish* (MS Thesis). Department of Zoology, University of Bergen, Bergen, Norway, pp. 1–77.

Estrada, C. and Murillo-Carretero, M. (2005). Nitric oxide and adult neurogenesis in health and disease. *Neuroscientist* 11,294–307.

Fantal, A. G., Nekani, N., Shepard, T., Cornel, L. M., Unis, A. S. and Lemire, R. J. (1997). The teratogenicity of NG-nitro-L-arginine methyl ester (L-NAME), a nitric oxide synthase inhibitor, in rats. *Reprod. Toxicol.* 11,709–717.

Feng, Q., Wei, S., Xiangru, L., Hamilton, J. A., Ming, L., Tianqing, P. and Siu-Pok, Y. (2002). Development of heart failure and congenital septal defects in mice lacing endothelial nitric oxide synthase. *Circulation* 106,873–879.

Ferrero, R., Rodriguez-Pascual, F., Miras-Portugal, M. T. and Torres, M. (1999). Comparative effects of several nitric oxide donors on intracellular cyclic GMP levels in bovine chromaffin cells: Correlation with nitric oxide production. *Br. J. Pharmacol.* 127,779–787.

Fritsche, R., Schwerte, T. and Pelster, B. (2000). Nitric oxide and vascular reactivity in developing zebrafish, *Danio rerio*. *Am. J. Physiol. Regul. Integr. Comp. Physiol.* 279,R2200–R2207.

Funakoshi, K., Kadota, T., Atobe, Y., Nakano, M., Goris, R. C. and Kishida, R. (1999). Nitric oxide synthase in the glossopharyngeal and vagal afferent pathway of a teleost, *Takifugu niphobles*. The branchial vascular innervation. *Cell Tissue Res.* 298,45–54.

Gage, F. (2000). Mammalian neural stem cells. *Science* 287,1433–1438.

Gallo, V. P. and Civinini, A. (2001). Immunohistochemical localization of nNOS in the head kidney of larval and juvenile rainbow trout, *Oncorhynchus mykiss*. *Gen. Comp. Endocrinol.* 124,21–29.

Ganster, R. W. and Geller, D. A. (2000). Molecular regulation of inducible nitric oxide synthase. In *Nitric Oxide, Biology and Pathobiology* (ed. L. J. Ignarro), pp. 129–156, Academic Press, San Diego.

Gibbins, I. L., Olsson, C. and Holmgren, S. (1995). Distribution of neurons reactive for NADPH-diaphorase in the branchial nerves of a teleost fish, *Gadus morhua*. *Neurosci. Lett.* 193,113–116.

Gibbs, S. M. (2003). Regulation of neuronal proliferation and differentiation by nitric oxide. *Mol. Neurobiol.* 27,107–120.

Grandel, H., Kaslin, J., Ganz, J., Wenzel, I. and Brand, M. (2006). Neural stem cells and neurogenesis in the adult zebrafish brain: Origin, proliferation dynamics, migration and cell fate. *Dev. Biol.* 295(1), 263–277.

Gratton, J.-P., Fontana, J. and Sessa, W. C. (2000). Molecular control of endothelial derived nitric oxide. A new paradigm for endothelial NOS regulation by posttranslational modifications. In *Nitric Oxide, Biology and Pathobiology* (ed. L. J. Ignarro), pp. 157–166, Academic Press, San Diego.

Green, K. and Campbell, G. (1994). Nitric oxide formation is involved in vagal inhibition of the stomach of the trout (*Salmo gairdneri*). *J. Auton. Nerv. Syst.* 50,221–229.

Gross, P. M., Weave, D. F., Bowers, R. J., Nag, S., Ho, L. T., Pang, J. J. and Espinosa, F. J. (1994). Neurotoxicity in conscious rats following intraventricular SNAP, a nitric oxide donor. *Neuropharmacology* 33,915–927.

269

Guix, F. X., Uribesalgo, I., Coma, M. and Munoz, F. J. (2005). The physiology and pathophysiology of nitric oxide in the brain. *Prog. Neurobiol.* 76,126–152.

Guo, S., Wilson, S. W., Cooke, S., Chitnis, A. B., Driever, W. and Rosenthal, A. (1999). Mutations in the zebrafish unmask shared regulatory pathways controlling the development of catecholaminergic neurons. *Dev. Biol.* 208,473–487.

Han, R. N. N. and Stewart, D. J. (2006). Defective lung vascular development in endothelial nitric oxide synthase-deficient mice. *Trends Cardiovasc. Med.* 16,29–34.

Handy, R. L., Wallace, P., Gaffen, Z. A., Whitehead, K. J. and Moore, P. K. (1995). The antinociceptive effect of 1-(2-trifluoromethylphenyl) imidazole (TRIM), a potent inhibitor of neuronal nitric oxide synthase in vitro, in the mouse. *Br. J. Pharmacol.* 116,2349–2350.

Hanneman, E. and Westerfield, M. (1989). Early expression of acetylcholinesterase activity in functionally distinct neurons of the zebrafish. *J. Comp. Neurol.* 284,350–361.

Hefler, L. A., Reyes, C. A., O'Brien, W. E. and Gregg, A. R. (2001). Perinatal development of endothelial nitric oxide synthase-deficient mice. *Biol. Reprod.* 64,666–673.

Higashijima, S., Mandel, G. and Fetcho, J. R. (2004). Distribution of prospective glutamatergic, glycinergic, and GABAergic neurons in embryonic and larval zebrafish. *J. Comp. Neurol.* 480,1–18.

Holmqvist, B. and Ekström, P. (1997). Subcellular localization of neuronal nitric oxide synthase in the brain of a teleost; an immunoelectron and confocal microscopical study. *Brain Res.* 745,67–82.

Holmqvist, B., Ellingsen, B., Alm, P., Forsell, J., Øyan, A. M., Goksøyr, A., Fjose, A. and Seo, H. C. (2000). Identification and distribution of nitric oxide synthase in the brain of adult zebrafish. *Neurosci. Lett.* 292,119–122.

Holmqvist, B., Ellingsen, B., Forsell, J., Zhdanova, I. and Alm, P. (2004). The early ontogeny of neuronal nitric oxide synthase systems in the zebrafish. *J. Exp. Biol.* 207,923–935.

Holmqvist, B., Falk Olsson, C., Larsson, B. and Alm, P. (2001). The ontogeny of nitric oxide synthase systems in the mouse. In *31st Annual Meeting of the Society for Neuroscience,* Abstract 693.4, San Diego, CA, USA.

Holmqvist, B., Falk Olsson, C., Svensson, M.-L., Svanborg, C., Forsell, J. and Alm, P. (2006). Expression of nitric oxide synthase isoforms in the mouse kidney: Cellular localization and influence by lipopolysaccharide and Toll-like receptor 4. *J. Mol. Histol.* 36(8-9), 499–516.

Holmqvist, B. I., Östholm, T., Alm, P. and Ekström, P. (1994). Nitric oxide synthase in the brain of a teleost. *Neurosci. Lett.* 171,205–208.

Ignarro, L. J. (2000). Introduction and overview. In *Nitric Oxide, Biology and Pathobiology* (ed. L. J. Ignarro), pp. 3–19, Academic Press, San Diego.

Jadhao, A. G. and Malz, C. R. (2003). Localization of the neuronal form of nitric oxide synthase (bNOS) in the diencephalon and pituitary gland of the catfish, *Synodontis multipunctatus*: An immunocytochemical study. *Gen. Comp. Endocrinol.* 132,278–283.

Jimenez, A., Esteban, F. J., Sanchez-Lopez, A. M., Pedrosa, J. A., Del Moral, M. L., Hernandez, R., Blanco, S., Barroso, J. B., Rodrigo, J. and Peinado, M. A. (2001). Immunohistochemical localisation of neuronal nitric oxide synthase in the rainbow trout kidney. *J. Chem. Neuroanat.* 21,289–294.

270

Kaslin, J. and Panula, P. (2001). Comparative anatomy of the histaminergic and other aminergic systems in zebrafish (*Danio rerio*). *J. Comp. Neurol.* 440,342–377.

Kimmel, C. B., Ballard, W. W., Kimmel, S. R., Ullmann, B. and Schilling, T. F. (1995). Stages of embryonic development of the zebrafish. *Dev. Dyn.* 203,253–310.

Kirsche, W. (1967). On postembryonic matrix zones in the brain of various vertebrates and their relationship to the study of the brain structure. *Z. Mikrosk. Anat. Forsch.* 77,313–406.

Kiss, J. Z. and Rougon, G. (1997). Cell biology of polysialic acid. *Curr. Opin. Neurobiol.* 7,640–646.

Kleinert, H., Biossel, J.-P., Schwarz, P. M. and Förstermann, U. (2000). Regulation of the expression of nitric oxide synthase isoforms. In *Nitric Oxide, Biology and Pathobiology* (ed. L. J. Ignarro), pp. 105–128, Academic Press, San Diego.

Krumenacker, J. S. and Murad, F. (2006). NO–cGMP signaling in development and stem cells. *Mol. Genet. Metab.* 87,311–314.

Kuzin, B., Roberts, I., Peunova, N. and Enikolopov, G. (1996). Nitric oxide regulates cell proliferation during *Drosophila* development. *Cell* 87,639–649.

Laing, K. J., Hardie, L. J., Aartsen, W., Grabowski, P. S. and Secombes, C. J. (1999). Expression of an inducible nitric oxide synthase gene in rainbow trout *Oncorhynchus mykiss*. *Dev. Comp. Immunol.* 23,71–85.

Lauder, J. M. (1988). Neurotransmitters as morphogens. *Prog. Brain Res.* 73,365–387.

Lauder, J. M. (1993). Neurotransmitters as growth regulatory signals: Role of receptors and second messengers. *Trends Neurosci.* 16,233–240.

Lauder, J. M., Liu, J., Devaud, L. and Morrow, A. L. (1998). GABA as a trophic factor for developing monoamine neurons. *Perspect. Dev. Neurobiol.* 5,247–259.

Lee, M. A., Cai, L., Habner, N., Lee, Y. A. and Lindpaintner, K. (1997). Tissue- and development-specific expression of multiple alternatively spliced transcripts of rat neuronal nitric oxide synthase. *J. Clin. Invest.* 100,1507–1512.

Lee, T. C., Zhao, Y. D., Courtman, D. W. and Stewart, D. J. (2000). Abnormal aortic valve development in mice lacking endothelial nitric oxide synthase. *Circulation* 101,2345–2348.

Li, Z. S. and Furness, J. B. (1993). Nitric oxide synthase in the enteric nervous system of the rainbow trout, *Salmo gairdneri*. *Arch. Histol. Cytol.* 56,185–193.

Macdonald, R., Xu, Q., Barth, K. A., Mikkola, I., Holder, N., Fjose, A., Krauss, S. and Wilson, S. W. (1994). Regulatory gene expression boundaries demarcate sites of neuronal differentiation in the embryonic zebrafish forebrain. *Neuron* 13,1039–1053.

Marx, M., Rutishauser, U. and Bastmeyer, M. (2001). Dual function of polysialic acid during zebrafish central nervous system development. *Development* 128,4949–4958.

Massman, G. A., Zhang, J., Sallah, J. and Figueroa, J. P. (2000). Developmental and regional expression patterns of type I nitric oxide synthase mRNA and protein in fetal sheep brain during the last third of gestation. *Dev. Brain Res.* 124,141–152.

Matarredona, E. R., Murillo-Carretero, M., Moreno-Lopez, B. and Estrada, C. (2005). Role of nitric oxide in subventricular zone neurogenesis. *Brain Res. Rev.* 49,355–366.

Matsumoto, T., Pollock, J. S., Nakane, M. and Förstermann, U. (1993). Developmental changes of cytosolic and particulate nitric oxide synthase in rat brain. *Dev. Brain Res.* 73,199–203.

271

Mauceri, A., Fasulo, S., Ainis, L., Licata, A., Lauriano, E. R., Martinez, A., Mayer, B. and Zaccone, G. (1999). Neuronal nitric oxide synthase (nNOS) expression in the epithelial neuroendocrine cell system and nerve fibers in the gill of the catfish, *Heteropneustes fossilis. Acta Histochem.* 101,437–448.

McLean, D. L. and Fetcho, J. R. (2004). Ontogeny and innervation patterns of dopaminergic, noradrenergic, and serotonergic neurons in larval zebrafish. *J. Comp. Neurol.* 480,38–56.

McNeill, B. and Perry, S. F. (2005). Nitric oxide and the control of catecholamine secretion in rainbow trout *Oncorhynchus mykiss. J. Exp. Biol.* 208,2421–2431.

Ming, G. L. and Song, H. (2005). Adult neurogenesis in the mammalian central nervous system. *Annu. Rev. Neurosci.* 28,223–250.

Moore, P. K., Handy, R. L. (1997). Selective inhibitors of neuronal nitric oxide synthase – Is no NOS really good NOS for the nervous system? *Trends Pharmacol. Sci.* 18,204–211.

Moreno-López, B., Noval, J. A., González-Bonet, L. and Estrada, C. (2000). Morphological basis for a role of nitric oxide in adult neurogenesis. *Brain Res.* 869,244–250.

Mungrue, I. N. Bredt, D. S., Stewart, D. J. Husain, M. (2003). From molecules to mammals: What's NOS got to do with it? *Acta Physiol. Scand.* 179,123–135.

Nasseem, K. M. (2005). The role of nitric oxide in cardiovascular diseases. *Mol. Aspects Med.* 26,33–65.

Neumann, N. F., Fagan, D. and Belosevic, M. (1995). Macrophage activating factor(s) secreted by mitogen stimulated goldfish kidney leukocytes synergize with bacterial lipopolysaccharide to induce nitric oxide production in teleost macrophages. *Dev. Comp. Immunol.* 19,473–482.

Northington, F. J., Koehler, R. C., Traystman, R. T. and Martin, L. J. (1996). Nitric oxide synthase 1 and nitric oxide synthase 3 protein expression is regionally and temporally regulated in fetal brain. *Dev. Brain Res.* 95,1–14.

Northington, F. J., Tobin, J. R., Harris, A. P., Traystman, R. T. and Koehler, R. C. (1997). Developmental and regional differences in nitric oxide synthase activity and blood flow in the sheep brain. *J. Cereb. Blood Flow Metab.* 17,109–115.

Ogilvie, P., Schilling, K., Billinsley, M. L. and Schmidt, H. H. W. (1995). Induction and variants of neuronal nitric oxide synthase type I during synaptogenesis. *FASEB J.* 9,799–806.

Olsson, C. and Holmgren, S. (1996). Involvement of nitric oxide in inhibitory innervation of urinary bladder of Atlantic cod, *Gadus morhua. Am. J. Physiol.* 270, R1380–R1385.

Olsson, C. and Holmgren, S. (1997). Nitric oxide in the fish gut. *Comp. Biochem. Physiol. A: Physiol.* 118,959–964.

Östholm, T., Holmqvist, B. I., Alm, P. and Ekström, P. (1994). Nitric oxide synthase in the CNS of the Atlantic salmon. *Neurosci. Lett.* 168,233–237.

Øyan, A. M., Nilsen, F., Goksøyr, A. and Holmqvist, B. (2000). Partial cloning of constitutive and inducible nitric oxide synthases and detailed neuronal expression of NOS mRNA in the cerebellum and optic tectum of adult Atlantic salmon (*Salmo salar*). *Mol. Brain. Res.* 78,38–49.

Pellegrino, D., Palmerini, C. A. and Tota, B. (2004). No hemoglobin but NO: The icefish (*Chionodraco hamatus*) heart as a paradigm. *J. Exp. Biol.* 207,3855–3864.

272

Pelster, B., Grillitsch, S. and Schwerte, T. (2005). NO as a mediator during the early development of the cardiovascular system in the zebrafish. *Comp. Biochem. Physiol. A: Mol. Integr. Physiol.* 142,215–220.

Peunova, N., Scheinker, V., Cline, H. and Enikolopov, G. (2001). Nitric oxide is an essential negative regulator of cell proliferation in *Xenopus* brain. *J. Neurosci.* 21,8809–8818.

Phung, Y. T., Bekker, J. M., Hallmark, O. G. and Black, S. M. (1999). Both neuronal NO synthase and nitric oxide are required for PC 12 cell differentiation: A cGMP independent pathway. *Mol. Brain Res.* 64,165–178.

Pierce, R. L., Pierce, M. R., Liu, H., Kadowitz, P. J. and Miller, M. J. (1995). Limb reduction defects after prenatal inhibition of nitric oxide synthase in rats. *Pediatr. Res.* 38,905–911.

Poon, K. L., Richardson, M., Lam, C. S., Khoo, H. E. and Korzh, V. (2003). Expression pattern of neuronal nitric oxide synthase in embryonic zebrafish. *Gene Expr. Patterns* 3,463–466.

Reif, A., Schmitt, A., Fritzen, S., Chourbaji, S., Bartsch, C., Urani, A., Wycislo, M., Mössner, R., Sommer, C., Gass, P. and Lesch, K.-P. (2004). Differential effect of endothelial nitric oxide synthase (NOS-III) on the regulation of adult neurogenesis and behaviour. *Eur. J. Neurosci.* 20,885–895.

Rivier, C. (2002). Role of nitric oxide and carbon monoxide in modulating the activity of the rodent hypothalamic–pituitary–adrenal axis. *Front. Horm. Res.* 29,15–49.

Seki, T. and Arai, Y. (1993). Distribution and possible roles of the highly polysialylated neural cell adhesion molecule (NCAM-H) in the developing and adult central nervous system. *Neurosci. Res.* 17,265–290.

Shin, D. H., Lim, H. S., Cho, S. K., Lee, H. Y., Lee, H. W., Lee, K. H., Chung, Y. H., Cho, S. S., Ik Cha, C. and Hwang, D. H. (2000). Immunocytochemical localization of neuronal and inducible nitric oxide synthase in the retina of zebrafish, *Brachydanio rerio. Neurosci. Lett.* 292,220–222.

Szabó, C. (2000). Pathophysiological roles of nitric oxide in inflammation. In *Nitric Oxide, Biology and Pathobiology* (ed. L. J. Ignarro), pp. 841–872, Academic Press, San Diego.

Tanaka, M., Yoshida, S., Yano, M. and Hanaoka, F. (1994). Roles of endogenous nitric oxide in cerebellar cortical development in slice cultures. *Neuroreport* 5,2049–2052.

Terada, H., Nagai, T., Okada, S., Kimura, H. and Kitahama, K. (2001). Ontogenesis of neurons immunoreactive for nitric oxide synthase in rat forebrain and midbrain. *Dev. Brain Res.* 128,121–137.

Tota, B., Amelio, D., Pellegrino, D., Ip, Y. K. and Cerra, M. C. (2005). NO modulation of myocardial performance in fish hearts. *Comp. Biochem. Physiol. A: Mol. Integr. Physiol.* 142,164–177.

Towler, P. K., Bennett, G. S., Moore, P. K. and Brain, S. D. (1998). Neurogenic oedema and vasodilatation: Effect of a selective neuronal NO inhibitor. *Neuroreport* 9,1513–1518.

Tranguch, S., Steuerwald, N. and Huet-Hudson, Y. M. (2003). Nitric oxide synthase production and nitric oxide regulation of preimplantation embryo development. *Biol. Reprod.* 68,1538–1544.

Unger, J. L. and Glasgow, E. (2003). Expression of isotocin-neurophysin mRNA in developing zebrafish. *Gene Expr. Patterns* 3,105–108.

Uretsky, A. D. and Chang, J. P. (2000). Evidence that nitric oxide is involved in the regulation of growth hormone secretion in goldfish. *Gen. Comp. Endocrinol.* 118,461–470.

Uretsky, A. D., Weiss, B. L., Yunker, W. K. and Chang, J. P. (2003). Nitric oxide produced by a novel nitric oxide synthase isoform is necessary for gonadotropin-releasing hormone-induced growth hormone secretion via a cGMP-dependent mechanism. *J. Neuroendocrinol.* 15,667–676.

Uttenthal, L. O., Alonso, D., Fernandez, A. P., Campbell, R. O., Moro, M. A., Leza, J. C., Lizazoain, I., Esteban, F. J., Barroso, J. B., Valderrama, R., Pedrosa, J. A., Peinado, M. A., Serrano, J., Richard, A., Bentura, M. L., Santacana, M., Martinez-Murillo, R. and Rodrigo, J. (1998). Neuronal and inducible nitric oxide synthase and nitrotyrosine immunoreactivities in the cerebral cortex of the aging rat. *Microsc. Res. Tech.* 43,75–88.

Villani, L., Minelli, D., Giuliani, A. and Quaglia, A. (2001). The development of NADPH-diaphorase and nitric oxide synthase in the visual system of the cichlid fish, *Tilapia mariae. Brain Res. Bull.* 54,569–574.

Virgili, M., Poli, A., Beraudi, A., Giuliani, A. and Villani, L. (2001). Regional distribution of nitric oxide synthase and NADPH-diaphorase activities in the central nervous system of teleosts. *Brain Res.* 901,202–207.

Wang, T., Ward, M., Grabowski, P. and Secombes, C. J. (2001). Molecular cloning, gene organization and expression of rainbow trout (*Oncorhynchus mykiss*) inducible nitric oxide synthase (iNOS) gene. *Biochem. J.* 358,747–755.

Wendland, B., Schweizer, F. E., Ryan, T. A., Nakane, M., Murad, F., Scheller, R. H. and Tsien, R. W. (1994). Existence of nitric oxide synthase in rat hippocampal pyramidal cells. *Proc. Natl. Acad. Sci. U.S.A.* 91,2151–2155.

Wingrove, J. A. and O'Farrell, P. H. (1999). Nitric oxide contributes to behavioral, cellular, and developmental responses to low oxygen in *Drosophila. Cell* 98,105–114.

Wullimann, M. F. and Knipp, S. (2000). Proliferation pattern changes in the zebrafish brain from embryonic through early postembryonic stages. *Anat. Embryol. (Berlin)* 202,385–400.

Yamamoto, T., Yao, Y., Harumi, T. and Suzuki, N. (2003). Localization of the nitric oxide/cGMP signaling pathway-related genes and influences of morpholino knockdown of soluble guanylyl cyclase on medaka fish embryogenesis. *Zool. Sci.* 20,181–191.

Zaccone, G., Ainis, L., Mauceri, A., Lo Cascio, P., Lo Giudice, F. and Fasulo, S. (2003a). NANC nerves in the respiratory air sac and branchial vasculature of the Indian catfish, *Heteropneustes fossilis. Acta Histochem.* 105,151–163.

Zaccone, G., Mauceri, A., Ainis, L., Licata, A. and Fasulo, S. (2003b). Nitric oxide synthase in the gill of air sac of the Indian catfish, *Heteopneustes fossilis.* In *Catfishes* (eds G. Arratia, B. G. Kapoor, M. Chardon and R. Diogo), pp. 771–778, Science Publishers, Inc., Plymouth, UK.

Zaccone, G., Mauceri, A. and Fasulo, S. (2006). Neuropeptides and nitric oxide synthase in the gill and the air-breathing organs of fishes. *J. Exp. Zool. A: Comp. Exp. Biol.* 305(5), 428–439.

Zhang, R., Zhang, L., Zhang, Z., Wang, Y., Lu, M., LaPointe, M. and Chopp, M. (2001). A nitric oxide donor induces neurogenesis and reduces functional deficits after stroke in rats. *Ann. Neurol.* 50,602–611.

Ziche, M. and Morbidelli, L. (2000). Nitric oxide and angiogenesis. *J. Neurooncol.* 50,139–148.

Zupanc, G. K. H. (2001). Adult neurogenesis and neuronal regeneration in the central nervous system of teleost fish. *Brain Behav. Evol.* 58,250–275.

NO in the development of fish

Bernd Pelster*

Institut für Zoologie, Leopold-Franzens-Universität Innsbruck, A-6020 Innsbruck, Austria

Abstract. Autonomic innervation, at least of the cardiovascular system, appears late during the development of vertebrate embryos. In the early stages of innervation, control may be achieved by the activity of hormones, including for example the messenger molecule nitric oxide (NO). While NO in adult vertebrates is known to play a key function in many physiological processes, such as control of vascular tone, neurotransmission, macrophage activity and angiogenesis, very little is known about the onset of NO responsiveness during development. In fish, the presence of neuronal NO synthase (nNOS) and inducible NOS (iNOS) have been established by cloning and sequencing. The presence of endothelial NOS (eNOS) is indicated by immunological data, but attempts to identify eNOS at the molecular level have failed so far. nNOS expression, in particular, has been shown to occur at very early developmental stages. Analysis of the effect of NO on the cardiovascular system in zebrafish embryos and larvae revealed almost no effects on cardiac activity during chronic exposure to NO-producing chemicals, whereas vascular reactivity was observed in veins and arteries of the zebrafish in early developmental stages (5–6 days post-fertilization). Chronic exposure to a NO donor also modified the development of the vascular system by inducing an earlier appearance of some blood vessels in the trunk region of the zebrafish larvae. The nervous system and the gut also appear to be organs in which the early expression of NOS is of functional importance.

Keywords: NO synthases; NADPH-diaphorase; circulatory system; cardiac activity; angiogenesis; vasculogenesis; enteric nervous system; gut motility; gills; catecholamines; zebrafish.

Introduction

Nitric oxide (NO) is a widespread and important messenger molecule, which can alter protein function and also modify gene expression in a large number of tissues. Accordingly, NO is involved in many different physiological processes, including cell proliferation, differentiation, smooth muscle contraction, apoptosis, macrophage activity and neurotransmission (Alderton *et al.*, 2001; Stuart-Smith, 2002).

A very important site of action of NO is the cardiovascular system, where it was first described as endothelium-derived relaxing factor (EDRF). For Stuart-Smith (2002), the discovery of NO was the greatest achievement of vascular biology in the latter part of the 20th century.

Corresponding author: Tel.: + +43-512-5076180. Fax: + +43-512-5072930.
E-mail: Bernd.Pelster@uibk.ac.at (B. Pelster).

ADVANCES IN EXPERIMENTAL BIOLOGY
VOLUME 01   ISSN 1872-2423
DOI: 10.1016/S1872-2423(07)01012-5

The presence of NO is extensively documented in mammals, where it is a potent regulator modifying the contractility of vascular smooth muscle cells and therefore influencing vascular tone and blood pressure. The influence on the vascular tone is due to its diffusion from endothelial cells into vascular smooth muscle cells, where it activates a soluble guanylyl cyclase (sGC), which generates cyclic guanosine monophosphate (cGMP). cGMP in turn mediates vasodilation (Denninger and Marletta, 1999). NO released from nitrergic nerves may also contribute to the control of vasoreactivity.

Nitric oxide is produced by NO synthases (NOS; EC 1.14.13.39) using arginine as a substrate (Alderton et al., 2001). Three different isoforms of NOS have been identified in mammals after the first description of NOS in 1989. These three isoforms typically are referred to as: NOS1, NOS I or neuronal NOS (nNOS); NOS2, NOS II or inducible NOS (iNOS); and NOS3, NOS III or endothelial NOS (eNOS) (Alderton et al., 2001). While nNOS has been identified in neuronal tissue, eNOS is typically located in the endothelium. iNOS was believed to be an inducible isoform present in a number of different tissues.

While the NO system and the structure and function of NO synthases have been most thoroughly characterized in mammals, a number of comparative studies have also revealed the presence of the NO system in non-mammalian vertebrates (McGeer and Eddy, 1996; Olson et al., 1997; Olsson and Holmgren, 1997; Schwerte et al., 1999). Cloning and sequencing studies have demonstrated the presence of nNOS and iNOS in a number of different fish species (Barroso et al., 2000; Oyan et al., 2000; Saeij et al., 2000; Wang et al., 2001), suggesting an early evolution of the vertebrate NOS system. Using monoclonal antibodies, eNOS could be detected in the retina of white bass (Haverkamp et al., 1999). Similar to mammals, NO appears to be an important signalling molecule in many fish cell types. In recent years several studies have indicated that NOS is expressed in early embryonic or larval stages of fish and that NO signalling is of major importance during early development. This chapter addresses the earliest presence of NOS in fish and attempts to summarize our current knowledge on the importance of NO during early development of fish.

Nitric oxide synthases in fish

Nitric oxide synthases exhibit a bidomain structure with an N-terminal oxygenase domain and a C-terminal reductase domain (Fig. 1). These two domains are connected by a calmodulin-binding domain (CaM)

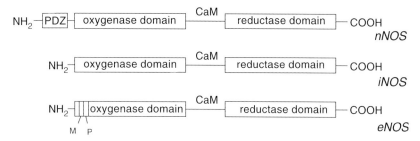

Fig. 1. General structure of NOS isoforms. NO synthases contain an oxygenase and a reductase domain, connected via a calmodulin-binding domain (CaM). nNOS is characterized by a PDZ domain, while eNOS contains a myristoylation (M) and a palmitoylation (P) site. Modified after Alderton *et al.* (2001).

suggesting that Ca^{2+} may have an influence on the activity. The three distinct isoforms of NOS mentioned above have been described in mammals: nNOS, mainly located in neuronal tissue; eNOS, found in endothelial cells; and iNOS, characterized by the fact that this isoform is inducible in a wide range of cells and tissues. It has also been reported that iNOS, in contrast to eNOS and nNOS, is not dependent on Ca^{2+}, but neither the inducibility nor the Ca^{2+} dependence appear to be unequivocal. In mammals all three isoforms contain a zinc-binding motif, which is located differently in the three isoforms. In addition, nNOS is characterized by a PSD-95, Discs-large, ZO-1 (PDZ) motif, *i.e.*, small protein–protein interaction modules that are thought to play a role in the clustering of submembranous signalling molecules. eNOS contains a myristoylation site and a palmitoylation site, which may locate the enzyme to membranes (Fig. 1).

The first attempts to demonstrate the presence of NOS in agnathans used NADPH-diaphorase histochemistry. These studies suggested the presence of NOS activity in the central nervous tissue of the lamprey (Schober *et al.*, 1994) and in the olfactory mucosa of larval sea lamprey (Zielinski *et al.*, 1996).

By cloning and sequencing studies, the presence of nNOS and iNOS was demonstrated in a number of fish species, including cyprinid, cichlid and salmonid fish, and also in the marine fish *Stenotomus chrysops* (the scup) (Barroso *et al.*, 2000; Cox *et al.*, 2001; Oyan *et al.*, 2000; Saeij *et al.*, 2000; Wang *et al.*, 2001). At the protein level about 60% sequence identity (or more) has been reported. iNOS of rainbow trout, for example, revealed a remarkable conservation in sequence and structure (Wang *et al.*, 2001). Recently nNOS of the euryhaline killifish has been described, confirming a high degree of conservation, including the PDZ motif (Hyndman *et al.*, 2006). Taken together these results suggest early evolution of the vertebrate NOS system.

An interesting and not unequivocally discussed question is the presence of eNOS in fish. To date no eNOS sequence has been reported for fish, although attempts have been made to identify this isoform in fishes (Donald *et al.*, 2004). The different isoforms of NOS have evolved by gene duplication, and iNOS appears to be the most ancient isoform (Wang *et al.*, 2001). A first gene duplication gave rise to the nNOS isoform. According to the sequence information all nNOS proteins of fish examined so far (*Oreochromis, Stenotomus, Oryzias, Fundulus, Takifugu* and *Danio* species) cluster together, and are separated from amphibian or other vertebrate nNOS proteins; the same is true for iNOS isoforms (Bordieri *et al.*, 2005; Hyndman *et al.*, 2006). This could easily be explained if we assume that the first gene duplication occurred prior to the divergence of fish. A second gene duplication gave rise to the eNOS isoforms, which again appear to be completely separated from all nNOS forms. A possible conclusion from these considerations is that the gene duplication separating nNOS from eNOS also occurred quite early in vertebrate evolution. But, as already mentioned, eNOS has not yet been identified in fish at the molecular level. Nevertheless, there is evidence from different groups indicating the presence of eNOS in fish on the basis of immunological data. Using monoclonal antibodies, eNOS-like immunoreactivity could be detected in the retina of white bass (Haverkamp *et al.*, 1999), and the presence of eNOS has been reported in developing zebrafish (Fritsche *et al.*, 2000). eNOS-like immunoreactivity has also been reported for gill cells of Atlantic salmon (Ebbesson *et al.*, 2005) and at least for a small population of gill cells of *Heteropneustes fossilis* (Zaccone *et al.*, 2006), and Tota *et al.* (2005) have provided immunohistochemical evidence for the presence of eNOS in the gill tissues of several adult fish species. Donald *et al.* (2004), however, could not detect eNOS in the dorsal aorta of the shovelnose ray, *Rhinobatus typus*. A recent study on the Australian eel indicated that nNOS may produce NO, which is involved in vascular control (Jennings *et al.*, 2004). These authors also could not detect eNOS and concluded that this isoform was absent in the Australian eel. Without molecular proof, it is difficult to accept the presence of eNOS in fish. A positive antibody reaction does not exclude the possibility of cross-reactions. In addition, although there obviously is a high degree of similarity within all vertebrates with respect to the function of NO, fish may not exactly follow the mammalian pattern in NOS isoform structure and function.

A recent study also provided physiological evidence for the presence of enOS in rainbow trout. Hypoxia resulted in the production of NO, which was abolished by removal of the endothelium (McNeill and Perry, 2005; see also below). Thus, the presence of enOS and its contribution to vascular control in fish appears to be a promising area for future research. Nevertheless, irrespective of the isoform of NOS present in fish, NO appears to be an important signalling molecule in many fish cell types.

Expression of NOS during early development

Several studies have addressed the early expression of nNOS in agnathans and in fish. In larval sea lamprey (*Petromyzon marinus*), immunohistochemistry of NADPH-diaphorase as well as physiological evidence indicated the involvement of NOS activity and NO in the functioning of the olfactory mucosa (Zielinski *et al.*, 1996). Similarly, in larval lamprey (*Lampetra planeri*), NOS activity has been inferred by the presence of NADPH-diaphorase activity in various parts of the central nervous tissue (Schober *et al.*, 1994).

In the tilapia, the presence of NADPH-diaphorase, indicating the presence of NOS, has been demonstrated in central nervous tissue as early as 20 h after fertilization (Villani, 1999). Expression of nNOS in very early developmental stages has also been demonstrated for the zebrafish *Danio rerio* (Holmqvist *et al.*, 2004; Poon *et al.*, 2003). nNOS mRNA expression could be localized in the ventral forebrain as early as 16 and 19 hpf (hours post-fertilization). The first expression of nNOS appears to be mainly in central nervous tissue, but some expression has also been reported for the skin of the zebrafish (Holmqvist *et al.*, 2004). Between 40 and 55 hpf, nNOS mRNA was detected in peripheral organs, namely in the gut and in the vicinity of the presumptive swimbladder tissue. At 72 hpf, the presence of nNOS was clearly detected in the gut and the swimbladder tissue of zebrafish larvae (Fig. 2). At the same stage, immunologically positive cells were detected in the enteric system of zebrafish larvae (Poon *et al.*, 2003), and Holmberg *et al.* (2006) detected nNOS-like immunoreactivity in neurons of the distal and middle intestine of zebrafish. Thus, expression of nNOS in enteric neurons of the zebrafish is observed before the onset of exogenous feeding. nNOS expression has also been reported for the head kidney of 12 dpf (days post-fertilization) larvae and juvenile trout *Oncorhynchus mykiss* (Gallo and Civinini, 2001).

280

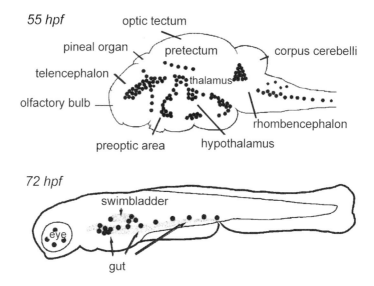

Fig. 2. Schematic representation of nNOS mRNA-expressing cell populations in zebrafish at 55 hpf (hours post-fertilization) and at 72 hpf. Modified after Holmqvist *et al.* (2004).

Using immunohistochemistry, the expression of eNOS was detected in cardiac muscle cells and in the dorsal vein of 3 and 5 dpf zebrafish larvae (Fritsche *et al.*, 2000). In the dorsal artery the distribution of eNOS immunoreactivity was weaker and somewhat diffuse. nNOS immuno-reactivity was not detected in peripheral cardiovascular tissues.

NO and vascular reactivity

Conflicting results have been reported regarding the importance of NO for the control of vascular smooth muscle cells in adult fish. While some studies could not detect any vasoactive influence of NO (Olson and Villa, 1991), it has also been proposed that NO functions only in some fish (Nilsson and Söderström, 1997) or may or may not be associated with air breathing (Cox *et al.*, 2001; Staples *et al.*, 1995).

Although not really unequivocal, it appears quite likely that at least some blood vessels of adult fish show NO-dependent responses in con-tractility. The next question yet unresolved is whether this response is related to the presence of eNOS in the vascular endothelium, or whether the observed vasoreactivity is due to perivascular nitrergic nerves. For the short-finned eel *Anguilla australis*, Donald and Broughton (2005) reported that nicotine induced an endothelium-independent vasodilation, which could be blocked by the sGC inhibitor ODQ (1H-[1,2,4] oxadiazolo[4,3-a]quinoxalin-1-one) and by the competitive NOS inhibitor L-NNA (N^G-nitro-L-arginine), and it was also reduced by N^ω-propyl-L-arginine, a more specific inhibitor for nNOS. It was therefore concluded that nicotine specifically activated nitrergic nerves, and that the vasoreactivity in this species was not attributable to the presence of eNOS (Donald and Broughton, 2005). A recent study on the Australian eel also indicated that the nNOS may produce NO which is involved in vascular control (Jennings *et al.*, 2004). These authors proposed the absence of eNOS in the Australian eel.

The early expression of NOS suggests that NO might contribute to cardiovascular control mechanisms during early development. This appears to be especially important because vascular control by the autonomic nervous system is known to be functional only quite late during larval development of fish and amphibians (Fritsche, 1997; Pelster, 1999; Protas and Leontieva, 1992). Application of the NO donor sodium nitroprusside (SNP) to the dorsal artery or dorsal vein of zebrafish larvae at 5 dpf indeed resulted in significant vasodilation, while injection of the L-arginine analogue and NOS inhibitor L-NAME (N^G-nitro-L-arginine methyl ester) close to the vessel resulted in vasoconstriction in both vessels (Fig. 3) (Fritsche *et al.*, 2000). Together with the immunohistochemical staining of eNOS in these vessels (see above), these results suggest that NO production by the eNOS significantly contributes to vascular control in these early stages of zebrafish development. The dorsal artery of fish supplies the intersegmental arteries with blood, and the administration of SNP significantly increased the number of erythrocytes passing through the intersegmental vessels in the zebrafish larvae.

The blood supply to these intersegmental vessels appears to be under a vasodilatory tonus, which is due to the presence of NO, as indicated by the combined application of catecholamines and inhibitors for NO production. Injection of epinephrine close to the dorsal vessels in the trunk of the larvae had no effect on the diameter of the dorsal artery or vein in 5 or 6 dpf larvae. However, when NO formation was inhibited by preincubation of the vessels with L-NAME, the application of epinephrine resulted in a significant vasoconstriction (Fritsche *et al.*, 2000).

These results thus suggest that eNOS is involved in vascular control in developing zebrafish.

282

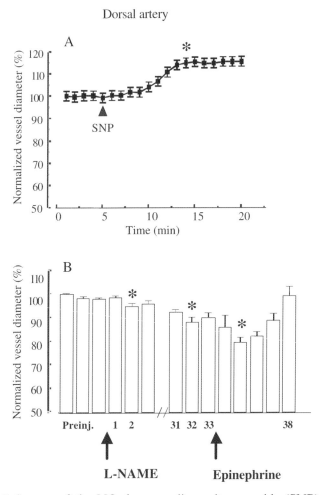

Dorsal artery

Fig. 3. (A) Influence of the NO donor sodium nitroprusside (SNP) on the diameter of the dorsal aorta in zebrafish larvae at 5 dpf and 6 dpf. (B) Effect of the NOS inhibitor L-NAME and of L-NAME plus epinephrine on the normalized diameter of the dorsal artery. Asterisks indicate significant differences ($p < 0.05$) Modified after Fritsche *et al.* (2000).

There also appear to be NO-dependent mechanisms indirectly related to control of the cardiovascular system. Adrenaline is an important effector controlling cardiac activity as well as peripheral resistance, and catecholamines are produced and released from chromaffin tissue. In adult rainbow trout, NO donors did not affect catecholamine secretion under resting conditions, but abolished the enhancement of catecholamine secretion of chromaffin cells by electrical field stimulation (McNeill and Perry, 2005). In addition, hypoxia evoked NO production in trout, which was abolished after removal of the endothelium by saponin treatment. These results clearly demonstrate that NO is an important player in the control of the cardiovascular system, either directly, by modifying the contractility of smooth muscle cells, or by influencing the release of other vasoactive hormones. If it is accepted that endothelial cells most likely produce eNOS, these results also indicate the potential involvement of eNOS in vascular control under hypoxic conditions, and therefore provide physiological evidence for the presence of eNOS in fish (McNeill and Perry, 2005).

Taken together these results suggest that there is indeed NO vasoreactivity in fish, and that NO-dependent control mechanisms are established quite early during development, possibly to compensate for the late appearance of the autonomic control of the vascular system. nNOS and perhaps also eNOS activity appear to be involved in the production of NO, but molecular proof for the existence of the latter is still missing. It therefore cannot be excluded that nNOS predominates in fish and that this NOS also serves the same functions in fish with respect to the control of vascular smooth muscle cells as eNOS does in mammals. The data also suggest that there are species-specific differences with respect to the contribution of the NO system to vascular control.

The influence of NO on cardiac activity in early stages

In addition to blood vessels, the heart may also be affected by NO, and NOS-related immunoreactivity has been found in cardiac muscle cells of the zebrafish (Fritsche *et al.*, 2000). Chronic exposure of eyed salmon eggs (*Salmo salar*) to the NO donors isosorbide dinitrate (ISDN; 0.1 mM) or SNP (0.1 mM) for two weeks resulted in bradycardia, reducing the heart rate by about 10% as compared with untreated control animals (Eddy *et al.*, 1999). Transfer of the ISDN-incubated eggs into freshwater resulted in transient tachycardia, while transfer of control eggs into the ISDN solution caused a decrease in heart rate by about 20% within four hours.

In our laboratory, zebrafish eggs were raised from 1 dpf until 12 dpf in 1 mM SNP. SNP is a NO donor, but may also result in the formation of cyanide. Therefore, solutions were exchanged daily, and aerobic and anaerobic metabolism of the embryos and larvae was followed throughout the experiment in order to avoid negative effects on the metabolism which might interfere with the results (Pelster *et al.*, 2005). Oxygen consumption of SNP-incubated animals was similar to the oxygen consumption of untreated controls until 8 dpf. In line with the unchanged rate of oxygen consumption, no difference in lactate content of the tissue was detected until 8 dpf. At 9 dpf, however, oxygen consumption of treated animals was reduced by more than 50% in comparison with control animals, indicating that at this stage metabolic activity might be affected by SNP.

Heart rate and also stroke volume of the larvae was measured by digital motion analysis of video images of the heart (Pelster *et al.*, 2005). The results showed some variability in SNP-treated animals as well as in control animals until 8 dpf, but there was no significant difference between the two groups. Accordingly, cardiac output was not affected by the presence of SNP until 8 dpf. At 12 dpf, however, stroke volume was strongly reduced, resulting in a significant decrease in cardiac output in treated animals. These results indicated that until 8 dpf SNP incubation had no effect on metabolism and on cardiac activity in developing zebrafish. Beyond 8 dpf, however, aerobic metabolism was affected by the NO donor SNP and finally cardiac output was severely reduced, which in turn would hamper aerobic metabolism even more because of the reduction in oxygen supply.

NO and tissue vascularization during development

The vasodilatory effect of NO contributes to changes in local blood flow. Local blood flow and even haemodilution in small blood vessels may cause a modification of sheer stress (Long *et al.*, 2004). Shear stress, in turn, represents an important stimulus enhancing the formation of blood vessels by an extension of existing vessels (angiogenesis), or it may also stimulate the formation of new blood vessels (vasculogenesis). Accordingly, NO is known as an important modulator of VEGF (vascular endothelial growth factor) induced vascular permeability and of VEGF induced angiogenesis (Bicknell and Harris, 2004; Conway *et al.*, 2001; Cooke and Losordo, 2002; Duda *et al.*, 2004; Fukumura *et al.*, 2001; Prior *et al.*, 2004). Based on these considerations, the vascular bed of the zebrafish larvae raised in the presence of 1 mM SNP or ISDN was carefully analysed and compared to untreated control larvae. The vascular bed was analysed by visualizing the movement of erythrocytes using digital motion analysis of video recordings (Schwerte and Fritsche, 2003; Schwerte and Pelster, 2000; Schwerte *et al.*, 2003). A limitation of this method is that small vessels, which are not yet penetrated by erythrocytes, cannot be detected. At first glance the overall appearance of the vascular bed in the trunk region of larvae was similar in the treated and control samples, but a closer look at defined areas revealed significant differences.

The main difference between SNP-treated and control animals was in the timing of blood vessel development. Three different blood vessels could be identified with enhanced development in the presence of the NO donor SNP (Pelster et al., 2005). The first example was the turning point between the dorsal artery and the dorsal vein, where a secondary caudal loop is formed, which finally is included in the main loop and contributes to the formation of the caudal fin. While this secondary caudal loop was well established in only 43% of the control animals at 3 dpf, it was established in 64% of the SNP-treated larvae at this stage of development. At 5 dpf, 92% of control animals showed this loop, and all of the SNP-treated animals.

A second example was the so-called caudal vascular tree, which develops on the ventral side of the caudal loop connecting the dorsal artery and vein. At 5 dpf this vessel was observed in 10% of the SNP-treated animals, but it could not be detected in control larvae. At 7 dpf, the vessel also showed up in 8% of the control animals, but by this time it was found in 50% of the treated larvae. Like the secondary caudal loop, the caudal tree gives rise to the blood vessel system of the caudal fin.

The third example was the parachordal vessels, running longitudinally along the myoseptum (Isogai et al., 2001, 2003) and contributing to the supply of muscle tissue in the trunk region. The first indication of the development of these vessels was observed in SNP-treated animals, and until 7 dpf these parachordal vessels were significantly better expressed in SNP-treated animals.

The results clearly demonstrated that raising the larvae in the presence of an NO donor did not change the pattern of the vascular bed, but it did induce an earlier appearance of some blood vessels in the trunk region of the zebrafish larvae. Previous studies have suggested that the gross anatomical patterning of the early trunk vessels of the zebrafish mainly follows the genetic program, and there is little or no room for deviations from this program (Isogai et al., 2003; Weinstein, 1999). Data from these incubation studies (Pelster et al., 2005) confirm the observation that the wiring of blood vessels is more-or-less fixed in the early developmental stages. In addition, neither hypoxia nor an increased oxygen demand of the muscle tissue induced by exposure of the larvae to a constant water current changed the vascularization of the trunk until two weeks after fertilization (Pelster et al., 2003; Schwerte et al., 2003).

286

On the other hand, the timing of the vessel appearance shows some flexibility (Pelster et al., 2005). In slightly later stages this flexibility not only extends to the timing of blood vessel formation, it also extends to the capillarization of tissues. Zebrafish larvae trained in a swim tunnel between 21 and 32 dpf showed significantly improved vascularization of the tail fin and of the segmental muscle tissue (Pelster et al., 2003). Thus, the initial wiring of the vascular bed appears to be largely driven by the genetic program. After about two or three weeks of development, however, the vascularization of the trunk regions becomes responsive to environmental parameters like hypoxia. It remains to be shown how this flexibility is achieved, and whether this pattern extends to other regions of the fish body.

NO in the developing enteric system

NOS-like immunoreactivity in the gut has been detected in very early developmental stages of the zebrafish prior to the onset of feeding (Holmberg et al., 2006; Holmqvist et al., 2004; Poon et al., 2003), and these studies focussed on the expression of nNOS in enteric nerves. In addition to the immunohistochemical localization, Holmberg et al. (2006) also addressed the functional importance of NO in the enteric system. Using L-NAME to inhibit NO production, and SNP as an artificial NO donor, it was demonstrated that gut motility was inhibited by the production of NO. While larvae at 3 dpf did not respond to the inhibition of internal NO production or the application of an NO donor, at 4 dpf the anterograde contraction wave frequency was increased by addition of the NOS inhibitor L-NAME. The NO donor SNP in turn reduced the frequency of the anterograde contraction wave. At 4 dpf the erratic and spontaneous contraction waves observed in the gut of the zebrafish at 3 dpf are replaced by more distinct contraction patterns (Holmberg et al., 2003). At 5 dpf the retrograde contraction wave frequency was similarly affected. The importance of NO for the enteric nervous system and its inhibitory effect on smooth muscle cells of adult teleosts has been reviewed by Olsson and Holmgren (2001), and these results reveal that NOS activity and the production of NO are important for the control of gut motility even before the onset of exogenous feeding. It is quite possible that the appearance of more distinct contraction waves in the gut at 4 dpf is at least in part related to the appearance of NOS activity and the production of NO, contributing to the control of smooth muscle cells in the enteric system.

NO in the gill cells and opercular tissue

The detection of NOS in gill cells launched the idea that NO might also be involved in gill functioning. nNOS immunoreactivity has been detected in the killifish opercular skin and in gill cells adjacent to chloride cells (Evans, 2002). The presence of nNOS in gill tissue has also been confirmed for neurons in the parenchyma beneath the gill epithelium of some catfish species (Zaccone *et al.*, 2003, 2006). In a small population of neuroepithelial cells in the gills of *Heteropneustes fossilis* eNOS immunoreactivity has been observed (Zaccone *et al.*, 2006), and in anadromous Atlantic salmon nNOS and eNOS immunoreactivity was found in large cells at the base and along the secondary lamellae. In contrast to the study of Evans (2002), however, in anadromous salmon NOS immunoreactivity was found in cells which were also positive for Na^+/K^+-ATPase (Ebbesson *et al.*, 2005). Chloride cells, in particular, appeared to be characterized by co-localization of NOS and Na^+/K^+-ATPase.

The application of NO donors resulted in a partial inhibition of Na^+/K^+-ATPase activity in gill tissue of anadromous salmon (Ebbesson *et al.*, 2005), but also in gill and kidney tissues of brown trout (Tipsmark and Madsen, 2003). In brown trout, the whole-tissue cGMP concentration increased in response to the NO donor SNP, and the lipid-soluble analogue db-cGMP inhibited Na^+/K^+-ATPase activity in a similar manner to NO. It was therefore concluded that NO provokes an increase in cGMP concentration, which in turn might activate cGMP-dependent protein kinase (Tipsmark and Madsen, 2003). In Atlantic salmon, preliminary observations suggested that NO decreased the short circuit current (I_{sc}) across the opercular epithelium. Accordingly, NO appears to be involved in the transport of Na^+ and Cl^- across the opercular epithelium and gill cells. The presence of nNOS has been confirmed in neuronal cells of gill tissue several times, and nNOS appears to be the dominating isoform in this tissue. It may therefore be speculated that nNOS, and not eNOS, is mainly responsible for this effect. Although the contribution of NOS to gill function in larval fish has not yet been addressed, it appears quite likely that NOS is an important player in gill function right from the start.

In summary, the data available so far clearly demonstrate the importance of NO as a mediator and signalling molecule during the early development of fish. Especially for the cardiovascular system and the enteric system, the importance of NO signalling has been shown for embryonic or at least early larval stages. In zebrafish NO has be shown to be responsible for a vasodilatory tone present in major blood vessels soon after hatching, and the presence or absence of NO contributes to a redirection of blood flow. In the gut NO appears to contribute to the control of contractility even before the onset of feeding. NO therefore appears to be an important signalling molecule and a major humoral factor during development.

Acknowledgements

Research of the author was funded by the Austrian Science Foundation (FWF, P14976).

References

Alderton, W., Cooper, C. E. and Knowles, R. G. (2001). Nitric oxide synthases: Structure, function and inhibition. *Biochem. J.* 357,593–615.

Barroso, J. B., Carreras, A., Esteban, F. J., Peinado, M. A., Martinez-Lara, E., Valderrama, R., Jimenez, A., Rodrigo, J. and Lupianez, J. A. (2000). Molecular and kinetic characterization and cell type location of inducible nitric oxide synthase in fish. *Am. J. Physiol.* 279,R650–R656.

Bicknell, R. and Harris, A. L. (2004). Novel angiogenic signaling pathways and vascular targets. *Annu. Rev. Pharmacol. Toxicol.* 44,219–238.

Bordieri, L., Bonaccorsi di Patti, M. C., Miele, R. and Cioni, C. (2005). Partial cloning of neuronal nitric oxide synthase (nNOS) cDNA and regional distribution of nNOS mRNA in the central nervous system of the Nile tilapia *Oreochromis niloticus. Mol. Brain Res.* 142,123–133.

Conway, E. M., Collen, D. and Carmeliet, P. (2001). Molecular mechanisms of blood vessel growth. *Cardiovasc. Res.* 49,507–521.

Cooke, J. P. and Losordo, D. W. (2002). Nitric oxide and angiogenesis. *Circulation* 105,2133–2135.

Cox, R. L., Mariano, T., Heck, D. E., Laskin, J. D. and Stegeman, J. J. (2001). Nitric oxide synthase sequences in the marine fish *Stenotomus chrysops* and the sea urchin *Arbacia punctulata*, and phylogenetic analysis of nitric oxide synthase calmodulin-binding domains. *Comp. Biochem. Physiol. B: Biochem. Mol. Biol.* 130,479–491.

Denninger, J. W. and Marletta, M. A. (1999). Gyanylate cyclase and the NO/cGMP signaling pathway. *Biochim. Biophys. Acta* 1411,334–350.

Donald, J. A. and Broughton, B. R. S. (2005). Nitric oxide control of lower vertebrate blood vessels by vasomotor nerves. *Comp. Biochem. Physiol. A: Mol. Integr. Physiol.* 142,188–197.

Donald, J. A., Broughton, B. R. S. and Bennett, M. B. (2004). Vasodilator mechanisms in the dorsal aorta of the giant shovelnose ray, *Rhinobatus typus* (Rajiformes; Rhinobatidae). *Comp. Biochem. Physiol. A: Mol. Integr. Physiol.* 137,21–31.

Duda, F. G., Fukumura, D. and Jain, R. K. (2004). Role of eNOS in neovascularization: NO for endothelial progenitor cells. *Trends Mol. Med.* 10,143–145.

Ebbesson, L. O. E., Tipsmark, C. K., Holmqvist, B., Nilsen, T., Andersson, E., Stefansson, S. O. and Madsen, S. S. (2005). Nitric oxide synthase in the gill of Atlantic salmon: Colocalization with and inhibition of Na^+,K^+-ATPase. *J. Exp. Biol.* 208, 1011–1017.

Eddy, F. B., McGovern, L., Acock, N. and McGeer, J. C. (1999). Cardiovascular responses of eggs, embryos and alevins of Atlantic salmon and rainbow trout to nitric oxide donors, sodium prusside and isosorbide dinitrate, and inhibitors of nitric oxide synthase, N-nitro-L-arginine methyl ester and aminoguanidine, following short- and long-term exposure. *J. Fish Biol.* 55(Suppl. A), 119–127.

Evans, D. H. (2002). Cell signaling and ion transport across the fish gill epithelium. *J. Exp. Zool.* 293,336–347.

Fritsche, R. (1997). Ontogeny of cardiovascular control in amphibians. *Am. Zool.* 37,23–30.

Fritsche, R., Schwerte, T. and Pelster, B. (2000). Nitric oxide and vascular reactivity in developing zebrafish, *Danio rerio*. *Am. J. Physiol. Regulato. Integr. Comp. Physiol.* 279,2200–2207.

Fukumura, D., Gohongi, T., Kadambi, A., Izumi, Y., Ang, J., Yun, C. O., Buerk, D. G., Huang, P. L. and Jain, R. K. (2001). Predominant role of endothelial nitric oxide synthase in vascular endothelial growth factor-induced angiogenesis and vascular permeability. *Proc. Natl. Acad. Sci. USA* 98,2604–2609.

Gallo, V. P. and Civinini, A. (2001). Immunohistochemical Localization of nNOS in the Head Kidney of Larval and Juvenile Rainbow Trout, *Oncorhynchus mykiss*. *Gen. Comp. Endocrinol.* 124,21–29.

Haverkamp, S., Kolb, H. and Cuenca, N. (1999). Endothelial nitric oxide synthase (eNOS) is localized to Müller cells in all vertebrate retinas. *Vision Res.* 39,2299–2303.

Holmberg, A., Olsson, C. and Holmgren, S. (2006). The effects of endogenous and exogenous nitric oxide on gut motility in zebrafish *Danio rerio* embryos and larvae. *J. Exp. Biol.* 209,2472–2479.

Holmberg, A., Schwerte, T., Fritsche, R., Pelster, B. and Holmgren, S. (2003). Ontogeny of intestinal motility in correlation to neuronal development in zebrafish embryos and larvae. *J. Fish Biol.* 63,318–331.

Holmqvist, B., Ellingsen, B., Forsell, J., Zhdanova, I. and Alm, P. (2004). The early ontogeny of neuronal nitric oxide synthase systems in the zebrafish. *J. Exp. Biol.* 207,923–935.

Hyndman, K. A., Choe, K. P., Havird, J. C., Rose, R. E., Piermarini, P. M. and Evans, D. H. (2006). Neuronal nitric oxide synthase in the gill of the killifish, *Fundulus heteroclitus*. *Comp. Biochem. Physiol. B: Biochem. Mol. Biol.* 144,510–519.

Isogai, S., Horiguchi, M. and Weinstein, B. M. (2001). The vascular anatomy of the developing zebrafish: An atlas of embryonic and early larval development. *Dev. Biol.* 230,278–301.

Isogai, S., Lawson, N. D., Torrealday, S., Horiguchi, M. and Weinstein, B. M. (2003). Angiogenic network formation in the developing vertebrate trunk. *Development* 130, 5281–5290.

Jennings, B. L., Broughton, B. R. S. and Donald, J. A. (2004). Nitric oxide control of the dorsal aorta and the intestinal vein of the Australian short-finned eel *Anguilla australis*. *J. Exp. Biol.* 207,1295–1303.

290

Long, D. S., Smith, M. L., Pries, A. R., Ley, K. and Damiano, E. R. (2004). Micro-viscometry reveals reduced blood viscosity and altered shear rate and shear stress profiles in microvessels after hemodilution. *Proc. Natl. Acad. Sci. USA* 101, 10060–10065.

McGeer, J. C. and Eddy, F. B. (1996). Effects of sodium nitroprusside on blood circulation and acid-base and ionic balance in rainbow trout: Indications for nitric oxide induced vasodilation. *Can. J. Zool.* 74,1211–1219.

McNeill, B. and Perry, S. F. (2005). Nitric oxide and the control of catecholamine secretion in rainbow trout *Oncorhynchus mykiss. J. Exp. Biol.* 208,2421–2431.

Nilsson, S. and Söderström, V. (1997). Comparative aspects on nitric oxide in brain and its role as a cerebral vasodilator. *Comp. Biochem. Physiol.* 118A,949–958.

Olson, K. R., Conklin, D. J., Farrell, A. P., Keen, J. E., Takei, Y., Weaver, L., Smith, M. P. and Zhang, Y. (1997). Effects of natriuretic peptides and nitroprusside on venous function in trout. *Am. J. Physiol.* 273,R527–R539.

Olson, K. R. and Villa, J. (1991). Evidence against nonprostanoid endothelium-derived relaxing factor(s) in trout vessels. *Am. J. Physiol.* 260,925–933.

Olsson, C. and Holmgren, S. (1997). Nitric oxide in the fish gut. *Comp. Biochem. Physiol.* 118A,959–964.

Olsson, C. and Holmgren, S. (2001). The control of gut motility. *Comp. Biochem. Physiol. A: Mol. Integr. Physiol.* 128,479–501.

Oyan, A. M., Nilsen, F., Goksoyr, A. and Holmqvist, B. (2000). Partial cloning of constitutive and inducible nitric oxide synthases and detailed neuronal expression of NOS mRNA in the cerebellum and optic tectum of adult Atlantic salmon (*Salmo salar*). *Mol. Brain Res.* 78,38–49.

Pelster, B. (1999). Environmental influences on the development of the cardiac system in fish and amphibians. *Comp. Biochem. Physiol. A* 124,407–412.

Pelster, B., Grillitsch, S. and Schwerte, T. (2005). NO as a mediator during the early development of the cardiovascular system in the zebrafish. *Comp. Biochem. Physiol. A: Mol. Integr. Physiol.* 142,215–220.

Pelster, B., Sänger, A. M., Siegele, M. and Schwerte, T. (2003). Influence of swim training on cardiac activity, tissue capillarization, and mitochondrial density in muscle tissue of zebrafish larvae. *Am. J. Physiol.* 285,R339–R347.

Poon, K. L., Richardson, M., Lam, C. S., Khoo, H. E. and Korzh, V. (2003). Expression pattern of neuronal nitric oxide synthase in embryonic zebrafish. *Gene Expression Patterns* 3,463–466.

Prior, B. M., Yang, H. T., Terjung, R. L., 2004. What makes vessels grow with exercise training? *J. Appl. Physiol.* 97,1119–1128.

Protas, L. L. and Leontieva, G. R. (1992). Ontogeny of cholinergic and adrenergic mechanisms in the frog (*Rana temporaria*) heart. *Am. J. Physiol.* 262,R150–R161.

Saeij, J. P. J., Stet, R. J. M., Groeneveld, A., Verburg-van Kemenade, L. B. M., Muiswinkel, W. B. and Wiegertjes, G. F. (2000). Molecular and functional characterization of a fish inducible-type nitric oxide synthase. *Immunogenetics* 51, 339–346.

Schober, A., Malz, C. R., Schober, W. and Meyer, D. L. (1994). NADPH-diaphorase in the central nervous system of the larval lamprey (*Lampetra planeri*). *J. Comp. Neurol.* 345,94–104.

Schwerte, T. and Fritsche, R. (2003). Understanding cardiovascular physiology in zebrafish and *Xenopus* larvae: The use of microtechniques. *Comp. Biochem. Physiol.* 135A,131–145.

Schwerte, T., Holmgren, S. and Pelster, B. (1999). Vasodilation of swimbladder vessels in the European eel (*Anguilla anguilla*) induced by vasoactive intestinal polypeptide, nitric oxide, adenosine and protons. *J. Exp. Biol.* 202,1005–1013.

Schwerte, T. and Pelster, B. (2000). Digital motion analysis as a tool for analysing the shape and performance of the circulatory system in transparent animals. *J. Exp. Biol.* 203,1659–1669.

Schwerte, T., Überbacher, D. and Pelster, B. (2003). Non-invasive imaging of blood cell concentration and blood distribution in hypoxic incubated zebrafish in vivo (*Danio rerio*). *J. Exp. Biol.* 206,1299–1307.

Staples, J. F., Zapol, W. M., Bloch, K. D., Kawai, N., Val, V. M. and Hochachka, P. W. (1995). Nitric oxide responses of air-breathing and water-breathing fish. *Am. J. Physiol.* 268,R816–R819.

Stuart-Smith, K. (2002). Demystified ... nitric oxide. *J. Clin. Pathol. Mol. Pathol.* 55,360–366.

Tipsmark, C. K. and Madsen, S. S. (2003). Regulation of Na^+/K^+-ATPase activity by nitric oxide in the kidney and gill of the brown trout (*Salmo trutta*). *J. Exp. Biol.* 206,1503–1510.

Tota, B., Amelio, D., Pellegrino, D., Ip, Y. K. and Cerra, M. C. (2005). NO modulation of myocardial performance in fish hearts. *Comp. Biochem. Physiol. A: Mol. Integr. Physiol.* 142,164–177.

Villani, L. (1999). Development of NADPH-diaphorase activity in the central nervous system of the cichlid fish, *Tilapia mariae*. *Brain Behav. Evol.* 54,147–158.

Wang, T., Ward, M., Grabowski, P. and Secombes, C. S. (2001). Molecular cloning, gene organization and expression of rainbow trout (*Oncorhynchus mykiss*) inducible nitric oxide synthase (iNOS) gene. *Biochem. J.* 358,747–755.

Weinstein, B. M., 1999. What guides early embryonic blood vessel formation? *Developmental Dynamics* 215,2–11.

Zaccone, G., Ainis, L., Mauceri, A., Lo Cascio, P., Lo Giudice, F. and Fasulo, S. (2003). NANC nerves in the respiratory air sac and branchial vasculature of the Indian catfish, *Heteropneustes fossilis*. *Acta Histochem.* 105,151–163.

Zaccone, G., Mauceri, A. and Fasulo, S. (2006). Neuropeptides and nitric oxide synthase in the gill and the air-breathing organs of fishes. *J. Exp. Zool.* 305A,428–439.

Zielinski, B. S., Osahan, J. K., Hara, T. J., Hosseini, M. and Wong, E. (1996). Nitric oxide synthase in the olfactory mucosa of the larval sea lamprey (*Petromyzon marinus*). *J. Comp. Neurol.* 365,18–26.

Role of nitric oxide in vascular regulation in fish

Claudio Agnisola[1,*] and Daniela Pellegrino[2]

[1]*Dipartimento delle Scienze Biologiche, Università di Napoli Federico II, Via Mezzocannone 8, 80134 Napoli, Italy*
[2]*Dipartimento Farmaco-Biologico, Università della Calabria, Cosenza, Italy*

Abstract. Nitric oxide (NO) is one of the oldest signaling molecules in animals, which acts as an intercellular and intracellular messenger in a multitude of cell types. Its role in the vascular biology of terrestrial vertebrates, particularly in mammals, is well established and extensively documented. This review article deals with the occurrence and effects of NO in the fish vascular system. In fish, the information regarding the roles of NO in the control of vascular resistance is surprisingly scanty; on the other hand, there is increasing evidence for a role for the NO synthase (NOS)/NO system in this highly diverse group of animals. Many authors have reported the occurrence and localization of both constitutive and inducible NOS (iNOS) isoforms in fish tissues, including gills and heart. Endothelial NOS (eNOS) has been detected in the vascular and endocardial endothelium of eel and some Antarctic fish, as well as in the endothelial cells of developing zebrafish. Evidence has been also reported for NOS-independent NO production, and particularly on the conversion of nitrite to NO by erythrocytes under conditions of hypoxia. The functional roles (vascular effects) of NO in developing and adult fish have been investigated. Studies on various fish species show results specific to the species or to the particular vascular preparation used. Fish appear to be an ideal model for studying the conservation and diversity of the functional roles of NO in the control of vascular resistance.

Keywords: (cardio)vascular system; acethylcholine; calmodulin; coronary arteries; deoxy-hemoglobin; endothelium; eNOS; fish gill; hypoxia; immunofluorescence; NADPH-diaphorase; nitrite; fish larvae; NOS localization; serotonin; shear stress; soluble guanylyl cyclase (sGC); trout; vascular resistance; zebrafish.

Introduction

Nitric oxide (NO) is a ubiquitous signaling molecule that mediates a large repertoire of physiological and pathophysiological effects in cardio-vascular, nervous and immunological systems. Because of the high membrane permeability and short life, this molecule is ideal as a short-distance intercellular and intracellular messenger molecule. In particular, NO represents one of the most important regulators of vascular resistance (Dattilo and Makhoul, 1997).

Corresponding author: Tel.:+ 390812535144 (office); + 390812534128 (lab). Fax: + 390812535090.
E-mail: agnisola@unina.it (C. Agnisola).

ADVANCES IN EXPERIMENTAL BIOLOGY
VOLUME 01   ISSN 1872-2423
DOI: 10.1016/S1872-2423(07)01013-7

NO is mainly produced from the guanidino group of L-arginine, and this reaction is catalysed by a family of nitric oxide synthases (NOSs). There are three isoforms of NOS, derived from separate genes, and with different localization, regulation, catalytic properties and inhibitor sensitivity: the neuronal NOS (nNOS; also called NOS1 or NOS I), the inducible NOS (iNOS; also called NOS2 or NOS II) and the endothelial NOS (eNOS; also called NOS3 or NOS III). The properties of these isoforms are summarized in Table 1. The human isoforms have 51–57% sequence homology with each other (Alderton et al., 2001). NOS utilizes L-arginine, oxygen and NADPH as substrates, and requires FAD, FMN, calmodulin and tetrahydrobiopterin as cofactors. The constitutive isoforms, nNOS and eNOS, are calcium dependent, while the inducible isoform is calcium independent. A mitochondrial isoform (mtNOS), corresponding to a variant of nNOS, has also been reported in various cell types (Giulivi, 2003).

Table 1. Synopsis of the main characteristics of the three isoforms of nitric oxide synthase (NOS).

Characteristic	nNOS	eNOS	iNOS
Monomer mass (human)	~160 kDa	~133 kDa	~131 kDa
Tissues	Nervous system	Endothelium	Macrophages
	Skeletal muscle	Brain	Heart
	Pancreatic islets	Epithelium	Liver
	Endometrium		Smooth muscle
	Macula densa		Endothelium
Subcellular localization	Mainly cytosolic	Mainly particulate (caveolae)	Both cytosolic and particulate
Main roles	Neurotransmission	Vascular relaxation	Antimicrobial
	Renal tubular glomerular interactions	Decreased platelet adhesion	Cytotoxic
	Intestinal motility	Angiogenesis	Inflammation
Some inhibitors	N-methyl-L-arginine	N-methyl-L-arginine	1400W[a]
	N-nitro-L-arginine	N-nitro-L-arginine	Aminoguanidine
	7-Nitroindazole	7-Nitroindazole	S-benzylisothiourea
			L-N^6-(1-iminoethyl)lysine

[a]Reported to be the most specific iNOS inhibitor (Garvey et al., 1997).

The functional NOS protein is a dimer formed of two identical sub-units. There are three distinct domains in each subunit: a reductase domain, a calmodulin-binding domain and an oxygenase domain. The reductase domains contain the FAD and FMN moieties. The reductase domain of one subunit transfers electrons from NADPH to the oxygen-ase domain of the opposite subunit of the dimer. Calmodulin is required for the activity of all NOS isoforms. It detects changes in intracellular calcium levels, although its precise function is slightly different in each of the three isoforms. nNOS and eNOS have a much greater calcium dependence than iNOS. Calmodulin-binding increases the rate of elec-tron transfer from NADPH to the reductase domain and triggers elec-tron transfer from the reductase domain to the heme center. nNOS and eNOS have a 40–50 amino acid insert in the middle of the FMN-binding subdomain, described as an autoinhibitory loop. This insert is absent in iNOS. Analysis of mutants of eNOS and nNOS with this loop deleted has shown that the insert acts by destabilizing calmodulin-binding at low calcium concentrations and by inhibiting electron transfer from FMN to the heme group in the absence of the calcium–calmodulin complex (Alderton *et al.*, 2001 and references therein). The oxygenase domain contains the binding sites for tetrahydrobiopterin, heme and arginine, and catalyses the conversion of arginine into citrulline and NO. From this synthetic description, it appears evident that the members of the NOS family are the most highly regulated enzymes. Regulation of NO synthesis is attained by cofactors and calcium availability, post-trans-ductional modifications and protein–protein interactions (Bredt, 2003). The requirement for oxygen as a substrate makes NOS-dependent NO production highly dependent on oxygen availability.

The regulation of eNOS, a key source of NO in the cardiovascular system, has been extensively studied (Dudzinski *et al.*, 2006). The most important physiological stimulus for the continuous eNOS-derived for-mation of NO in the vascular system is shear stress, the frictional force exerted on the vessel surface by blood. NO-mediated responses to shear stress include vessel relaxation, inhibition of apoptosis and inhibition of platelet and monocyte adhesion (Boo and Jo, 2003). A step increase in shear stress stimulates eNOS in two phases: an initial burst phase (lasting from seconds up to 30 min), which is Ca^{2+}–calmodulin-dependent, fol-lowed by a Ca^{2+}-independent phase in which NO production is main-tained at a lower rate; this second phase involves protein kinases and eNOS phosphorylation (Boo and Jo, 2003; Dimmeler *et al.*, 1999). Chronic changes in shear stress may also modify the eNOS expression level by both transcriptional induction and stabilization of mRNA (Davis *et al.*, 2001a).

NOS-independent pathways for NO release have also been reported. In particular, the nitrite anion (NO_2^-), which is relatively abundant (concentrations of 100–1,000 nM) in the blood and tissues of mammalian species, including humans (Gladwin *et al.*, 2000; Rassaf *et al.*, 2003; Rodriguez *et al.*, 2003), has been proposed as the largest intravascular and tissue storage form of NO. Mechanisms for the *in vivo* conversion of nitrite to NO may involve enzymatic reduction, non-enzymatic disproportionation or acidic reduction (Cosby *et al.*, 2003). Several proteins have been shown to have nitrite reductase capacity: glutathione-*S*-transferases (Hill *et al.*, 1992), xanthine oxidoreductase (Millar *et al.*, 1998), deoxyhemoglobin (Doyle *et al.*, 1981), cytochrome P-450 enzymes (Delaforge *et al.*, 1993), and, recently, eNOS (Gautier *et al.* 2006). These mechanisms would be relevant mainly during pathological hypoxia and acidosis, when NOS activity is blocked (Duranski *et al.*, 2005; Webb *et al.*, 2004), but recently some authors have reported that nitrite is a signaling molecule also under physiological conditions (Bryan *et al.*, 2005). NOS-dependent and NOS-independent mechanisms of NO production are summarized in Fig. 1.

The principal target of NO is soluble guanylyl cyclase (sGC), resulting in an increase in the intracellular pool of cGMP, which in turn may activate a variety of effectors such as cyclic-nucleotide-gated channels, protein kinases and phosphodiesterases (Shah and MacCarthy, 2000). NO can also exert cGMP-independent actions, reacting directly with ion channels, enzymes, heme-proteins and reactive oxygen species (Davis *et al.*, 2001b).

Nitric oxide appeared as a signaling molecule before the radiation of the metazoans. Accumulating evidence reveals that NO is used in a wide variety of invertebrate and vertebrate animals as an orthograde

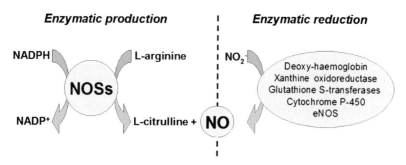

Fig. 1. Scheme of the NOS-dependent and NOS-independent mechanisms of NO production.

transmitter and co-transmitter, and as a modulator of conventional transmitter release, suggesting that these NO functions have been highly conserved during evolution (Feelisch and Martin, 1995; Jacklet, 1998; Torreilles, 2001).

NOS isoforms have been identified in echinoderms (Bishop and Brandhorst, 2001; Elphick and Melarange, 2001), coelenterates (Colasanti and Venturini, 1999), nematodes (Bascal et al., 2001; Pfarr et al., 2001) and annelids (Jacklet, 1998). In insects, NO signaling is involved in many physiological processes, including nervous and immune responses, development and integration (Kim et al., 2004, and references therein). An interesting example of NO versatality is the hematophagous insect *Rhodnius prolixus*, which produces NO to dilate blood vessels and to inhibit platelet aggregation in the host, thereby allowing more efficient blood sucking (Ribeiro and Nussenzveig, 1993). Insect NOSs have the greatest sequence identity with the nNOS isoform of mammals, and in insects NOS is expressed in a variety of adult and embryonic tissues (Kim et al., 2004, and references therein). In crustaceans, NO/cGMP signaling plays a role in neuronal development and in neuron, skeletal muscle and cardiac muscle regulation (Kim et al., 2004, and references therein). In the cephalopod *Sepia officinalis*, Chichery and Chichery (1994) localized NOS in the brain using histochemical techniques, and Palumbo and co-workers (1997, 1999, 2000) detected NOS activity both in the brain and in the ink gland, demonstrating that NO is involved in the ink defense system. NO has been reported to be essential for both tactile and visual learning in *Octopus vulgaris* (Robertson et al., 1994, 1996). The presence of NOS was also revealed in molluscan hemocytes, in which inducibility of this enzyme was also reported (Franchini et al., 1995), and in the hemocytes of the American horseshoe crab (*Limulus polyphemus*) (Radomski et al., 1991). The hemoglobin–NO interactions in invertebrates have also been reviewed (Gow et al. 1999).

Very few studies have investigated the possible role of NO in the cardiovascular system of invertebrates. eNOS has been detected in hemocytes of the scallop *Chlamys farreri* (Sun et al., 2005) and in the heart endothelium of *Nautilus pompilius* (Springer et al., 2004). In *Sepia officinalis*, NO regulates blood pressure by acting as a vasodilatory mediator (Schipp and Gebauer, 1999), as occurs in mammals.

The data on the occurrence and role of NO in invertebrates, while generating a wealth of comparative information, and underlining the importance of invertebrate studies to our knowledge of NO biology, appear insufficient to clarify the evolutionary story of eNOS and its role in the control of vascular function. More study is necessary, particularly in those invertebrate groups characterized by a closed circulatory system, such as cephalopods. These conclusions can be extended to lower vertebrates, and particularly to fish, where only in the last few years evidence is beginning to accumulate that NO may play a significant role in the control of vascular resistance, and that this role is at least in part endothelium dependent and may be related to the ecophysiology of different species.

Nitric oxide and the fish vascular system

The role of NO in mammalian vascular biology has been well and extensively documented (Bredt and Snyder, 1994; Ignarro, 1989, 1993; Moncada *et al.*, 1991; Radomski and Moncada, 1993; Schmidt and Walter, 1994). The vasodilatory action of NO is considered essential to maintain the matching between blood flow and tissue oxygen demand in several regions of the body. In non-mammalian terrestrial vertebrates the importance of NO in vasoregulation has been also established (Aksulu *et al.*, 2000; Hylland *et al.*, 1996; Rea and Parsons, 2001).

Fish are a group with great diversity, and have colonized a great variety of environments. Although experimental data are relatively scanty, there is increasing evidence that the NOS/NO system plays a role in fish circulation. Data include studies of NOS localization and evidence for a functional role for NO in the control of vascular resistance (Table 2).

NOS localization in fish circulation

The techniques used to detect NOSs in both cells and tissues are mainly NADPH-diaphorase and immunofluorescence. The NADPH-diaphorase histochemical method, based on dark blue formazan formation (Virgili *et al.*, 2001) localizes the occurrence and evaluates the total NOS activity, without discriminating among the various isoforms. The immunofluorescence technique allows the detection of specific isoforms (Pollock *et al.*, 1995).

The presence of NOS activity in fish tissues was reported for the first time by Li and Furness (1993) in the nervous enteric system. The occurrence of nNOS in various fish species is now well established both in the central nervous system (Anken and Rahmann, 1996; Bordieri and Cioni, 2004; Bordieri *et al.*, 2003; Cioni *et al.*, 1998; Conte, 2001, 2003; Cox *et al.*, 2001; Holmqvist *et al.*, 1994; Ostholm *et al.*, 1994; Oyan *et al.*,

Table 2. Synopsis of eNOS occurrence and NO effects in the vascular system of fish.

Species	eNOS localization	
	Site	Reference
Trematomus bernacchii	Sub-epicardial vessels	Amelio *et al.* (2006)
Salmo salar L.	Gills	Ebbesson *et al.* (2005)
Danio rero (larvae)	Dorsal vein	Fritsche *et al.* (2000)

	NO effects	
	Effect and site	
	Vasodilation:	
Salmo salar L.	Whole body	McGeer and Eddy (1996)
Chionodraco hamatus	Gills	Pellegrino *et al.* (2003)
Oncorhynchus mykiss	Gills	Smith *et al.* (2000)
Oncorhynchus mykiss	Coronary arteries	Mustafa *et al.* (1997)
Danio rero (larvae)	Dorsal artery and vein	Fritsche *et al.* (2000)
	Vasoconstriction:	
Anguilla anguilla	Gills	Pellegrino *et al.* (2002)

	NO-mediated mechanisms	
	Process	
Oncorhynchus mykiss	Serotonin vasodilation	Mustafa *et al.* (1997)
Oncorhynchus mykiss	Acetylcholine vasodilation	Mustafa *et al.* (1997); Soderstrom *et al.* (1995)
Carassius carassius	Acetylcholine vasodilation	Hylland and Nilsson (1995)
Oncorhynchus mykiss	Galanin vasodilation	Le Mével *et al.* (1998)
Oncorhynchus mykiss	Oxytocin vasodilation	Haraldsen *et al.* (2002)
Anguilla anguilla	Angiotensin effects on heart	Imbrogno *et al.* (2003)
Oncorhynchus mykiss	Stretch-dependent vasodilation	Mustafa and Agnisola (1998)

2000; Schober *et al.*, 1994a; Smith *et al.*, 2001; Virgili *et al.*, 2001) and peripheral nervous system (Donald and Broughton, 2005; Funakoshi *et al.*, 1999; Gibbins *et al.*, 1995; Holmqvist and Ekström, 1997; Karila *et al.*, 1997; Morlá *et al.*, 2003; Olsson and Karila, 1995; Radaelli, 1998; Schober *et al.*, 1994b).

The first report on the occurrence of constitutive NOS activity outside the nervous system in fish was from the gas bladder of the air-breathing teleost *Hoplerythrinus unitaeniatus* (Staples *et al.*, 1995). Several studies have reported the occurrence of iNOS in fish tissues, induced by infection and lipopolysaccharide (Laing *et al.*, 1999; Schoor and Plumb, 1994) or hypoxia (Sollid *et al.*, 2006). By biochemical, immunohistochemical and immunoblotting analyses, Barroso and co-workers (2000) showed iNOS activity in the head, kidney and liver of rainbow trout (*Oncorhynchus mykiss*), at both the cellular and the molecular level.

Amelio *et al.* (2006) have localized eNOS at the level of vascular endothelium in the Antarctic notothenioid *Trematomus bernacchii*. eNOS is mainly associated with the plasmalemma, as in the endocardial endothelium of *Chionodraco hamatus*, *T. bernacchii* and *Anguilla anguilla* (Tota *et al.*, 2005). This membrane targeting appears similar to that identified in mammals, in which eNOS is present in the caveolae associated with caveolin-3 (Cohen *et al.*, 2004; Feron *et al.*, 1996). The NOS spatial confinement (both at the tissue and the cellular level) adds a fundamental regulatory mechanism to NO signaling. In fact, the highly reactive and diffusive nature of NO requires the enzyme be localized close to NO targets (Barouch *et al.*, 2002).

NOSs are present and active in the gill of Atlantic salmon (Ebbesson *et al.*, 2005). Antibodies against the two constitutive NOS isoforms, nNOS and eNOS, both produced immunoreactivity restricted to large cells at the base and along the secondary lamellae. NADPH-diaphorase positive cells showed a similar distribution. Antibodies against the iNOS isoform only labeled small cells located deep in the filament. These results show that NO systems are abundant in the gill of Atlantic salmon and that NO may be produced preferentially by a constitutive NOS isoform. In addition, by *in vitro* and co-localization studies, the same authors found that NOSs and Na^+/K^+-ATPase have a similar distribution in the gill and that NO influences gill function via the inhibition of Na^+/K^+-ATPase activity (Ebbesson *et al.*, 2005).

The gills are also an important site of iNOS expression in rainbow trout, *Oncorhynchus mykiss*, following pathogen infection (Campos-Perez *et al.*, 2000).

The presence of NOS activity, revealed by NADPH-diaphorase, was found in about 55–85% of the neurons in the branchial nerve of the codfish, *Gadus morhua* (Gibbins *et al.*, 1995).

By immunohistochemistry, NOS immunoreactivity (rabbit primary antibodies) was also demonstrated in endothelial cells of the dorsal vein of developing zebrafish (*Danio rerio*) (Fritsche *et al.*, 2000).

Effects of NO on fish vasculature

Hylland and Nilsson (1995) first reported results suggesting that in crucian carp NO is an endogenous vasodilator that mediates the effects of acetylcholine. The use of NO donors has allowed demonstration of the responsiveness of fish body vasculature to NO. Fuentes *et al.* (1996) demonstrated that the NO donor SNP (sodium nitroprusside) induces dilation of body vasculature in *Salmo salar* alevins. In the salmon, NO-releasing compounds cause vasodilation *in vivo* by activating sGC (McGeer and Eddy, 1996).

NO exerts a basal vasodilator tone in the unstimulated branchial circulation of the Antarctic icefish, *Chionodraco hamatus*. In fact, NOS inhibition by L-NIO [N^5-(1-iminoethyl)-L-ornithine] induced a consistent vasoconstriction (20%), while the exogenous NO donor SIN-1 (morpholinosydnonimine) elicited a dose-dependent vasodilation (Pellegrino *et al.*, 2003). Interestingly, this NO-dependent vasodilation in the icefish, which is typical of mammals (see Moncada and Higgs, 1995), contrasts the NO-mediated branchial vasoconstriction that the same authors have detected in the European eel (*Anguilla anguilla*) (Pellegrino *et al.*, 2002). In this freshwater teleost, both endogenously derived (L-arginine) and exogenously derived (SNP, SIN-1) NO caused dose-dependent vasoconstriction under basal conditions, and vasodilation in acetylcholine pre-contacted preparations. This vasoconstrictor effect exerted by NO is cGMP-dependent. In fact, pre-treatment with the sGC inhibitor ODQ (1*H*-[1,2,4]oxadiazolo[4,3-α]quinoxalin-1-one) inhibited the effects of both SIN-1 and SNP, while a stable cGMP analogue (8-Br cGMP) induced dose-dependent vasoconstriction (Pellegrino *et al.*, 2002). It has been also reported that the NO donor SNP vasodilates branchial vessels of rainbow trout (*Oncorhyncus mykiss*) (Smith *et al.*, 2000). Whether this opposite response to NO of the gill vasculature from different species reflects diversities either in phylogeny and adaptation or in the variety of structural vascular sites remains unknown.

Evidence of the L-arginine–NO pathway in trout coronary arteries has been also shown, using both endogenously (L-arginine) and exogenously (SNP) derived NO. L-arginine induces vasodilation in non-working rainbow trout heart preparations perfused under constant pressure conditions (Mustafa and Agnisola, 1998). This vasodilatory response to L-arginine disappeared when the vascular endothelium was functionally destroyed by chemical denudation (Mustafa *et al.*, 1997). The NO donors nitroprusside and SNAP (*S*-nitroso-*N*-acetyl-1,1-penicillamine) displayed vasodilatory effects, while methylene blue, an inhibitor of the L-arginine–NO pathway, was vasoconstrictory (Mustafa et al., 1997). Finally, the NOS inhibitors L-NAME (N^G-nitro-L-arginine methyl ester) and L-NNA (N^G-nitro-L-arginine) induced both vasoconstrictory effects (Mustafa et al., 1997).

The dorsal artery and the dorsal vein of zebrafish larvae in vivo vasodilate in response to the NO donor SNP and vasoconstrict in response to L-NAME, suggesting that an endogenous NO production is affecting the larval vasculature and that there are NO-sensitive second-messenger systems in smooth muscles of the main dorsal blood vessels (Fritsche et al., 2000). As previously mentioned, immunohistological data revealed the presence of NOS in the endothelial cells of the dorsal vein at these early stages. These results indicate that, in larval tissue, the vascular tone is regulated locally by an eNOS/NO system before a functional autonomic innervation of the peripheral vascular system is developed (Fritsche et al., 2000).

NO-mediated vasodilatory mechanisms have also been documented in trout. Serotonin and acetylcholine NO-dependent (L-NAME and L-NNA inhibited) vasodilation has been reported in trout coronary arteries (Mustafa et al., 1997). A NO-dependent mechanism for the acetylcholine-induced vasodilation has been also reported in crucian carp (Hylland and Nilsson, 1995) and trout (Soderstrom et al., 1995). Le Mével et al. (1998) report that NO is possibly involved in the vasodilatory effect of galanin in trout. Oxytocin induces NO-mediated vasodilation in rainbow trout (Haraldsen et al., 2002). Finally, NO signaling has been reported to be involved in the effects of angiotensin II on the eel heart (Imbrogno et al., 2003).

The control of vascular resistance in fish may be of particular significance under conditions of hypoxia. Interest has focused mainly on the role of NO under such conditions, as NO vasodilation could help to maintain an adequate metabolic environment for tissue survival during exposure to hypoxia. On the other hand, reduced oxygen levels would reduce NOS-dependent NO production. Mustafa and Agnisola (1998) have shown that adenosine, a compound that is implicated in the fish hypoxic response (Bernier et al., 1996), has vasodilatory effects on the trout coronary vascular bed which are in part NO-mediated and which involve stretch receptors. Renshaw and Dyson (1999) have reported that hypoxia tolerance in epaulette sharks is associated with enhanced NOS production in the brain microvasculature. On the other hand, a NOS-independent mechanism may also be involved under hypoxic conditions. Jensen and Agnisola (2005) have recently demonstrated, in the coronary

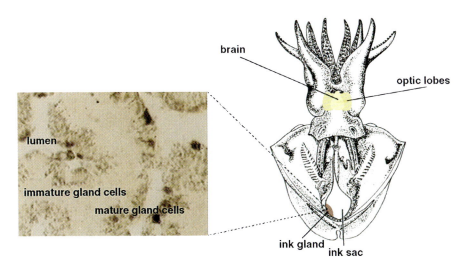

Plate 1. The cuttlefish *S. officinalis.* The central nervous system, comprising the brain and optic lobes, is shown together with the ink-producing system. Note the ink gland and the ink sac. Inset: microscopic structure of the ink gland. (For Black and White version, see page 51).

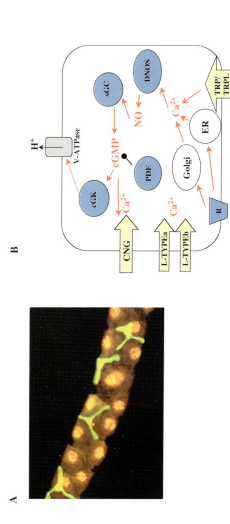

A

B

Plate 2. (A) Main segment of the *Drosophila* Malpighian tubule. Large nuclei of principal cells are stained orange with ethidium bromide, and star-shaped stellate cells are picked out with green fluorescent protein. Photo courtesy of J. A. T. Dow (Rosay *et al.*, 1997). (B) NO/cGMP signalling in the *Drosophila* tubule principal cell. Binding of capa peptides (capa-1, capa-2 and the closely related CAP$_{2b}$) to their cognate receptor, R, causes elevated intracellular calcium levels (Kean *et al.*, 2002). Calcium influx in the principal cell occurs *via* L-type calcium channels, Dmca1A and Dmca1D (represented here as a and b), the cyclic nucleotide-gated (CNG) channels (MacPherson *et al.*, 2001) and transient receptor potential (TRP)/transient receptor potential-like (TRPL) channels (MacPherson *et al.*, 2005). Major intracellular calcium stores in the tubule are the Golgi and the endoplasmic reticulum (ER) (Southall *et al.*, 2006), as well as the apical mitochondria (Terhzaz *et al.*, 2006). DNOS, a calcium-activated protein, produces NO, which in turn stimulates a soluble guanylyl cyclase (sGC). Cyclic GMP activates the cGMP-dependent protein kinase(s) (cGK), encoded by *dg1* and *dg2* genes (MacPherson *et al.*, 2004). cGMP can also activate calcium influx via CNG channels. The cGMP signal is efficiently terminated by a family of phosphodiesterases (PDE) (Day *et al.*, 2005). cGMP increases the transepithelial potential of the tubule, suggesting that a target of cGMP (or cGK) may be the apical vacuolar V-ATPase (Davies *et al.*, 1995); however, direct phosphorylation of V-ATPase subunits has yet to be demonstrated. (For Black and White version, see page 87).

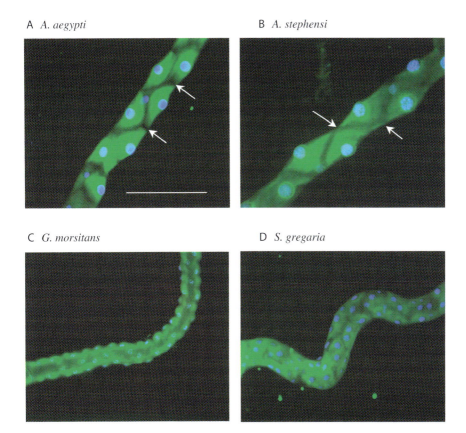

A *A. aegypti*

B *A. stephensi*

C *G. morsitans*

D *S. gregaria*

Plate 3. Tubules from (A) *Aedes aegypti*, (B) *Anopheles stephensi*, (C) *Glossina morsitans* and (D) *Schistocerca gregaria*, based on Pollock *et al.* (2004). Anti-uNOS antibody (Broderick *et al.*, 2003, 2004; Dow and Davies, 2001; Gibbs and Truman, 1998) was used to visualise NOS immunoreactivity (green); cell nuclei were visualised with the nuclear stain 4′,6′-diamidino-2-phenylindole hydro-chloride (DAPI; blue) (Broderick *et al.*, 2004). In (A) and (B), no green staining is observed in stellate cells (arrowed). *Glossina* (C) and *Schistocerca* (D) lack this cell type. Scale bar indicates 100 μm in (A) and (B); 200 μm in (C) and (D). (For Black and White version, see page 93).

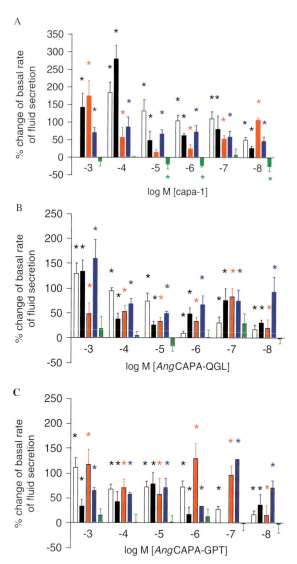

Plate 4. Fluid secretion rates in tubules from *D. melanogaster* (unshaded), *Aedes aegypti* (black), *Anopheles stephensi* (red), *G. morsitans* (blue) and *S. gregaria* (green) stimulated by either (A) capa-1, (B) *Ang*CAPA-QGL or (C) *Ang*CAPA-GPT at the concentrations shown (expressed as log molar, M). Basal rates of secretion were measured for 30 min prior to the addition of peptides. Secretion rates were measured for a further 40 min. Secretion rates are expressed as the percentage change of unstimulated tubules for each species, $\pm$SEM ($n = 6-8$). Asterisks denote statistically significant differences from basal values, $P<0.05$, determined using the Student's *t*-test (unpaired samples). (For Black and White version, see page 95).

A   B

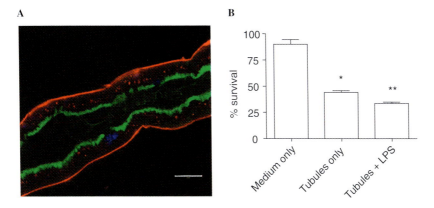

Plate 5. (A) Lipopolysaccharide (LPS) binds to tubule. Intact tubules from the *vhaSFD* gene-trap line, which express green fluorescent protein (GFP)-tagged *vhaSFD* (Morin *et al.*, 2001), were used in order to delineate the tubule apical membrane. These were incubated with AlexaFluor-LPS (*E. coli*) for up to 10 min. Principal cell nuclei are stained blue using DAPI; tubule apical membrane is defined by green vhaSFD:GFP (Torrie *et al.*, 2004); LPS binding and internalisation is indicated by red staining. Note the presence of red vesicles in the cytosol of the tubule cells. Scale bar = 15 μm. (B) Bacterial killing by excised, intact tubules. Sterile Schneider's medium was either left untreated, incubated with tubules only (6 h), or incubated with LPS-treated tubules (6 h). The medium was subsequently assessed for the effect on *E. coli* populations. Data are expressed as percentage survival of *E. coli* ± SEM (*n* = 4) from four different biological replicates. Data significantly different from controls are indicated by * and significant differences from tubules only are indicated by ** ($P < 0.05$), Student's *t*-test for unpaired samples. From McGettigan *et al.*, 2005. (For Black and White version, see page 97).

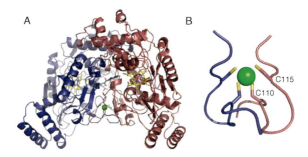

Plate 6. Structure of the human iNOS oxygenase domain and zinc tetrathiolate center. The oxygenase domain of NOS is homodimeric. (A) The entire oxygenase domain. The heme cofactor is shown as yellow lines, H_4B as light blue lines, and Zn^{2+} as a green sphere with the ligating cysteine thiols (Cys110 and Cys115 in human iNOS) shown as yellow lines. The zinc tetrathiolate and H_4B moieties make numerous interchain contacts to promote dimeric stability. Arginine was omitted for clarity. (B) Closeup view of the zinc tetrathiolate center. Figure was constructed with PyMol (DeLanoScientific LLC, San Francisco, USA) using Protein DataBank entry 1NSI. (For Black and White version, see page 156).

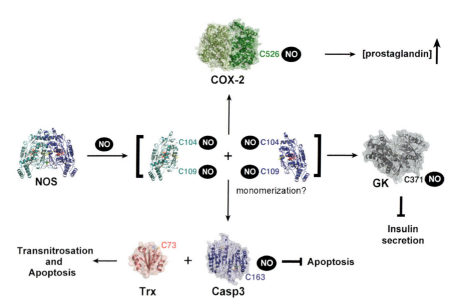

Plate 7. Potential NOS involvement in cellular nitrosothiol relay. NOS is the source of mammalian NO and undergoes spontaneous S-nitrosation *in vivo*. Through many known protein–protein interactions, it is plausible that NOS directly transfers nitrosothiols to cysteine residues of other proteins with disparate physiological outcomes. Dissociation of the NOS dimer after S-nitrosation *in vivo* requires further study, but *in vitro* reports indicate that it is plausible. This figure was constructed with PyMol (DeLanoScientific LLC, San Francisco, USA) using Protein DataBank entries 1NSI, 1CVU, 1V4S, 1NME, and 1ERU. Casp3, caspase-3. COX-2, cyclooxygenase-2. GK, glucokinase. Trx, thioredoxin-1. (For Black and White version, see page 169).

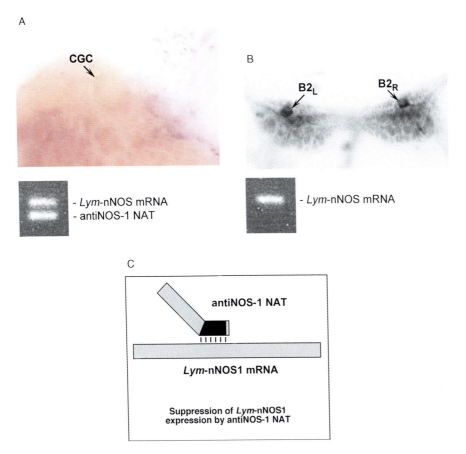

Plate 8. AntiNOS-1 natural antisense transcript (NAT)-mediated negative regulation of NOS expression in the CGCs. (A) The vast majority of the CGCs shows no NADPH-diaphorase activity (upper image) (Korneev *et al.*, 1999). Single-cell RT-PCR demonstrates the presence in the CGCs of *Lym*-nNOS mRNA as well as antiNOS-1 NAT (lower image). (B) B2 motoneurons localised in the buccal ganglia are strongly NADPH-diaphorase positive (upper image). Single-cell RT-PCR demonstrates the presence in the B2 motoneurons of *Lym*-nNOS mRNA. Note that the antiNOS-1 NAT is not detected (lower image). (C) AntiNOS-1 NAT suppresses translation of *Lym*-nNOS1 mRNA through the formation of RNA–RNA duplex molecules. The antisense region is shown in black (not to scale). (For Black and White version, see page 185).

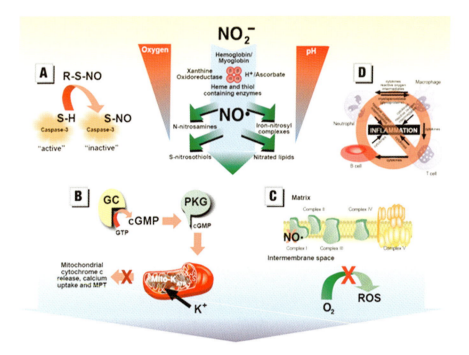

Plate 9. Potential mechanisms of ischemia/reperfusion (I/R) cytoprotection by nitrite-derived nitric oxide. Nitrite (NO_2^-), in the setting of acidosis and hypoxia present during ischemia, forms nitric oxide ($NO^\bullet$) catalysed by hemoglobin and other pathways. The derived nitric oxide or nitrite itself may modify cellular proteins or lipids, or may directly participate in actions which mitigate the injury occurring at reperfusion. Four potential mechanisms for this cytoprotection are presented: (A) S-nitrosation of critical regulatory thiols on proteins in the apoptotic pathway, such as caspase-3. (B) Stimulation of guanylyl cyclase (GC) to form cGMP, which in turn activates protein kinase G (PKG), resulting in stabilization of mitochondrial K_{ATP} channels. This stabilization is associated with decreased mitochondrial calcium accumulation, prevention of membrane permeability transition (MPT), and release of cytochrome c, actions linked to the development of cellular necrosis and apoptosis. (C) Direct inhibition of complex I of the electron transport chain with resultant decrease in the amount of cytotoxic reactive oxygen species (ROS) formed at the time of reperfusion. (D) Inhibition of acute inflammation, mediated mainly by polymorphonuclear granulocytes. (For Black and White version, see page 220).

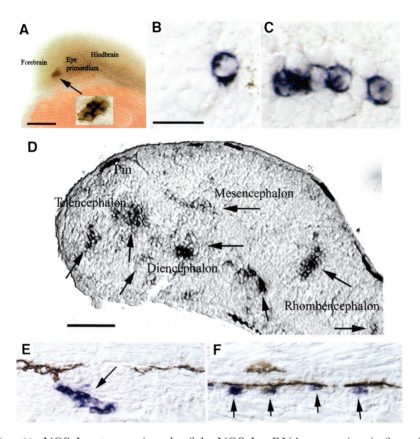

Plate 10. NOS I ontogeny in zebrafish. NOS I mRNA expression is first visualized in the brain ventro-rostral cell cluster at (A) 19 hpf (see arrow; see also
magnified inset at 23 hpf) from whole-mount preparations and (B) 24 hpf (bluish
labelling) from a cryosection. (C) NOS I mRNA expression in the ventrocaudal
cell cluster at 34 hpf. (D) NOS I mRNA expression at 55 hpf, demonstrating the
widespread distribution of NOS I populations in all major brain areas (arrows;
"Pin" indicates the pigmented pineal organ, which lacks NOS I expression). (E
and F) NOS I mRNA expression in the body at 72 hpf, in a putative enteric
ganglia (E) and along the nephritic and alimentary tracts (F). Scale bars represent 100 μm in (A) and (D), 20 μm in (B) and (C), and 10 μm in (E) and (F).
(For Black and White version, see page 243).

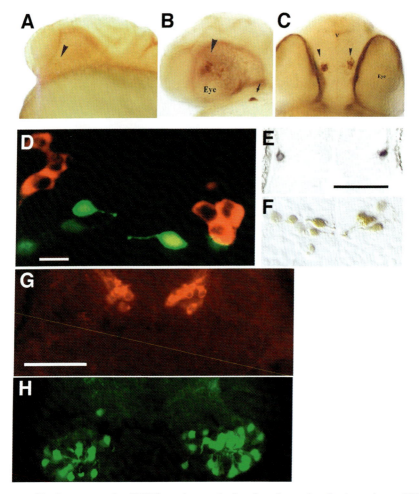

Plate 11. Brain serotonin (5HT) and catecholamine (tyrosine hydroxylase; TH) ontogeny coincide with that of NOS I in the zebrafish. Images demonstrate the first immunoreactive 5HT-expressing cells (A) and TH-expressing cells (B and C) in whole-mount preparations at 24 hpf and in (D) cryosection (5HT is green and TH is red) at 24 hpf, and in chromogen-immunolabelled cryosections at 40 hpf (E shows serotonin; F shows TH). Note the TH population in the arch-associated cell cluster (arrow in B). (G and H) Double labelling of the same section – (G) shows TH labelled in red and (H) shows 5HT labelled in green – from embryos at 55 hpf, with TH and 5HT cells located in the posterior tubercular region and hypothalamus, respectively, which also contain NOS I at this stage and in adults. Note the processes indicating cerebrospinal fluid-contacting properties of 5HT cells in (D) and (F). Scale bars represent 10 μm in (D), 100 μm in (E) and (F), and 100 μm in (G) and (H). (For Black and White version, see page 245).

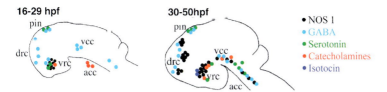

Plate 12. Schematic representation of the spatial relation between the early differentiating NOS I- and transmitter-expressing cell populations in the zebrafish brain, represented by coloured dots: NOS I (black), GABA (light blue), serotonin (green), catecholamines (tyrosine hydroxylase; red) and isotocin (dark blue). NOS I- and transmitter-expressing cell differentiations coincide in embryonic cell clusters, and follow the general pattern of embryonic brain neurogenesis. Images of embryos represent NOS I- and transmitter-expressing cells that are differentiated during two developmental periods (16–29 hpf and 30–50 hpf). drc, dorsoventral cell cluster. hc, hindbrain cell cluster. pin, pineal organ. vcc, ventrocaudal cell cluster. vrc, ventrorostral cell cluster. (For Black and White version, see page 246).

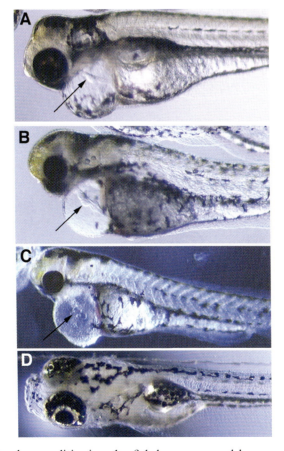

Plate 13. Body abnormalities in zebrafish larvae, caused by exposure of embryonic life stages to TRIM (A and B) or SNAP (C), in comparison to a non-treated control animal (D). Abnormal development of both head (tilted) and heart (tube-like; arrows) is demonstrated. (For Black and White version, see page 251).

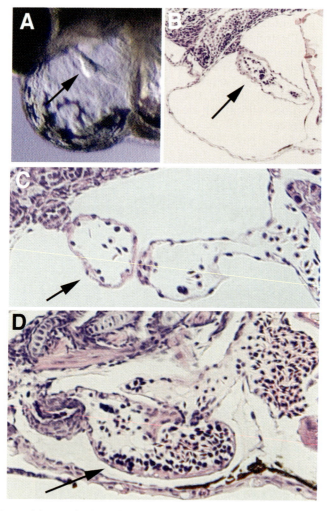

Plate 14. Heart histopathology in zebrafish larvae caused by exposure of embryonic life stages to TRIM (A–C). Animals exposed to TRIM possess an abnormal, tube-like heart, compared to the normal heart (D). (For Black and White version, see page 253).

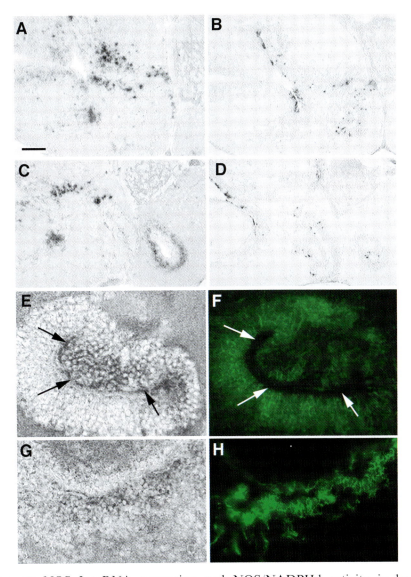

Plate 15. NOS I mRNA expression and NOS/NADPHd activity in hypothalamic proliferation zones of adult zebrafish. NOS I mRNA expression (A and
C) and proliferation nuclear antigen (PCNA) immunoreactivity (B and D) in
adjacent sections of the hypothalamic proliferation zones. Note the coincidence
of NOS I and PCNA located in different cell populations of periventricular and
ependymal cell layers, respectively. NOS/NADPHd activity (E and G) and polysialylated neural cell adhesion molecule (PSA-NCAM) immunofluorescence
(green in F and H) double labelling indicate different degrees of cellular co-
expression of NOS and PSA-NCAM, and that NOS activity is also present in
the inner periventricular and ependymal cell layers (arrows in E and F). Scale
bar in (A) represents 100 μm in (A)–(D), and 50 μm in (E)–(H). PSA-NCAM
antiserum used was a kind gift from Prof. Urs Rutishauser. (For Black and
White version, see page 260).

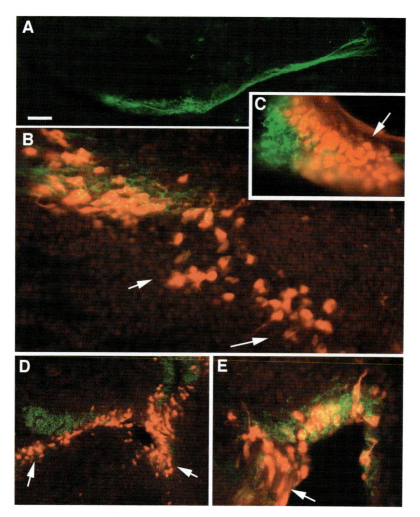

Plate 16. Proliferation zones and regions of plasticity and cell differentiation in the brain of adult zebrafish. Polysialylated neural cell adhesion molecule (PSA-NCAM; green) immunofluorescence labels the extracellular matrix within and between proliferation zones throughout the brain, demonstrated in the ventral telencephalon (A). Immunofluorescence double labelling of PCNA (red) and PSA-NCAM (green) demonstrates the cellular relation between cell proliferation activity and the putative migration routes of proliferation zones in the ventral telencephalon (B and C), hypothalamus (D) and dorsal thalamus/habenula (E). Note that the labelling of cells ranges from single PCNA expression (arrows), which prevails close to the areas of active cell mitosis of the ependymal and subependymal cell layers (see also Fig. 15), to areas of cellular co-expression of PCNA and PSA-NCAM (green around red cells) and single PSA-NCAM expression, which prevails in periventricular cell layers, *i.e.*, in regions of putative cell migration and differentiation (see Fig. 15). All images show sagittal views. Scale bar represents 100 μm in (A) and (D), and 50 μm in (B), (C) and (E). PSA-NCAM antiserum used was a kind gift from Prof. Urs Rutishauser. (For Black and White version, see page 262).

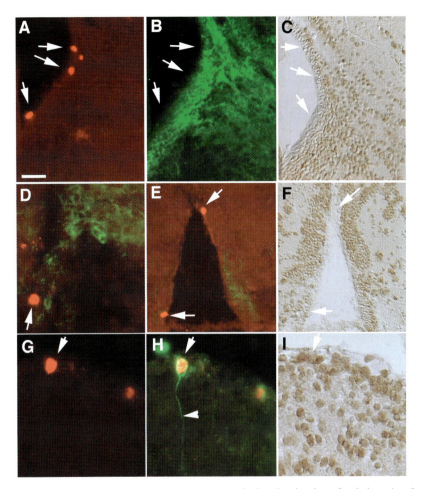

Plate 17. Proliferation zones and neurogenesis in the brain of adult zebrafish. Distribution of active cell mitosis, regions of plasticity and neuronal differentiation in proliferation zones of adult zebrafish, depicted in the hypothalamus (A–C) and rostral preoptic area (D–F) by immunohistochemical labelling of histon 3 (red), polysialylated neural cell adhesion molecule (PSA-NCAM; green) and HU/elav (weak to strong brownish). Note in (A)–(F) the histon 3-expressing cells in the ependymal cell layers, in relation to prevailing PSA-NCAM and HU labelling in periventricular and central brain regions. Note also the successive HU labelling intensity, ranging from weak in the inner periventricular layers to strong in the outer periventricular layers, in which the majority of NOS I cells are distributed. In contrast, at the same location as NOS I-expressing cells in the dorsal telencephalon (G), newly divided cells with processes into the brain (arrowheads) are co-expressing PSA-NCAM (H) and are located in the same cell layer as HU-positive cells (I), with varying labelling intensities. Scale bar represents 100 μm in (A)–(F), and 50 μm in (G)–(I). PSA-NCAM antiserum used was a kind gift from Prof. Urs Rutishauser. (For Black and White version, see page 264).

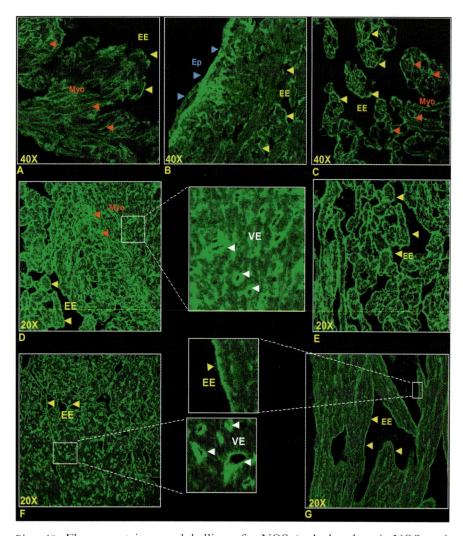

Plate 18. Fluorescent immunolabelling of e-NOS (polyclonal anti-eNOS antibody; 1:100) in the ventricular myocardium of (A) *Chionodraco hamatus*, (B) *Trematomus bernacchii*, (C) *Protopterus dolloi*, (D, E) the eel *Anguilla anguilla*, and (F, G) the tuna *Thunnus thynnus thynnus*. (D) and (F) show the compact myocardium; (E) and (G) show the trabecular myocardium (unpublished data). Method as described in Pellegrino *et al.*, 2004. Insets show details of immunofluorescent vessels of the eel and the tuna, and of the EE of the tuna. EE, endocardial endothelium. Myo, myocardiocytes. VE, vascular endothelium. Ep, epicardium. From Tota *et al.*, 2005. (For Black and White version, see page 317).

303

circulation of the isolated trout heart perfused with erythrocyte suspensions, that nitrite is converted to NO in a process that is not inhibited by the NOS inhibitor L-NNA. This supports the possible involvement of a deoxyhemoglobin-mediated reduction of nitrite to NO, a mechanism that would be significant under conditions of hypoxia to help maintain oxygen supply to tissues.

Conclusions

It is not possible to have an overall picture of the occurrence of endothelium-derived NO and its role in the control of vascular resistance in fish, because of the relatively sparse information available and the high diversity of this group of vertebrates. On the other hand, it is evident that eNOS is present in the vasculature of several fish species, and that NO is likely to play a significant role in the control of vascular tone under conditions of normoxia and hypoxia. Research on this topic is growing. Indeed, the high diversity of fish makes them an ideal study model for relating the occurrence and function of NO to the ecophysiological characteristics of a species.

References

Aksulu, H. E., Bingol, I., Karatas, F., Sagmanligil, H. and Ustundag, B. (2000). Changes in plasma angiotensin-converting enzyme activity and noradrenaline responses to long-term nitric oxide inhibition vary depending on their basal values in chickens. *Physiol. Res.* 49,175–182.
Alderton, W. K., Cooper, C. E. and Knowles, R. G. (2001). Nitric oxide synthases: Structure, function and inhibition. *Biochem. J.* 357,593–615.
Amelio, D., Garofano, F., Pellegrino, D., Giordano, F., Tota, B. and Cerra, M. C. (2006). Cardiac expression and distribution of nitric oxide synthases in the ventricle of the cold-adapted Antarctic teleosts, the hemoglobinless *Chionodraco hamatus* and the red-blooded *Trematomus bernacchii*. *Nitric Oxide* 15,190–198.
Anken, R. H. and Rahmann, H. (1996). An atlas of the distribution of NADPH-diaphorase in the brain of the highly derived swordtail fish *Xiphophorus helleri* (Atheriniformes: Teleostei). *J. Hirnforsch.* 37,421–449.
Barouch, L. A., Harrison, R. W., Skaf, M. W., Rosas, G. O., Cappola, T. P., Kobeissi, Z.A., Hobai, I. A., Lemmon, C. A., Burnett, A. L., O'Rourke, B., Rodriguez, E. R., Huang, P. L., Lima, J. A., Berkowitz, D. E. and Hare, J. M. (2002). Nitric oxide regulates the heart by spatial confinement of nitric oxide synthase isoforms. *Nature* 416,337–339.
Barroso, J. B., Carreras, A., Esteban, F. J., Peinado, M. A., Martinez-Lara, E., Valderrama, R., Jimenez, A., Rodrigo, J. and Lupianez, J. A. (2000). Molecular and

kinetic characterization and cell type location of inducible nitric oxide synthase in fish. *Am. J. Physiol. Regulat. Integr. Comp. Physiol.* 279,R650–R656.

Bascal, Z. A., Cunningham, J. M., Holden-Dye, L., O'Shea, M. and Walker, R. J. (2001). Characterization of a putative nitric oxide synthase in the neuromuscular system of the parasitic nematode, *Ascaris suum. Parasitology* 122,219–231.

Bernier, N. J., Harris, J., Lessard, J. and Randall, D. J. (1996). Adenosine receptor blockade and hypoxia-tolerance in rainbow trout and pacific hagfish- I effects on anaerobic metabolism. *J. Exp. Biol.* 199,485–495.

Bishop, C. D. and Brandhorst, B. P. (2001). NO/cGMP signaling and HSP90 activity represses metamorphosis in the sea urchin *Lytechinus pictus. Biol. Bull.* 201,394–404.

Boo, Y. C. and Jo, H. (2003). Flow-dependent regulation of endothelial nitric oxide synthase: Role of protein kinases. *Am. J. Physiol. Cell Physiol.* 285,C499–C508.

Bordieri, L. and Cioni, C. (2004). Co-localization of neuronal nitric oxide synthase with arginine-vasotocin in the preoptic-hypothalamo-hypophyseal system of the teleost *Oreochromis niloticus. Brain Res.* 1015,181–185.

Bordieri, L., Persichini, T., Venturini, G. and Cioni, C. (2003). Expression of nitric oxide synthase in the preoptic-hypothalamo-hypophyseal system of the teleost *Oreochromis niloticus. Brain Behav. Evol.* 62,43–55.

Bredt, D. S. (2003). Nitric oxide signaling specificity — the heart of the problem. *J. Cell Sci.* 116,9–15.

Bredt, D. S. and Snyder, S. H. (1994). Nitric oxide, a physiologic messenger molecule. *Annu. Rev. Bioche.* 63,175–195.

Bryan, N. S., Fernandez, B. O., Bauer, S. M., Garcia-Saura, M. F., Milsom, A. B., Rassaf, T., Maloney, R. E., Bharti, A., Rodriguez, J. and Feelisch, M. (2005). Nitrite is a signaling molecule and regulator of gene expression in mammalian tissues. *Nat. Chem. Biol.* 1,290–297.

Campos-Perez, J. J., Ward, M., Grabowski, P. S., Ellis, A. E. and Secombes, C. J. (2000). The gills are an important site of iNOS expression in rainbow trout Oncorhynchus mykiss after challenge with the gram-positive pathogen *Renibacterium salmoninarum. Immunology* 99,153–161.

Chichery, R. and Chichery, M. P. (1994). NADPH-diaphorase in a cephalopod brain (Sepia): Presence in an analogue of the cerebellum. *Neuroreport* 5,1273–1276.

Cioni, C., Francia, N., Fabrizi, C., Colasanti, M. and Venturini, G. (1998). Partial biochemical characterization of nitric oxide synthase in the caudal spinal cord of the teleost *Oreochromis niloticus. Neurosci. Lett.* 253,68–70.

Cohen, A. W., Hnasko, R., Schubert, W. and Lisanti, M. P. (2004). Role of caveolae and caveolins in health and disease. *Physiol. Rev.* 84,1341–1379.

Colasanti, M. and Venturini, G. (1999). Nitric oxide in invertebrates. *Mol. Neurobiol.* 17,157–174.

Conte, A. (2001). Role of pH on the calcium ion dependence of the nitric oxide synthase in the carp brain. *Brain Res. Bull.* 56,67–71.

Conte, A. (2003). Physiologic pH changes modulate calcium ion dependence of brain nitric oxide synthase in *Carassius auratus. Biochim. Biophys. Acta* 1619,29–38.

Cosby, K., Partovi, K. S., Crawford, J. H., Patel, R. P., Reiter, C. D., Martyr, S., Yang, B. K., Waclawiw, M. A., Zalos, G., Xu, X., Huang, K. T., Shields, H., Kim-Shapiro, D. B., Schechter, A. N., Cannon, R. O. and Gladwin, M. T. (2003). Nitrite reduction

305

to nitric oxide by deoxyhemoglobin vasodilates the human circulation. *Nat. Med.* 9,1498–1505.

Cox, R. L., Mariano, T., Heck, D. E., Laskin, J. D. and Stegeman, J. J. (2001). Nitric oxide synthase sequences in the marine fish *Stenotomus chrysops* and the sea urchin *Arbacia punctulata*, and phylogenetic analysis of nitric oxide synthase calmodulin-binding domains. *Comp. Biochem. Physiol. B: Biochem. Mol. Biol.* 130,479–491.

Dattilo, J. B. and Makhoul, R. G. (1997). The role of nitric oxide in vascular biology and pathobiology. *Ann. Vascular Surgery* 11,307–314.

Davis, M. E., Cai, H., Drummond, G. R. and Harrison, D. G. (2001a). Shear stress regulates endothelial nitric oxide synthase expression through c-Src by divergent signaling pathways. *Circ. Res.* 89,1073–1080.

Davis, K. L., Martin, E., Turko, I. V. and Murad, F. (2001b). Novel effects of nitric oxide. *Annu. Rev. Pharmacol. Toxicol.* 41,203–236.

Delaforge, M., Servent, D., Wirsta, P., Ducrocq, C., Mansuy, D. and Lenfant, M. (1993). Particular ability of cytochrome P-450 CYP3A to reduce glyceryl trinitrate in rat liver microsomes: Subsequent formation of nitric oxide. *Chem. Biol. Interact.* 86, 103–117.

Dimmeler, S., Fleming, I., Fisslthaler, B., Hermann, C., Busse, R. and Zeiher, A. M. (1999). Activation of nitric oxide synthase in endothelial cells by Akt-dependent phosphorylation. *Nature* 399,601–605.

Donald, J. A. and Broughton, B. R. S. (2005). Nitric oxide control of lower vertebrate blood vessels by vasomotor nerves. *Comp. Biochem. Physiol. A: Mol. Integr. Physiol.* 142,188–197.

Doyle, M. P., Pickering, R. A., De Weert, T. M., Hoekstra, J. W. and Pater, D. (1981). Kinetics and mechanism of the oxidation of human deoxyhemoglobin by nitrites. *J. Biol. Chem.* 256,12393–12398.

Dudzinski, D. M., Igarashi, J., Greif, D. and Michel, T. (2006). The regulation and pharmacology of endothelial nitric oxide synthase. *Annu. Rev. Pharmacol. Toxicol.* 46,235–276.

Duranski, M. R., Greer, J. J., Dejam, A., Jaganmohan, S., Hogg, N., Langston, W., Patel, R. P., Yet, S. F., Wang, X., Kevil, C. G., Gladwin, M. T. and Lefer, D. J. (2005). Cytoprotective effects of nitrite during in vivo ischemia-reperfusion of the heart and liver. *Clin. Invest.* 115,1232–1240.

Ebbesson, L. O., Tipsmark, C. K., Holmqvist, B., Nilsen, T., Andersson, E., Stefansson, S. O. and Madsen, S. S. (2005). Nitric oxide synthase in the gill of Atlantic salmon: Colocalization with and inhibition of Na$^+$,K$^+$-ATPase. *J. Exp. Biol.* 208,1011–1017.

Elphick, M. R. and Melarange, R. (2001). Neural control of muscle relaxation in echinoderms. *J. Exp. Biol.* 204,875–885.

Feelisch, M. and Martin, J. F. (1995). The early role of nitric oxide in evolution. *Trends Ecol. Evol.* 10,496–499.

Feron, O., Belhassen, L., Kobzik, L., Smith, T. W., Kelly, R. A. and Michel, T. (1996). Endothelial nitric oxide synthase targeting to caveolae. Specific interactions with caveolin isoforms in cardiac myocytes and endothelial cells. *J. Biol. Chem.* 271, 22810–22814.

Franchini, A., Conte, A. and Ottaviani, E. (1995). Nitric oxide: An ancestral immunocyte effector molecule. *Adv. Neuroimmunol.* 5,463–478.

306

Fritsche, R., Schwerte, T. and Pelster, B. (2000). Nitric oxide and vascular reactivity in developing zebrafish. *Danio rerio. Am. J. Physiol. Regulat. Integr. Comp. Physiol.* 279,R2200–R2207.

Fuentes, J., McGeer, J. C. and Eddy, F. B. (1996). Drinking rate in juvenile Atlantic salmon, *Salmo salar L* fry in response to a nitric oxide donor, sodium nitroprusside and an inhibitor of angiotensin converting enzyme, enalapril. *Fish Physiol. Biochem.* 15,65–69.

Funakoshi, K., Kadota, T., Atobe, Y., Nakano, M., Goris, R. C. and Kishida, R. (1999). Nitric oxide synthase in the glossopharyngeal and vagal afferent pathway of a teleost, *Takifugu niphobles*. The branchial vascular innervation. *Cell Tissue Res.* 298,45–54.

Garvey, E. P., Oplinger, J. A., Furfine, E. S., Kiffi, R. J., Laszloi, F., Whittlei, B. J. R. and Knowles, R. G. (1997). 1400W is a slow, tight binding, and highly selective inhibitor of inducible nitric-oxide synthase in vitro and in vivo. *J. Biol. Chem.* 272,4959–4963.

Gautier, C., van Faassen, E., Mikula, I., Martasek, P. and Slama-Schwok, A. (2006). Endothelial nitric oxide synthase reduces nitrite anions to NO under anoxia. *Biochem. Biophys. Res. Commun.* 341,816–821.

Gibbins, I. L., Olsson, C. and Holmgren, S. (1995). Distribution of neurons reactive for NADPH-diaphorase in the branchial nerves of a teleost fish, *Gadus morhua. Neurosci. Lett.* 193,113–116.

Giulivi, C. (2003). Characterization and function of mitochondrial nitric-oxide synthase. *Free Radic. Biol. Med.* 34,397–408.

Gladwin, M. T., Shelhamer, J. H., Schechter, A. N., Pease-Fye, M. E., Waclawiw, M. A., Panza, J. A., Ognibene, F. P. and Cannon, R. O. (2000). Role of circulating nitrite and *S*-nitrosohemoglobin in the regulation of regional blood flow in humans. *Proc. Natl. Acad. Sci. USA* 97,11482–11487.

Gow, A. J., Luchsinger, B. P., Pawloski, J. R., Singel, D. J. and Stamler, J. S. (1999). The oxyhemoglobin reaction of nitric oxide [see comments]. *Proc. Natl. Acad. Sci. USA* 96,9027–9032.

Haraldsen, L., Soderstrom-Lauritzsen, V. and Nilsson, G. E. (2002). Oxytocin stimulates cerebral blood flow in rainbow trout (*Oncorhynchus mykiss*) through a nitric oxide dependent mechanism. *Brain Res.* 929,10–14.

Hill, K. E., Hunt Jr, R. W., Jones, R., Hoover, R. L. and Burk, R. F. (1992). Metabolism of nitroglycerin by smooth muscle cells. Involvement of glutathione and glutathione *S*-transferase. *Biochem. Pharmacol.* 43,561–566.

Holmqvist, B. and Ekström, P. (1997). Subcellular localization of neuronal nitric oxide synthase in the brain of a teleost; an immunoelectron and confocal microscopical study. *Brain Res.* 745,67–82.

Holmqvist, B. I., Ostholm, T., Alm, P. and Ekstrom, P. (1994). Nitric oxide synthase in the brain of a teleost. *Neurosci. Lett.* 171,205–208.

Hylland, P. and Nilsson, G. E. (1995). Evidence that acetylcholine mediates increased cerebral blood flow velocity in crucian carp through a nitric oxide-dependent mechanism. *J. Cereb. Blood Flow Metab.* 15,519–524.

Hylland, P., Nilsson, G. E. and Lutz, P. L. (1996). Role of nitric oxide in the elevation of cerebral blood flow induced by acetylcholine and anoxia in the turtle. *J. Cereb. Blood Flow Metab.* 16,290–295.

307

Ignarro, L. J. (1989). Biological actions and properties of endothelium-derived nitric oxide formed and released from artery and vein. *Circ. Res.* 65,1–21.

Ignarro, L. J. (1993). Nitric oxide-mediated vasorelaxation. *Thromb. Haemost.* 70,148–151.

Imbrogno, S., Cerra, M. C. and Tota, B. (2003). Angiotensin II-induced inotropism requires an endocardial endothelium-nitric oxide mechanism in the in-vitro heart of *Anguilla anguilla. J. Exp. Biol.* 206,2675–2684.

Jacklet, J. W. (1998). Nitric oxide signaling in invertebrates. *Invert. Neurosci.* 3,1–14.

Jensen, F. B. and Agnisola, C. (2005). Perfusion of the isolated trout heart coronary circulation with red blood cells effects of oxygen supply and nitrite on coronary flow and myocardial oxygen consumption. *J. Exp. Biol.* 208,3665–3674.

Karila, P., Messenger, J. and Holmgren, S. (1997). Nitric oxide synthase- and neuropeptide Y-containing subpopulations of sympathetic neurons in the coeliac ganglion of the Atlantic cod, *Gadus morhua*, revealed by immunohistochemistry and retrograde tracing from the stomach. *J. Auton. Nerv. Syst.* 66,35–45.

Kim, H.-W., Batista, L. A., Hoppes, J. L., Lee, K. J. and Mykles, D. L. (2004). A crustacean nitric oxide synthase expressed in nerve ganglia, Y-organ, gill and gonad of the tropical land crab, *Gecarcinus lateralis. J. Exp. Biol.* 207,2845–2857.

Laing, K. J., Hardie, L. J., Aartsen, W., Grabowski, P. S. and Secombes, C. J. (1999). Expression of an inducible nitric oxide synthase gene in rainbow trout *Oncorhynchus mykiss. Dev. Comp. Immunol.* 23,71–85.

Le Mével, J. C., Mabin, D., Hanley, A. M. and Conlon, J. M. (1998). Contrasting cardiovascular effects following central and peripheral injections of trout galanin in trout. *Am. J. Physio. Regulat. Integr. Compar. Physiol.* 275,R1118–R1126.

Li, Z. S. and Furness, J. B. (1993). Nitric oxide synthase in the enteric nervous system of the rainbow trout, *Salmo gairdneri. Arch. Histol. Cytol.* 56,185–193.

McGeer, J. C. and Eddy, F. B. (1996). Effects of sodium nitroprusside on blood circulation and acid-base and ionic balance in rainbow trout: Indications for nitric oxide induced vasodilation. *Can. J. Zool.* 74,211–1219.

Millar, T. M., Stevens, C. R., Benjamin, N., Eisenthal, R., Harrison, R. and Blake, D. R. (1998). Xanthine oxidoreductase catalyses the reduction of nitrates and nitrite to nitric oxide under hypoxic conditions. *FEBS Lett.* 427,225–228.

Moncada, S. and Higgs, E. A. (1995). Molecular mechanisms and therapeutic strategies related to nitric oxide. *FASEB J.* 9,1319–1330.

Moncada, S., Palmer, R. M. and Higgs, E. A. (1991). Nitric oxide: Physiology, pathophysiology, and pharmacology. *Pharmacol. Rev.* 43,109–142.

Morlá, M., Agustí, A. G. N., Rahman, I., Motterlini, R., Saus, C., Morales-Nin, B., Company, J. B. and Busquets, X. (2003). Nitric oxide synthase type I (nNOS), vascular endothelial growth factor (VEGF) and myoglobin-like expression in skeletal muscle of Antarctic icefishes (Notothenioidei: Channichthyidae). *Polar Biol.* 26,458–462.

Mustafa, T. and Agnisola, C. (1998). Vasoactivity of adenosine in the trout (*Oncorhynchus mykiss*) coronary system: Involvement of nitric oxide and interaction with noradrenaline. *J. Exp. Biol.* 201,3075–3083.

Mustafa, T., Agnisola, C. and Hansen, J. K. (1997). Evidence for NO-dependent vasodilation in the trout (*Onchorhynchus mykiss*) coronary system. *J. Compar. Physiol. B* 167,98–104.

308

Olsson, C. and Karila, P. (1995). Coexistence of NADPH-diaphorase and vasoactive intestinal polypeptide in the enteric nervous system of the Atlantic cod (*Gadus morhua*) and the spiny dogfish (*Squalus acanthias*). *Cell Tissue Res.* 280,297–305.

Ostholm, T., Holmqvist, B. I., Alm, P. and Ekström, P. (1994). Nitric oxide synthase in the CNS of the Atlantic salmon. *Neurosci. Lett.* 168,233–237.

Oyan, A. M., Nilsen, F., Goksoyr, A. and Holmqvist, B. (2000). Partial cloning of constitutive and inducible nitric oxide synthases and detailed neuronal expression of NOS mRNA in the cerebellum and optic tectum of adult Atlantic salmon (*Salmo salar*). *Mol. Brain Res.* 78,38–49.

Palumbo, A., Di Cosmo, A., Gesualdo, I., d'Ischia, M. (1997). A calcium-dependent nitric oxide synthase and NMDA R1 glutamate receptor in the ink gland of *Sepia officinalis*: A hint to a regulatory role of nitric oxide in melanogenesis? *Biochem. Biophys. Res. Commun.* 235, 429–432

Palumbo, A., Di Cosmo, A., Poli, A., Di Cristo, C. and d'Ischia, M. (1999). A calcium/calmodulin-dependent nitric oxide synthase, NMDAR2/3 receptor subunits, and glutamate in the CNS of the cuttlefish *Sepia officinalis*: Localization in specific neural pathways controlling the inking system. *J. Neurochem.* 73,1254–1263.

Palumbo, A., Poli, A., Di Cosmo, A. and d'Ischia, M. (2000). N-Methyl-D-aspartate receptor stimulation activates tyrosinase and promotes melanin synthesis in the ink gland of the cuttlefish *Sepia officinalis* through the nitric Oxide/cGMP signal transduction pathway. A novel possible role for glutamate as physi. *J. Biol. Chem.* 275,16885–16890.

Pellegrino, D., Acierno, R. and Tota, B. (2003). Control of cardiovascular function in the icefish *Chionodraco hamatus*: Involvement of serotonin and nitric oxide. *Comp. Biochem. Physiol. A: Mol. Integr. Physiol.* 134,471–480.

Pellegrino, D., Sprovieri, E., Mazza, R., Randall, D. J. and Tota, B. (2002). Nitric oxide-cGMP-mediated vasoconstriction and effects of acetylcholine in the branchial circulation of the eel. *Comp. Biochem. Physiol. A: Mol. Integr. Physiol.* 132,447–457.

Pfarr, K. M., Qazi, S. and Fuhrman, J. A. (2001). Nitric oxide synthase in filariae: Demonstration of nitric oxide production by embryos in *Brugia malayi* and *Acanthocheilonema viteae*. *Exp. Parasitol.* 97,205–214.

Pollock, J. S., Forstermann, U., Tracey, W. R. and Nakane, M. (1995). Nitric oxide synthase isozymes antibodies. *Histochemistry* 27,738–744.

Radaelli, G. (1998). Different putative neuromodulators are present in the nerves which distribute to the teleost skeletal muscle. *Histol. Histopathol.* 13,939–947.

Radomski, M. W., Martin, J. F. and Moncada, S. (1991). Synthesis of nitric oxide by the hemocytes of the American horseshoe crab (*Limulus polyphemus*). *Phil. Trans. R. Soc. London B* 334,129–133.

Radomski, M. W. and Moncada, S. (1993). Regulation of vascular homeostasis by nitric oxide. *Thromb. Haemost.* 70,36–41.

Rassaf, T., Bryan, N., Maloney, R., Specian, V., Kelm, M., Kalyanaraman, B., Rodriguez, J., Feelisch, M. (2003). NO adducts in mammalian red blood cells: Too much or too little? *Nat. Med.* 9, 481–483

Rea, M. S. and Parsons, R. H. (2001). Evidence of nitric oxide and angiotensin II regulation of circulation and cutaneous drinking in *Bufo marinus*. *Physiol. Biochem. Zool.* 74,127–133.

Renshaw, G. M. and Dyson, S. E. (1999). Increased nitric oxide synthase in the vasculature of the epaulette shark brain following hypoxia. *Neuroreport* 10,1707–1712.

Ribeiro, J. M. and Nussenzveig, R. H. (1993). Nitric oxide synthase activity from a hematophagous insect salivary gland. *FEBS Lett.* 330,165–168.

Robertson, J. D., Bonaventura, J. and Kohm, A. P. (1994). Nitric oxide is required for tactile learning in *Octopus vulgaris*. *Proc. Biol. Sci.* 256,269–273.

Robertson, J. D., Bonaventura, J., Kohm, A. and Hiscat, M. (1996). Nitric oxide is necessary for visual learning in *Octopus vulgaris*. *Proc. Biol. Sci.* 263,1739–1743.

Rodriguez, J., Maloney, R., Rassaf, T., Bryan, N. and Feelisch, M. (2003). Chemical nature of nitric oxide storage forms in rat vascular tissue. *Proc. Natl. Acad. Sci. USA* 100,336–341.

Schipp, R. and Gebauer, M. (1999). Nitric oxide: A vasodilatatory mediator in the cephalic aorta of *Sepia officinalis* (L.) (Cephalopoda). *Invert. Neurosci.* 4,9–15.

Schmidt, H. H. and Walter, U. (1994). NO at work. *Cell* 78,919–925.

Schober, A., Malz, C. R., Schober, W. and Meyer, D. L. (1994a). NADPH-diaphorase in the central nervous system of the larval lamprey (*Lampetra planeri*). *J. Comp. Neurol.* 345,94–104.

Schober, A., Meyer, D. L. and Von Bartheld, C. S. (1994b). Central projections of the nervus terminalis and the nervus praeopticus in the lungfish brain revealed by nitric oxide synthase. *J. Comp. Neurol.* 349,1–19.

Schoor, W. P. and Plumb, J. A. (1994). Induction of nitric oxide synthase in channel catfish *Ictalurus punctatus* by *Edwarsiella ictaluri*. *Dis. Aquat. Org.* 19,153–155.

Shah, A. M. and MacCarthy, P. A. (2000). Paracrine and autocrine effects of nitric oxide on myocardial function. *Pharmacol. Ther.* 86,49–86.

Smith, M. P., Takei, Y. and Olson, K. R. (2000). Similarity of vasorelaxant effects of natriuretic peptides in isolated blood vessels of salmonids. *Physiol. Biochem. Zool.* 73,494–500.

Smith, G. T., Unguez, G. A. and Reinauer, R. M. J. (2001). NADPH-diaphorase activity and nitric oxide synthase-like immunoreactivity colocalize in the electromotor system of four species of gymnotiform fish. *Brain Behav. Evol.* 58,122–136.

Soderstrom, V., Hylland, P. and Nilsson, G. E. (1995). Nitric oxide synthase inhibitor blocks acetylcholine induced increase in brain blood flow in rainbow trout. *Neurosci. Lett.* 197,191–194.

Sollid, J., Rissanen, E., Tranberg, H. K., Thorstensen, T., Vuori, K. A. M., Nikinmaa, M. and Nilsson, G. E. (2006). HIF-1alpha and iNOS levels in crucian carp gills during hypoxia-induced transformation. *J. Comp. Physiol. B* 176,359–369.

Springer, J., Ruth, P., Beuerlein, K., Westermann, B. and Schipp, R. (2004). Immunohistochemical localization of cardioactive neuropeptides in the heart of a living fossil, *Nautilus pompilius* L. (Cephalopoda, Tetrabranchiata). *Histochem. J.* 35, 21–28.

Staples, J. F., Zapol, W. M., Bloch, K. D., Kawai, N., Val, V. M. and Hochachka, P. W. (1995). Nitric oxide responses of air-breathing and water-breathing fish. *Am. J. Physiol.* 268,R816–R819.

Sun, H., Wang, Y., Wang, X., Ge, L. and Sun, I. (2005). Study of nitric oxide and nitric oxide synthase in haemolymph of scallop *Chlamys farreri*. *Oceanol. Limnol. Sin.* 36,343–348.

310

Torreilles, J. (2001). Nitric oxide: One of the more conserved and widespread signaling molecules. *Front. Biosci.* 6,D1161–D1172.

Tota, B., Amelio, D., Pellegrino, D., Ip, Y. K. and Cerra, M. C. (2005). NO modulation of myocardial performance in fish hearts. *Comp. Biochem. Physiol. A* 142,164–177.

Virgili, M., Poli, A., Beraudi, A., Giuliani, A. and Villani, L. (2001). Regional distribution of nitric oxide synthase and NADPH-diaphorase activities in the central nervous system of teleosts. *Brain Res.* 901,202–207.

Webb, A., Bond, R., McLean, P., Uppal, R., Benjamin, N. and Ahluwalia, A. (2004). Reduction of nitrite to nitric oxide during ischemia protects against myocardial ischemia-reperfusion damage. *Proc. Natl. Acad. Sci. USA* 101,13683–13688.

311

NOS distribution and NO control of cardiac performance in fish and amphibian hearts

Bruno Tota*, Sandra Imbrogno, Rosa Mazza and Alfonsina Gattuso

Department of Cell Biology, University of Calabria, Via P. Bucci, 87030 Arcavacata (CS), Italy

Abstract. Nitric oxide (NO), generated endogenously by a family of NO synthases (NOS) in the heart, has important autocrine–paracrine effects on cardiac function, modulating the inotropic state, excitation–contraction coupling, diastolic function, heart rate and β-adrenergic responsiveness. Fish and amphibian hearts share common structural and functional aspects with higher vertebrates, while differing in relevant ultrastructural, myoarchitectural, vascular and pumping features. This synopsis deals with cardiac NOS expression and localization in phylogenetically and eco-physiologically different teleost species, as well as in lungfish and frog, thus documenting the long evolutionary history of cardiac NO. In particular, the role of NO in the mechanical performance of teleost and frog hearts, both in the absence (*i.e.,* unstimulated heart preparations) and in the presence of physical (*i.e.,* load changes) and chemical (inotropic agonists) stimuli, is analysed. Using teleost and amphibian hearts as natural models in which the coronary system is absent, or scarcely present, the importance of an endocardial endothelium (EE) NO-mediated intracavitary control of mechanical performance is emphasized. This highlights the ancient autocrine–paracrine role of the cardiac NOS/NO system during the evolution of the poikilotherm vertebrate heart.

Keywords: amphibians; avascular heart; chemical stimulation; endocardial endothelium; Frank-Starling response; lungfish; nitric oxide; NOS isoforms; stroke volume; teleosts.

Introduction

Nitric oxide (NO), the free radical gas produced by NO synthase (NOS), exerts a wide range of physiological and pathophysiological effects extended to almost every cell type within the cardiocirculatory system. In the mammalian heart, NO modulates several aspects of myocardial function, such as excitation–contraction coupling, Frank–Starling response, force–frequency relationship, β-adrenergic and cholinergic effects, as well as oxygen consumption, hypertrophic remodelling and myocardial regeneration (Massion *et al.*, 2005, and references therein).

Corresponding author: Tel.: +39-984-492907. Fax: +39-984-492906.
E-mail: tota@unical.it (B. Tota).

ADVANCES IN EXPERIMENTAL BIOLOGY
VOLUME 01   ISSN 1872-2423
DOI: 10.1016/S1872-2423(07)01014-9

The three NOS isoforms (NOS1/NOS I or nnos, NOS2/NOS II or iNOS and NOS3/NOS III or eNOS) are widely distributed in cardiac tissues. eNOS and nNOS are constitutively expressed in myocytes, where eNOS is enriched in plasmalemmal and T-tubular caveolae, and nNOS is localized in the sarcoplasmic reticulum. nNOS is also expressed in both adrenergic and cholinergic nervous fibres and eNOS in both endothelial and endocardial cells (Andries et al., 1998; Brutsaert, 2003). In pathological conditions, such as septic shock (Jung et al., 2000; Kleinert et al., 2003) or heart failure (Massion et al., 2003; Paulus and Bronzwaer, 2004), the calcium-independent iNOS is induced in both the cardiomyocytes and the inflammatory cells infiltrating the myocardium. However, beyond its established role in response to pathophysiological conditions, the identification of a constitutive iNOS has also been demonstrated in mammalian cardiac cells under normal conditions (Buchwalow et al., 2001; Cohen et al., 2003). More recently, a mitochondrial NOS (mtNOS), corresponding to a variant of nNOS, has been reported in various cell types, including cardiomyocytes (Giulivi, 2003). The apparent redundancy of the expression of NOS isoforms within the myocardium, their precise spatial subcellular and tissue confinements, together with the limited diffusibility of NO, appear crucial to the ability of NO to regulate heart performance, restricting its effects to closely co-localized targets (Hare and Stamler, 2005).

The classical view holds that the most important second messenger transducing the effects of NO is cyclic guanidine monophosphate (cGMP), NO being a powerful activator of soluble guanylyl cyclase (sGC). However, cGMP-independent mechanisms operating through NO interaction with ion channels, haem proteins, iron/non-haem complexes, free thiol residues, or superoxide anions, have been also shown (Balligand and Cannon, 1997; Hare, 2003).

After a brief outline of the basic functional morphology of the fish and amphibian hearts, in this review we will summarize present knowledge regarding the intracardiac NOS system in these cold-blooded vertebrates, which, in the absence of other data, is mostly based on our own studies. We will first consider the expression and zonal localization of NOS in the heart of teleost, lungfish and frog species, then the role of NO on the mechanical performance of teleost and frog cardiac preparations both in the absence (i.e., unstimulated heart) and in the presence of physical (i.e., changes in load) or chemical (inotropic agonists) stimulation. Finally, we will discuss the paracrine endocardial endothelium (EE)-NO intracavitary modulation of ventricular performance, highlighting the ancestral role of this system, which appears to be particularly important in lower vertebrates.

Basic functional morphology of the fish and amphibian hearts

The fish heart is made up of four chambers coupled in series: the sinus venosus, the atrium, the ventricle, and the outflow tract, which consists of a proximal conus arteriosus and a distal bulbus arteriosus. In a typical water-breathing fish in which the gills are the respiratory organ, the peripheral venous blood is driven into the sinus venosus, the atrium and the ventricle, from where it is pumped to the gills to be oxygenated; thence it is distributed to the body and back to the heart.

Most teleosts and the lungfish (belonging to the sarcopterygians, which gave rise to the very first terrestrial vertebrates, the amphibians) have an entirely trabeculated ventricle, *i.e.,* the *spongiosa*, supplied by the intertrabecular lacunary system; other teleosts have a mixed type of ventricle, *i.e.,* an external *compacta* and an inner *spongiosa* (Icardo *et al.*, 2005a; Tota, 1983; Tota *et al.*, 1983). In this article, the hearts of the Antarctic teleosts and the African lungfish *Protopterus dolloi* exemplify the entirely trabeculated ventricle, while the eel and tuna hearts exemplify the mixed type of ventricular myoarchitecture. The entirely trabeculated ventricle can only be supplied by the lacunary system (avascular heart) or, in addition, can have a vascular supply (usually coronary arteries); the *compacta* is always vascularized (Tota, 1989; Tota *et al.*, 1983). From a phylogenetic point of view, the fish heart is the prototype of the higher vertebrate hearts, as illustrated by the recent development of zebrafish embryology (Stainier and Fishman, 1994). The fish heart is a venous heart since it is supplied only, or mostly, by venous blood. In comparison with the higher vertebrates, it is designed as a low-pressure region exposed to relatively low and variable pO_2 levels (Farrell and Jones, 1992; Olson, 1998). However, among the vertebrates, fish, particularly the teleosts, show the highest interspecific variation not only in myocardial architecture and cardiac blood supply, as epitomized by the different types of ventricle arrangement reported (Farrell and Jones, 1992; Tota and Gattuso, 1996; Tota *et al.*, 1983), but also in haemodynamic capabilities. In fact, a variety of ventricular pumps exists in fish, including pumps that mainly produce high pressures with relatively modest volume flow (pressure pumps) and pumps that invest the same power mainly in a high stroke volume with small increments in pressure (volume pumps). This highlights the functional plasticity of the heart chamber (see for

314

references Tota and Gattuso, 1996), which has been related to eco-physiological features and to differences in the factors determining systemic oxygen transport capacities, e.g., cardiac performance, blood flow, blood oxygen content and haematocrit (Agnisola et al., 1997).

The amphibian heart consists of a sinus venosus, right and left atria divided by an anatomically complete internal septum, a ventricle lacking any internal subdivision (with the exception of the salamander species *Siren*), and a conus arteriosus with a spiral valve. Due to the impressive variety of respiratory and circulatory patterns exhibited by the amphibians, it is impossible to generalize the mode of action of the "amphibian heart" (Foxon, 1964) and here we will only consider the frog (*Rana esculenta*) heart. The right atrium drains "venous" blood, while the left atrium drains more oxygenated "arterial" blood. The blood from the two atria remains to a large extent unmixed in the single ventricle, with a selective-like distribution during ventricular systole (according to the so-called "classical hypothesis"). In fact, the "venous" blood during early systole, under the influence of the spiral valve of the conus, leaves first, being injected in the low-pressure pulmo-cutaneous arch, hence perfusing the lungs and the skin; in the middle phase of ventricular systole the more mixed blood from the centre of the ventricle is ejected on the other side of the spiral valve of the conus, being distributed to the systemic arches. The last to leave the ventricle is the "oxygenated" blood, which under the greatest blood pressure is canalized to the brain through the high-resistance carotid arches (Foxon, 1964). It is worth noting that the patterns of atrial and, especially, ventricular myocardial trabeculations seen in air-breathing fish and amphibian hearts play a role in the evolutionary remodelling of intracardiac blood flow.

The use of teleost and amphibian hearts as natural models in which the coronary system is absent, or poorly present, allows, unlike mammalian hearts, examination of the cardiac nitrergic influences on myocardial performance independently from the vascular changes. Furthermore, the great surface area of the EE covering the internal cavity of these trabeculated hearts makes them suitable natural models for analysing the NO-mediated cross-talk between the EE and the myocardium.

Due to their phylogenetic and morpho-functional aspects, the hearts of fish and amphibians have been very helpful in developing knowledge about basic and comparative cardiac physiology, as emphasized by their use in recent studies on cardiac development (Stainier and Fishman, 1994), epigenetic remodelling (Hove et al., 2003) and adult heart regeneration (Becker et al., 1974; Poss et al., 2002).

In this context, studies of the presence and function of the NOS system in the hearts of phylogenetically distant species, which disclose an under-exploited number of natural cardiac models, could highlight aspects of evolutionary unity and diversity regarding the role of NO in the control of cardiac function in vertebrates. At the same time, studies of cardiac NO in lower vertebrates inject new life into our understanding of the autocrine–paracrine control of the vertebrate heart.

NOS distribution in cardiac tissues

Teleosts and lungfish

Although early comparative studies in avian (Hasegawa and Nishimura, 1991), reptilian (Knight and Burnstock, 1993) and amphibian (Knight and Burnstock, 1996; Rumbaut et al., 1995) species provided evidence that the NOS system is present, very few up-to-date studies have examined NOS localization and distribution in the cardiovascular system of non-mammalian vertebrates. Fish are known to express the three isoforms of NOS; however, studies on heart and blood vessel preparations show variable results depending on the species and the particular preparation used (for reviews see Eddy, 2005; Tota et al., 2005).

Recently, Tota et al. (2005) and Amelio et al. (2006) reported cardiac NOS localization in teleosts with distinct phylogenies and ecophysiological habitats, i.e., the Antarctic stenotherm notothenioids (the icefish Chionodraco hamatus and the red-blooded Trematomus bernacchii), the temperate eurytherms eel (Anguilla anguilla) and tuna (Thunnus thynnus thynnus), as well as the dipnoan air-breathing fish (the African lungfish, Protopterus dolloi) (Table 1). The presence of eNOS has been reported both in the spongiosa of the African lungfish P. dolloi and in the mixed-type ventricular myocardium of the tuna and eel, where it is expressed in both the vascular and the EE (Fig. 1). Of note, despite the remarkably different phylogenetic, ecophysiological and cardiac morphofunctional traits that characterize these fish, eNOS always localizes in the ventricular endothelial cells, suggesting that, as in mammals (Andries et al., 1998), also in fish NO is released by both the EE and endothelium of coronary vessels. Moreover, to a lesser extent, eNOS is also localized in the ventricular myocardiocytes of the spongiosa (lungfish, eel, tuna) and the compacta (eel and tuna) (Fig. 1). In both endothelial and myocardial cells, eNOS is prevalently associated with the plasmalemma. This membrane targeting appears similar to that identified in mammalian cardiac

316

Table 1. eNOS and iNOS immulolocalization pattern in the ventricle of *Chionodraco hamatus*, *Trematomus bernacchii*, *Anguilla anguilla*, *Protopterus dolloi* and *Thunnus thynnus thynnus* (Tota *et al.*, 2005).

	eNOS				INOS			
	EP	EE	VE	Myo	EP	EE	VE	Myo
C. hamatus	+ +	+ +	NA	+	–	–	NA	+ +
T. bernacchii	+ +	+ +	+ +	+	–	–	–	+ +
A. anguilla	+ +	+ +	+ +	+	–	–	–	+ +
P. dolloi	+ +	+ +	+	+	?	?	?	?
T. thynnus thynnus	+ +	+ +	+ +	+	?	?	?	?

EP, epicardium. EE, endocardial endothelium. VE, vascular endothelium. Myo, myocardiocytes.
+, presence. –, absence. ?, not yet detected.
NA, not applicable (*C. hamatus* lacks a VE).

cells, in which eNOS is present in the caveolae associated with the protein caveolin-3 (Cohen *et al.*, 2004; Feron *et al.*, 1996).

The detection, by both immunofluorescence and Western blotting analysis, of eNOS and iNOS enzymes in the heart ventricle of the cold-adapted Antarctic teleosts, *i.e.,* the haemoglobinless *C. hamatus* and the red-blooded *T. bernacchii*, is of interest from an ecophysiological point of view. The eNOS isoform localizes at the level of the EE (*C. hamatus* and *T. bernacchii*) cells and in the endothelium of the subepicardial vessels (*T. bernacchii*), while iNOS is exclusively expressed within the cytoplasm of the myocardiocytes, being absent from the EE and the endothelium of the subepicardial vessels of both teleosts (Amelio *et al.*, 2006). Interestingly, in the two Antarctic teleosts, iNOS expression is susceptible to septic stimulation by bacterial lipopolysaccharide (LPS), suggesting the importance and the maintenance of the NOS system in the general mechanism of defence against pathogens (Amelio *et al.*, 2006). The conservation of the NOS system in these two notothenioid species, which have evolved in the frigid waters of the Southern Ocean and are characterized by notable stenothermia, suggests that neither the evolutionary geographic isolation nor the species-specific cold-elicited adaptations have affected the ventricular NOS patterns in these teleosts.

Of note, eNOS is expressed also in the visceral pericardium of all fish examined (Fig. 1) (Amelio *et al.*, 2006). In mammals, this tissue may affect several aspects of cardiac physiology, including pericardial permeability and fluid production, by the release of a number of cardiomodulatory substances such as atrial natriuretic peptides, endothelin-1 (ET-1),

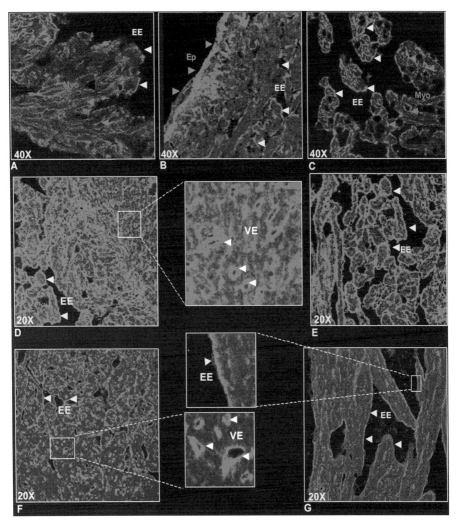

Fig. 1. Fluorescent immunolabelling of e-NOS (polyclonal anti-eNOS antibody; 1:100) in the ventricular myocardium of (A) *Chionodraco hamatus,* (B) *Trematomus bernacchii,* (C) *Protopterus dolloi,* (D, E) the eel *Anguilla anguilla,* and (F, G) the tuna *Thunnus thynnus thynnus.* (D) and (F) show the compact myocardium; (E) and (G) show the trabecular myocardium (unpublished data). Method as described in Pellegrino *et al.,* 2004. Insets show details of immunofluorescent vessels of the eel and the tuna, and of the EE of the tuna. EE, endocardial endothelium. Myo, myocardiocytes. VE, vascular endothelium. Ep, epicardium. From Tota *et al.,* 2005. (See Colour Plate Section in this book).

prostaglandins and growth factors into the pericardial fluid (Mebazaa *et al.*, 1998). In fish also, the visceral pericardium shows morpho-functional traits indicative of autocrine–paracrine activity (*e.g.*, atrial natriuretic peptides in *T. bernacchii* and *C. hamatus*; Cerra *et al.*, 1997). For example, the subepicardial tissue of *T. bernacchii* heart displays many of the morphological criteria that define it as a site specialized in the production of humoral immune response, showing, at the same time, the tissue organization found in germinal centres of higher verte-brates (Icardo *et al.*, 1999). Accordingly, the presence of eNOS in the epicardium of these teleosts strongly supports an autocrine–paracrine role for fish pericardium.

Amphibians

Studies of the localization of NOS isoforms in amphibian hearts have, until now, been limited to frogs. Clark *et al.* (1994) first demonstrated that an nNOS, found in nerve cells in the frog heart, occurs only at the atrial level. Later, in the frog (*R. esculenta*), Sys *et al.* (1997) showed that the expression of an immunoreactive constitutive NOS was exclusively located in the EE; no cNOS staining was observed in the myocardium, although the authors did not exclude the possibility of its presence in the epicardium. Recently, using Western blot analysis, the expression of eNOS in the EE of frog heart has been confirmed by Adler *et al.* (2004). Taken together, these results suggest that in the frog heart EE is the major NOS-expressing cell type. However, the precise localization of the different NOS isoforms in the cardiac tissues of amphibian hearts re-quires further investigations on more species.

NO modulation of cardiac function in teleost and frog hearts

While in mammals the role of endogenously produced NO in the control of mechanical cardiac function is well established and has been exten-sively reviewed (Schultz *et al.*, 2005), in non-mammalian vertebrates the few available studies appear contentious because of contradictory results. The discrepancy observed between different studies may be explained by either the different organ/tissue preparations used, or the lack of rigorous standardization in the experimental conditions or species specificity.

To uncover intracardiac NO-cGMP signalling, we have used *in vitro* isolated and perfused whole heart preparations working at "physiologi-cal" conditions, which allow the evaluation of heart performance free from extrinsic neuro-humoral stimuli (Shah, 1996; Tota *et al.*, 2005).

Teleosts

NO and basal cardiac performance

Studies examining the effects of both NOS inhibitors and NO donors in working teleost hearts (*Anguilla anguilla*: Imbrogno *et al.*, 2001; *Salmo salar*: Gattuso *et al.*, 2002) have shown that the presence of an endogenous nitrergic tone negatively modulates basal cardiac performance (Fig. 2). In the eel heart, this basal nitrergic tone involves a cGMP-dependent transduction mechanism (Fig. 2) (Imbrogno *et al.*, 2001).

The detection of a NO-cGMP-dependent positive inotropism in the heart of the cold-adapted Antarctic teleost, the icefish *C. hamatus* (Pellegrino *et al.*, 2004), characterized by the evolutionary loss of haemoglobin (Hb) (Ruud, 1954), is of particular interest from an evolutionary and ecophysiological perspective. Hb appears not to be crucial for the fitness of these endemic inhabitants of the subzero frigid, thermally stable and highly oxygenated Antarctic waters, which represent a paradigm of the "blind cave fish phenomenon" (Somero *et al.*, 1998). The haemoglobin-less condition has stimulated in these organisms multiple cardiocirculatory and subcellular compensations to achieve efficient oxygen delivery (Pellegrino *et al.*, 2004). At the same time, the lack of one of the major mechanisms for affecting NO bioactivity (the NO-Hb reaction) makes the icefish a particularly intriguing model for studying NO function in fish. Therefore, the presence of a functional NOS system in the hearts of eel and icefish, *i.e.,* two teleosts that differ remarkably in their evolutionary history and ecophysiology, stresses the paramount importance of cardiac NO in fish. However, in clear contrast with the negative inotropic effect observed in the eel heart preparations, the icefish heart is under the tonic influence of a NO-cGMP-dependent positive inotropism. The reasons underlying such different behaviour remain to be elucidated. It seems unlikely that they are related either to interspecific differences in cardiac structure, or to the experimental hierarchic level of investigation, since all these hearts show the same type of myoarchitecture and intracardiac blood supply and have been studied under the same *in vitro* experimental conditions. Therefore, we suppose that these different results may reflect important species-specific differences in the NO-elicited inotropic effects, as already emphasized in mammalian and amphibian hearts (for a review see Fischmeister *et al.*, 2005). Moreover, it may be of relevance that the red-blooded stenothermal counterpart of the icefish, *T. bernacchii*, shows the same NO-dependent pattern detected in temperate species, as well as in the frog. Nothing is known about the functional

320

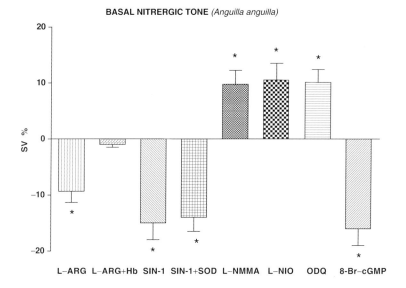

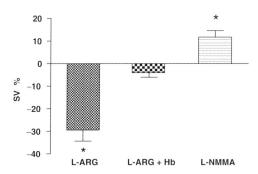

Fig. 2. Basal nitrergic tone in *Anguilla anguilla* and *Salmo salar*. Upper graph: Effects of the NO donor L-arginine (L-ARG; 10^{-7} M) alone and plus haemoglobin (Hb; 10^{-6} M), the NO donor SIN-1 (3-morpholino-sydnonimine; 10^{-9} M) alone and plus superoxide dismutase (SOD; $10\,\text{Uml}^{-1}$), the NOS inhibitor L-NMMA (L-N^G-monomethyl-arginine; 10^{-5} M), the NOS inhibitor L-NIO [N^5-(1-iminoethyl)-L-ornithine; 10^{-5} M], the guanylyl cyclase inhibitor ODQ (1*H*-[1,2,4]oxadiazolo[4,3-*a*]quinoxalin-1-one; 10^{-5} M) and the cGMP analogue 8-Br-cGMP (10^{-7} M) on stroke volume (SV) in *A. anguilla* (Imbrogno *et al.*, 2001). Lower graph: Effects of L-ARG (10^{-7} M) alone and plus Hb (10^{-6} M) and L-NMMA (10^{-5} M) on stroke volume (SV) in *S. salar* (Gattuso *et al.*, 2002). Percentage changes were evaluated as means $\pm$ SEM of the 4–5 experiments for each drug. Asterisks indicate significant differences ($P < 0.05$) from rates in untreated controls normalized to 0% SV.

behaviour, *e.g.,* kinetics of binding and dissociation, of NOS at 0°C; however, the comparison between icefish and *T. bernacchii* suggests that temperature *per se* cannot be responsible for the opposite NO-induced cardiac responses exhibited by the icefish.

NO and the Frank–Starling mechanism

According to the Frank–Starling Law of the heart (heterometric regulation), which is common to all vertebrate hearts, the end-diastolic volume and the consequent stretch of the myocardial fibres is a major determinant of the stroke volume, and hence cardiac output. When the return of venous blood to the heart (preload) increases, the lengthened atrial and ventricular myocardial fibres will contract more vigorously and perform more work, hence increasing the stroke volume.

In mammals, mechanical stimuli increase cardiac NO release from both the vascular endothelium and cardiomyocytes, playing a major role in the modulation of heart function (Petroff *et al.*, 2001; Pinsky *et al.*, 1997; Prendergast *et al.*, 1997). In fish, also the end-diastolic volume and consequent stretch of the myocardial fibres may contribute to the regulation of cardiac performance. However, unlike mammals, fish, in response to different haemodynamic loads, are able to increase cardiac output mainly through increased stroke volume rather than increased heart rate (Farrell and Jones, 1992; Olson, 1998).

We have reported that the hearts of both temperate eurytherms and cold-adapted teleosts are very sensitive to filling pressure changes (icefish: Tota *et al.*, 1991; Agnisola *et al.*, 1997; eel: Imbrogno *et al.*, 2001; salmon: Gattuso *et al.*, 2002; gilthead seabream: Icardo *et al.*, 2005b). Moreover, in eel and salmon we have documented that a basal release of endogenous NO modulates the Frank–Starling response by making the heart more sensitive to preload (Fig. 3), thus emphasizing the ubiquitous role of the cardiac NOS system in the control of vertebrate cardiac mechanical function.

The mechanism underlying the effect of NO on the stretch-induced increase in the force of contraction is, to date, not fully understood. Influences of NO on either systolic function (*e.g.,* mechanisms involving a modulation of intracellular calcium transient; see Méry *et al.*, 1993; Shah *et al.*, 1994) or diastolic function (e.g., reduction in diastolic stiffness; see Paulus *et al.*, 1994) have been postulated in mammalian heart preparations. It has been recently proposed that, in mammals, both constitutive NOS isoforms (i.e,. eNOS and nNOS) are involved in NO modulation of the Frank–Starling response, through a cGMP-independent mechanism.

322

The nNOS-derived NO promotes myocyte relaxation, through a facilitatory effect on calcium reuptake into the sarcoplasmic reticulum by SERCA-ATPase (sarcoplasmic/endoplasmic reticulum Ca^{2+}-ATPase), while the eNOS-derived NO sustains the subsequent increase in calcium transient, and thus in the force of contraction, by an activating S-nitrosylation of the ryanodine receptor Ca^{2+} release channels (Massion et al., 2005). The subcellular compartmentation of NOS isoforms, such as the differences in their mode of activation, has been proposed to be implicated in targetting NO to specific intracellular effectors; this in turn may make it possible to exert diverse and specific actions within the same cell type (Casadei and Sears, 2003; Hare, 2004).

In mammals, various pathophysiological conditions, such as endothelial damage, in vivo endotoxin injection, or activation of pro-inflammatory cytokines, induce the calcium-independent iNOS to produce high amounts of cytosolic NO, which may elicit organ and tissue dysfunction (Schultz et al., 2005). In salmonids, iNOS activity has been demonstrated by biochemical, immunohistochemical and immunoblotting evidence (Barroso et al., 2000), while iNOS transcript expression has been detected in the gills after injection challenge with the gram-positive pathogen Renibacterium salmoninarum (Campos-Perez et al., 2000). In Atlantic salmon (S. salar) affected by infectious salmon anaemia (ISA), an endothelial-endocardial viral disease, we have documented cardiac dysfunction, measured in terms of deterioration in the Frank–Starling response, correlated with the severity of the disease (Gattuso et al., 2002). Although these studies provided no direct evidence for iNOS in the pathogenesis of cardiac dysfunction, it was of interest that the depressed contractile responsiveness to the Frank–Starling mechanism, observed in the diseased hearts was completely reverted by the iNOS-specific inhibitor L-NIL [L-N^6-(1-iminoethyl)lysine] (Fig. 3). This suggests that in ISA-infected fish the induction and activation of iNOS may play a relevant role in the pathogenesis of heart failure.

NO and chemical stimulation

Experimental work carried out in the past 15 years has provided considerable advances in understanding the autocrine–paracrine role of NO in modulating the performance of the mammalian heart. In particular, intracardiac NO appears to optimize and fine-tune the heart responses to numerous cardioactive agents, such as neurotransmitters, hormones, autacoids, etc. Several cardioactive agents exert their effects by modulating cardiac ion channels, which in turn are regulated by NO through cGMP-

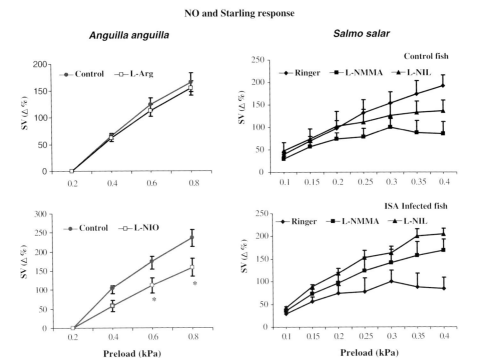

NO and Starling response

Fig. 3. NO modulation of the Starling response in teleost hearts. Left panels: Effects of preload on stroke volume (SV) under control conditions (filled circles) and after treatment with the NOS inhibitor L-NIO (10^{-5} M) (lower panel; open squares) or the NO donor L-Arg (10^{-7} M) (upper panel; open squares) in *Anguilla anguilla* (Imbrogno *et al.*, 2001). Right panels: Effects of preload on SV after treatment with the NOS inhibitor L-NMMA (10^{-6} M) (squares) or the iNOS inhibitor L-NIL (10^{-6} M) (triangles) or mock treatment (Ringer solution; diamonds) in control and ISA-infected *Salmo salar* (Gattuso *et al.*, 2002). Percentage changes were evaluated as mean $\pm$SEM of 4–6 experiments for each group. Paired Student's *t*-test was used for comparison within groups; two-way ANOVA was used for comparison between groups. Asterisks indicate significant differences ($P < 0.05$) between groups.

dependent or cGMP-independent mechanisms. Although many studies have focussed on the effects of NO and cGMP on Ca^{2+} channels (Hare, 2003), other cardiac ion channels have also been shown to be regulated by this pathway, such as ATP-sensitive K^+ channels (Han *et al.*, 2002), pacemaker *f*-channels (Herring *et al.*, 2001) and voltage-dependent fast Na^+ current channels (Ahmmed *et al.*, 2001). The spatial NOS compartmentalization in the proximity of cell membrane receptors and ion channels appears crucial in determining specific couplings of various extracellular chemical stimuli with appropriate intracellular signal effectors (Barouch et al., 2002; Cohen et al., 2004). In particular, the involvement of NO in adrenergic (sympathetic) and cholinergic (parasympathetic) neuromodulation has been extensively studied. For example, constitutive NOS activity linked to the muscarinic cholinergic signal transduction cascade has been described in several mammalian hearts (for a review see Balligand, 2000). Similarly, interactions between angiotensin II (ANG II) and eNOS have been reported to take part in the downstream transduction cascade activated by the angiotensin type 1 (AT_1) receptor (Li et al., 2002; Paton et al., 2001). In the absence of other data regarding fish heart, we will report below some examples of the NO signal mediating inotropic effects induced by chemical stimuli, such as acetylcholine (ACh) and ANG II in the in vitro working eel heart.

ACh generally produces negative chronotropic and inotropic effects in the heart; however, positive inotropic responses have been sporadically reported both in mammalian and nonmammalian vertebrates (Buccino et al., 1966; Biegon et al., 1980; Chan and Chow, 1976). In the isolated and perfused eel heart, exogenous ACh elicits a biphasic concentration-dependent inotropic action; i.e., a positive response at nanomolar concentrations, mediated by M_1 muscarinic receptors, and a negative one at micromolar concentrations, which involves M_2 muscarinic receptors (Imbrogno et al., 2001). The ACh-mediated positive inotropism occurs through a NO-cGMP signal transduction mechanism. In contrast to the ACh-NO signalling in the frog heart (see below), in the heart of A. anguilla the NO-cGMP mechanism appears to be involved only in the positive cholinergic response. In fact, pre-treatment with blockers of various steps of the NO-cGMP signalling abolished the positive effects of ACh without influencing the negative one (Fig. 4).

In mammals, ANG II, the principal effector of the renin–angiotensin system, modulates various cardiac functions, including chronotropic and inotropic properties (Sekine et al., 1999), myocardial growth (Grinstead and Young, 1992) and coronary vasoconstriction (Timmermans et al., 1993). On the eel heart, ANG II elicits a direct negative inotropic effect (Imbrogno et al., 2003). Although ANG II has been found to exert both direct and indirect (via cardiac adrenoceptors) effects on the heart of the American eel Anguilla rostrata and of the trout Oncorhynchus mykiss (Bernier and Perry, 1999; Oudit and Butler, 1995), no data are available regarding the signal transduction mechanisms involved. Recently, Imbrogno et al. (2003) reported that the ANG II-mediated inotropism involves a NO-cGMP transduction pathway (Imbrogno et al., 2003; Fig. 4).

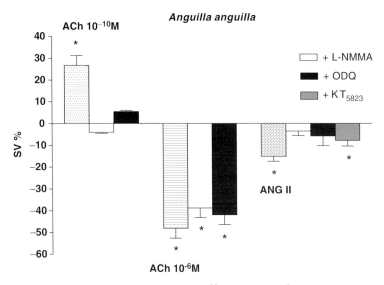

Fig. 4. Effects of acetylcholine (ACh; 10^{-10} M and 10^{-6} M) on stroke volume (SV) before and after treatment with the NOS inhibitor L-NMMA (10^{-5} M) and the guanylyl cyclase inhibitor ODQ (10^{-5} M), and effects of angiotensin II (ANG II; 10^{-8} M) on SV before and after treatment with L-NMMA (10^{-5} M), ODQ (10^{-5} M) and the protein kinase A inhibitor KT_{5823} (10^{-7} M) in *Anguilla anguilla*. Percentage changes from pre-treatment values were evaluated as means $\pm$ SEM of 4–5 experiments for each drug. Asterisks indicate significant differences ($P < 0.05$) from rates in untreated controls normalized to 0% SV. For details see Imbrogno *et al.* (2001, 2003, 2004).

Frogs

In contrast to the impressive biodiversity and variety of anatomical and physiological adaptations exhibited by the amphibians, very few data, mostly limited to frogs, are available on cardiac NOS in these vertebrates. Due to its morpho-functional characteristics, which make it a powerful natural tool for cardiac physio-pharmacological research, the frog heart has paved the way for basic discoveries in cardiac physiology, as illustrated by the classical studies carried out by Frank and Loewi on the length–tension relationships (the Frank–Starling law of the heart; Frank, 1895) and on neurochemical transmission (the "Vagus-Stoff" and the "Sympatikus-Stoff"; Loewi, 1921). However, to date no studies are available on the nervous and autocrine–paracrine control of heart performance exerted by the cardiac NOS system. Using an *in vitro* isolated and perfused frog (*R. esculenta*) heart working at physiological haemodynamic loads,

326

we have analysed aspects of NO modulation of cardiac performance both under basal conditions and chemical stimulation. The extensive inter-trabecular lacunary system supplying the completely trabeculated frog ventricle makes this heart particularly suitable for exploring the EE-NO-mediated modulation of cardiac function (see below).

NO under basal conditions and chemical stimulation

Basal conditions
In the frog heart, an endogenous nitrergic tone negatively modulates mechanical cardiac performance (Sys *et al.*, 1997). In fact, the admin-istration of NOS inhibitors induced a positive inotropic effect, while NO donors elicited negative inotropy via activation of sGC (Sys *et al.*, 1997) (Fig. 5). Moreover, the NO donor SIN-1 generated a biphasic dose-response curve with a positive effect at lower doses and a negative one at

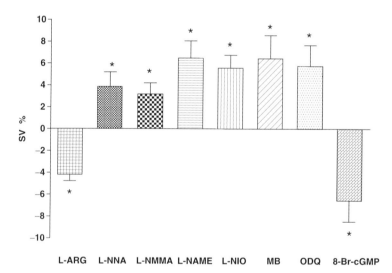

Fig. 5. Basal nitrergic tone in *Rana esculenta*. Effects of the NO donor L-ARG (10^{-7}M), the NOS inhibitors L-NNA (nitro-L-arginine; 10^{-5}M), L-NMMA (10^{-4}M), L-NAME (N^G-nitro-L-arginine methyl ester; 10^{-4}M) and L-NIO (10^{-5}M), the guanylyl cyclase inhibitors methylene blue (MB; 10^{-6}M) and ODQ (10^{-5}M) and the cGMP analogue 8-Br-cGMP (10^{-6}M) on stroke volume (SV). Percentage changes were evaluated as means $\pm$SEM of 4–5 experiments for each drug. Asterisks indicate significant differences ($P<0.05$) from rates in untreated controls normalized to 0% SV. For details see Gattuso *et al.*, 1999.

higher doses (data not shown) (Gattuso *et al.*, 1999). Interestingly, in frog (*R. esculenta*) isolated ventricular myocytes, Méry *et al.* (1993) demonstrated a dose-dependent trans-sarcolemmal calcium current (I_{Ca}) response to NO-cGMP activation, which involves the modulation of PDE2 and PDE3 phosphodiesterase isoforms (Fischmeister *et al.*, 2005).

Chemical stimulation
NO modulation of frog cardiac performance has been studied in relation to stimulation with ACh, isoproterenol (ISO) and ET-1.

ACh induces a biphasic effect, with a positive (lower concentrations) and a negative (higher concentrations) inotropism. Both cholinergic responses are mediated by NOS and sGC activation since they are blocked by pre-treatment with both NOS inhibitors and GC inhibitors (Fig. 6). Infusion of the cGMP analogue 8-Br-cGMP shifted the positive inotropism at lower doses to negative. Milrinone, which specifically blocks the cGMP-inhibited cAMP-phosphodiesterase (cG_i-PDE or PDE3), abolished the positive effect of ACh but not the negative one (Gattuso *et al.*, 1999), suggesting that nanomolar and micromolar doses of cGMP can

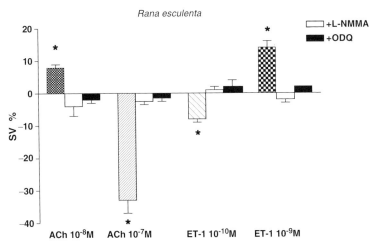

Fig. 6. Effects of acetylcholine (ACh; 10^{-8}M and 10^{-7}M) on stroke volume (SV) before and after treatment with NOS inhibitor L-NMMA (10^{-4}M) and the guanylyl cyclase inhibitor ODQ (10^{-5}M), and effects of endothelin-1 (ET–1; 10^{-10} M and 10^{-9}M) on SV before and after treatment with L-NMMA (10^{-4}M) and ODQ (10^{-5}M) in *R. esculenta*. Percentage changes from untreated controls were evaluated as means $\pm$ SEM of 5–6 experiments for each drug. Asterisks indicate significant differences (P < 0.05) from rates in untreated controls normalized to 0% SV. For details see Gattuso *et al.*, 1999 and Tota *et al.*, 2000.

play a role in the fine-tuning of cardiac cAMP concentration by positive and negative controls via inhibition of the cG_i-PDE and stimulation of the cG_s-PDE, respectively, in agreement with the findings of Méry et al. (1993). In their study on frog (R. esculenta) isolated ventricular myocytes, Méry and colleagues demonstrated a biphasic trans-sarcolemmal I_{Ca} response to the NO donor SIN-1, which appears to be either excitatory or inhibitory depending on its nanomolar or micromolar range, with a consequent dose-dependent stimulation of guanylyl cyclase activity not only via the "soluble" NO-sensitive guanylyl cyclase, but possibly also via the "membrane-bound" isoform of the enzyme (Méry et al., 1993). Thus, neurotransmitters responsible for stimulating cardiac NO production (e.g., ACh; Balligand et al., 1993) are likely to affect cardiac contractility via this pathway. Perfusion with the NO donor SIN-1, which mimics the biphasic dose–response of ACh, confirms that both these cGMP-elevating agents act via an NO-dependent mechanism (Gattuso et al., 1999). It remains to be clarified what are the species-specific biochemical (e. g., myocardial PDE pattern) and environmental factors that underlie the different cardiac response between eel and frog.

In the frog heart, ISO, which acts mainly through β_2-adrenergic receptors (Jurevicius et al., 2003), induces a dose-dependent positive inotropism mediated by NO-cGMP signalling, which is blocked by inhibitors of this pathway (Tota et al., 2000; data not shown). The presence of an intrinsic adrenergic tone is also revealed by the finding that, as shown in various mammalian ventricular preparations, the negative inotropic action of muscarinic agonists appears to be most evident under conditions of elevated cAMP concentration (Balligand et al., 1993; Gattuso et al., 1999; MacDonnell et al., 1995).

ET-1, a powerful vasoconstrictor agent, is the predominant isoform of isopeptides (ETs) containing 21 amino acids, which, as local humoral factors, take part in the regulation of cardiovascular homeostasis. In the mammalian heart, ET-1 regulates normal cardiac function, modulating inotropic and chronotropic effects, diastolic relaxation and response to preload (Brunner et al., 2006). ET-1 effects are mediated by ET_A and ET_B receptors, which are expressed in several tissues, including vascular endothelium, smooth muscle cells, cardiomyocytes and EE (Leite-Moreira and Bras-Silva, 2004). It has been demonstrated that ET-1, by interacting with endothelial ET_B receptors, regulates NO synthase activity and NO production (Brunner et al., 2006). On the frog (R. esculenta) heart, ET-1 (human or porcine) induced a significant negative inotropic effect in the concentration range from 10^{-11} M to 10^{-8} M. However, at higher doses of ET-1 (10^{-9} M and 10^{-8} M), 60% of hearts showed a positive inotropic response. An NO-cGMP signal transduction mechanism was involved both in the negative (lower doses) and in the positive (higher doses) inotropism of ET-1, since both were abolished by NOS and sGC inhibitors (Tota *et al.*, 2000; Fig. 6).

EE-NO integrated cardiac modulation

During gastrulation, the embryonic endoderm cells adjacent to the precardiac mesoderm play a role in heart development, modulating the formation of cardiac mesoderm, primary myocardium and endocardium. Arising from the same cardiac mesodermal precursors after formation of the primitive cardiac tube, these endocardial and myocardial cells become separated by a dense layer of extracellular matrix, the cardiac jelly (Lough and Sugi, 2000). During further development, these two distinct cardiac cell layers reciprocally interact through autocrine and paracrine signalling. For example, cardiomyocyte development, survival and contractility appear to be modulated by signalling mediators secreted by endothelial cells (neuregulin, NO, ET-1), while cardiomyocytes may promote endothelial cell survival and assembly through VEGF-A (vascular endothelial growth factor A) and angiopoietin-1 (Hsieh *et al.*, 2006). Furthermore, it has been shown that during embryogenesis, EE cells are able to regulate cardiac morphogenesis by sensing and transducing biomechanical stimuli caused by pulsatile blood flow (Hove *et al.*, 2003). This aspect of morphogenetic plasticity is of evolutionary interest in relation to the intraluminal blood re-routing, which occurred in the heart of lower vertebrates during one of the most momentous events in their phylogeny, *i.e.,* their transition from water-breathing to air-breathing. This intracardiac redistribution, which accommodates a time-related shift of the cardiac output between the aerial gas exchanger and the systemic arterial circuits, is due to the myocardial trabeculation and partitioning of both the atrium and the ventricle, as epitomized by some extant lungfish (*e.g., Lepidosiren*) and early amphibians (Johansen, 1985). The specific arrangements of myocardial trabeculation and the rheological properties of the perfusing blood may allow selective laminar or streamlined flow, which avoids mixing of the oxygenated blood from the aerial gas exchanger with the systemic venous blood (Johansen, 1985). The early ontogenetic and phylogenetic functions of EE are retained in the adult heart, in which, similar to the regulation of vascular smooth muscle contraction by the vascular endothelium, EE plays a pivotal role in the regulation of myocardial performance, as shown in various mammalian species (Brutsaert, 2003). In fact, located between luminal blood and subjacent cardiac muscle, EE is ideally suited to act as a sensor–integrator system, which transduces the intracavitary physical and chemical stimuli, thereby directly regulating the performance of the subjacent myocardium through the synthesis and release of auto-crine–paracrine substances like NO, prostacyclin, ANG II and endo-thelin (for a review see Brutsaert, 2003). It may be expected that in the avascular, or poorly vascularized, hearts of fish and amphibians, the EE-induced autocrine–paracrine modulation of myocardial performance is more relevant than in the compact and vascularized heart of the homeo-therms. In fact, in the avascular hearts, in which EE lines the extensive and complex myocardial trabeculation, being at the same time the only barrier between the cardiac lumen (lacunary spaces) and the subjacent myocardium, there is a much higher endothelial surface-to-myocardial volume ratio than in the compact type of higher vertebrate hearts. For example, in *Rana pipiens,* where the distended ventricle has a wall thick-ness of 2 mm, the endocardial surface area is between 500 and $1000 \, cm^2/$ g, estimated from perfused sections of the ventricle. It is assumed that the total cellular surface area is $6600 \, cm^2/g$ of the distended ventricle (Brady, 1964). Such a comparatively greater amount of EE covering the internal cavity of the cold-blooded heart offers a good opportunity to investigate the role of this tissue as a key actor in the paracrine NO-mediated in-tracavitary regulation of cardiac function.

This function may be uncovered by exposing the luminal surface of the fish and amphibian ventricle to low concentrations of the detergent Triton X-100, so that no structural or ultrastructural changes in EE are detectable and pacemaker activity – as well as atrial, myocardial, me-chanical and endothelial secretory performance – remains intact (frog: Sys *et al.,* 1997; eel: Imbrogno *et al.,* 2001). This functional damage of EE induces a positive inotropic effect probably due to an interruption of the signal transduction pathway that normally activates eNOS in EE (Gattuso *et al.,* 1999; Imbrogno *et al.,* 2001; Sys *et al.,* 1997) (Table 2). These data, together with the identification of eNOS on the EE of both eel (Tota *et al.,* 2005) and frog (Sys *et al.,* 1997) hearts, strongly support the idea that in these animals EE is an important cellular source of NO, which, in turn, acts as a diffusible messenger between EE and the sub-jacent myocardium. The role of NO-EE in sensing and transducing chemical stimuli is illustrated by its involvement in inotropic cholinergic responses, both in eel and frog hearts. While in the eel heart, only the positive inotropism induced by nanomolar concentrations of ACh is abolished when EE is functionally damaged (Imbrogno *et al.,* 2001), in the frog, EE integrity appears to be a prerequisite for transducing both

331

Table 2. EE-NO involvement in the cardiac performance modulation of eel and frog heart.

Cardiac performance modulator	EE-NO involvement	
	Eel	Frog
Acetylcholine (positive inotropism)	+	+
Acetylcholine (negative inotropism)	–	+
Angiotensin II	+	?
Endothelin-1 (positive inotropism)	?	+
Endothelin-1 (negative inotropism)	?	+

For details, see text.
+, involvement. –, no involvement. ?, not yet detected.

the positive and negative inotropic effects of ACh (Gattuso et al., 1999; Table 2). Of note, the positive effects of ACh in eel, and both cholinergic responses in frog are mediated by the M_1 receptor subtype (Gattuso et al., 1999; Imbrogno et al., 2001). The EE-dependent transduction of M_1-mediated cholinergic response suggests that in eel and frog this receptor subtype is located at the EE level, resembling the situation in mammals, i.e., M_2 and M_4 muscarinic receptors are preferentially located on the myocardiocytes, and M_1, M_3 and M_5 are principally expressed by endothelial cells (Brodde and Michel, 1999). The findings that functional damage of EE and the inhibition of M_1 receptors mimic the situation obtained when the NO-cGMP mechanism is abolished (see Figs. 4 and 6), support the role of an EE-NO-cGMP-dependent pathway in the signal transduction activated by M_1 stimulation. Furthermore, the intracavitary ANG II signal in the eel, and the biphasic effect of ET-1 in the frog, also appear to be mediated by EE, since they are abolished by EE impairment by Triton X-100 treatment. Again, the NO-cGMP inhibition reproduces the situation obtained when EE is functionally damaged (Table 2) (eel: Imbrogno et al., 2003, 2004; frog: Tota et al., 2000). Taken together, these facts provide compelling evidence that in eel and frog hearts the EE, through the release of NO, orchestrates in an autocrine–paracrine manner the signal transduction pathways interposed between the luminal chemical stimuli and the subjacent myocardium. This signalling module, through protein–protein interactions involving several G-protein-coupled receptors and other allosteric modulators, may provide a rationale for the importance of the spatial compartmentation of NOS in the regulation of cardiac function by NO (Feron and

332

Balligand, 2006). The EE caveolae are the site of many proteins involved in signal transduction cascades, including cell membrane receptors, ion channels, G-proteins, protein kinases, etc., and appear to be the major candidate as the domain where multiple extracellular stimuli converge to modulate eNOS (and potentially other NOS isoforms) activity.

Conclusions and perspectives

This review has shown that cardiac NO signalling is highly conserved phylogenetically in fish and amphibians. The supposed multidimensional role played by the EE-NO system in the avascular trabeculated hearts of lower vertebrates could also be extended to other aspects of cardiac physiology. For example, in mammals it has been shown that endogenous NO, through a reversible inhibition of cytochrome c oxidase, reduces cellular respiration, and, as suggested by Thomas et $al.$ (2001), this inhibition of O_2 consumption may improve extension of the O_2 gradient to cells further away from the vessel. Conceivably, this EE-NO-induced modulation of cell respiration and oxygen gradient may exert a relevant ischaemic cardioprotection in avascular "venous"-type hearts, in which the blood is retained for a longer period in the lacunae, thereby exposing the cardiomyocytes to relatively large fluctuations of pO_2 and, in the case of increased tissue work, to hypoxic conditions (Tota, 1983). The close proximity of EE, as a NO source, to the underlying myocardium may represent the morphological counterpart of this homeostatic function. Moreover, NO is known to prevent both platelet adhesion to EE and platelet aggregation (Radomski and Moncada, 1993). Therefore, in trabeculated hearts, where the turbulent flow in the myriad of lacunary spaces may represent a potent stimulus for thrombotic processes, EE-derived NO could exert a more pronounced protective role. To verify the EE-NO involvement in such novel aspects of poikilotherm myocardial homeostasis is a stimulating challenge for future studies.

References

Adler, A., Huang, H., Wang, Z., Conetta, J., Levee, E., Zhang, X. and Hintze, T. (2004). Endocardial endothelium in the avascular frog heart: Role for diffusion of NO in control of cardiac O_2 consumption. $Am.$ $J.$ $Physiol.$ 287,H14–H21.

Agnisola, C., Acierno, R., Calvo, J., Farina, F. and Tota, B. (1997). In vitro cardiac performance in the sub-antarctic Notothenoids $Eleginops$ $maclovinus$ (subfamily Eleginopinae), $Paranotothenia$ $magellanica$ and $Patagonotothen$ $tassellata$ (subfamily Nototheniinae). $Comp.$ $Biochem.$ $Physiol.$ 118A,1437–1445.

Ahmmed, G.U., Xu, Y., Hong Dong, P., Zhang, Z., Eiserich, J. and Chiamvimonvat, N. (2001). Nitric oxide modulates cardiac Na$^+$ channel via protein kinase A and protein kinase G. *Circ. Res.* 89,1005–1013.

Amelio, D., Garofalo, F., Pellegrino, D., Giordano, F. and Tota B., Cerra,M.C. (2006). Cardiac expression and distribution of nitric oxide synthases in the ventricle of the cold-adapted Antarctic teleosts, the hemoglobinless Chionodraco hamatus and the red-blooded *Trematomus bernacchii*. *Nitric Oxide* 15,190–198.

Andries, L.J., Brutsaert, D.L. and Sys, S.U. (1998). Nonuniformity of endothelial constitutive nitric oxide synthase distribution in cardiac endothelium. *Circ. Res.* 82,195–203.

Balligand, J.L. (2000). Regulation of cardiac function by nitric oxide. In *Nitric Oxide. Handbook of Experimental Pharmacology* (ed. B. Mayer), Vol. 143, pp. 206–234, Springer-Verlag, Berlin.

Balligand, J.L. and Cannon, P.J. (1997). Nitric oxide synthases and cardiac muscle: Autocrine and paracrine influences. *Arterioscler. Thromb. Vasc. Biol.* 17,184–185.

Balligand, J.L., Kelly, R.A., Marsden, P.A., Smith, T.W. and Michel, T. (1993). Control of cardiac muscle cell function by an endogenous nitric oxide signaling system. *Proc. Natl. Acad. Sci. USA* 90,347–351.

Barouch, L.A., Harrison, R.W., Skaf, M.W., Rosas, G.O., Cappola, T.P., Kobeissi, Z.A., Hobai, I.A., Lemmon, C.A., Burnett, A.L., O'Rourke, B., Rodriguez, E.R., Huang, P.L., Lima, J.A., Berkowitz, D.E. and Hare, J.M. (2002). Nitric oxide regulates the heart by spatial confinement of nitric oxide synthase isoforms. *Nature* 416,337–339.

Barroso, J.B., Carreras, A., Esteban, F.J., Peinado, M.A., Martinez-Lara, E., Valderrama, R., Jumenez, A., Rodrigo, J. and Lupianez, J.A. (2000). Molecular and kinetic characterization and cell type location of inducible nitric oxide syinthase in fish. *Am. J. Physiol.* 279,R650–R656.

Becker, R.O., Chapin, S. and Sherry, R. (1974). Regeneration of the ventricular myocardium in amphibians. *Nature* 248,145–147.

Bernier, N.J. and Perry, S.F. (1999). Cardiovascular effects of angiotensin-II-mediated adrenaline release in rainbow trout *Oncorhynchus mykiss*. *J. Exp. Biol.* 202,55–66.

Biegon, R.L., Epstein, P.M. and Pappano, A.J. (1980). Muscarinic antagonism of the effects of phosphodiesterase inhibitor (methylisobutylxanthine) in embryonic chick ventricle. *J. Pharmacol. Exp. Ther.* 215,348–356.

Brady, A.J. (1964). Physiology of the amphibian heart. In *Physiology of the Amphibia* (ed. J.A. Moore), pp. 211–250, Academic Press, New York and London.

Brodde, O.E. and Michel, M.C. (1999). Adrenergic and muscarinic receptors in the human heart. *Pharmacol. Rev.* 51,651–690.

Brunner, F., Bras-Silva, C., Cerdeira, A.S. and Leite-Moreira, A.F. (2006). Cardiovascular endothelins: Essential regulators of cardiovascular homeostasis. *Pharmacol. Ther.* 111,508–531.

Brutsaert, D.L. (2003). Cardiac endothelial-myocardial signaling: Its role in cardiac growth, contractile performance, and rhythmicity. *Physiol. Rev.* 83,59–115.

Buccino, R.A., Sonnenblick, E.H., Cooper, T. and Braunwald, E. (1966). Direct positive inotropic effect of acetylcholine on myocardium. Evidence for multiple cholinergic receptors in the heart. *Circ. Res.* 19,1097–1108.

334

Buchwalow, I.B., Schulze, W., Karczewski, P., Kostic, M.M., Wallukat, G., Morwinski, R., Krause, E.G., Muller, J., Paul, M., Slezak, J., Luft, F.C. and Haller, H. (2001). Inducible nitric oxide synthase in the myocard. *Mol. Cell Biochem.* 217,73–82.

Campos-Perez, J.J., Ellis, A.E. and Secombes, C.J. (2000). Toxicity of nitric oxide and peroxynitrite to bacterial pathogens of fish. *Dis. Aquat. Org.* 43,109–115.

Casadei, B. and Sears, C.E. (2003). Nitric-oxide-mediated regulation of cardiac contractility and stretch responses. *Prog. Biophys. Mol. Biol.* 82,67–80.

Cerra, M.C., Canonaco, M., Acierno, R. and Tota, B. (1997). Different binding activity of A- and B-type natriuretic hormones in the heart of two Antartic teleosts, the red blooded *Trematomus bernacchii* and the haemoglobinless *Chionodraco hamatus*. *Comp. Biochem. Physiol.* 118,993–999.

Chan, D.K. and Chow, P.H. (1976). The effects of acetylcholine, biogenic amines and other vasoactive agents on the cardiovascular functions of the del, *Anguilla japonica. J. Exp. Zool.* 196,13–26.

Clark, R.B., Kinsberg, E.R. and Giles, W.R. (1994). Histochemical localization of nitric oxide synthase in the bullfrog intracardiac ganglion. *Neurosci. Lett.* 182,255–258.

Cohen, A.W., Hnasko, R., Schubert, W. and Lisanti, M.P. (2004). Role of caveolae and caveolins in health and disease. *Physiol. Rev.* 84,1341–1379.

Cohen, R.I., Hassell, A.M., Ye, X., Marzouk, K. and Liu, S.F. (2003). Lipopolysaccharide down-regulates inducible nitric oxide synthase expression in swine heart in vivo. *Biochem. Biophys. Res. Commun.* 307,451–458.

Eddy, F.B. (2005). Role of nitric oxide in larval and juvenile fish. *Comp. Biochem. Physiol.* 142A,221–230.

Farrell, A.P. and Jones, D.R. (1992). The heart. In *Fish Physiology* (eds W.S. Hoar and D.R. Randall), pp. 1–88, Academic Press, London.

Feron, O. and Balligand, J.L. (2006). Caveolins and the regulation of endothelial nitric oxide synthase in the heart. *Cardiovasc. Res.* 69,788–797.

Feron, O., Belhassen, L., Kobzik, L., Smith, T.W., Kelly, R.A. and Michel, T. (1996). Endothelial nitric oxide synthase targeting to caveolae. Specific interactions with caveolin isoforms in cardiac myocytes and endothelial cells. *J. Biol. Chem.* 271,22810–22814.

Fischmeister, R., Castro, L., Abi-Gerges, A., Rochais, F. and Vandecasteele, G. (2005). Species- and tissue-dependent effects of NO and cyclic GMP on cardiac ion channels. *Comp. Biochem. Physiol.* 142,136–143.

Foxon, G.E.H. (1964). Blood and respiration. In *Physiology of the Amphibia* (ed. J.A. Moore), pp. 151–209, Academic Press, New York and London.

Frank, O. (1895). Zur Dynamik des Herzmuskels. *Z. Biol.* 32,370–447.

Gattuso, A., Mazza, R., Imbrogno, S., Sverdrup, S., Tota, B. and Nylund, A. (2002). Cardiac performance in *Salmo salar* with infectious salmon anaemia (ISA): Putative role of nitric oxide. *Dis. Aquat. Org.* 52,11–20.

Gattuso, A., Mazza, R., Pellegrino, D. and Tota, B. (1999). Endocardial endothelium mediates luminal ACh-NO signaling in the isolated frog heart. *Am. J. Physiol.* 276,H633–H641.

Giulivi, C. (2003). Characterization and function of mitochondrial nitric-oxide synthase. *Free Radic. Biol. Med.* 34,397–408.

Grinstead, W.C. and Young, J.B. (1992). The myocardial renin-angiotensin system: Existence, importance, and clinical implications. *Am. Heart J.* 123,1039–1045.

Han, J., Kim, N., Joo, H., Kim, E. and Earm, Y.E. (2002). ATP-sensitive K^+ channel activation by nitric oxide and protein kinase G in rabbit ventricular myocytes. *Am. J. Physiol.* 283,H1545–H1554.

Hare, J.M. (2003). Nitric oxide and excitation-contraction coupling. *J. Mol. Cell. Cardiol.* 35,719–729.

Hare, J.M. (2004). Spatial confinement of isoform of cardiac nitric-oxide synthase: unravelling the complexities of nitric oxide's cardiobiology. *Lancet* 363,1338–1339.

Hare, J.M. and Stamler, J.S. (2005). NO/redox disequilibrium in the failing heart and cardiovascular system. *J. Clin. Invest.* 115,509–517.

Hasegawa, K. and Nishimura, H. (1991). Humoral factors mediates acetylcholine-induced endothelium-dependent relaxation of chicken aorta. *Gen. Comp. Endocrinol.* 84,164–169.

Herring, N., Rigg, L., Terrar, D.A. and Paterson, D.J. (2001). NO–cGMP pathway increases the hyperpolarisation-activated current, I_f, and heart rate during adrenergic stimulation. *Cardiovasc. Res.* 52,446–453.

Hove, J.R., Koster, R.W., Forouhar, A.S., Acevedo-Bolton, G., Fraser, S.E. and Gharib, M. (2003). Intracardiac fluid forces are an essential epigenetic factor for embryonic cardiogenesis. *Nature* 421,172–177.

Hsieh, P.C.H., Davis, M.E., Lisowski, L.K., LeeHsieh, R.T. *et al.* (2006). Endothelial-cardiomyocyte interactions in cardiac development and repair. *Annu. Rev. Physiol.* 68,51–66.

Icardo, J.M., Colvee, E., Cerra, M.C. and Tota, B. (1999). Bulbus arteriosus of the Antarctic teleosts II: The red-blooded *Trematomus bernacchii*. *Anat. Rec.* 256,116–126.

Icardo, J.M., Imbrogno, S., Gattuso, A., Colvee, E. and Tota, B. (2005b). The heart of *Sparus auratus*: A reappraisal of cardiac functional morphology in teleosts. *J. Exp. Zool.* 303,665–675.

Icardo, J.M., Ojeda, J.L., Colvee, E., Tota, B., Wong, W.P. and Ip, Y.K. (2005a). Heart inflow tract of the African lungfish *Protopterus dolloi*. *J. Morphol.* 263,30–38.

Imbrogno, S., Angelone, T., Corti, A., Adamo, C., Helle, K.B. and Tota, B. (2004). Influence of vasostatins, the chromogranin A-derived peptides, on the working heart of the eel (*Anguilla anguilla*): Negative inotropy and mechanism of action. *Gen. Comp. Endocrinol.* 139,20–28.

Imbrogno, S., Cerra, M.C. and Tota, B. (2003). Angiotensin II-induced inotropism requires an endocardial endothelium-nitric oxide mechanism in the in vitro heart of *Anguilla anguilla*. *J. Exp. Biol.* 206,2675–2684.

Imbrogno, S., De Iuri, L., Mazza, R. and Tota, B. (2001). Nitric oxide modulates cardiac performance in the heart of *Anguilla anguilla*. *J. Exp. Biol.* 204,1719–1727.

Johansen, K. (1985). A phylogenetic overview of cardiovascular shunts. In *Cardiovascular Shunts: Phylogenetic, Ontogenetic and Clinical Aspects* (eds K. Johansen and W. Burggren), pp. 1–37, Raven Press, New York.

Jung, F., Palmer, L.A., Zhou, N. and Johns, R.A. (2000). Hypoxic regulation of inducible nitric oxide synthase via hypoxia inducible factor-1 in cardiac myocytes. *Circ. Res.* 86,319–325.

Jurevicius, J., Skeberdis, V.A. and Fischmeister, R. (2003). Role of cyclic nucleotide phosphodiesterase isoforms in cAMP compartmentation following beta2-adrenergic stimulation of ICa,L in frog ventricular myocytes. *J. Physiol.* 551,239–252.

Kleinert, H., Schwarz, P.M. and Forstermann, U. (2003). Regulation of the expression of inducible nitric oxide synthase. *Biol. Chem.* 384,1343–1364.

Knight, G.E. and Burnstock, G. (1993). Acetylcholine induces relaxation via the release of nitric oxide from endothelial cells of the garter snake (*Thamnophis sirtalis parietalis*) aorta. *Comp. Biochem. Physiol.* 106C,383–388.

Knight, G.E. and Burnstock, G. (1996). The involvement of the endothelium in the relaxation of the leopard frog (*Rana pipiens*) aorta in response to acetylcholine. *Br. J. Pharmacol.* 118,1518–1522.

Leite-Moreira, A.F. and Bras-Silva, C. (2004). Inotropic effects of ETB receptor stimulation and their modulation by endocardial endothelium, NO, and prostaglandins. *Am. J. Physiol.* 287(3), H1194–H1199.

Li, H., Wallerath, T. and Forstermann, U. (2002). Physiological mechanisms regulating the expression of endothelial-type NO synthase. *Nitric Oxide* 7,132–147.

Loewi, O. (1921). Uber humorale Ubertragbarkeit der Herznervenwirkung. *Pflugers Arch. Ges. Physiol.* 482,167–178.

Lough, J. and Sugi, Y. (2000). Endoderm and heart development. *Dev. Dyn.* 217,27–342.

MacDonnell, K., Tibbits, G.F. and Diamond, J. (1995). cGMP elevation does not mediate muscarinic agonist-induced negative inotropy in rat ventricular cardiomyocytes. *Am. J. Physiol.* 269,H1905–H1912.

Massion, P.B., Feron, O., Dessy, C. and Balligand, J.-L. (2003). Nitric oxide and cardiac function: Ten years after, and continuing. *Circ. Res.* 93(5), 388–398.

Massion, P.B., Pelat, M., Belge, C. and Balligand, J.-L. (2005). Regulation of the mammalian heart function by nitric oxide. *Comp. Biochem. Physiol.* 142A,144–150.

Mebazaa, A., Wetzel, R.C., Dodd-o, J.M., Redmond, E.M., Shah, A.M., Maeda, K., Maistre, G., Lakatta, E.G. and Robotham, J.L. (1998). Potential paracrine role of the pericardium in the regulation of cardiac function. *Cardiovasc. Res.* 40(2), 332–342.

Méry, P.F., Pavoine, C., Belhassen, L., Pecker, F. and Fischmeister, R. (1993). Nitric oxide regulates cardiac Ca^{2+} current. Involvement of cGMP inhibited and cGMP-stimulated phosphodiesterases through guanylyl cyclase activation. *J. Biol. Chem.* 268(35), 26286–26295.

Olson, R.K. (1998). The cardiovascular system. In *The Physiology of Fishes* (ed. H.D. Evans), pp. 129–154, CRC Press, Boca Raton, New York.

Oudit, G.Y. and Butler, D.G. (1995). Angiotensin II and cardiovascular regulation in a freshwater teleost, *Anguilla rostrata* Le Sueur. *Am. J. Physiol.* 269,R726–R735.

Paton, J.F., Deuchars, J., Ahmad, Z., Wong, L.F., Murphy, D. and Kasparov, S. (2001). Adenoviral vector demonstrates that angiotensin II-induced depression of the cardiac baroreflex is mediated by endothelial nitric oxide synthase in the nucleus tractus solitarii of the rat. *J. Physiol.* 531,445–458.

Paulus, W.J. and Bronzwaer, J.G. (2004). Nitric oxide's role in the heart: Control of beating or breathing?. *Am. J. Physiol.* 287,H8–H13.

Paulus, W.J., Vantrimpont, P.J. and Shah, A.M. (1994). Acute effects of nitric oxide on left ventricular relaxation and diastolic distensibility in man. *Circulation* 89,2070–2078.

Pellegrino, D., Palmerini, C.A. and Tota, B. (2004). No haemoglobin but NO: The icefish (*Chionodraco hamatus*) heart as a paradigm. *J. Exp. Biol.* 207,3855–3864.

Petroff, M.G.V., Kim, S.H., Pepe, S., Dessy, C., Marban, E., Balligand, J.-L. and Sollott, S.J. (2001). Endogenous nitric oxide mechanisms mediate the stretch dependence of Ca^{2+} release in cardiomyocytes. *Nat. Cell. Biol.* 3,867–873.

Pinsky, D.J., Patton, S., Mesaros, S., Brovkovych, V., Kubaszewski, E., Grunfeld, S. and Malinski, T. (1997). Mechanical transduction of nitric oxide synthesis in the beating heart. *Circ. Res.* 81,372–379.

Poss, K.D., Wilson, L.G. and Keating, M.T. (2002). Heart regeneration in zebrafish. *Science* 298,2188–2190.

Prendergast, B.D., Sagach, V.F. and Shah, A.M. (1997). Basal release of nitric oxide augments the Frank–Starling response in the isolated heart. *Circulation* 96,1320–1329.

Radomski, M.W. and Moncada, S. (1993). Regulation of vascular homeostasis by nitric oxide. *Thromb. Haemost.* 70,36–41.

Rumbaut, R.E., McKay, M.K. and Huxley, V.H. (1995). Capillary hydraulic conductivity is decreased by nitric oxide synthase inhibition. *Am. J. Physiol. Heart Circ. Physiol.* 37,H1856–H1861.

Ruud, J.T. (1954). Vertebrates without erythrocytes and blood pigment. *Nature* 173,848–850.

Schultz, R., Rassaf, T., Massion, P.B., Kelm, M. and Balligand, J.-L. (2005). Recent advances in the understanding of the role of nitric oxide in cardiovascular homeostasis. *Pharmacol. Ther.* 108,225–256.

Sekine, T., Kusano, H., Nishimaru, K., Tanaka, Y., Tanaka, H. and Shigenobu, K. (1999). Developmental conversion of inotropism by endothelin I and angiotensin II from positive to negative in mice. *Eur. J. Pharmacol.* 374,411–415.

Shah, A.M. (1996). Paracrine modulation of heart cell function by endothelial cells. *Cardiovasc. Res.* 31,847–867.

Shah, A.M., Spurgeon, H., Sollott, S.J., Talo, A. and Lakatta, E.G. (1994). 8-Bromo-cGMP reduces the myofilament response to Ca^{2+} in intact cardiac myocytes. *Circ. Res.* 74,970–978.

Somero, G.N., Fields, P.A., Hofman, G.E., Weinstein, R.B. and Kawall, H. (1998). Cold adaptation and stenothermy in Antarctic notothenoid fishes: What has been gained and what has been lost?. In *Fishes of Antartica, A Biological Overview* (eds G. Di Prisco, E. Pisano and A. Clarke), pp. 97–109, Springer-Verlag, Italy.

Stainier, D.Y. and Fishman, M.C. (1994). The zebrafish as a model system to study cardiovascular development. *Trends Cardiovasc. Med.* 4,207–212.

Sys, S.U., Pellegrino, D., Mazza, R., Gattuso, A., Andries, L.J. and Tota, B. (1997). Endocardial endothelium in the avascular heart of the frog: Morphology and role of nitric oxide. *J. Exp. Biol.* 200,3109–3118.

Thomas, D.D., Liu, X., Kantrow, S.P. and Lancaster, J.R. (2001). The biological lifetime of nitric oxide: Implications for the perivascular dynamics of NO and O_2. *Proc. Natl. Acad. Sci. USA* 98,355–360.

Timmermans, P.B., Wong, P.C., Chiu, A.T., Herblin, W.F., Benfield, P., Carini, D.J., Lee, R.J., Wexler, R.R., Saye, J.A. and Smith, R.D. (1993). Angiotensin II receptors and angiotensin II receptor antagonists. *Pharmacol. Rev.* 45(2), 205–251.

Tota, B. (1983). Vascular and metabolic zonation in the ventricular myocardium of mammals and fishes. *Comp. Biochem. Physiol.* 76,423–437.

338

Tota, B. (1989). Vascular and metabolic zonation in the ventricular myocardium of mammals and fishes. *Comp. Biochem. Physiol.* 76,423–427.

Tota, B., Acierno, R. and Agnisola, C. (1991). Mechanical performance of the isolated and perfused heart of the haemoglobinless antarctic icefish *Chionodraco hamatus* (Lonnberg): Effects of loading conditions and temperature. *Phil. Trans. R. Soc. Lond.* 332,191–198.

Tota, B., Amelio, D., Pellegrino, D., Ip, Y.K. and Cerra, M.C. (2005). NO modulation of myocardial performance in fish hearts. *Comp. Biochem. Physiol.* 142,164–177.

Tota, B., Cimini, V., Salvatore, G. and Zummo, G. (1983). Comparative study of the arterial and lacunary systems of the ventricular myocardium of elasmobranchs and teleost fishes. *Am. J. Anat.* 167,15–32.

Tota, B. and Gattuso, A. (1996). Heart ventricle pumps in teleosts and elasmobranchs. A morphodynamic approach. *J. Exp. Zool.* 275,162–171.

Tota, B., Pellegrino, D., Gattuso, A., Mazza, R. and Imbrogno, S. (2000). The avascular heart of the frog provides the key to endocardial endothelium nitric oxide signal-transduction. In *The Biology of Nitric Oxide* (eds S. Moncada, L.E. Gustafsson, N.P. Wiklund and E.A. Higgs)61, Portland Press, London.

Nitric oxide and histamine in hibernation and neuroprotection

Pertti Panula[1], Giuseppina Giusi[2], Rosa Maria Facciolo[2], Tina Sallmen[3], Minamaija Lintunen[3] and Marcello Canonaco[2]

[1]*Neuroscience Center, Institute of Biomedicine/Anatomy, Biomedicum Helsinki, University of Helsinki, POB 63, Haartmaninkatu 8, 00014 Helsinki, Finland*
[2]*Comparative Neuroanatomy and Cytology Laboratory, Ecology Department, University of Calabria, Arcavacata di Rende (Cosenza), 87030, Italy*
[3]*Department of Biology, Abo Akademi University, Turku, Finland*

Abstract. The short-lived free radical gas nitric oxide (NO) and the histaminergic neuronal systems are widely distributed in most brain regions, thereby being involved in various homeostatic and neurobiological activities as well as neurodegenerative processes. In the case of the first neuronal system, its production relies on three specifically and dimerically active NO synthase (NOS) enzyme isoforms. An enzymatic system related to the activation of its two major neuromediators (glutamate and gamma-aminobutyric acid [GABA]), which are co-localized to NOergic fibers throughout the different brain regions. Consequently, NOergic signaling deriving from this complex neuronal system tends to strengthen its role in the successful execution of determinant behaviors such as sensori-motor tasks. It is worthwhile noting that the activation of endothelial NOS in blood vessels seems to be beneficial for the maintenance of cerebral blood flow not only in pathological syndromes but also in physiological conditions, such as the torpor state of hibernators, that manifest ischemic-like damage. Although the free radical gas modifies the circadian sleep–wake cycle, this cycle, which in hibernators is replaced by ultradian rhythms, does not appear to be linked to NOergic influences. It seems that the phylo-genetically old group of histaminergic neurons, which activates three distinct G-protein-coupled receptors in the brain, is more directly involved in the regulation of hibernation. This article reviews the modulatory role of NOergic and histaminergic neuronal systems in hibernation and in other physiological states that involve neuronal plasticity. Potentially important protective roles of these systems in neurodegenerative diseases and cerebral ischemia are also discussed.

Keywords: nitric oxide; NO synthase; histamine; hibernation; cerebral ischemia; homeostasis; arousal; torpor; histaminergic subtype receptors; neuroprotection; neurogenesis; sleep-wake cycle; mitogen-activated protein kinases; tuberomamillary nucleus; N-methyl-D-aspartate; hypothalamic suprachiasmatic nucleus; GABAA receptor; hippocampus; amygdala; neurodegenerative diseases.

Introduction

The discovery of the intercellular messenger nitric oxide (NO), an endothelium-derived relaxing factor (Ignarro *et al.*, 1987) capable of

E-mail: pertti.panula@helsinki.fi (P. Panula).

ADVANCES IN EXPERIMENTAL BIOLOGY
VOLUME 01   ISSN 1872-2423
DOI: 10.1016/S1872-2423(07)01015-0

promoting numerous cerebral neuro-signaling functions (Garthwaite and Boulton, 1995), has opened a new dimension in the concept of neural communication. NO is an inorganic, unstable, and short-lived free radical gas capable of modulating a variety of biological processes (Ohkuma and Katsura, 2001). Studies have shown that neuronal NO is coupled to local blood flow (Iadecola, 1997) and synaptic plastic functions such as long-term potentiation (LTP) and depression (Haley, 1998; Shibuki and Okada, 1991). Because of its gaseous nature and short lifetime, NO displays several unusual features – this gas is not stored in vesicles, it is freely diffusible after synthesis, and its action is confined to the site of production. In addition, the formation of NO is related to neuronal phosphorylative cascade processes as indicated by the anatomical segregation of brain sites involved with its production as well as the expression of its second messenger, cyclic guanosine monophosphate (cGMP) (Ignarro, 2002).

The synthesis of this gas is regulated by the dimerically active NO synthase (NOS) enzyme, which is associated with calmodulins (CaMs). These isoforms are distributed in a heterogeneous manner and show different catalytic and inhibitory properties (Alderton et al., 2001). The first type of synthase enzyme to be synthesized was termed neuronal NOS (nNOS; also called NOS1 or NOS I), which was shown to be the predominating isoform in neuronal tissue. The others are the inducible type (iNOS; also called NOS2 or NOS II), which is characteristic of a wide range of cells and tissues, and the endothelial type (eNOS; also called NOS3 or NOS III), which is the major isoform of vascular cells (Bredt and Snyder, 1990). The heterogeneous distribution of NOS-containing neurons in the central nervous system (CNS) indicates that NO is able to act as either a neurotransmitter or neuromodulator in cGMP-dependent target cells (Garbers, 1992). Recently, NO has also proven to be a principal modulator of synapse-specific plasticity events (Namiki et al., 2005), which appear to be tightly linked to the promotion of neurogenesis – a phenomenon implicated in the recovery of CNS neurodegenerative processes (Reif et al., 2004). In this case, it is the suppression of NO production that activates its neurogenic role, as displayed by the large number of cells generated especially in the olfactory subependyma and dentate gyrus of the hippocampus (Packer et al., 2003) when the activity of iNOS and nNOS is blocked. At the same time, similar studies, in combination with the detection of dense hypothalamic NO-producing neurons, have led to the proposal of a NOergic modulatory role linked to homeostatic functions. This correlation is strongly supported by the finding that hypothalamic NO-enriched neurons maintain in constant equilibrium hypoxia-inducible factors (Hirota et al., 2004) that are important for cellular energy supply and demand (Ruiz-Stewart et al., 2004) via the activation of mitogen-activated protein kinase (MAPK) pathways, especially when torpor terminates and a normometabolic rate is again established (Zhu et al., 2005).

On the other hand, few data have been reported on the role of NO in hibernation, probably because greater attention has been directed to other neuromediators, i.e., the histaminergic system as the main neuronal system of hibernators (Sallmen et al., 1999). It is well known that brain histaminergic neurons send widespread, diffuse axons from the tuberomammillary nucleus (TM) neurons to the different brain areas (Ericson et al., 1991a). Histamine is able to interact with the four subtypes of histamine receptors: H_1, H_2, H_3, and H_4, of which the H_1 and H_2 receptors are located postsynaptically. The H_3 receptors are located on the somata and axon terminals of histaminergic neurons, where they serve as either autoreceptors, modulating synthesis and release of histamine, or as pre- and postsynaptic sites of the numerous brain regions (Hill et al., 1997). The H_4 receptor, which was only recently identified, is homologous to the H_3 receptor and displays mostly chemotaxis activities. In addition to activating several coupled receptors, the first three receptor subtypes are coupled to G-proteins and hence are able to activate cyclic adenosine monophosphate (cAMP) and MAPK pathways (Drutel et al., 2001), in many cases via interaction with other neurotransmitter systems such as gamma-aminobutyric acid (GABA) and/or glutamate (Xu et al., 2004). On the basis of the widespread projections of the histaminergic pathways, histamine appears to be involved in numerous neurophysiological functions, namely brain energy metabolism, locomotor activity, feeding behaviors, vestibular function, and analgesia (Brown et al., 2001; Haas and Panula, 2003). Moreover, changes in brain histamine levels are evident in cerebral ischemia, as shown by occlusion of the middle cerebral artery in the rat being responsible for neuronal disorders in the whole brain (Adachi, 2005). Consequently, the major neuronal role of H_1 and H_3 receptors levels in ischemic and hibernating conditions may represent a relevant physiological response to neuronal stress and anoxia, and opens the possibility for the therapeutic use of histamine and/or histamine receptor ligands to treat some of the more serious motor impairments such as Alzheimer's and Parkinson's diseases.

342

NO synthesis and localization in the CNS

Synthesis

NO is synthesized via the activation of NOS enzymes, and due to its rather short half-life (only a few seconds) it does not accumulate in specialized cellular vesicles (Bredt and Snyder, 1991). The tightly bound cofactors of these widely distributed specific enzymes (Marletta, 1994) allow them to catalyse the aerobic transformation of L-arginine, in the presence of NADPH, to L-citrulline plus NO (Fig. 1). The promotion of this reaction is controlled by the distinct chemical and physical properties of the different NOS isoforms.

Moreover, a widespread nomenclature based on earlier observations has classified nNOS and eNOS as constitutive isoforms, whereas inducible reactions have been attributed to iNOS expression. This nomenclature remains somewhat questionable, since under physiological conditions iNOS might serve as a constitutive isoform in some cells, while in damaged tissues it may function as an inducible element. This is similar to the formation of nNOS and eNOS, in which the genes coding for them may be expressed not only under physiological conditions but also under pathological states. Interestingly, the activity of these NOSs is tightly correlated to the presence of Ca^{2+}. Ca^{2+} is required for nNOS and eNOS functions, while it is not essential for iNOS since CaM spontaneously binds tightly to this isoform (Abu-Soud and Stuehr, 1993; Sase and Michel, 1997). Although all three isoforms are typical of neurons (Nathan and Xie, 1994), it is the very high activity of nNOS associated with NO production in the different CNS areas that constitutes the main target of this tissue. nNOS-positive cells are densely distributed in cerebral regions such as the hypothalamus, cerebellar cortex, olfactory bulb, and specific nuclei in the brain stem (Nazli and Thippeswamy, 2002), which is in-line with a possible role for nNOS in the processing of specific sensory and/or motor signals, neuroprotection, and the regulation of

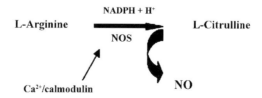

Fig. 1. Aerobic oxidation of L-arginine through a two-step electron transfer catalysed reaction, supplying the end products L-citrulline and NO.

cerebral blood flow in neurotoxic and ischemic phenomena in these regions (Estrada et al., 1993; Moro et al., 2004).

The activation of NOS throughout the CNS is assured by its effects on receptors for glutamate, a major presynaptically released neuromediator. NOS binds to both postsynaptic N-methyl-D-aspartate-type glutamate receptors (NMDAR) and non-NMDA glutamate receptors – interactions that lead to an influx of Ca^{2+} through receptor-operated ion channels (Garthwaite, 1991). However, the role of NMDA-dependent phosphorylation on NOS activity still remains to be defined, because glutamate has been shown to be related to NMDAR-induction of nNOS phosphorylation in a dose-dependent manner in most brain regions in the past few years only (Cardenas et al., 2000). GABA, another major neuronal system with inhibitory functions, has been shown to coexist with glutamatergic and NOergic neurons in different brain areas such as the striatum and cortex. Glutamate and NO exert excitatory effects on firing activity, while GABA exerts an inhibitory effect (Liu et al., 2005). The coexistence of these three neuronal systems in distinct cerebral sites seems to underlie not only the abnormal NO-induced locomotor behaviors that follow different forms of chemically induced seizures (Shih et al., 2004), but also the potentiation of neuroendocrine secretory processes via cGMP-independent mechanisms (Ozaki et al., 2000). In blood vessels, it is the activation of acetylcholine that, through the binding of endothelial muscarinic receptors, facilitates the entry of Ca^{2+} and subsequently promotes eNOS activity. Hence, the early formation of this enzyme seems to be notably beneficial for maintaining normal cerebral blood flow, as shown by low eNOS transcriptional levels in cerebral ischemic injuries (Huang et al., 1996) following hypoxic conditions. These events are typical of pathological syndromes and of hibernation, which manifest similar injuries during the traumatic fluctuation of cerebral blood flow during torpor–arousal states (Stenzel-Poore et al., 2003).

Localization

Identification of the major NO-producing sites in the different areas of the CNS was established through the evaluation of NOS-positive neurons. The greatest amount of nNOS-positive cells is typical of hippocampal and cerebellar areas during the postnatal developmental period (Chung et al., 2004), when they are required to promote neurogenic events. In the case of the cerebellum, the deep molecular layers, namely the basket and granular cells, contain the highest levels of nNOS-positive somata and dendrites (Baader and Schilling, 1996). A similar trend is also typical of the hippocampus, in which consistently dense numbers of stained nNOS-positive cells appear to characterize the pyramidal and granular cell layers of the dentate gyrus and oriens-pyramidalis areas (Weiss et al., 1998), respectively. nNOS-positive cells have also been reported for mesencephalic areas, pons, medulla, and laterodorsal and pedunculopontine tegmental nuclei (Dun et al., 1994; Yousef et al., 2004). In these areas, dense nNOS fibers characterize all cortical layers and the mesencephalic granular fields. Even the hypothalamus, from a functional point of view, displays a strong labeling of nNOS-immunoreactive cell bodies in the preoptic region, the supraoptic and paraventricular nuclei, the lateral hypothalamic area, the ventromedial and arcuate nuclei, and areas of the mammillary region (Nylen et al., 2001; Yamada et al., 1996). The other two isoforms (eNOS and iNOS) display a widespread distribution pattern, with eNOS being primarily located in endothelial cells of cerebral vessels (Siles et al., 2002), while iNOS is typical of glial cells and damaged neurons of inflammatory and neurodegenerative processes (Heneka and Feinstein, 2001).

Neurobiological role of NO

NO and cerebral ischemia

Cerebral ischemia is a peculiar phenomenon that causes an overactivation of membrane receptors, which in turn leads to an extracellular accumulation of glutamate plus an increase in the intracellular Ca^{2+} concentration. This in turn induces an nNOS activity that is responsible for damage to lipids, proteins, and nucleic acids, *i.e.,* an overall reduction of cellular energy sources that is typical of necrosis (Liu, 2003). The role of nNOS in similar cellular events is supported by the elevated levels of NO in specific CNS sites following cerebral ischemia/reperfusion injury, traumatic head injury, and spinal cord damage (Cherian et al., 2000). In this context, while nNOS and iNOS are mainly responsible for the traumatic profiles of the ischemic-induced events, eNOS instead exerts a protective role (Moro et al., 2004). Recent studies have also suggested that eNOS assures such a role by enhancing vasodilatatory activities as well as reducing the number of blood elements undergoing aggregation, which are regarded as posing significant risks to cardio-circulatory performance. eNOS-knockout mice that are devoid of this type of isoenzyme seemed to suffer significant damage following reversible ischemic events after treatment with a selective eNOS inhibitor (Chen et al., 2005). Conversely, an nNOS-dependent NO overproduction during the early stages of ischemia may be detrimental to neurons by accelerating cell metabolism and/or damaging DNA. In particular, the activation of this specific isoform in brain regions such as cerebellar granule cells by oxygen and glucose deprivation, which is typical of anoxic and glucopenic conditions, is associated with brain injury (Scorziello *et al.*, 2004). Only after decreased NO production following treatment with its inhibitor L-NAME (N^G-nitro-L-arginine methyl ester), an increase in mitochondrial oxidation and an improvement in cell survival detected.

At present, studies are beginning to show that during brain insults, NO is capable of inducing an important recovery of damaged brain areas by switching from a proliferation to a differentiation state or neurogenesis (Gibbs, 2003). This phenomenon appears to be under the influence of physiological factors such as environmental stimulation, aging, and physical exercise, which appears to be a prime candidate for the onset of neurogenesis. Stress and ischemia rather appear to be the prime pathological stimuli (Gage *et al.*, 1998; Kee *et al.*, 2001). Thus, while neurogenic mechanisms mainly operate during CNS developmental stages, Gage *et al.* (1998) have unveiled functional evidence of neurogenesis in adults also, especially after brain injuries such as seizures or stroke (Parent, 2003). Inhibitory cellular mechanisms blocking neurogenesis in the subgranular zone of dentate gyrus, one of the two main neurogenic loci expressing nNOS (Moreno-López *et al.*, 2000) in the adult brain, have been correlated with pathogenic events of depression and Alzheimer's disease (Wen *et al.*, 2004). Although most neurogenesis studies in adults have focused on proliferation, the analysis of survival rates could also be important because NO can act as a bifunctional regulator of apoptosis (Kim *et al.*, 1999). In fact, nNOS-dependent NO production appears to decrease neurogenesis or to act as an antiproliferative molecule, whereas NO derived from iNOS and eNOS seems to stimulate this activity (Reif *et al.*, 2004; Zhu *et al.*, 2003).

NO and homeostasis

At the hypothalamus–pituitary–adrenal axis, it is the modulatory activity of certain hormones, such as corticotropin-releasing hormone, that conserves normal homeostatic parameters during stressful conditions (McCann *et al.*, 2000). The actions of corticotropin-releasing hormone on homeostasis seem to be explained via the interaction of c-fos in discrete hypothalamic magnocellular neurons of paraventricular and supraoptic nuclei (PVN and SON). In this context, NO assumes a determinant role: these hypothalamic sites are not only key targets of

346

NO production, as shown by the high concentration of NOergic fibers in PVN and SON, but are also involved in the expression of the stress-induced hormones vasopressin and oxytocin (Vincent *et al.*, 1994). Blockade of these fibers by NOS antagonists strongly conditioned corticotropin (ACTH) and corticosterone responses in stress-induced paradigms (Bugajski *et al.*, 2004).

In line with these neuroendocrine-dependent homeostatic functions, the sleep–wake cycle also constitutes an important physiological condition. Hypothalamic centers are involved in non-rapid eye movement sleep (NREM) via NOergic-dependent modifications of fluid balance as well as energy production and thermoregulatory processes (Steiner *et al.*, 2002), making NO an important element in such homeostatic functions. It is known that the synchronization of extrinsic cellular activities within the multicellular intrinsic oscillator of the hypothalamic suprachiasmatic nucleus and of other sleeping centers (retino-hypothalamic tract and pontine areas) plays a key regulatory role in vigilance and wakefulness of paradoxical sleeping states (Hars, 1999; Morin, 1994) – a role that heavily depends on NMDA receptors activating hypothalamic sleeping centers in a Ca^{2+}-dependent manner (Hars, 1999; Hauser *et al.*, 2005). Moreover, a beneficial role of the NOergic system was reported for some syndromes that involve altered hypothalamic effects and REM sleeping disorders (such as epilepsy during the absence of seizures or convulsions, and Parkinson's disease) – as shown by the antiepileptic actions of high NO levels (Faradji *et al.*, 2000). Previously our laboratory demonstrated that using L-NAME to block NO production resulted in elevated levels of benzodiazepine receptor (Fig. 2), a major $GABA_A$ receptor component involved in the suppression of anxiogenic and epileptic behaviors that favor sleeping activities (Facciolo *et al.*, 1996).

NO and hibernation

Hibernation is a unique physiological condition that allows several mammals to survive throughout periods of food shortage by a series of metabolic adaptations. As a consequence, these animals display highly elastic morphological, physiological, and behavioral phenotypic features. Neuroprotective adaptations such as hypothermia, metabolic suppression, immunosuppression, and increased antioxidant defense may be essential for successful hibernation, since cerebral blood flow fluctuates dramatically, falling 80–90% during torpor (*i.e.,* to ischemic-like levels) and returning to normal levels in a reperfusion-like manner, every 2 to 3 weeks during brief, periodic arousals (Osborne and Hashimoto, 2003).

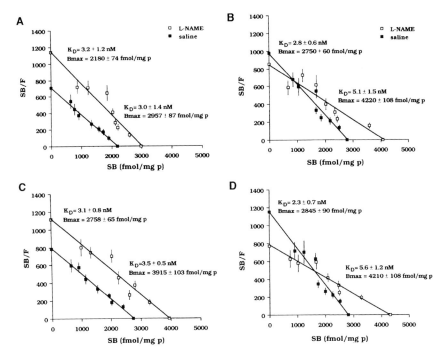

Fig. 2. The inhibiting effects of N^G-nitro-L-arginine methyl ester (L-NAME) on substrate binding activity of the benzodiazepine receptor in the absence (A, B) or presence (C, D) of 20λμM GABA in some amygdalar (A, C) and hippocampal (B, D) areas of YOS (Yoshida) rats. Error bars indicate ±SEM (standard error of mean). F, free substrate. SB, bound substrate (femtomol per mg receptor protein). B_{max}, maximum binding. K_D, dissociation constant. (The authors would like to thank Facciolo *et al.*, 1996 for having granted permission to use their data.)

A typical hibernating session of mammals is characterized by extended bouts of entry, torpor, and arousal states (Ueda and Ibuka, 1995) that can last from a few days to up to 5 weeks in some species. Torpor seems to depend strongly on the overall metabolic activity, and if the basal body temperature rises to approximately 36°C, the animals are maintained in these arousal bouts for up to 24λh before reentry into the torpor state (Carey *et al.*, 2003). The fact that some seasonal hibernators undergo frequent synaptogenesis processes underlies the necessity for the hibernating animals to renew neuronal molecular elements. This activity very likely derives from extracellular stimuli, such as environmental stressors, which are capable of eliciting the interplay of multiple signaling pathways such as those belonging to MAPK (Zhu *et al.*, 2005). The brain

348

of hibernators is able to tolerate traumatic neuronal injury *in vivo* (Zhou *et al.*, 2001) as well as oxygen glucose deprivation *in vitro* (Frerichs and Hallenbeck, 1998). Hence, hibernation is considered a useful model to study neuroprotection mechanisms and may also lead to the discovery of novel strategies and therapeutics for the treatment of stroke, traumatic brain injury, and neurodegenerative diseases (Drew *et al.*, 2001). It is surprising that only scarce data regarding the role of NO on hibernation is available, especially given the known role of NO in sleeping behavior, as discussed above. However, apart from studies suggesting that NOergic fibers control the circadian sleep–wake cycle of hibernators (Zhu *et al.*, 2005), the few data on this role derive from the fact that, in spite of NO being able to modify the sleep–wake cycle, this cycle, which in hibernators is replaced by ultradian rhythms characterized by NREM sleep (Palchykova *et al.*, 2002; Ruby, 2003), is very likely regulated by other neuroreceptor systems, *i.e.*, histaminergic (Sallmen *et al.*, 1999), through a histamine-dependent variation of hypothalamus-pituitary-adrenal axis responses (Lantoine *et al.*, 1998).

The histaminergic system and its roles in the brain

The histaminergic system is a phylogenetically old group of neurons that projects to all parts of the brain of almost all animal species studied. Its histamine content varies greatly between species (Almeida and Beaven, 1981; Reite, 1972). In vertebrates, such as fish (Brodin *et al.*, 1990; Ekstrom *et al.*, 1995, 1998), rodents (Panula *et al.*, 1984; Takeda *et al.*, 1984), and humans (Panula *et al.*, 1990), the histamine-producing neurons are located in the TM of the posterior hypothalamus.

The major terminal areas of the histaminergic projections originating from the TM are also slightly different in different species, but they cover essentially all areas of the CNS (Inagaki *et al.*, 1988; Panula *et al.*, 1989). This is in agreement with the widespread distribution of the three histamine receptors in the brain (Haas and Panula, 2003). In all mammals studied, some areas are almost always well innervated by histaminergic fibers. Thus, for example, the cerebral cortex, amygdala, substantia nigra, and striatum receive at least moderate or dense innervation. The density of nerve fibers in the mammalian hippocampus and thalamus varies, and the most distal parts of the CNS, the retina and spinal cord, also receive histaminergic fibers from TM. Reciprocal innervations of the histaminergic and the other aminergic cell groups are evident in most animal species (Kaslin and Panula, 2001).

The projections of TM neurons are widespread, and afferents to this area also arrive from many different brain areas (Ericson *et al.*, 1991a). As a consequence, the histaminergic neuronal system is well positioned to manage general regulatory events in the whole brain, which are important in *e.g.*, sleep, hibernation, and shortages of nutrients and/or oxygen. The brainstem innervation to TM arises mainly from the adrenergic cell groups C1–C3, noradrenergic groups A1–A3, and serotonergic groups B5–B9, whereas the locus coeruleus and the dopaminergic neuron groups of the substantia nigra and ventral tegmental area send only few fibers to TM (Ericson *et al.*, 1989). In this context, the histaminergic system is well connected to other major neuronal systems and hence is able to regulate whole-brain activity.

The histaminergic TM neurons contain several co-transmitters and neuromodulatory peptides. The GABA-synthesizing enzyme glutamate decarboxylase (GAD; Takeda *et al.*, 1984; Vincent *et al.*, 1982) and GABA itself are found in TM neurons (Airaksinen *et al.*, 1992; Ericson *et al.*, 1991b), and this renders TM neurons one of the few types of GABAergic cells that reach almost all areas throughout the CNS. The neuropeptides galanin, thyrotropin-releasing hormone (TRH), pro-enkephalin-derived peptides, and substance P are also found co-localized in histamine-producing TM neurons, although species differences exist in their co-localization. The functional significance of TM neuropeptides is not clear (Airaksinen *et al.*, 1992; Staines *et al.*, 1986), but cells expressing them form a heterogeneous subgroup among TM neurons. The majority of TM neurons also express adenosine deaminase (Senba *et al.*, 1987; Staines *et al.*, 1986), which may be an important factor responsible for the inactivation of adenosine, which is formed during some physiological conditions such as wakefulness.

The essential amino acid L-histidine is taken into multipolar TM neurons, and decarboxylated to histamine by the specific enzyme histidine decarboxylase (HDC). HDC is present not only in the cell bodies of TM neurons but also in the distal branches of axons, suggesting that the whole process of histamine synthesis can take place in target areas of the CNS. Histamine is then taken up in storage vesicles by vesicular monoamine transporter 2 (VMAT-2). After being released, histamine is methylated by histamine *N*-methyltransferase to *tele*-methylhistamine, a metabolite that does not show any affinity to the known histamine receptors, and does not have histamine-like activity at ion channels. Although some reports indicate that histamine is taken up by glial cells (Huszti, 1998), but high-affinity uptake of histamine in neurons has been difficult to demonstrate. In the absence of a high-affinity uptake system for histamine, extracellular methylation is the principal inactivation mechanism, as histaminase activity in the CNS is low (Haas and Panula, 2003).

The histaminergic neurons in TM are pacemaker cells that fire at a slow but regular rate (Haas and Panula, 2003). The firing pattern of TM neurons is dependent on the behavioral state in cats and rodents. Their activity is high during waking and attention, and low or absent during sleep. The blocking of histaminergic activity during sleep is thought to be mediated mainly by GABAergic input from the ventrolateral preoptic area (VLPO), which is active during slow-wave sleep (Sherin et al., 1996). The TM nucleus is mutually interconnected with all the aminergic, orexinergic, and presumably other nuclei in the mesencephalon and diencephalon.

Histamine acts through ionotropic receptors in the insect eye (Hardie, 1989) and in molluscs (Gisselmann et al., 2002; Zheng et al., 2002), but specific ion-channel receptors for histamine have not been identified with certainty in mammals. Of the four known G-protein-coupled histamine receptors, the H_1 receptor is important in allergy, the H_2 receptor regulates gastric acid secretion, and the H_3 receptor is a presynaptic auto- and heteroreceptor that regulates the release of many transmitters, especially in the brain. It was only recently found that the H_4 receptor is based on homology with H_3 receptor. Its functions include chemotaxis, and many H_3 receptor ligands have also affinity for the H_4 receptor.

The H_1 receptor is a 486–491 amino acid protein encoded by an intronless gene (Yamashita et al., 1991). It is coupled to the $G_{q/11}$ protein and phospholipase C (PLC). The H_2 receptor was first cloned in dogs (Gantz et al., 1991), and then in a number of other species. Like the H_1 receptor, it is coupled to the G_s protein and protein kinase A (PKA) and is encoded by an intronless gene of 358–359 amino acids. Several isoforms of the H_3 receptor (Coge et al., 2001; Drutel et al., 2001) have been identified. They consist of 326–445 amino acids and are derived from a single gene (Lovenberg et al., 1999) by alternative splicing. As can be deduced from the H_3 receptor heterogeneity, its gene structure is more complex than that of the previously known H_1 and H_2 receptors. Interestingly, the various isoforms of the histamine H_3 receptor are differently coupled to second messenger systems (Drutel et al., 2001). The H_3 receptor is coupled to $G_{i/o}$, it displays significant constitutive activity, and consequently it controls histamine release and synthesis (Morisset et al., 2000). The pharmacology of histamine receptors and their transduction mechanisms have been discussed in several extensive reviews (Bakker et al., 2002; Hill et al., 1997; Schwartz et al., 1991). H_1 receptors are known to mediate excitatory actions and the classic antihistamines have been shown to act as H_1 antagonists, so that their sedative effects are now known to be mediated through specific receptor actions that are particularly abundant in telencephalic areas like the cortex and the hippocampus. The excitatory actions are mediated by $G_{q/11}$-protein activation of phospholipases, particularly PLC, which leads to the formation of diacylglycerol (DAG) and inositol trisphosphate (IP$_3$). The latter releases Ca^{2+} from intracellular stores, which activates several important Ca^{2+}-dependent processes in neurons. H_2 receptors are coupled to the G_s protein, adenylyl cyclase, and PKA, which phosphorylates proteins and activates the transcription factor CREB (cyclic adenosine monophosphate response element binding protein), suggesting their possible involvement in plasticity through gene regulation. The direct effect of H_2 activity on neuronal membranes is usually excitatory. H_3 receptors are located on neuronal cell bodies, dendrites, and axon terminals, where they inhibit histamine synthesis/release as well as the release of other transmitters, *e.g.*, glutamate, noradrenaline (norepinephrine), and GABA, very likely through innervation of the acetylcholine system (Xu *et al.*, 2004). H_3 receptors are coupled to G_q and high-voltage-activated Ca^{2+} channels, which regulate transmitter release (Haas and Panula, 2003). H_3 receptors can also be coupled negatively to cAMP, and activate the MAPK pathway (Drutel *et al.*, 2001), a mechanism that is important in plasticity activities that are associated with neuronal damage and recovery.

Histamine, hibernation, and neuroprotection

In addition to activating several G-protein-coupled receptors, histamine causes direct facilitation of the NMDA glutamate receptor channel through its polyamine modulatory site (Bekkers, 1993; Vorobjev *et al.*, 1993). This pH-sensitive action is inhibited by spermidine, an interaction that allows histamine to clearly modulate neuronal function in such a manner that it plays a central role in synaptic plasticity. Interestingly, synchronous bursting in the hippocampus is strongly promoted by histamine (Yanovsky and Haas, 1998), a natural stimulus that is specific for (LTP) and as a consequence turns out to be a major cellular model of synaptic plasticity (Bliss and Collingridge, 1993). H_1 receptor stimulation increases intracellular Ca^{2+} concentration and activates protein kinase C (PKC), which regulate the induction threshold of long-term changes in synaptic efficacy. H_2 receptors activate cAMP formation and PKA, which are involved in the maintenance of this phenomenon (Selbach

352

et al., 1997). H$_3$ receptor activation reduces the release of many neuro-transmitters, including glutamate, which in this case turns out to be the main regulator of synaptic plasticity in telencephalic areas, particularly in the hippocampus and the striatum (Brown and Haas, 1999; Doreulee *et al.*, 2001).

Histamine is required for arousal states (for references, see Haas and Panula, 2003). For example, lesioning of the posterior hypothalamus causes hypersomnia, and injections of the GABAergic agonist muscimol in this area strongly affect wakefulness in cats (Lin *et al.*, 1989). The releasing of histamine in target brain areas varies in parallel with his-taminergic neuron firing (Mochizuki *et al.*, 1991) during sleep and wake-fulness states. Studies have shown that H$_1$ receptor antagonists can increase slow-wave sleep. In fact, mice devoid of the histamine-synthe-sizing enzyme HDC, and thus neuronal histamine, display deficits in waking, attention, and exploratory behavior when placed in a new en-vironment (Parmentier *et al.*, 2002). In addition to its direct effects, his-tamine can also regulate higher cortical and hippocampal activity through the stimulation of the nucleus basalis cholinergic neurons (and subsequently cortical areas) as well as the serotonergic neurons in the dorsal raphe nucleus. These effects are mediated by H$_1$ receptors (Brown *et al.*, 2002; Cecchi *et al.*, 2001) that are expressed by the target neurons (Lintunen *et al.*, 1998). The histaminergic neuron activity and/or brain histamine release show differences in several states that involve altered consciousness, like sleep (Parmentier *et al.*, 2002), narcolepsy (John *et al.*, 2004), and hibernation (Sallmen *et al.*, 1999). By using golden-mantled ground squirrels, Sallmen *et al.* (1999) showed that histamine levels in all studied brain areas were strongly elevated in mid-bout hibernating an-imals as compared to euthermic control animals. In order to show whether this was due to accumulation of the transmitter in the nerve terminals or whether the active release and metabolism of histamine were instead responsible for such hibernating events, the first histamine me-tabolite *tele*-methylhistamine was also evaluated. Levels of this metabo-lite, which have been closely correlated with histamine turnover rate in the brain, were also clearly elevated in hibernating animals. Many of the brain areas proposed to play key roles in hibernation (*e.g.*, the septum, hippocampus, hypothalamus, cortex, and raphe nuclei) displayed a greatly increased density of histamine-immunoreactive nerve fibers in hibernating animals. No new cell groups were found to synthesize his-tamine in hibernating animals, and the expression level of HDC mRNA in the TM neurons was not different from that of euthermic animals. Increased histamine levels occurred in neurons rather than in, for example, mast cells, which can also be found in the brain. The global increase of brain histamine level might suggest a general, rather than circuit-specific function for histamine in hibernation. However, when applied to the hippocampus, one of the critical sites involved in the neural circuit believed to control hibernation, just before the expected ending of the hibernation bout, histamine significantly prolonged the remaining bout length (Sallmen et al., 2003a). Although the mechanism of action of histamine during hibernation is still open, changes in receptor function might provide clues to the importance of a particular mechanism involving only some of the known histamine receptor(s). In golden-mantled ground squirrels, the expression of H_3 receptor mRNA is increased in the caudate nucleus, putamen, and cortex concomitantly with an increase in receptor ligand binding (Fig. 3), whereas in the hippocampus H_3 mRNA expression is decreased (Sallmen et al., 2003b, c).

The expression of H_1 and H_2 receptors in the hippocampus is increased during hibernation in this species (Sallmen et al., 2003b). The results lend support to a working hypothesis that histamine in the hippocampus is involved in maintaining hibernation by activating H_1 and H_2 receptors. Experiments with specific receptor agonists and inverse agonists would be good experimental tools to cast light on the detailed mechanism, but no data are available yet. The role of endogenous histamine could be investigated by inhibiting histamine synthesis with α-fluoromethyl-histidine, an irreversible inhibitor of histamine synthesis. The possibility also exists that the status of L-histidine transport through the blood–brain barrier, entry into the neurons, and subsequent decarboxylation to histamine under the conditions of hibernation are altered. The availability of L-histidine regulates histamine synthesis, as HDC is not saturated during normal conditions.

It is worthwhile noting that changes in brain histamine levels are also evident in brain ischemia: occlusion of the middle cerebral artery in the rat increases histamine levels in the striatum and cortex (Adachi, 2005), and histamine levels are also elevated in brain ischemia in primates (Subramanian et al., 1981). In kainic acid (KA)-induced limbic seizures, histamine levels show a more complex pattern of changes: $6\lambda h$ after systemic KA injection, there is a rapid elevation of histamine level in the piriform cortex, amygdala, hippocampus, and striatum followed by a return to normal levels (Lintunen et al., 2005). Subsequently, a second delayed increase in histamine level, which was also detected in neurons, seemed to occur within a week. Lowering the brain histamine level increases neuronal death in hippocampal area CA1 in a four-vessel ischemia model in rat (Adachi, 2005), suggesting that adequate brain

354

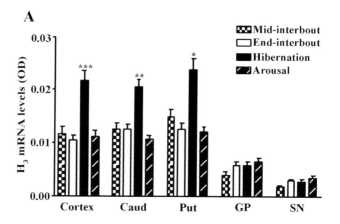

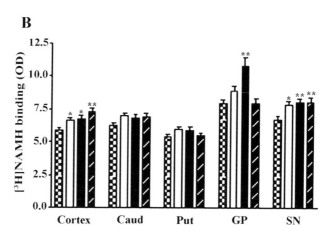

Fig. 3. Comparisons of H_3 receptor mRNA expression and binding across the hibernation bout. Graphs showing changes in (A) H_3 receptor mRNA expression and (B) H_3 receptor binding to tritiated N-α-methyl-histamine (NAMH) in the ground squirrel brain across the hibernation bout. One-way ANOVA (analysis of variance) with the Bonferroni *post hoc* test (*$p < 0.05$, **$p < 0.01$, ***$p < 0.001$). Mid-interbout $n = 5$, end-interbout $n = 4$, hibernation $n = 4$, arousal $n = 3$. Caud, caudate nucleus. Put, putamen. GP, globus pallidus. SN, substantia nigra, pars reticulata. OD, optic density. Error bars indicate $\pm$SEM (standard error of mean). (The authors would like to thank Sallmen *et al.*, 2003b for having granted permission to use their data.)

355

histamine levels may be beneficial for neuroprotection in ischemia. Post-ischemic treatment of gerbils with histamine failed to protect from neu-ronal death, whereas in one study postischemic treatment of the rats with L-histidine, a precursor of histamine, had a beneficial effect on neuronal death in striatum in a middle cerebral artery occlusion study (Adachi, 2005). Due to L-histidine being rapidly decarboxylated to histamine, this beneficial effect may be consequent to the increased brain histamine lev-els. A fairly complex pattern of changes in central histamine receptor expression and ligand binding follows a four-vessel ischemia in some brain areas of the rat, such as the striatum. In this case lower H_1 receptor mRNA expression, and higher H_2 and H_3 receptor levels than in the controls seem to be typical of these animals (Lozada et al., 2005). The increase of H_3 receptor expression appears to be associated with a con-comitant increase in receptor ligand binding, suggesting that the role of the H_3 receptor may be significant in the striatum. The possible role of the H_3 receptor in neuronal plasticity seems to be suggested by the rapid, transient pattern of changes in H_3 mRNA expression in the hippocam-pus, piriform cortex, and amygdala (Fig. 4) following systemic KA

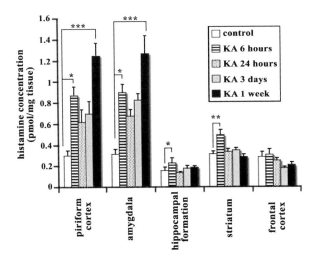

Fig. 4. Histamine concentrations in some rat brain areas after saline injection (control) and 6λh, 24λh, 3 days, and 1 week after kainic acid (KA) injection. Statistical results from Bonferroni post hoc tests are shown (*$p<0.05$, **$p<0.01$, ***$p<0.001$). Error bars indicate $\pm$SEM (standard error of mean). (The au-thors would like to thank Lintunen et al., 2005 for having granted permission to use their data.)

administration (Lintunen *et al.*, 2005) as well as in 3-nitropropionic-dependent neurodegenerative effects (Canonaco *et al.*, 2005).

It is particularly significant that the major changes concerned an isoform of H_3 (H_{3A}) that is effectively coupled to the MAPK pathway (Drutel *et al.*, 2001). The neuroprotective effect of histamine was recently demonstrated *in vitro* using a coculture system, in which hippocampal and hypothalamic explants were maintained together (Kukko-Lukjanov *et al.*, 2006). In this system, the posterior hypothalamic explant innervates the hippocampal explant and provides it with histaminergic innervation. A control explant from anterior hypothalamus does not provide the histaminergic innervation. The presence of a posterior hypothalamic explant was associated with significant neuroprotection against KA-induced cell death in the hippocampus, an effect that could be reversed with α-fluoromethylhistidine (Kukko-Lukjanov *et al.*, 2006). To confirm that the effect was due to histamine rather than an unknown substance that might be present in the posterior hypothalamus, exogenous histamine was also tested. From the experimental testing, histamine appeared to also inhibit KA-induced cell death in hippocampal CA3 field. The inhibition of H_1 receptor activity by the application of two well-known histamine H_1 antagonists (triprolidine or mepyramine) reduced the ability of posterior hypothalamic explants to protect the hippocampus from KA-induced cell death. Thus, in this *in vitro* model, where not all connections of the hippocampus are intact, histamine seems to have significant neuroprotective effects. It is thus possible that the increased histamine level and histamine release observed in both ischemic conditions and hibernation represent a relevant physiological response to neuronal stress and anoxia, opening up the possibility of the therapeutic use of histamine and/or histamine receptor ligands to treat related clinical conditions.

Concluding remarks

The fact that histaminergic neurons are concentrated in a single nucleus with widespread fiber projections in the brain, while NOergic neurons are rather widespread in most cerebral regions, may account for distinct neurobiological roles. The large number of hypothalamic neurons producing the free radical gas NO suggests that the hypothalamus is the main brain region involved in NO-dependent neuronal functions, a view supported *e.g.*, by the dense NOergic fibers found in the PVN and SON. Their presence in these hypothalamic sites underlies the major neuroendocrine regulatory activities (Bugajski *et al.*, 2004) linked to homeostatic functions such as the sleep–wake cycle. In this context, these hypo-thalamic-controlled functions, together with others such as energy production and thermoregulatory processes, tend to strengthen the importance of the role exerted by the coexistence of NO with GABA and NMDA (Segovia and Mora, 1998). Seasonal hibernators undergo profound plastic changes in CNS circuitries and regulation of excitability, and so their brain is protected against different types of insults that are harmful for euthermic organisms. For this reason, hibernation turns out to be a useful model to study neuroprotection mechanisms. The NOergic system seems to preferentially exert influences on neurodegenerative processes and on sleeping behaviors. At the level of the CNS, studies have begun to show a direct histaminergic-dependent regulation of the NO system as demonstrated by TM neuron stimulation of NOS via the modulatory effects of the H_1 receptor subtype on G-protein-coupled pathways, resulting in an activation of SON vasopressinergic neurons (Yang and Hatton, 2002). There is currently little evidence for a role for NO in brain neuronal activities, aside the NOergic influence on glomerular functions (Sandovici et al., 2004), the maintenance of blood flow (Kudej and Vatner, 2003) of true hibernators, and the ability of cerebral neurons to tolerate hypoxia-induced cellular stress during the arousal state (Ma et al., 2005). Although many transmitters regulate hibernation, current evidence suggests that histaminergic circuits may be crucial for the execution of hibernation. Such a role is substantially evident in the hippocampus of hibernators, in which the activation of H_1 and H_2 receptors in this brain region may maintain the animals in a torpor state (Sallmen et al., 2003b). Interactions of brain histamine with the H_1, H_2, and different isoforms of the H_3 receptor are also involved in other conditions related to plasticity and neurodegeneration, such as those associated with brain ischemia and limbic seizures. The in vitro application of histamine seems to strongly protect against KA-induced hippocampal neuronal damage, whereas H_1 receptor antagonists attenuate such a protective effect. The details of the mechanisms by which this phylogenetically old neuronal system operates in hibernation, plasticity, and in neurodegenerative processes are still open. As it continues to receive greater attention in the field of neuronal plasticity and repair, it will be necessary that future studies be directed towards the possible heterogeneity of TM histaminergic neurons and the significance of the circuitries that either suffer from similar damage or that instead display evident plastic changes. Moreover, the possible functional difference between subpopulations of this phylogenetically old group of neurons during the different phases of hibernation could provide further helpful

358

information about the possible mechanisms of neuroprotection in, for example, seizures and ischemia, as well as in neurodegenerative diseases (Alguacil and Perez-Garcia, 2003).

References

Abu-Soud, H.M. and Stuehr, D.J. (1993). Nitric oxide synthases reveal a role for calmodulin in controlling electron transfer. *Proc. Natl. Acad. Sci. USA* 90,10769–10772.

Adachi, N. (2005). Cerebral ischemia and brain histamine. *Brain Res. Rev.* 50,275–286.

Airaksinen, M.S., Alanen, S., Szabat, E., Visser, T.J. and Panula, P. (1992). Multiple neurotransmitters in the tuberomammillary nucleus: Comparison of rat, mouse, and guinea pig. *J. Comp. Neurol.* 323,103–116.

Alderton, W.K., Cooper, C.E. and Knowles, R.G. (2001). Nitric oxide synthases: Structure, function and inhibition. *J. Biochem.* 357,593–615.

Alguacil, L.F. and Perez-Garcia, C. (2003). Histamine H3 receptor: A potential drug target for the treatment of central nervous system disorders. *Curr. Drug Targets CNS Neurol. Disord.* 2,303–313.

Almeida, A.P. and Beaven, M.A. (1981). Phylogeny of histamine in vertebrate brain. *Brain Res.* 208,244–250.

Baader, S.L. and Schilling, K. (1996). Glutamate receptors mediate dynamic regulation of nitric oxide synthase expression in cerebellar granule cells. *J. Neurosci.* 16,1440–1449.

Bakker, R.A., Timmerman, H. and Leurs, R. (2002). Histamine receptors: Specific ligands, receptor biochemistry, and signal transduction. *Clin. Allergy Immunol.* 17,27–64.

Bekkers, J.M. (1993). Enhancement by histamine of NMDA-mediated synaptic transmission in the hippocampus. *Science* 261,104–106.

Bliss, T.V.P. and Collingridge, G.L. (1993). A synaptic model of memory: Long-term potentiation in the hippocampus. *Nature* 361,31–39.

Bredt, D.S. and Snyder, S.H. (1990). Isolation of nitric oxides synthetase, a calmodulin-requiring enzyme. *Proc. Natl. Acad. Sci. USA* 87,682–685.

Bredt, D.S. and Snyder, S.H. (1991). NO, a novel neuronal messenger. *Neuron* 8,3–11.

Brodin, L., Hokfelt, T., Grillner, S. and Panula, P. (1990). Distribution of histaminergic neurons in the brain of the lamprey *Lampetra fluviatilis* as revealed by histamine-immunohistochemistry. *J. Comp. Neurol.* 292,435–442.

Brown, R.E. and Haas, H.L. (1999). On the mechanism of histaminergic inhibition of glutamate release in the rat dentate gyrus. *J. Physiol. (London)* 515,777–783.

Brown, R.E., Sergeeva, O.A., Eriksson, K.S. and Haas, H.L. (2002). Convergent excitation of dorsal raphe serotonin neurons by multiple arousal systems (orexin/hypocretin, histamine and noradrenaline). *J. Neurosci.* 22,8850–8859.

Brown, R.E., Stevens, D.R. and Haas, H.L. (2001). The physiology of brain histamine. *Prog. Neurobiol.* 63,637–672.

Bugajskı, J., Gadek-Michalska, A. and Bugajski, A. (2004). Nitric oxide and prostaglandin systems in the stimulation of hypothalamic-pituitary-adrenal axis by neurotransmitters and neurohormones. *J. Physiol. Pharmacol.* 55,679–703.

Canonaco, M., Madeo, M., Alò, R., Giusi, G., Granata, T., Canonaco, A., Carelli, A. and Facciolo, R.M. (2005). The histaminergic signaling system exerts a neuroprotective role against neurodegenerative-induced processes in the hamster. *J. Pharmacol. Exp. Ther.* 315,188–195.

Cardenas, A., Moro, M.A., Hurtado, O., Leza, J.C., Lorenzo, P., Castrillo, A., Bodelòn, O.G., Boscà, L. and Lizasoian, I. (2000). Implication of glutamate in the expression of inducible nitric oxide synthase after oxygen and glucose deprivation in rat forebrain slices. *J. Neurochem.* 74,2041–2048.

Carey, H.V., Andrews, M.T. and Martin, S.L. (2003). Mammalian hibernation: Cellular and molecular responses to depressed metabolism and low temperature. *Physiol. Rev.* 83,1153–1181.

Cecchi, M., Passani, M.B., Bacciottini, L., Mannaioni, P.F. and Blandina, P. (2001). Cortical acetylcholine release elicited by stimulation of histamine H1 receptors in the nucleus basalis magnocellularis: A dual-probe microdialysis study in the freely moving rat. *Eur. J. Neurosci.* 13,68–78.

Chen, J., Zacharek, A., Zhang, C., Jiang, H., Li, Y., Roberts, C., Lu, M., Kapke, A. and Chopp, M. (2005). Endothelial nitric oxide synthase regulates brain-derived neurotrophic factor expression and neurogenesis after stroke in mice. *J. Neurosci.* 25,2366–2375.

Cherian, L., Goodman, J.C. and Robertson, C.S. (2000). Brain nitric oxide changes after controlled cortical impact injury in rats. *J. Neurophysiol.* 83,2171–2178.

Chung, Y.H., Kim, Y.S. and Lee, W.B. (2004). Distribution of neuronal nitric oxide synthase-immunoreactive neurons in the cerebral cortex and hippocampus during postnatal development. *J. Mol. Histol.* 35,765–770.

Coge, F., Guenin, S.P., Audinot, V., Renouard-Try, A., Beauverger, P., Macia, C., Ouvry, C., Nagel, N., Rique, H., Boutin, J.A. and Galizzi, J.P. (2001). Genomic organization and characterization of splice variants of the human histamine H3 receptor. *J. Biochem.* 355,279–288.

Doreulee, N., Yanovsky, Y., Flagmeyer, I., Stevens, D.R., Haas, H.L. and Brown, R.E. (2001). Histamine H(3) receptors depress synaptic transmission in the corticostriatal pathway. *Neuropharmacology* 40,106–113.

Drew, K.L., Rice, M.E., Kuhn, T.B. and Smith, M.A. (2001). Neuroprotective adaptations in hibernation: Therapeutics implications for ischemia-reperfusion, traumatic brain injury and neurodegenerative diseases. *Free Radical Biol. Med.* 31,563–573.

Drutel, G., Peitsaro, N., Karlstedt, K., Wieland, K., Smit, M.J., Timmerman, H., Panula, P. and Leurs, R. (2001). Identification of rat H_3 receptor isoforms with different brain expression and signaling properties. *Mol. Pharmacol.* 59,1–8.

Dun, N.J., Dun, S.L. and Förstermann, U. (1994). Nitric oxide synthase immunoreactivity in rat pontine medullary neurons. *Neuroscience* 59,429–445.

Ekstrom, P., Holmqvist, B.I. and Panula, P. (1995). Histamine-immunoreactive neurons in the brain of the teleost *Gasterosteus aculeatus L.* Correlation with hypothalamic tyrosine hydroxylase- and serotonin-immunoreactive neurons. *J. Chem. Neuroanat.* 8,75–85.

Ericson, H., Blomqvist, A. and Köhler, C. (1989). Brainstem afferents to the tuberomammillary nucleus in the rat brain with special reference to monoaminergic innervation. *J. Comp. Neurol.* 281,169–192.

Ericson, H., Blomqvist, A. and Köhler, C. (1991a). Origin of neuronal inputs to the region of the tuberomammillary nucleus of the rat brain. *J. Comp. Neurol.* 311,45–64.

Ericson, H., Köhler, C. and Blomqvist, A. (1991b). GABA-like immunoreactivity in the tuberomammillary nucleus: An electron microscopic study in rat. *J. Comp. Neurol.* 305,462–469.

Estrada, C., Mengual, E. and Gonzáles, C. (1993). Local NADPH-diaphorase neurons innervate pial arteries and lie close or project to intracerebral blood vessels: A possible role for NO in the regulation of cerebral blood flow. *J. Cereb. Blood Flow Metab.* 13,984–987.

Facciolo, R.M., Tavolaro, R., Chinellato, A., Ragazzi, E., Canonaco, M. and Fassina, G. (1996). Effects of *N*-nitro-L-arginine methyl ester on benzodiazepine binding in some limbic areas of hyperlipidaemic rats. *Pharmacol. Biochem. Behav.* 54,431–437.

Faradji et al., (2000)Faradji, H., Rousset, C., Debilly, G., Vergnes, M. and Cespuglio, R. (2000). Sleep and epilepsy: A key role for nitric oxide? *Epilepsia* 41,794–801.

Frerichs, K.U. and Hallenbeck, J.M. (1998). Hibernation in ground squirrels induces state and species-specific tolerance to ipoxia and aglycemia: An in vitro study in hippocampal slices. *J. Cereb. Blood Flow Metab.* 18,168–175.

Gage, F.H., Kempermann, G., Palmer, T.D., Peterson, D.A. and Ray, J. (1998). Multipotent progenitor cells in the adult dentate gyrus. *J. Neurobiol.* 36,49–266.

Gantz, I., Schaffer, M., DelValle, J., Logsdon, C., Campbell, V., Uhler, M. and Yamada, T. (1991). Molecular cloning of a gene encoding the histamine H2 receptor. *Proc. Natl. Acad. Sci. USA* 88,5937.

Garbers, D.L. (1992). Guanylyl cyclase receptors and their endocrine paracrine and autocrine ligands. *Cell* 71,1–4.

Garthwaite, J. (1991). Glutamate, nitric oxide and cell-cell signalling in the nervous system. *Trends Neurosci.* 14,60–67.

Garthwaite, J. and Boulton, C.L. (1995). Nitric oxide signaling in the central nervous system. *Annu. Rev. Physiol.* 57,683–706.

Gibbs, S.M. (2003). Regulation of neuronal proliferation and differentiation by nitric oxide. *Mol. Neurobiol.* 27,107–120.

Gisselmann, G., Pusch, H., Hovemann, B.T. and Hatt, H. (2002). Two cDNAs coding for histamine-gated ion channels in *D. melanogaster*. *Nat. Neurosci.* 5,11–12.

Haas, H.L. and Panula, P. (2003). The role of histamine and the tuberomamillary nucleus in the nervous system. *Nat. Rev. Neurosci.* 4,121–130.

Haley, J.E. (1998). Gases as neurotransmitters. *Essays Biochem.* 33,79–91.

Hardie, R.C. (1989). A histamine-activated chloride channel involved in neurotransmission at a photoreceptor synapse. *Nature* 339,704–706.

Hars, B. (1999). Endogenous nitric oxide in the rat pons promotes sleep. *Brain Res.* 816,209–219.

Hauser, W., Sassmann, A., Qadri, F., Johren, O. and Dominiak, P. (2005). Expression of nitric oxide synthase isoforms in hypothalamo-pituitary-adrenal axis during the development of spontaneous hypertension in rats. *Mol. Brain Res.* 138,198–204.

Heneka, M.T. and Feinstein, D.L. (2001). Expression and function of inducible nitric oxide synthase in neurons. *J. Neuroimmunol.* 114,8–18.

Hill, S.J., Ganellin, C.R., Timmerman, H., Schwartz, J.C., Shankley, N.P., Young, J.M., Schunack, W., Levi, R. and Haas, H.L. (1997). International Union of Pharmacology. XIII. Classification of histamine receptors. *Pharmacol. Rev.* 49,253–278.

361

Hirota, K., Fukuda, R., Takabuchi, S., Kisaka-Kondok, S., Adachi, T., Fukuda, K. and Semenza, G.L. (2004). Induction of hypoxia-inducible factor 1 activity by muscarinic acetylcholine receptor signaling. J. Biol. Chem. 279,41521–41528.

Huang, Z., Huang, P.L., Ma, J., Meng, W., Ayara, C., Fishman, M.C. and Moskowitz, M.A. (1996). Enlarged infarcts in endothelial nitric oxide synthase knockout mice are attenuated by nitro-L-arginine. J. Cereb. Blood Flow Metab. 16,981–987.

Huszti, Z. (1998). Carrier-mediated high affinity uptake system for histamine in astroglial and cerebral endothelial cells. J. Neurosci. Res. 51,551–558.

Iadecola, C. (1997). Bright and dark sides of nitric oxide in ischemic brain injury. Trends Neurosci. 20,132–139.

Ignarro, L.J. (2002). Nitric oxide as a unique signaling molecule in the vascular system: A historical overview. J. Physiol. Pharmacol. 53,503–514.

Ignarro, L.J., Buga, G.M., Wood, K.S., Byrns, R.E. and Chaudhuri, G. (1987). Endothelium-derived relaxing factor produced and released from artery and vein is nitric oxide. Proc. Natl. Acad. Sci. USA 84,9265–9269.

Inagaki, S., Yamatodani, A., Ando-Yamamoto, M., Tohyama, M., Watanabe, T. and Wada, H. (1988). Organization of histaminergic fibers in the rat brain. J. Comp. Neurol. 273,283–300.

John, J., Wu, M.F., Boehmer, L.N. and Siegel, J.M. (2004). Cataplexy-active neurons in the hypothalamus: Implications for the role of histamine in sleep and waking behavior. Neuron 42,619–634.

Kaslin, J. and Panula, P. (2001). Comparative anatomy of the histaminergic and other aminergic systems in zebrafish (Danio rerio). J. Comp. Neurol. 440,342–377.

Kee, N.J., Preston, E. and Wojtowicz, J.M. (2001). Enhanced neurogenesis after transient global ischemia in the dentate gyrus of the rat. Exp. Brain Res. 831,238–287.

Kim, Y.M., Bombeck, C.A. and Billiar, T.R. (1999). Nitric oxide as a bifunctional regulator of apoptosis. Circ. Res. 84,253–256.

Kudej, R.K. and Vatner, S.F. (2003). Nitric oxide-dependent vasodilation mantains blood flow in true hibernating myocardium. J. Mol. Cell. Cardiol. 35,931–935.

Kukko-Lukjanov, T.-K., Soini, S., Michelsen, K.A., Panula, P. and Holopainen, I.E. (2006). Histaminergic neurons protect the developing hippocampus from kainic acid-induced neuronal damage in an organotypic co-culture system. J. Neurosci. 26,1088–1097.

Lantoine, F., Iouzalen, N., Devynck, M.A. and Millanvoye-van Brussel, E. (1998). Nitric oxide production in human endothelial cells stimulated by histamine requires Ca^{2+} influx. J. Biochem. 330,695–699.

Lin, J.S., Sakai, K., Vanni, M.G. and Jouvet, M. (1989). A critical role of the posterior hypothalamus in the mechanisms of wakefulness determined by microinjection of muscimol in freely moving cats. Brain Res. 479,225–240.

Lintunen, M., Sallmen, T., Karlstedt, K., Fukui, H., Eriksson, K.S. and Panula, P. (1998). Postnatal expression of H_1 receptor mRNA in the rat brain: Correlation to L-histidine decarboxylase expression and local upregulation in limbic seizures. Eur. J. Neurosci. 10,2287–2301.

Lintunen, M., Sallmen, T., Karlstedt, K. and Panula, P. (2005). Systemic kainic acid induces profound transient changes in the limbic histaminergic system. Neurobiol. Dis. 20,155–169.

362

Liu, C.N., Liu, X., Gao, D. and Li, S. (2005). Effects of SNP, GLU and GABA on the neuronal activity of striatum nucleus in rats. *Pharmacol. Res.* 51,547–551.

Liu, K.P. (2003). Ischemia-reperfusion-related repair deficit after oxidative stress: Implications of faulty transcripts in neuronal sensitivity after brain injury. *J. Biomed. Sci.* 10,4–13.

Lovenberg, T.W., Roland, B.L., Wilson, S.J., Jiang, X., Pyati, J., Huvar, A., Jackson, M.R. and Erlander, M.G. (1999). Cloning and functional expression of the human histamine H_3 receptor. *Mol. Pharmacol.* 55,1101–1107.

Lozada, A., Munyao, N., Sallmen, T., Lintunen, M., Leurs, R., Lindsberg, P.J. and Panula, P. (2005). Postischemic regulation of central histamine receptors. *Neuroscience* 136,371–379.

Ma, Y.L., Zhu, X., Rivera, P.M., Toien, O., Barnes, B.M., LaManna, J.C., Smith, M.A. and Drew, K.L. (2005). Absence of cellular stress in brain after hypoxia induced by arousal from hibernation in Arctic ground squirrels. *Am. J. Physiol. Regul. Integr. Comp. Physiol.* 289,R1297–R1306.

Marletta, M.A. (1994). Nitric oxide synthase: Aspects concerning structure and catalysis. *Cell* 78,927–930.

McCann, S.M., Antunes-Rodrigues, J., Franci, C.R., Anselmo-Franci, J.A., Karanth, S. and Rettori, V. (2000). Role of the hypothalamic pituitary adrenal axis in the control of the response to stress and infection. *Braz. J. Med. Biol. Res.* 33,1121–1131.

Mochizuki, T., Yamatodani, A., Okakura, K., Takemura, M., Inagaki, N. and Wada, H. (1991). In vivo release of neuronal histamine in the hypothalamus of rats measured by microdialysis. *Naunyn Schmiedebergs Arch. Pharmacol.* 343,190–195.

Moreno-López, B., Noval, J.A., González-Bonet, L. and Estrada, C. (2000). Morphological bases for a role of nitric oxide in adult neurogenesis. *Brain Res.* 869,244–250.

Morin, L.P. (1994). The circadian visual system. *Brain Res. Rev.* 67,102–127.

Morisset, S., Rouleau, A., Ligneau, X., Gbahou, F., Tardivel-Lacombe, J., Stark, H., Schunack, W., Ganellin, C.R., Schwartz, J.C. and Arrang, J.M. (2000). High constitutive activity of native H3 receptors regulates histamine neurons in brain. *Nature* 408,860–864.

Moro, M.A., Cárdenas, A., Hurtado, O., Leza, J.C. and Lizasoain, I. (2004). Role of nitric oxide after brain ischemia. *Cell Calcium* 36,265–275.

Namiki, S., Kakizawa, S., Hirose, K. and Iino, M. (2005). NO signaling decodes frequency of neuronal activity and generates synapse-specific plasticity in mouse cerebellum. *J. Physiol.* 566,849–863.

Nathan, C. and Xie, Q.W. (1994). Nitric oxide synthases: Roles, tolls, and controls. *Cell* 78,915–918.

Nazli, M. and Thippeswamy, T. (2002). Nitric oxide signaling system in rat brain stem: Immunocytochemical studies. *Anat. Histol. Embryol.* 31,252–256.

Nylen, A., Skagerberg, G., Alm, P., Larsson, B., Holmqvist, B.I. and Andersson, K.E. (2001). Detailed organization of nitric oxide synthase, vasopressin and oxytocin immunoreactive cell bodies in the supraoptic nucleus of the female rat. *Anat. Embryol. (Berl.)* 203,309–321.

Ohkuma, S. and Katsura, M. (2001). Nitric oxide and peroxynitrite as factors to stimulate neurotransmitter release in the CNS. *Prog. Neurobiol.* 64,97–108.

Osborne, P.G. and Hashimoto, M. (2003). State-dependent regulation of cortical blood flow and respiration in hamsters: Response to hypercapnia during arousal from hibernation. *J. Physiol.* 547,963–970.

Ozaki, M., Shibuya, I., Kabashima, N., Isse, T., Noguchi, J., Ueta, Y., Inoue, Y., Shigematsu, A. and Yamashita, H. (2000). Preferential potentiation by nitric oxide of spontaneous inhibitory postsynaptic currents in rat supraoptic neurones. *J. Neuroendocrinol.* 12,273–281.

Packer, M., Stasiv, Y., Benraiss, A., Chmielnick, I.E., Grinberg, A., Westphal, H., Goldman, S.A. and Enikolopov, G. (2003). Nitric oxide negatively regulates mammalian adult neurogenesis. *Proc. Natl. Acad. Sci. USA* 100,9566–9571.

Palchykova, S., Deboer, T. and Tobler, I. (2002). Selective sleep deprivation after daily torpor in the Djungarian hamster. *J. Sleep Res.* 11,313–319.

Panula, P., Airaksinen, M.S., Pirvola, U. and Kotilainen, E. (1990). A histamine-containing neuronal system in human brain. *Neuroscience* 34,127–132.

Panula, P., Pirvola, U., Auvinen, S. and Airaksinen, M.S. (1989). Histamine-immunoreactive nerve fibres in the rat brain. *Neuroscience* 28,585–610.

Panula, P., Yang, H.Y. and Costa, E. (1984). Histamine-containing neurons in the rat hypothalamus. *Proc. Natl. Acad. Sci. USA* 81,2572–2576.

Parent, J.M. (2003). Injury induced neurogenesis in the adult mammalian brain. *Neuroscientist* 9,261–272.

Parmentier, R., Ohtsu, H., Djebbara-Hannas, Z., Valatx, J.L., Watanabe, T. and Lin, J.S. (2002). Anatomical, physiological, and pharmacological characteristics of histidine decarboxylase knock-out mice: Evidence for the role of brain histamine in behavioral and sleep-wake control. *J. Neurosci.* 22,7695–7711.

Reif, A., Schmitt, A., Fritzen, S., Chourbaji, C., Bartsch, A., Urani, M., Wycislo, R., Mössner, R., Sommer, C., Gass, P. and Lesch, K.P. (2004). Differential effect of endothelial nitric oxide synthase (NOS-III) on the regulation of adult neurogenesis and behaviour. *Eur. J. Neurosci.* 20,885–895.

Reite, O.B. (1972). Comparative physiology of histamine. *Physiol. Rev.* 52,778–819.

Ruby, N.F. (2003). Hibernation: When good clocks go cold. *J. Biol. Rhythms* 18,275–286.

Ruiz-Stewart, I., Tiyyagura, S.R., Lin, J.E., Kazerounian, S., Pitari, G.M., Schulz, S., Martin, E., Murad, F. and Waldman, S.A. (2004). Guanylyl cyclase is an ATP sensor coupling nitric oxide signaling to cell metabolism. *Proc. Natl. Acad. Sci. USA* 101,37–42.

Sallmen, T., Beckman, A.L., Stanton, T.L., Eriksson, K.S., Tarhanen, J., Tuomisto, L. and Panula, P. (1999). Major changes in the brain histamine system of the ground squirrel *Citellus lateralis* during hibernation. *J. Neurosci.* 19,1824–1835.

Sallmen, T., Lozada, A.F., Anichtchik, O.V., Beckman, A.L., Leurs, R. and Panula, P. (2003c). Changes in hippocampal histamine receptors across the hibernation cycle in ground squirrels. *Hippocampus* 13,745–754.

Sallmen, T., Lozada, A.F., Anichtchik, O.A., Michelsen, K.A., Beckman, A.L. and Panula, P. (2003b). Increased brain histamine is associated with increased histamine H3 receptor expression and signaling in hibernation. *BMC Neurosci.* 4,24–33.

Sallmen, T., Lozada, A.F., Beckman, A.L. and Panula, P. (2003a). Intrahippocampal histamine delays arousal from hibernation. *Brain Res.* 966,317–320.

364

Sandovici, M., Henning, R.H., Hut, R.A., Strijkstra, A.M., Epema, A.H., van Goor, H. and Deelman, L.E. (2004). Differential regulation of glomerular and interstitial endothelial nitric oxide synthase expression in kidney of hibernating ground squirrel. *Nitric Oxide* 11,194–200.

Sase, K. and Michel, T. (1997). Expression and regulation of endothelial nitric oxide synthase. *Trends Cardiovasc. Med.* 7,28–37.

Schwartz, J.C., Arrang, J.M., Garbarg, M., Pollard, H. and Ruat, M. (1991). Histaminergic transmission in the mammalian brain. *Physiol. Rev.* 71,1–51.

Scorziello, A., Pellegrini, C., Secondo, A., Sirabella, R., Formisano, L., Sibaud, L., Amoroso, S., Canzoniero, L.M.T., Annunziato, L. and Di Renzo, G.F. (2004). Neuronal NOS activation during oxygen and glucose deprivation triggers cerebellar granule cell death in the later reoxygenation phase. *J. Neurosci. Res.* 76,812–821.

Segovia, G. and Mora, F. (1998). Role of nitric oxide in modulating the release dopamine, glutamate and GABA in striatum of the freely moving rat. *Brain Res. Bull.* 45,275–279.

Selbach, O., Brown, R.E. and Haas, H.L. (1997). Long-term increase of hippocampal excitability by histamine and cyclic AMP. *Neuropharmacology* 36,1539–1548.

Senba, E., Daddona, P.E. and Nagy, J.I. (1987). Adenosine deaminase-containing neurons in the olfactory system of the rat during development. *Brain Res. Bull.* 18,635–648.

Sherin, J.E., Shiromani, P.J., McCarley, R.W. and Saper, C.B. (1996). Activation of ventrolateral preoptic neurons during sleep. *Science* 271,216–219.

Shibuki, K. and Okada, D. (1991). Endogenous nitric oxide release required for long-term synaptic depression in the cerebellum. *Nature* 349,326–328.

Shih, Y.H., Lee, A.W., Huang, Y.H., Ko, M.H. and Fu, Y.S. (2004). GABAergic neuron death in the striatum following kainate-induced damage of hippocampal neurons: Evidence for the role of NO in locomotion. *Int. J. Neurosci.* 114,1119–1132.

Siles, E., Martinez-Lara, E., Canuelo, A., Sanchez, M., Hernandez, R., Lopez-Ramos, J.C., Del Moral, M.L., Esteban, F.J., Blanco, S., Pedrosa, J.A., Rodrigo, J. and Peinado, M.A. (2002). Age-related changes of the nitric oxide system in the rat brain. *Brain Res.* 956,385–392.

Staines, W.A., Yamamoto, T., Daddona, P.E. and Nagy, J.I. (1986). Neuronal colocalization of adenosine deaminase, monoamine oxidase, galanin and 5-hydroxytryptophan uptake in the tuberomammillary nucleus of the rat. *Brain Res. Bull.* 17,351–365.

Steiner, A.A., Antunes-Rodrigues, J., McCann, S.M. and Branco, L.G. (2002). Antipyretic role of the NO-cGMP pathway in the anteroventral preoptic region of the rat brain. *Am. J. Physiol. Regul. Integr. Comp. Physiol.* 282,584–593.

Stenzel-Poore, M.P., Stevens, S.L., Xiong, Z., Lessov, N.S., Harrington, C.A., Mori, M., Meller, R., Rosenzweig, H.L., Tobar, E., Shaw, T.E., Chu, X. and Simon, R.P. (2003). Effect of ischaemic preconditioning on genomic response to cerebral ischaemia: Similarity to neuroprotective strategies in hibernation and hypoxia-tolerant states. *Lancet* 362,1028–1037.

Subramanian, N., Theodore, D. and Abraham, J. (1981). Experimental cerebral infarction in primates: Regional changes in brain histamine content. *J. Neural Transm.* 50,225–232.

Takeda, N., Inagaki, S., Shiosaka, S., Taguchi, Y., Oertel, W.H., Tohyama, M., Watanabe, T. and Wada, H. (1984). Immunohistochemical evidence for coexistence of histidine decarboxylase like and glutamate decarboxylase-like immunoreactivities in nerve cells of the magnocellular nucleus of the posterior hypothalamus of rats. *Proc. Natl. Acad. Sci. USA* 81,7647–7650.

Takeda, N., Inagaki, S., Taguchi, Y., Tohyama, M., Watanabe, T. and Wada, H. (1984). Origins of histamine-containing fibers in the cerebral cortex of rats studied by immunohistochemistry with histidine decarboxylase as a marker and transection. *Brain Res.* 323,55–63.

Ueda, S. and Ibuka, N. (1995). An analysis of factors that induce hibernation in Syrian hamsters. *Physiol. Behav.* 58,653–657.

Vincent, S.R., Das, S. and Maines, M.D. (1994). Brain heme oxygenase isoenzymes and nitric oxide synthase are co-localized in select neurons. *Neuroscience* 63,223–231.

Vincent, S.R., Hokfelt, T. and Wu, J.Y. (1982). GABA neuron systems in hypothalamus and the pituitary gland, Immunohistochemical demonstration using antibodies against glutamate decarboxylase. *Neuroendocrinology* 34,117–125.

Vorobjev, V.S., Sharonova, I.N., Walsh, I.B. and Haas, H.L. (1993). Histamine potentiates N-methyl-D-aspartate responses in acutely isolated hippocampal neurons. *Neuron* 11,837–844.

Weiss, S.W., Albers, D.S., Iadarola, M.J., Dawson, T.M., Dawson, V.L. and Standaert, D.G. (1998). NMDAR1 glutamate receptor subunit isoforms in neostriatal, neocortical, and hippocampal nitric oxide synthase neurons. *J. Neurosci.* 18,1725–1734.

Wen, P.H., Patrick, R.H., Xiaoping, C., Gluck, K., Gregory, A., Younkin, S.G., Younkin, L.H., DeGasperi, R., Gama Sosa, M.A., Robakis, N.L., Haroutunian, V. and Elder, G.A. (2004). The presenilin-1 familial Alzheimer disease mutant P117L impairs neurogenesis in the hippocampus of adult mice. *Exp. Neurol.* 188,224–237.

Xu, C., Michelsen, K.A., Wu, M., Morozova, E., Panula, P. and Alreja, M. (2004). Histamine innervation and activation of septohippocampal GABAergic neurons: involvement of local ACh release. *J. Physiol.* 56,657–670.

Yamada, K., Emson, P. and Hokfelt, T. (1996). Immunohistochemical mapping of nitric oxide synthase in the rat hypothalamus and colocalization with neuropeptides. *J. Chem. Neuroanat.* 10,295–316.

Yamashita, M., Fukui, H., Sugama, K., Horio, Y., Ito, S., Mizuguchi, H. and Wada, H. (1991). Expression cloning of a cDNA encoding the bovine histamine H1 receptor. *Proc. Natl. Acad. Sci. USA* 88,11515–11519.

Yang, Q.Z. and Hatton, G.I. (2002). Histamine H-1 receptor modulation of inter-neuronal coupling among vasopressinergic neurons depends on nitric oxide synthase activation. *Brain Res.* 955,115–122.

Yanovsky, Y. and Haas, H.L. (1998). Histamine increases the bursting activity of pyramidal cells in the CA3 region of mouse hippocampus. *Neurosci. Lett.* 240,110–112.

Yousef, T., Neubacher, U., Eysel, U.T. and Volgushev, M. (2004). Nitric oxide synthase in rat visual cortex: An immunohistochemical study. *Brain Res. Protoc.* 13,57–67.

Zheng, Y., Hirschberg, B., Yuan, J., Wang, A.P., Hunt, D.C., Ludmerer, S.W., Schmatz, D.M. and Cully, D.F. (2002). Identification of two novel *Drosophila melanogaster* histamine-gated chloride channel subunits expressed in the eye. *J. Biol. Chem.* 277,2000–2005.

Zhou, F., Zhu, X., Castellani, R.J., Stimmelmayr, R., Perry, G., Smith, M.A. and Drew, K. (2001). Hibernation, a model of neuroprotection. *Am. J. Physiol.* 158,2145–2151.

Zhu, D.Y., Liu, S.H., Sun, H.S. and Lu, Y.M. (2003). Expression of inducible nitric oxide synthase after focal cerebral ischemia stimulates neurogenesis in the adult rodent dentate gyrus. *J. Neurosci.* 23,223–229.

Zhu, X., Smith, M.A., Perry, G., Wang, Y., Ross, A.P., Zhao, H.W., LaManna, J.C. and Drew, K.L. (2005). MAPKs are differentially modulated in arctic ground squirrels during hibernation. *J. Neurosci. Res.* 80,862–868.

Nitric oxide, peroxynitrite and matrix metalloproteinases: Insight into the pathogenesis of sepsis

Jonathan Cena[1], Ava K. Chow[2] and Richard Schulz[1,2,*]

[1]*Department of Pharmacology, Cardiovascular Research Group, 4-62 Heritage Medical Research Centre, University of Alberta, Edmonton, Alberta T6G 2S2, Canada*
[2]*Department of Pediatrics, Cardiovascular Research Group, 4-62 Heritage Medical Research Centre, University of Alberta, Edmonton, Alberta T6G 2S2, Canada*

Abstract. Sepsis remains a significant cause of morbidity and mortality in North America. Clinical trials in the past have produced only modest reductions in mortality. Part of the reason for the failure of these trials is a general lack of understanding of the pathogenesis of sepsis. Gram-negative sepsis is initiated by lipopolysaccharide, a component in the outer wall of Gram-negative bacteria. It is important to understand the cardiovascular pathophysiology of sepsis, as cardiovascular symptoms predominate. These symptoms include altered blood coagulation, as well as vascular and myocardial complications. During sepsis, there is an enhanced state of coagulation due to the activation of extrinsic and intrinsic clotting pathways. The vascular complications involve the development of two interacting factors: overproduction of vasodilatory substances and vascular hyporeactivity to vasoconstrictors. The cardiac complications are believed to be due to an intrinsic decrease in myocardial contractile function. This review will focus on three important mechanisms for the pathogenicity of sepsis: nitric oxide, oxidative stress and matrix metalloproteinases. An enhanced amount of nitric oxide is generated via increased expression of inducible nitric oxide synthase during sepsis. Oxidative stress is also enhanced and is increased primarily by the formation of peroxynitrite, the toxic reaction product of nitric oxide and superoxide. A new emerging field in sepsis pathophysiology is the activation of matrix metalloproteinases as a result of enhanced oxidative stress. An upregulation of these enzymes during sepsis has been demonstrated in both clinical and basic science models. Moreover, studies have shown a beneficial effect of pharmacological inhibition and genetic ablation of matrix metalloproteinases. This review provides an overview of the cardiovascular abnormalities of sepsis as well as various mechanisms of its pathogenicity, with particular emphasis on matrix metalloproteinases.

Keywords: matrix metalloproteinase; nitric oxide; peroxynitrite; sepsis; heart; vasculature; oxidative stress.

Defining sepsis

Sepsis, a fatal condition arising through the body's response to an infection, is the leading cause of death in intensive care units (ICUs) in North

Corresponding author: Fax: +1-780-492-9753.
E-mail: richard.schulz@ualberta.ca (R. Schulz).

ADVANCES IN EXPERIMENTAL BIOLOGY
VOLUME 01   ISSN 1872-2423
DOI: 10.1016/S1872-2423(07)01016-2

America (Kumar *et al.*, 2001a). In the USA sepsis and its associated syndromes account for 2.9% of all hospital admissions and 10% of admissions into the ICU (Rivers *et al.*, 2005). In 1995, an estimated 9.3% of all deaths in the USA were attributed to sepsis. Mortality rates vary among the population, from 3.2% in children to 43% in the elderly. Additionally, the annual expenditure for sepsis in the US is estimated at US$16.7 billion.

Sepsis is characterized by both an infection and a systemic inflammatory response (Levy *et al.*, 2003). Clinically, sepsis indicates what has also been termed "severe sepsis", which is defined as sepsis complicated by global organ dysfunction. Septic shock, a subset of severe sepsis, is defined as sepsis manifested by circulatory failure characterized by persistent arterial hypotension. Additionally, widespread intravascular coagulation results from a downregulation of fibrinolytic mechanisms (Bone, 1992). Research into the pathogenesis of sepsis has revealed many potential therapeutic targets such as inflammatory mediators, bacterial toxins and specific enzymes; however, no completely effective pharmacotherapy has yet been discovered. Part of the problem in developing treatments for this syndrome is dealing with its enormous complexity (Marshall, 2003).

Clinically, sepsis is typically characterized by fever, hyperventilation, tachycardia and persistent hypotension. Laboratory tests frequently show altered white blood cell counts and decreased platelet count, and the patient is frequently acidotic. Bloodborne infection is confirmed by a positive blood culture, though the condition may progress to death before the results of the test are available.

Treatment of sepsis typically requires admission into the ICU. Therapy is typically supportive: oxygen, fluid resuscitation and vasopressor agents to improve blood pressure. Administration of broad-spectrum antibiotics intravenously to combat the infection is hindered by the persistent systemic hypotension. The complication and mortality rate of sepsis remains high. Consequently, it is essential that the mechanisms that underlie the severe and persistent hypotension observed in septic patients be more fully elucidated in order to facilitate more positive treatment outcomes.

Unravelling the underlying mechanisms is difficult because the natural immunological response to bacterial infection releases an unrelenting cascade of mediators, thus complicating the pathophysiology of sepsis. This review will focus on the cardiovascular pathophysiology of sepsis, with particular emphasis on the involvement of nitric oxide (NO), peroxynitrite (ONOO⁻) and matrix metalloproteinases (MMPs).

Initiation of sepsis

The role of lipopolysaccharide

Though the origin of bacterial infection can vary widely, the body's reaction to systemic invasion by bacterial pathogens follows a common course. Initiation of the septic cascade first begins with exposure to immunogens in the bloodstream. Lipopolysaccharide (LPS), a cell wall component of Gram-negative bacteria (Fig. 1), can spark the inflammatory cascade, leading to sepsis-like conditions. LPS, also called endotoxin, consists of a membrane-anchoring lipid A domain and a covalently linked polysaccharide portion (Alexander and Rietschel, 2001). The polysaccharide region consists of up to 50 repeating oligosaccharide units and has extreme structural variability among bacterial species. The terminal end of the LPS molecule, called the O-specific chain, protects the bacteria from phagocytosis. Interestingly, the polysaccharide moiety in Gram-negative bacteria has evolved to mimic human glycolipids and thus allow them a consequentially increased resistance to the immune system defences. The lipid A domain is shown to be the primary immunoreactive centre of LPS. It is highly sensitive to various components in the immune system (Medzhitov and Janeway, 2000). LPS first binds to CD14 receptors in the immune system via its lipid A portion (Antal-Szalmas, 2000; Landmann et al., 2000). The LPS–CD14 complex is then recognized by toll-like receptor-4 (TLR-4) located on neutrophils, macrophages and endothelial cells (Modlin et al., 1999; Vasselon and Detmers, 2002).

Occupancy of the TLR activates multiple signal transduction pathways and mobilizes transcription factors such as nuclear factor-κB (NF-κB) (Li et al., 2003) which stimulate the transcription of many genes that encode for immunomodulatory molecules such as the pro-inflammatory cytokines tumour necrosis factor-α (TNF-α) and interleukin-1β (IL-1β) (Fig. 1). Many of these downstream mediators may stimulate the release of other factors, potentially amplifying the inflammatory cascade. Accompanying the release of pro-inflammatory mediators is the release of anti-inflammatory cytokines such as transforming growth factor-β and others. In sepsis, an imbalance occurs which shifts the balance towards inflammation. Continuous generation of these mediators makes

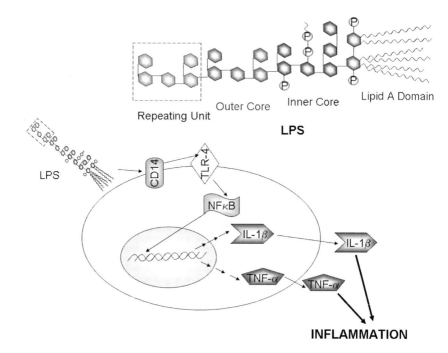

Fig. 1. (Top) General structure of lipopolysaccharide (LPS) from Gram-negative bacteria. LPS consists of a membrane-anchoring lipid A domain and a co-valently linked polysaccharide portion. The terminal end of the LPS molecule contains up to 50 repeating oligosaccharide units. The central core domains are more structurally conserved among bacterial species. (Bottom) Simplified inflammatory pathway in LPS signaling. LPS stimulates intracellular pathways downstream of toll-like receptor-4 (TLR-4) via nuclear factor-κB (NF-κB), leading to the transcription of pro-inflammatory cytokines such as interleukin-1β (IL-1β) and tumour necrosis factor-α (TNF-α).

the pathogenesis of sepsis self-perpetuating and independent of the initial exposure to endotoxin.

Lipoteichoic acid (LTA) and peptidoglycan are two major cell wall components in Gram-positive bacteria, both of which stimulate inflammatory responses in various *in vitro* and *in vivo* models (Wang *et al.*, 2003). Approximately 50% of all cases of sepsis are initiated by Gram-positive bacteria, *Staphylococcus aureus* being the most commonly associated microbe. Although there are many parallels between Gram-positive and Gram-negative sepsis, this review will focus on the pathogenesis of Gram-negative sepsis.

Immune response to sepsis

The inflammatory cascade in sepsis involves the generation of a plethora of powerful pro-inflammatory cytokines such as interleukin-2 (IL-2), IL-4, IL-6, IL-8, IL-10, interferon-γ (IFN-γ) and TNF-α (Borrelli *et al.*, 1996; Casey *et al.*, 1993; Endo *et al.*, 1994; Girardin *et al.*, 1988; Marchant *et al.*, 1995). Release of this diverse array of cytokines complicates the search for the primary mediators involved in the septic cascade. However, recent evidence brings to light the importance of TNF-α and IL-1β in the inflammatory process. Natanson *et al.* (1988) have shown that in dogs, many cardiovascular manifestations of sepsis coincide with peak TNF-α serum concentrations. Moreover, several groups have demonstrated a dose-dependent correlation between the administration of TNF-α and the pathophysiological changes of sepsis in canine models (Eichenholz *et al.*, 1992; Natanson *et al.*, 1989; Walley *et al.*, 1994). Despite promising experimental evidence, clinical trials involving blockade of TNF-α have been unsuccessful. IL-1β is another important culprit of the septic cascade based on recent studies. Experimentally, infusion of this cytokine into animals and humans results in cardiovascular dysfunction as well as other sepsis-related abnormalities. Interestingly, blockade of IL-1β by receptor antagonists was found to reduce cardiovascular dysfunction and mortality in animal models (Fisher *et al.*, 1994; Ohlsson *et al.*, 1990; Wakabayushi *et al.*, 1991); however, human trials produced more modest results. A possible explanation for the failure of these anti-cytokine trials may lie in the similarity of effects seen among the cytokine family. The redundancy of action observed in the cytokine cascade may compensate for the inhibition of any one particular cytokine, rendering inhibition of a specific pro-inflammatory cytokine an ineffective therapeutic strategy.

Therapies involving the inhibition of cytokine synthesis or their mechanism of action may prove to be detrimental. Animal studies demonstrate that antibody neutralization of TNF-α activity increases mortality, and combination therapy (blockade of both TNF-α and IL-1β receptors) was actually fatal in various models of sepsis (Echtenacher *et al.*, 2001; Eskandari *et al.*, 1992). Possible reasons for these results may be that suppression of the immune system allows pathogenic substances to exert their effects unhindered, and some suggest that clinical trials involving immunosuppressive agents were conducted inappropriately (Opal, 2003). A crucial problem concerning these failed clinical trials appears to be the heterogeneity of the patient population involved. The categorization of septic patients enrolled in these trials exhibited discrepancies leading to unreliable or misleading conclusions (Riedemann *et al.*, 2003).

Manifestations of sepsis

Coagulation

Another severe consequence of septic shock is widespread activation of blood coagulation. A shift in the balance from anti-coagulation to pro-coagulation represents a deleterious consequence of the septic cascade. Activation of the extrinsic and intrinsic pathways in the coagulation cascade is the major contributor to the cardiovascular complications observed in septic patients. Physiologically, tissue factor (TF) is normally contained beneath the intact vascular endothelium. However, during injury, TF is exposed and binds and activates factor VII (FVII), initiating the extrinsic coagulation pathway. This complex (TF–FVIIa) functions to activate factor IX in the intrinsic coagulation pathway. These two events synergistically promote the formation of factor Xa, which catalyzes the activation of prothrombin to thrombin. Lastly, thrombin cleaves the inactive zymogen fibrinogen, forming fibrin monomers which act to form the clot.

In parallel with the facilitation of the coagulation cascade is the existence of anti-coagulant factors to prevent widespread thrombosis (Aird, 2001). Antithrombin, TF pathway inhibitor and activated protein C are the major physiological anti-coagulants that act to oppose thrombus formation. Antithrombin functions to inhibit thrombin formation as well as binding with factors Xa, IXa, XIa and XIIa (Mammen, 1998). TF pathway inhibitor inhibits factor Xa directly as well as the TF–VIIa complex. Protein C is produced in an inactive state. Circulating thrombin binds to thrombomodulin, an endothelial surface protein. This complex then activates protein C (Faust *et al.*, 2001), which inhibits factors V and VIII as well as decreasing further thrombin formation (Grinnell and Joyce, 2001). Activated protein C has been accepted as a powerful tool in combating the deleterious effects of the septic cascade (Levi, 2003).

Vascular complications

The development of persistent systemic vasodilation in sepsis involves two interacting factors. The first is the hyporeactivity of vascular smooth muscle to vasoconstrictors. The second is the enhanced biosynthesis of vasodilatory substances from both the endothelium (Crowley, 1996) and other cells.

Normally, endogenous adrenergic agonists such as noradrenaline (norepinephrine) maintain vascular tone by occupying α-receptors on vascular smooth muscle. Very high doses of vasoconstrictors which would cause severe hypertension in the healthy individual are necessary in the septic patient in order to at least partially reverse the marked hypotension. This suggests that the vasculature is insensitive to adrenergic vasoconstriction. Possible explanations for this phenomenon may be an alteration of adrenoceptor affinity or adrenoceptor number, or various intracellular signaling mechanisms.

Although the cellular pathogenesis of vascular hyporeactivity is unclear, several theories have been presented. The emergence of cytokines is a central factor in the pathogenesis of vascular hyporeactivity. TNF-α, IL-1β and IL-6 are three pro-inflammatory cytokines that are important mediators of hyporeactivity (Vila and Salaices, 2005). It has also been suggested that superoxide ($O_2 \bullet^-$), produced by activated neutrophils and macrophages associated with tissue damage and inflammatory response, can oxidize catecholamines, rendering them incapable of vasoconstriction; therefore, benefits of such vasoconstrictor agents are limited (Macarthur et al., 2000). There is emerging evidence which implicates a disruption in ion homeostasis across the vascular smooth muscle cell membrane in sepsis. Chen et al. (2005) demonstrated an abnormal activation of the Na^+/K^+-ATPase pump in aortae from endotoxemic rats. This results in a rapid depletion of cellular energy as well as a decrease in contractility due to the ionic imbalance. The adenosine triphosphate (ATP)-sensitive potassium channel (K_{ATP}) has also been implicated in the pathogenesis of vascular hyporeactivity. Studies have demonstrated an increase in K_{ATP} channel activity in the vascular smooth muscle of LPS-treated rats (Chen et al., 2005; Sorrentino et al., 1999). This results in vascular smooth muscle cell hyperpolarization, thereby further limiting contractility.

The vascular endothelium lines the luminal side of blood vessels and serves as an interface between circulating blood and tissue. It performs a variety of physiological functions which are essential for homeostasis. These include the regulation of organ perfusion, vascular tone and permeability, as well as the adhesion of platelets and bloodborne cells and the modulation of coagulation (Bassenge, 1996). By virtue of its contact with the bloodstream, the endothelium is the first to come in contact with pathogens. A large body of evidence supports the role of the endothelium in orchestrating the septic immune response. Endothelial activation leads to the release of vasodilators, allowing immune cells to infiltrate the infected tissue. Under normal conditions, the endothelium functions as an anti-coagulant surface; however, during sepsis it is believed to undertake a pro-coagulatory phenotype (Grandel and Grimminger, 2003) by expressing TF, leading to the activation of FVII (Grignani and Maiolo, 2000; Rosenberg and Aird, 1999) and consequential activation of the extrinsic coagulation pathway.

The decrease in vascular tone observed can be explained by enhanced production of several vasodilatory substances. Cytokines such as TNF-α, IFN-γ and IL-1β have been shown to produce excessive amounts of two potent vasodilators, prostaglandin I_2 and NO (see below) (Crowley, 1996). These vasodilators may be endothelial derived or upregulated in the vascular wall (Hernanz et al., 2004; Stoclet et al., 1999).

Myocardial complications

The cardiovascular manifestations of sepsis dominate the clinical presentation of affected patients (Kumar et al., 2001a). Cardiac complications of sepsis arise as the course of septic shock progresses. Parker et al. (1994) have shown that patients who survive septic shock have a reversible depression in left ventricular ejection fraction of <0.4, while paradoxically, those that died had a value of >0.4, although these values normalize 10 days after the onset of sepsis. Additionally, the same group also revealed that septic patients have an abnormal response in left ventricular stroke work index (Ognibene et al., 1988). The correlation between the impairment of cardiovascular function and the poor prognosis of septic patients underscores the importance of understanding the mechanisms for cardiovascular dysfunction associated with sepsis (Clowes et al., 1966, 1970; Kwaan and Weil, 1969; MacLean et al., 1967). Patient presentation is characterized by an elevated cardiac output, reduced afterload and low preload (Kumar et al., 2001a). These signs reflect several pathophysiological changes in the cardiovascular system; one important change is intrinsic contractile dysfunction of the heart.

The cellular mechanisms involved in the development of myocardial dysfunction have been studied for decades. Several studies have shown that cardiac myocyte contractile depression occurs within 10 min of their exposure to pro-inflammatory cytokines such as TNF-α and IL-1β (Kumar et al., 2001b). In contrast to the acute effects of these pro-inflammatory cytokines, prolonged depressant activity appears to involve a change in protein expression in the cardiac myocyte. A large body of evidence supports the involvement of nitric oxide synthase (NOS) as a central mechanism (Krishnagopalan et al., 2002; Schulz et al., 1992, 1995). Pathological generation of NO and its downstream effector, cyclic guanosine monophosphate (cGMP), are believed to contribute to myocardial dysfunction associated with sepsis. Furthermore, recent evidence suggests that symptoms of sepsis include a cardiovascular state of increased oxidative stress (Salvemini and Cuzzocrea, 2002). Various studies have demonstrated the involvement of reactive oxygen species, as well as the efficacy of antioxidant treatment to relieve these symptoms (Salvemini and Cuzzocrea, 2002). Accompanying the increase in oxidative stress is the emergence of another adverse complication, an increased proteolytic state within heart tissue.

Mechanisms of pathogenicity

Nitric oxide

Since the discovery of NO, scientific research has unravelled its existence in many physiological systems. Moreover, NO has been a central component of various cardiovascular abnormalities, including sepsis. NO is upregulated during sepsis, and overproduction of this free radical is a well-established indicator of, and contributor to, the pathogenesis of septic shock.

A link between cytokine generation and oxidative stress appears to be the biosynthesis of NO, a small, short-lived molecule that has received tremendous attention in recent years. NO is a labile gas with a half-life in the time-frame of seconds in physiological buffer at 37°C. The effects of NO were first observed in aortic rings. Light mechanical rubbing of the lumen of a segment of rabbit aorta (a protocol for removing the vascular endothelium) completely inhibited the vasorelaxant properties of acetylcholine (Furchgott and Zawadzki, 1980). At the time, the chemical identity of this relaxant factor was unknown, and it was termed endothelium-derived relaxing factor (EDRF). It was later determined to be NO (Ignarro et al., 1987; Palmer et al., 1987). The generation of NO is catalyzed by three distinct isoforms of NOS: endothelial NOS (eNOS), inducible NOS (iNOS) and neuronal NOS (nNOS). All isoforms of NOS catalyze the oxidation of the amino acid L-arginine to produce NO and citrulline. NO is primarily generated in the cardiovascular system by a Ca^{2+}-dependent NOS in cardiac myocytes (Schulz et al., 1992), endocardial endothelial cells (Schulz et al., 1991) and vascular endothelial cells (Pollock et al., 1991). This NOS isoform was later identified to be eNOS (Balligand et al., 1995). nNOS activity has also been localized to the endoplasmic reticulum in cardiac myocytes (Xu et al., 1999), as well as to cardiac neurons (Calupca et al., 2000; Sawada et al., 1997) and blood vessels innervated with non-adrenergic non-cholinergic nerves, where it is released as a neurotransmitter (Toda and Okamura, 2003). NO exerts a number of regulatory and cytoprotective effects, such as promoting vaso-dilation (Moncada et al., 1991), and decreasing the adhesion of platelets (Radomski et al., 1987) and neutrophils (Kubes et al., 1991) to the en-dothelium, as well as inhibiting platelet aggregation (Radomski, 1989).

Oxidation of one of the guanidine nitrogen atoms of L-arginine forms NO and citrulline. In this process, 1.5 molecules of NADPH per molecule of NO are consumed and molecular oxygen is reduced (Knowles and Moncada, 1994). NO exerts many but not all of its biological effects via its direct activation of soluble guanylyl cyclase, which catalyzes the formation of cGMP from guanosine triphosphate (GTP) (Murad et al., 1987). Soluble guanylyl cyclase contains a haem moiety which is essential for the binding of NO and the activation of the enzyme. cGMP can be acted upon by phosphodiesterases which render it biologically inactive. Of the phosphodiesterase family, phosphodiesterase V is mainly respon-sible for the enzymatic cleavage of cGMP in vascular smooth muscle cells.

The main downstream action of cGMP generation is the activation of protein kinase G, its associated protein kinase. This kinase is involved in the regulation of various enzymes via phosphorylation (Munzel et al., 2003). In vascular smooth muscle, this ultimately promotes vasodila-tion via reducing intracellular Ca^{2+} levels. Specifically, protein kinase G acts to attenuate contraction through a number of mechanisms in-cluding the indirect inhibition of myosin phosphorylation by inhibi-tion of the GTPase protein RhoA as well as activation of K^+ channels and inhibition of the inositol triphosphate receptor (Birschmann and Walter, 2004).

Under inflammatory conditions, NO can be produced in higher con-centrations following the expression of iNOS in the endocardial endothe-lium (Smith et al., 1993), vascular endothelial cells (Radomski et al., 1990), cardiac myocytes (Balligand et al., 1994; Schulz et al., 1992), vascular smooth muscle (Rees et al., 1990) and neutrophils (McCall et al., 1991). Pro-inflammatory cytokines such as those involved in the septic cascade (e.g., TNF-α, IL-1β and IFN-γ) are capable of inducing iNOS (Busse and Mulsch, 1990) in these cell types. This active isoform of NOS continuously produces NO at high concentrations which can be sustained over several hours. Its ability to generate larger amounts of NO can be attributed to its ability to catalyze NO formation in the absence of stimu-lation via Ca^{2+}/calmodulin (Lirk et al., 2002). Additionally, iNOS has been shown to generate $O_2 \bullet^-$ in the absence of substrates such as L-arginine and other NOS cofactors (Xia et al., 1998).

Evidence of enhanced NO and/or peroxynitrite (ONOO$^-$; see below) production is found in the plasma of septic patients as their metabolites,

NO_2^- and NO_3^- (Groeneveld *et al.*, 1996; Ochoa *et al.*, 1991). This overproduction of NO contributes to cardiac dysfunction and systemic vasodilation. This deleterious combination imparts a fatal decrease in cardiac output, resulting in an impairment of tissue perfusion and oxygen extraction (Thiemermann, 1997). The potential relevance of NO in these conditions was supported by *in vivo* animal studies demonstrating the attenuation of cardiovascular effects of cytokines and LPS by NOS inhibitors (Kilbourn *et al.*, 1990a,b; Nava *et al.*, 1992).

During endotoxemia or sepsis, pro-inflammatory cytokines stimulate the expression of iNOS in a variety of cell types including endothelial cells, macrophages, Kupffer cells, cardiac myocytes and vascular smooth muscle cells (Naseem, 2005; Rees *et al.*, 1990; Schulz *et al.*, 1992; Titheradge, 1999). Vascular smooth muscle, for example, is capable of expressing iNOS and this is believed to be responsible for the excessive production of NO and subsequent vasodilation seen in sepsis (Titheradge, 1999). Additionally excessive production of NO by iNOS has been shown to be cytotoxic to the endothelium (Rees *et al.*, 1990; Whittle, 1995), resulting in peripheral oedema and a loss of intravascular blood volume. iNOS knockout mice are resistant to endotoxin-induced hypotension, indicating the involvement of this NOS isoform in the pathogenesis of sepsis (MacMicking *et al.*, 1995). However, various animal studies and clinical trials involving the specific pharmacological inhibition of NOS have revealed difficulties in this strategy (Cobb, 2001; Cobb *et al.*, 1999; Nava *et al.*, 1991). This may be due to the variable induction of iNOS at different anatomical sites. Excessive inhibition of NOS may result in augmented microvascular vasoconstriction, cell hypoxia and lactic acidosis (Li *et al.*, 1992; Rackow and Astiz, 1991). Moreover, it is evident that NO also plays some protective roles in the setting of septic shock, and a careful titration of NOS inhibitor is required to only partially block excess NO production in this setting (Nava *et al.*, 1991; Schulz *et al.*, 1995). These observations bring to light a possible therapeutic strategy targeting the inhibition of iNOS induction; however, as previously mentioned, clinical trials designed to block various inflammatory mediators participating in iNOS induction have produced only modest results (Remick, 2003).

Oxidative stress – Peroxynitrite

A considerable body of evidence suggests that enhanced oxidative stress is a major contributor to endotoxic shock. Moreover, antioxidant therapy has proven beneficial in various models of sepsis. A key link between NO and oxidative stress in sepsis is the discovery of ONOO⁻ as a mediator in this process.

Oxidative stress results from an imbalance between oxidant and anti-oxidant species. A considerable body of evidence demonstrates the involvement of oxidative stress as a central component of various cardiovascular pathologies, including sepsis (Fig. 2). H_2O_2, $O_2 \bullet^-$, hydroxyl radical (OH•) and ONOO⁻ represent the best known reactive species generated from oxygen. $O_2 \bullet^-$ is normally reduced to H_2O_2 via superoxide dismutase (SOD) (McCord and Fridovich, 1969) and the H_2O_2 is then metabolized intracellularly by either glutathione peroxidase or catalase. However, in some scenarios it decomposes to OH• via the iron-dependent Fenton reaction (Reaume et al., 1996). Both $O_2 \bullet^-$ and OH• are reactive oxygen species; the latter is several orders of magnitude more reactive than the former (de Groot, 1994; Kehrer, 1993). These radicals initiate a chain reaction, particularly with membrane lipids, further perpetuating their damaging effects. NADPH oxidase and xanthine oxidase, two major sources of $O_2 \bullet^-$, are also increased in the vascular wall

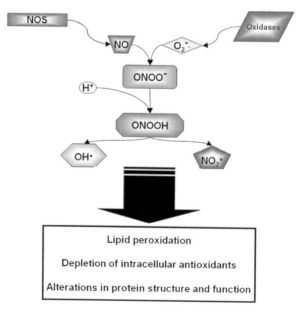

Fig. 2. Pathway for the generation of peroxynitrite and its toxic decomposition products. Excess NO can combine with $O_2 \bullet^-$ to form ONOO⁻. At physiological pH, ONOO⁻ is protonated and spontaneously decomposes into the toxic metabolites OH• and $NO_2 \bullet$.

(Brandes et al., 1999) and heart (Ferdinandy et al., 2000; Khadour et al., 2002) during endotoxemia or exposure of these tissues to pro-inflammatory cytokines.

The mechanism by which NO exerts damaging effects throughout the cardiovascular system involves the formation of $ONOO^-$, the toxic reaction product of NO and $O_2 \bullet^-$ (Beckman et al., 1990) (Fig. 2). During sepsis, the upregulation of NO biosynthesis coincides with an increase in the generation of $O_2 \bullet^-$. At physiological pH, $ONOO^-$ is protonated to form the unstable intermediate peroxynitrous acid. This then readily decomposes into several products, including the highly reactive free radicals nitrogen dioxide ($NO_2 \bullet$) and hydroxyl radical ($OH \bullet$).

These species are far more chemically reactive than $ONOO^-$ itself and are therefore partially responsible for the detrimental effects of $ONOO^-$. Their effects include protein and DNA damage, lipid oxidation, ion channel and transporter malfunction, finally leading to cell death. $ONOO^-$ can alter the structure and function of proteins via its reaction with susceptible amino acid residues (Alvarez and Radi, 2003). This includes the nitration of tyrosine residues, which, for example, inactivates SOD (Yamakura et al., 1998). $ONOO^-$ causes both the peroxidation of lipid membranes and the DNA strand breakage (Szabo, 1999). Previous studies have confirmed the involvement of $ONOO^-$ in human and animal models of sepsis by a variety of experimental methods (Khadour et al., 2002; Kooy et al., 1997; Oyama et al., 1998).

$ONOO^-$ has been shown to deplete cellular glutathione, an endogenous antioxidant in all cells including endothelial (Phelps et al., 1995) and smooth muscle cells (Szabo et al., 1996) as well as the plasma (Van der Vliet et al., 1994). Concerning vascular hyporeactivity, it has been suggested that this phenomenon may be partially attributed to the capacity of $O_2 \bullet^-$ to inactivate endogenous norepinephrine (Macarthur et al., 2000), resulting in the production of adrenochromes (Graham, 1978) which have been implicated in various cardiovascular abnormalities (Singal et al., 1982; Yates et al., 1981).

Activation of poly(ADP-ribose) polymerase (PARP) by $ONOO^-$ has emerged as a significant mechanism in the reduction of cardiac contractility and vascular hyporeactivity in sepsis (Evgenov and Liaudet, 2005; Tasatargil et al., 2005). PARP is a highly conserved enzyme found in nuclei. Cleavage of PARP creates an active enzyme that is involved in the repair of single-strand DNA breaks by recruiting and activating DNA repair enzymes. Increased PARP activation during sepsis, however, can impart a disturbance in cellular metabolism by depleting intracellular stores of its substrate, nicotinamide adenine dinucleotide (NAD^+)

(Pacher *et al.*, 2005). This results in an impairment of glycolysis, Kreb's cycle activity and mitochondrial electron transport, thus resulting in ATP depletion. Treatment of animals with a PARP inhibitor reduced the hyporesponsiveness observed in LPS-treated rats. PARP activation is also involved in the regulation of various inflammatory proteins, including iNOS as well as other cytokines and chemokines.

Matrix metalloproteinases

Recent evidence indicates that $ONOO^-$ as well as various cytokines can enhance the activity and/or expression of MMPs, targeting them as important culprits for the cardiovascular dysfunction associated with sepsis. A growing body of evidence supports the involvement of MMPs, as well as the beneficial effects of MMP inhibition in various models of sepsis.

Classification and structure

MMPs are a large family of zinc-dependent endopeptidases which were first discovered as a collagenolytic activity released from the tail of a tadpole undergoing metamorphosis (Gross and Lapiere, 1962). They are best known as proteolytic enzymes which degrade extracellular matrix proteins, including collagen, necessary for tissue remodelling processes. MMPs are classified by numerical designation (MMP-1 through MMP-28) and are also categorized by their *in vitro* substrate specificity for certain extracellular matrix substrates. Groups of MMPs include the collagenases (MMP-1, MMP-8 and MMP-13), the stromelysins (MMP-3 and MMP-10), the matrilysins (MMP-7 and MMP-26), membrane-type MMPs (MT-MMPs, 1 through 8), and the gelatinases (MMP-2 and MMP-9).

All MMPs are initially synthesized in an inactive zymogen form (pro-MMP) (Woessner, 1998). Structurally, MMPs have a signaling peptide at the N-terminus allowing secretion into the endoplasmic reticulum and eventual transport out of the cell. Beside the signal peptide lies a hydrophobic propeptide domain involved in shielding the catalytic domain next to it. This catalytic domain is present in all MMPs and is known as the "matrixin fold", which forms substrate binding pockets. The catalytic Zn^{2+} is coordinated to a cysteinyl sulphydryl group on the propeptide domain. This intermolecular association is termed the "cysteine switch" and is conserved across the MMP family (Van Wart and Birkedal-Hansen, 1990).

The key feature of the collagenases is their ability to cleave a specific site of interstitial collagens I, II and III (Visse and Nagase, 2003). The stromelysins can activate a number of pro-MMPs, including pro-MMP-1 (Suzuki *et al.*, 1990). Matrilysins are characterized by the lack of an intact haemopexin domain (Visse and Nagase, 2003). MT-MMPs are transmembrane and glycosyl phosphatidyl inositol anchored proteins. They are all capable of activating pro-MMP-2 as well as digesting a number of extracellular matrix substrates. The gelatinases are known for their ability to digest denatured collagens (gelatin). These enzymes have three repeats of a type II fibronectin domain contained within the catalytic domain which bind gelatin, collagens and laminin (Allan *et al.*, 1995). Those MMPs not classified into the above categories are involved in other biological functions, including macrophage migration (MMP-12) (Shipley *et al.*, 1996), enamel formation (MMP-20) (Li *et al.*, 2001) and tissue homeostasis (Lohi *et al.*, 2001).

The propeptide domain must be perturbed in order for the MMP to become proteolytically active. Various mechanisms have been elucidated concerning activation of this class of enzymes and each is distinct in its own way. One mechanism involves the activation of MMPs in the extracellular space; this occurs in two distinct steps. First, activation is initiated by other proteases (*e.g.*, trypsin or other MMPs) that cleave the propeptide at specific sites. Upon cleavage, the shielding of the catalytic cleft is withdrawn, exposing the catalytic Zn^{2+} ion. The propeptide, however, is not entirely removed and the newly active MMP undergoes intermolecular autocatalysis which cleaves the remaining propeptide, thus generating a lower molecular weight enzyme (Morgunova *et al.*, 1999). MMPs can also be activated by other members of its class. One important example involves the activation of MMP-2. MT1-MMP, which is located in the plasma membrane, associates with two molecules of tissue inhibitor of matrix metalloproteinase-2 (TIMP-2). This complex binds pro-MMP-2, allowing for proteolytic activation of MMP-2 (Strongin *et al.*, 1995). Another mechanism of activation involves the proprotein convertase, furin (Kang *et al.*, 2002; Pei and Weiss, 1995; Sato *et al.*, 1996). In contrast to extracellular activation, this action takes place intracellularly as furin is present in the Golgi network. However, the proteolytic action of furin on MMP-2 in the Golgi can also result in an inactive MMP-2 (Cao *et al.*, 2005). After proteolytic activation, the MMP is either targeted to the cell membrane for insertion (in the case of MT-MMPs) or secreted from the cell.

Function
In the context of cardiovascular diseases involving enhanced oxidative stress, the most important activation mechanism that can occur both

intracellularly and extracellularly involves a direct posttranslational modification of a cysteine residue in the autoinhibitory propeptide domain (Fig. 3). In this pathway, oxidative species such as H_2O_2, $O_2 \bullet^-$, OH• and $ONOO^-$ oxidize the sulphydryl bond of the cysteinyl group involved in coordinating the catalytic Zn^{2+} ion. This causes a conformational change in the enzyme which exposes the catalytic Zn^{2+} ion and produces an active enzyme to which the propeptide domain is still attached (Okamoto et al., 1997; Rajagopalan et al., 1996). There is inaccuracy in the current nomenclature of MMPs in that pro-MMPs are commonly referred as the inactive zymogen form of MMPs solely due to

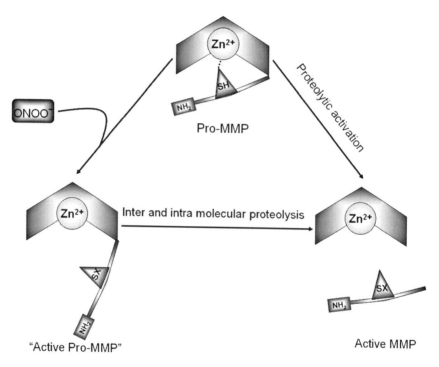

Fig. 3. Activation of matrix metalloproteinases (MMPs) via proteolysis and ONOO−. MMPs are synthesized as inactive zymogens (pro-MMPs) and are classically activated on the luminal side of the plasma membrane via proteolytic removal of the propeptide by other proteases, including MMPs (right-hand pathway). However, under conditions of oxidative stress, $ONOO^-$ is capable of oxidizing the sulphydryl moiety (SX) of the cysteinyl group coordinated to the catalytic Zn^{2+} ion. This "active pro-MMP" can cleave susceptible protein targets and can also undergo inter- and intramolecular catalysis to yield active MMP without the propeptide domain.

their higher molecular weight, as seen in sodium dodecyl sulphate poly-acrylamide gel electrophoresis (SDS-PAGE) – whereas MMP activation by oxidative stress results in a proteolytically active "pro-MMP" form which is only distinguishable in molecular weight from its zymogen form with mass spectrometry, and not SDS-PAGE.

Despite the large variability in function, MMPs are best known to degrade extracellular matrix proteins and are involved in both physio-logical and pathological processes including embryogenesis, organo-genesis, angiogenesis, wound healing and platelet aggregation (Galis and Khatri, 2002; Sawicki et al., 1997). Recent evidence implicates this class of enzymes in the pathogenesis of cancer, inflammatory arthritis, and pulmonary and cardiovascular diseases (Chakraborti et al., 2003; Overall et al., 2002).

Inhibitors of MMPs

The tissue inhibitors of matrix metalloproteinases (TIMPs) are endog-enous proteins involved in the regulation and inhibition of MMP activity. To date four TIMPs have been identified and each binds to a MMP in a 1:1 stoichiometric ratio (Brew et al., 2000). Structurally, TIMPs have an N-terminal inhibitory domain and a smaller C-terminal domain. TIMP-1 to TIMP-4 have a broad range of inhibitory activity against several MMPs. These TIMPs differ in that TIMP-2 is constitu-tively expressed, whereas TIMP-1 can be induced by pro-inflammatory cytokines (Li et al., 1999). TIMP-3 is less characterized but has been shown to be involved in angiogenesis as well as lung abnormalities (Martin et al., 2005). In the heart, TIMP-3 is found in the extracellular matrix (Leco et al., 1994; Young et al., 2002). TIMP-4 is perhaps the most widespread TIMP in the cardiovascular system and has been locali-zed to the sarcomere of cardiac myocytes (Schulze et al., 2003). It has been suggested to have cardioprotective effects (Dollery et al., 1999).

Pharmacological inhibitors of MMP activity have been utilized and exhibit different mechanisms of action. The tetracycline class of antibio-tics is recognized to have MMP inhibitory activity distinct from its anti-microbial effects (Golub et al., 1998). Specifically, doxycycline has been shown to interact with the structural Zn^{2+} ions of MMPs (Garcia et al., 2005). GM6001 is a hydroxamic acid based MMP inhibitor designed to act as a bidentate ligand for the catalytic Zn^{2+} in the active site (Galardy et al., 1994). Pharmaceutical companies have produced several proprietary compounds in the last two decades as promising treat-ments for inflammatory disease (e.g., rheumatoid arthritis) and cancer (Peterson, 2006).

MMPs and sepsis

Although few studies have dealt with the relationship between MMPs and sepsis, strong evidence has been provided by researchers using isolated cell culture models, human and animal models, and septic patients. Focus has been centred on the gelatinases (MMP-2 and MMP-9) since they are abundant (MMP-2) or can be induced by cytokines (MMP-9) in a variety of cardiovascular cell types. Xie *et al.* (1994) demonstrated the dose-dependent relationship between the addition of LPS and MMP-2 and MMP-9 activities in isolated murine macrophages. Pugin *et al.* (1999) performed similar experiments using human blood. In this model, MMP-9 was shown to have increased activity after stimulation by LPS. Albert *et al.* (2003) found that circulating MMP-9 activity increased significantly within 2 h after administration of LPS to human volunteers. These studies, among others, demonstrate the relationship between LPS and MMP-2 and MMP-9 activities, providing a possible link between MMPs and clinical sepsis.

Myocardial depression seen in sepsis ultimately involves a reversible attenuation of the contractile efficiency of the myofilaments in cardiac myocytes (Crowley, 1996). This may be due to decreased cytosolic Ca^{2+} release, dysregulation of intracellular Ca^{2+} and/or decreased myofilament sensitivity to Ca^{2+}. These mechanisms have been shown in *in vivo* models of sepsis-induced myocardial dysfunction. Indeed, upregulation of proteolytic MMPs occurs in models of sepsis, with the concomitant degradation of key contractile elements in the cardiac sarcomere such as troponin I (Felten *et al.*, 2004; Gao *et al.*, 2003; Lalu *et al.*, 2003).

In baboons subjected to *E. coli*-induced sepsis, an increase in MMP-9 activity was found in serum (Paemen *et al.*, 1997). Dubois *et al.* (2002) conducted a study using MMP-9 knockout mice and found that they were significantly more resistant than controls to lethal doses of LPS. Carney *et al.* (2001) studied the utility of MMP inhibition using a chemically modified tetracycline (devoid of anti-bacterial activity, yet retaining MMP inhibitory action) in pigs which were administered LPS. The LPS-treated group exhibited a dramatic reduction in blood pressure which was abolished in the group treated with a MMP inhibitor, demonstrating the involvement of MMPs in the development of severe hypotension induced by LPS. Steinberg *et al.* (2005) showed that the same chemically modified tetracycline significantly reduced morbidity in a pig model of sepsis induced by the introduction of a faecal blood clot into the peritoneal cavity and accompanied by mesenteric artery occlusion. Lalu *et al.* (2003) demonstrated that the MMP inhibitors Ro 31-9790 or doxycycline attenuated LPS-induced myocardial contractile dysfunction in rats. In the same rat model of endotoxemia, Lalu *et al.* (2004) found that the symptoms of endotoxic shock peaked 6–12 h after LPS administration and were accompanied by a subsequent loss in ventricular MMP-2 activity. Likewise, plasma levels of MMP-2 were found to be significantly depressed 3–12 h after LPS administration. Plasma MMP-9 activity and protein levels, however, peaked 1 h after LPS administration. The same group showed activation of MMP-2 in aortae isolated from rats treated with LPS *in vivo* or exposed to IL-1β *in vitro* (Lalu *et al.*, 2006). There was a loss of TIMP-4 protein in the aortae from LPS-treated rats, suggesting a TIMP/MMP imbalance. Importantly, MMP inhibitors prevented the LPS- or IL-1β-induced hyporeactivity to vasoconstrictors, a hallmark of the severe hypotension seen in sepsis. The balance between MMPs and TIMPs is an important factor to consider in dealing with MMP activity. One study by Martin *et al.* (2003) demonstrated that TIMP-3 knockout mice are more prone to the detrimental effects of sepsis due to the increased activity of MMPs.

Additionally, cytokines have been shown not only to upregulate MMP activity and expression, but to in fact be regulated by MMPs themselves (Overall *et al.*, 2002). Certain members of the MMP family have been shown to be involved in the proteolytic processing of both pro- and anti-inflammatory cytokines and chemokines. Zhang *et al.* (2003) revealed a mechanism in which the chemokine stromal cell derived factor-1 is cleaved by MMP-2 into a highly neurotoxic substance. Conversely, MMP-2 mediated cleavage of monocyte chemoattractant protein-3 results in a product which is an antagonist of the receptors for this protein, thus providing a mechanism by which a MMP may attenuate inflammation (McQuibban *et al.*, 2002). These complexities add both a new dimension and possible cautions at the frontier of MMP drug design.

Conclusions

Despite convincing evidence implicating the contribution of MMPs in the cardiovascular manifestations of sepsis, there is still much to be discovered. The exact list of proteolytic targets of MMPs in sepsis, whether in the heart, vascular wall or the bloodstream, has yet to be revealed. There is increasing evidence of the involvement of MMPs in cardiovascular diseases involving enhanced oxidative stress, including sepsis. Targeting MMPs for the treatment of sepsis presents a novel and promising therapeutic approach which requires careful study.

386

Acknowledgements

JC and AKC acknowledge financial support from the 75th Anniversary Award, Faculty of Medicine & Dentistry, University of Alberta. Results from the authors' laboratory are supported by grants from the Heart and Stroke Foundation of Alberta, NWT and Nunavut and the Canadian Institutes of Health Research. RS is a scientist of the Alberta Heritage Foundation for Medical Research.

References

Aird, W. C. (2001). Vascular bed-specific hemostasis: Role of endothelium in sepsis pathogenesis. *Crit. Care Med.* 29,S28–S35.

Albert, J., Radomski, A., Soop, A., Sollevi, A., Frostell, C. and Radomski, M. W. (2003). Differential release of matrix metalloproteinase-9 and nitric oxide following infusion of endotoxin to human volunteers. *Acta Anaesthesiol. Scand.* 47,407–410.

Alexander, C. and Rietschel, E. T. (2001). Bacterial lipopolysaccharides and innate immunity. *J. Endotoxin Res.* 7,167–202.

Allan, J. A., Docherty, A. J., Barker, P. J., Huskisson, N. S., Reynolds, J. J. and Murphy, G. (1995). Binding of gelatinases A and B to type-I collagen and other matrix components. *Biochem. J.* 309,299–306.

Alvarez, B. and Radi, R. (2003). Peroxynitrite reactivity with amino acids and proteins. *Amino Acids* 25,295–311.

Antal-Szalmas, P. (2000). Evaluation of CD14 in host defence. *Eur. J. Clin. Invest.* 30,167–179.

Balligand, J. L., Kobzik, L., Han, X., Kaye, D. M., Belhassen, L., O'Hara, D. S., Kelly, R. A., Smith, T. W. and Michel, T. (1995). Nitric oxide-dependent parasympathetic signaling is due to activation of constitutive endothelial (type III) nitric oxide synthase in cardiac myocytes. *J. Biol. Chem.* 270,14582–14586.

Balligand, J. L., Ungureanu-Longrois, D., Simmons, W. W., Pimental, D., Malinski, T. A., Kapturczak, M., Taha, Z., Lowenstein, C. J., Davidoff, A. J. and Kelly, R. A. (1994). Cytokine-inducible nitric oxide synthase (iNOS) expression in cardiac myocytes. Characterization and regulation of iNOS expression and detection of iNOS activity in single cardiac myocytes in vitro. *J. Biol. Chem.* 269,27580–27588.

Bassenge, E. (1996). Endothelial function in different organs. *Prog. Cardiovasc. Dis.* 39,209–228.

Beckman, J. S., Beckman, T. W., Chen, J., Marshall, P. A. and Freeman, B. A. (1990). Apparent hydroxyl radical production by peroxynitrite: Implications for endothelial injury from nitric oxide and superoxide. *Proc. Natl. Acad. Sci. USA* 87,1620–1624.

Birschmann, I. and Walter, U. (2004). Physiology and pathophysiology of vascular signaling controlled by guanosine 3′,5′-cyclic monophosphate-dependent protein kinase. *Acta Biochim. Pol.* 51,397–404.

Bone, R. C. (1992). Modulators of coagulation. A critical appraisal of their role in sepsis. *Arch. Intern. Med.* 152,1381–1389.

Borrelli, E., Roux-Lombard, P., Grau, G. E., Girardin, E., Ricou, B., Dayer, J. and Suter, P. M. (1996). Plasma concentrations of cytokines, their soluble receptors, and antioxidant vitamins can predict the development of multiple organ failure in patients at risk. *Crit. Care Med.* 24,392–397.

Brandes, R. P., Koddenberg, G., Gwinner, W., Kim, D., Kruse, H. J., Busse, R. and Mugge, A. (1999). Role of increased production of superoxide anions by NAD(P)H oxidase and xanthine oxidase in prolonged endotoxemia. *Hypertension* 33,1243–1249.

Brew, K., Dinakarpandian, D. and Nagase, H. (2000). Tissue inhibitors of metalloproteinases: Evolution, structure and function. *Biochem. Biophys. Acta* 1477,267–283.

Busse, R. and Mulsch, A. (1990). Induction of nitric oxide synthase by cytokines in vascular smooth muscle cells. *FEBS Lett.* 275,87–90.

Calupca, M. A., Vizzard, M. A. and Parsons, R. L. (2000). Origin of neuronal nitric oxide synthase (NOS)-immunoreactive fibers in guinea pig parasympathetic cardiac ganglia. *J. Comp. Neurol.* 426,493–504.

Cao, J., Rehemtulla, A., Pavlaki, M., Kozarekar, P. and Chiarelli, C. (2005). Furin directly cleaves proMMP-2 in the trans-Golgi network resulting in a nonfunctioning proteinase. *J. Biol. Chem.* 280,10974–10980.

Carney, D. E., McCann, U. G., Schiller, H. J., Gatto, L. A., Steinberg, J., Picone, A. L. and Nieman, G. F. (2001). Metalloproteinase inhibition prevents acute respiratory distress syndrome. *J. Surg. Res.* 99,245–252.

Casey, L. C., Balk, R. A. and Bone, R. C. (1993). Plasma cytokine and endotoxin levels correlate with survival in patients with the sepsis syndrome. *Ann. Intern. Med.* 199,771–778.

Chakraborti, S., Mandal, M., Das, S., Mandal, A. and Chakraborti, T. (2003). Regulation of matrix metalloproteinases: An overview. *Mol. Cell. Biochem.* 253,269–285.

Chen, S. J., Chen, K. H. and Wu, C. C. (2005). Nitric oxide–cyclic GMP contributes to abnormal activation of Na^+–K^+-ATPase in the aorta from rats with endotoxic shock. *Shock* 23,179–185.

Clowes, G. H., Jr., Farrington, G. H., Zuschneid, W., Cossette, G. R. and Saravis, C. (1970). Circulating factor in the etiology of pulmonary insufficiency and right heart failure accompanying severe sepsis (peritonitis). *Ann. Surg.* 171,663–678.

Clowes, G. H., Jr., Vucinic, M. and Weidner, M. G. (1966). Circulatory and metabolic alterations associated with survival or death in peritonitis: Clinical analysis of 25 cases. *Ann. Surg.* 163,866–885.

Cobb, J. P. (2001). Nitric oxide synthase inhibition as therapy for sepsis: A decade of promise. *Surg. Infect. (Larchmt.)* 2,93–101.

Cobb, J. P., Hotchkiss, R. S., Swanson, P. E., Chang, K., Qiu, Y., Laubach, V. E., Karl, I. E. and Buchman, T. G. (1999). Inducible nitric oxide synthase (iNOS) gene deficiency increases the mortality of sepsis in mice. *Surgery* 126,438–442.

Crowley, S. R. (1996). The pathogenesis of septic shock. *Heart Lung* 25,124–134.

de Groot, H. (1994). Reactive oxygen species in tissue injury. *Hepatogastroenterology* 41,328–332.

Dollery, C. M., McEwan, J. R., Wang, M., Sang, Q. A., Liu, Y. E. and Shi, Y. E. (1999). TIMP-4 is regulated by vascular injury in rats. *Ann. N. Y. Acad. Sci.* 878,740–741.

Dubois, B., Starckx, S., Pagenstecher, A., Oord, J., Arnold, B. and Opdenakker, G. (2002). Gelatinase B deficiency protects against endotoxin shock. *Eur. J. Immunol.* 32, 2163–2171.

Echtenacher, B., Weigl, K., Lehn, N. and Mannel, D. N. (2001). Tumor necrosis factor-dependent adhesions as a major protective mechanism early in septic peritonitis in mice. *Infect. Immun.* 69,3550–3555.

Eichenholz, P. W., Eichaker, P. Q., Hoffman, W. D., Banks, S. M., Parrillo, J. E., Danner, R. L. and Natanson, C. (1992). Tumor necrosis factor challenges in canines: Patterns of cardiovascular dysfunction. *Am. J. Physiol.* 263,H668–H675.

Endo, S., Inada, K., Yamada, Y., Takauwa, T., Kasai, T., Nakae, H., Yoshida, M. and Ceska, M. (1994). Plasma endotoxin and cytokine concentrations in patients with hamorrhagic shock. *Crit. Care Med.* 22,949–955.

Eskandari, M. K., Bolgos, G., Miller, C., Nguyen, D. T., DeForge, L. E. and Remick, D. G. (1992). Anti-tumor necrosis factor antibody therapy fails to prevent lethality after cecal ligation and puncture or endotoxemia. *J. Immunol.* 148,2724–2730.

Evgenov, O. V. and Liaudet, L. (2005). Role of nitrosative stress and activation of poly(ADP-ribose) polymerase-1 in cardiovascular failure associated with septic and hemorrhagic shock. *Curr. Vasc. Pharmacol.* 3,293–299.

Faust, S. N., Levin, M., Harrison, O. B., Goldin, R. D., Lockhart, M. S., Kondaveeti, S., Laszik, Z., Esmon, C. T. and Heyderman, R. S. (2001). Dysfunction of endothelial protein C activation in severe meningococcal sepsis. *N. Engl. J. Med.* 345, 408–416.

Felten, M. L., Cosson, C., Charpentier, J., Paradis, V., Benhamou, D., Mazoit, J. X. and Edouard, A. R. (2004). Effect of isoproterenol on the cardiac troponin I degradation and release during early TNFalpha-induced ventricular dysfunction in isolated rabbit heart. *J. Cardiovasc. Pharmacol.* 44,532–538.

Ferdinandy, P., Danial, H., Ambrus, I., Rothery, R. A. and Schulz, R. (2000). Peroxynitrite is a major contributor to cytokine-induced myocardial contractile failure. *Circ. Res.* 87,241–247.

Fisher, C. J., Slotman, G. H., Opal, S. M., Pribble, J. P., Bone, R. C., Emmanuel, G., Ng, D., Bloedow, D. C. and Catalano, M. A. (1994). Initial evaluation of human recombinant interleukin-1 receptor antagonist in the treatment of sepsis syndrome: A randomized, open-label, placebo-controlled multicenter trial. The IL-1RA Sepsis Syndrome Study Group. *Crit. Care Med.* 22,12–21.

Furchgott, R. F. and Zawadzki, J. V. (1980). The obligatory role of endothelial cells in the relaxation of arterial smooth muscle by acetylcholine. *Nature* 288,373–376.

Galardy, R. E., Cassabonne, M. E., Giese, C., Gilbert, J. H., Lapierre, F., Lopez, H., Schaefer, M. E., Stack, R., Sullivan, M., Summers, B., Tressler, R., Tyrrell, D., Wee, J., Allen, S. D., Castellot, J. J., Barletta, J. P., Schultz, G. S., Fernandez, L. A., Fisher, S., Cui, T., Foellmer, H. G., Grobelny, D. and Holleran, W. M. (1994). Low molecular weight inhibitors in corneal ulceration. *Ann. N. Y. Acad. Sci.* 732,315–323.

Galis, Z. S. and Khatri, J. J. (2002). Matrix metalloproteinases in vascular remodeling and atherogenesis: the good, the bad, and the ugly. *Circ. Res.* 90,251–262.

Gao, C. Q., Sawicki, G., Suarez-Pinzon, W. L., Csont, T., Wozniak, M., Ferdinandy, P. and Schulz, R. (2003). Matrix metalloproteinase-2 mediates cytokine-induced myocardial contractile dysfunction. *Cardiovasc. Res.* 57,426–433.

Garcia, R. A., Pantazatos, D. P., Gessner, C. R., Go, K. V., Woods, V. L., Jr. and Villarreal, F. J. (2005). Molecular interactions between matrilysin and the matrix metalloproteinase inhibitor doxycycline investigated by deuterium exchange mass spectrometry. *Mol. Pharmacol.* 67,1128–1136.

Girardin, E., Grau, G. E., Dayer, J. M., Roux-Lombard, P. and Lambert, P. H. (1988). Tumor necrosis factor and interleukin-1 in the serum of children with severe infection purpura. *N. Engl. J. Med.* 319,397–400.

Golub, L. M., Lee, H. M., Ryan, M. E., Giannobile, W. V., Payne, J. and Sorsa, T. (1998). Tetracyclines inhibit connective tissue breakdown by multiple non-antimicrobial mechanisms. *Adv. Dent. Res.* 12,12–26.

Graham, D. G. (1978). Oxidative pathways for catecholamines in the genesis of neuro-melanin and cytotoxic quinones. *Mol. Pharmacol.* 14,633–643.

Grandel, U. and Grimminger, F. (2003). Endothelial responses to bacterial toxins in sepsis. *Crit. Rev. Immunol.* 23,267–299.

Grignani, G. and Maiolo, A. (2000). Cytokines and hemostasis. *Haematologica* 85,967–972.

Grinnell, B. W., Joyce, D. (2001). Recombinant human activated protein C: A system modulator of vascular function for treatment of severe sepsis. *Crit. Care Med.*, 29, S53–S60 discussion S60–S61.

Groeneveld, P. H., Kwappenberg, K. M., Langermans, J. A., Nibbering, P. H. and Curtis, L. (1996). Nitric oxide (NO) production correlates with renal insufficiency and multiple organ dysfunction syndrome in severe sepsis. *Intensive Care Med.* 22,1197–1202.

Gross, J. and Lapiere, C. M. (1962). Collagenolytic activity in amphibian tissues: A tissue culture assay. *Proc. Natl. Acad. Sci. USA* 48,1014–1022.

Hernanz, R., Alonso, M. J., Zibrandtsen, H., Alvarez, Y., Salaices, M. and Simonsen, U. (2004). Measurements of nitric oxide concentration and hyporeactivity in rat superior mesenteric artery exposed to endotoxin. *Cardiovasc. Res.* 62,202–211.

Ignarro, L. J., Buga, G. M., Wood, K. S., Byrns, R. E. and Chaudhuri, G. (1987). Endothelium-derived relaxing factor produced and released from artery and vein is nitric oxide. *Proc. Natl. Acad. Sci. USA* 84,9265–9269.

Kang, T., Nagase, H. and Pei, D. (2002). Activation of membrane-type matrix metal-loproteinase 3 zymogen by the proprotein convertase furin in the trans-Golgi network. *Cancer Res.* 62,675–681.

Kehrer, J. P. (1993). Free radicals as mediators of tissue injury and disease. *Crit. Rev. Toxicol.* 23,21–48.

Khadour, F. H., Panas, D., Ferdinandy, P., Schulze, C., Csont, T., Lalu, M. M., Wildhirt, S. M. and Schulz, R. (2002). Enhanced NO and superoxide generation in dysfunctional hearts from endotoxemic rats. *Am. J. Physiol. Heart Circ. Physiol.* 283,H1108–H1115.

Kilbourn, R. G., Gross, S. S., Jubran, A., Adams, J., Griffith, O. W., Levi, R. and Lodato, R. F. (1990a). N^G-Methyl-L-arginine inhibits tumor necrosis factor-induced hypotension: Implications for the involvement of nitric oxide. *Proc. Natl. Acad. Sci. USA* 87,3629–3632.

Kilbourn, R. G., Jubran, A., Gross, S. S., Griffith, O. W., Levi, R., Adamn, J. and Lodato, R. F. (1990b). Reversal of endotoxin-mediated shock by N^G-methyl-L-arginine, an inhibitor of nitric oxide synthesis. *Biochem. Biophys. Res. Commun.* 172,1132–1138.

Knowles, R. G. and Moncada, S. (1994). Nitric oxide synthases in mammals. *Biochem. J.* 298(2), 249–258.

Kooy, N. W., Lewis, S. J., Royall, J. A., Ye, Y. Z., Kelly, D. R. and Beckman, J. S. (1997). Extensive tyrosine nitration in human myocardial inflammation: Evidence for the presence of peroxynitrite. *Crit. Care Med.* 25,812–819.

Krishnagopalan, S., Kumar, A. and Parrillo, J. E. (2002). Myocardial dysfunction in the patient with sepsis. *Curr. Opin. Crit. Care.* 8,376–388.

Kubes, P., Suzuki, M. and Granger, D. N. (1991). Nitric oxide: An endogenous modulator of leukocyte adhesion. *Proc. Natl. Acad. Sci. USA* 88,4651–4655.

Kumar, A., Cameron, H. and Parrillo, J. (2001a). Myocardial dysfunction in septic shock: Part I. Clinical manifestation of cardiovascular dysfunction. *J. Cardiothorac. Vasc. Anesth.* 15,364–376.

Kumar, A., Cameron, H. and Parrillo, J. (2001b). Myocardial dysfunction in septic shock: Part II. Role of cytokines and nitric oxide. *J. Cardiothorac. Vasc. Anesth.* 15,485–511.

Kwaan, H. M. and Weil, M. H. (1969). Differences in the mechanism of shock caused by bacterial infections. *Surg. Gynecol. Obstet.* 128,37–45.

Lalu, M. M., Cena, J., Chowdhury, R., Lam, A. and Schulz, R. (2006). Matrix metalloproteinases contribute to endotoxin and interleukin-1β induced vascular dysfunction. *Br. J. Pharmacol.* 149,31–42.

Lalu, M. M., Csont, T. and Schulz, R. (2004). Matrix metalloproteinase activities are altered in the heart and plasma during endotoxemia. *Crit. Care Med.* 32, 1332–1337.

Lalu, M. M., Gao, C. Q. and Schulz, R. (2003). Matrix metalloproteinase inhibitors attenuate endotoxemia induced cardiac dysfunction: A potential role for MMP-9. *Mol. Cell. Biochem.* 251,61–66.

Landmann, R., Muller, B. and Zimmerli, W. (2000). CD14, new aspects of ligand and signal diversity. *Microbes Infect.* 2,295–304.

Leco, K. J., Khokha, R., Pavloff, N., Hawkes, S. P. and Edwards, D. R. (1994). Tissue inhibitor of metalloproteinases-3 (TIMP-3) is an extracellular matrix-associated protein with a distinctive pattern of expression in mouse cells and tissues. *J. Biol. Chem.* 269,9352–9360.

Levi, M. (2003). Benefit of recombinant human activated protein C beyond 28-day mortality: There is more to life than death. *Crit. Care Med.* 31,984–985.

Levy, M. M., Fink, M. P., Marshall, J. C., Abraham, E., Angus, D., Cook, D., Cohen, J. and Opal, S. M. (2003). SCCM/ESICM/ACCP/ATS/SIS International Sepsis Definitions Conference. *Crit. Care Med.* 31,1250–1256.

Li, T., Croce, K. and Winquist, R. J. (1992). Regional differences in the effects of septic shock on vascular reactivity in the rabbit. *J. Pharmacol. Exp. Ther.* 261, 959–963.

Li, W., Gibson, C. W., Abrams, W. R., Andrews, D. W. and DenBesten, P. K. (2001). Reduced hydrolysis of amelogenin may result in X-linked amelogenesis imperfecta. *Matrix Biol.* 19,755–760.

Li, X., Tupper, J. C., Bannerman, D. D., Winn, R. K., Rhodes, C. J. and Harlan, J. M. (2003). Toll-like receptor 4-induced activation of NF-κB in endothelial cells. *Infect. Immun.* 71,4414–4420.

Li, Y. Y., McTiernan, C. F. and Feldman, A. M. (1999). Proinflammatory cytokines regulate tissue inhibitors of metalloproteinases and disintegrin metalloproteinase in cardiac cells. *Cardiovasc. Res.* 42,162–172.

Lirk, P., Hoffmann, G. and Rieder, J. (2002). Inducible nitric oxide synthase – Time for reappraisal. *Curr. Drug Targets Inflamm. Allergy* 1,89–108.

Lohi, J., Wilson, C. L., Roby, J. D. and Parks, W. C. (2001). Epilysin, a novel human matrix metalloproteinase (MMP-28) expressed in testis and keratinocytes and in response to injury. *J. Biol. Chem.* 276,10134–10144.

Macarthur, H., Westfall, T. C., Riley, D. P., Misko, T. P. and Salvemini, D. (2000). Inactivation of catecholamines by superoxide gives new insights on the pathogenesis of septic shock. *Proc. Natl. Acad. Sci. USA* 97,9753–9758.

MacLean, L. D., Mulligan, W. G., McLean, A. P. and Duff, J. H. (1967). Patterns of septic shock in man – A detailed study of 56 patients. *Ann. Surg.* 166,543–562.

MacMicking, J. D., Nathan, C., Hom, G., Chartrain, N., Fletcher, D. S., Trumbauer, M., Stevens, K., Xie, Q. W., Sokol, K., Hutchinson, N., Chen, H. and Mudgett, J. S. (1995). Altered responses to bacterial infection and endotoxic shock in mice lacking inducible nitric oxide synthase. *Cell* 81,641–650.

Mammen, E. F. (1998). Antithrombin: Its physiological importance and role in DIC. *Semin. Thromb. Hemost.* 24,19–25.

Marchant, A., Alegre, M. L., Hakim, A., Pierard, G., Marecaux, G., Friedman, G., DeGroote, D., Khan, R. J., Vincent, J. L. and Goldman, M. (1995). Clinical and biological significance of interleukin-10 plasma levels in patients with septic shock. *J. Clin. Immunol.* 15,266–273.

Marshall, J. (2003). Such stuff dreams are made on: Mediator-directed therapy in sepsis. *Nat. Rev. Drug Discov.* 2,391–405.

Martin, E. L., McCaig, L. A., Moyer, B. Z., Pape, M. C., Leco, K. J., Lewis, J. F. and Veldhuizen, R. A. (2005). Differential response of TIMP-3 null mice to the lung insults of sepsis, mechanical ventilation, and hyperoxia. *Am. J. Physiol. Lung Cell. Mol. Physiol.* 289,L244–L251.

Martin, E. L., Moyer, B. Z., Pape, M. C., Starcher, B., Leco, K. J. and Veldhuizen, R. A. (2003). Negative impact of tissue inhibitor of metalloproteinase-3 null mutation on lung structure and function in response to sepsis. *Am. J. Physiol. Lung Cell. Mol. Physiol.* 285,L1222–L1232.

McCall, T. B., Palmer, R. M. and Moncada, S. (1991). Induction of nitric oxide synthase in rat peritoneal neutrophils and its inhibition by dexamethasone. *Eur. J. Immunol.* 21,2523–2527.

McCord, J. M. and Fridovich, I. (1969). Superoxide dismutase. An enzymic function for erythrocuprein (hemocuprein). *J. Biol. Chem.* 244,6049–6055.

McQuibban, G. A., Gong, J. H., Wong, J. P., Wallace, J. L., Clark-Lewis, I. and Overall, C. M. (2002). Matrix metalloproteinase processing of monocyte chemoattractant proteins generates CC chemokine receptor antagonists with anti-inflammatory properties in vivo. *Blood* 100,1160–1167.

Medzhitov, R. and Janeway, C., Jr. (2000). Innate immunity. *N. Engl. J. Med.* 343,338–344.

Modlin, R. L., Brightbill, H. D. and Godowski, P. J. (1999). The toll of innate immunity on microbial pathogens. *N. Engl. J. Med.* 340,1834–1835.

Moncada, S., Palmer, R. M. and Higgs, E. A. (1991). Nitric oxide: Physiology, patho-physiology, and pharmacology. *Pharmacol. Rev.* 43,109–142.

Morgunova, E., Tuuttila, A., Bergmann, U., Isupov, M., Lindqvist, Y., Schneider, G. and Tryggvason, K. (1999). Structure of human pro-matrix metalloproteinase-2: Activation mechanism revealed. *Science* 284,1667–1670.

Munzel, T., Feil, R., Mulsch, A., Lohmann, S. M., Hofmann, F. and Walter, U. (2003). Physiology and pathophysiology of vascular signaling controlled by guanosine $3',5'$-cyclic monophosphate-dependent protein kinase. *Circulation* 108,2172–2183.

Murad, F., Waldman, S., Molina, C., Bennett, B. and Leitman, D. (1987). Regulation and role of guanylate cyclase-cyclic GMP in vascular relaxation. *Prog. Clin. Biol. Res.* 249,65–76.

Naseem, K. M. (2005). The role of nitric oxide in cardiovascular diseases. *Mol. Aspects Med.* 26,33–65.

Natanson, C., Danner, R. L., Fink, M. P., MacVittie, T. J., Walker, R. I., Conklin, J. J. and Parrillo, J. E. (1988). Cardiovascular performance with *E. coli* challenges in a canine model of human sepsis. *Am. J. Physiol.* 254,H558–H569.

Natanson, C., Eichenholz, P. W., Danner, R. L., Eichacker, P. Q., Hoffman, W. D., Kuo, G. C., Banks, S. M., MacVittie, T. J. and Parrillo, J. E. (1989). Endotoxin and tumor necrosis factor challenges in dogs simulate the cardiovascular profile of human septic shock. *J. Exp. Med.* 169,823–832.

Nava, E., Palmer, R. M., Moncada, S. (1991). Inhibition of nitric oxide synthesis in septic shock: How much is beneficial? *Lancet* 338, 1555–1557.

Nava, E., Palmer, R. M. and Moncada, S. (1992). The role of nitric oxide in endotoxic shock: Effects of N^{G}-monomethyl-L-arginine. *J. Cardiovasc. Pharmacol.* 20(Suppl. 12), S132–S134.

Ochoa, J. B., Udekwu, A. O., Billiar, T. R., Curran, R. D., Cerra, F. B., Simmons, R. L. and Peitzman, A. B. (1991). Nitrogen oxide levels in patients after trauma and during sepsis. *Ann. Surg.* 214,621–626.

Ognibene, F. P., Parker, M. M., Natanson, C., Shelhamer, J. H. and Parrillo, J. E. (1988). Depressed left ventricular performance. Response to volume infusion in patients with sepsis and septic shock. *Chest* 93,903–910.

Ohlsson, K., Bjork, P., Bergenfeldt, M., Hageman, R. and Thompson, R. C. (1990). Interleukin-1 receptor antagonist reduces mortality from endotoxin shock. *Nature* 348,550–552.

Okamoto, T., Akaike, T., Nagano, T., Miyajima, S., Suga, M., Ando, M., Ichimori, K. and Maeda, H. (1997). Activation of human neutrophil procollagenase by nitrogen dioxide and peroxynitrite: A novel mechanism for procollagenase activation involving nitric oxide. *Arch. Biochem. Biophys.* 342,261–274.

Opal, S. M. (2003). Severe sepsis and septic shock; defining the clinical problem. *Scand. J. Infect. Dis.* 35,529–534.

Overall, C. M., McQuibban, G. A. and Clark-Lewis, I. (2002). Discovery of chemokine substrates for matrix metalloproteinases by exosite scanning: A new tool for de-gradomics. *Biol. Chem.* 383,1059–1066.

Oyama, J., Shimokawa, H., Momii, H., Cheng, X., Fukuyama, N., Arai, Y., Egashira, K., Nakazawa, H. and Takeshita, A. (1998). Role of nitric oxide and peroxynitrite in the cytokine-induced sustained myocardial dysfunction in dogs in vivo. *J. Clin. Invest.* 101,2207–2214.

Pacher, P., Schulz, R., Liaudet, L. and Szabo, C. (2005). Nitrosative stress and pharmacological modulation of heart failure. *Trends Pharmacol. Sci.* 26,302–310.

Paemen, L., Jansen, P. M., Proost, P., Van Damme, J., Opdenakker, G., Hack, E. and Taylor, F. B. (1997). Induction of gelatinase B and MCP-2 in baboons during sublethal and lethal bacteraemia. *Cytokine* 9,412–415.

Palmer, R. M., Ferrige, A. G. and Moncada, S. (1987). Nitric oxide release accounts for the biological activity of endothelium-derived relaxing factor. *Nature* 327,524–526.

Parker, M. M., Ognibene, F. P. and Parrillo, J. E. (1994). Peak systolic pressure/end-systolic volume ratio, a load-independent measure of ventricular function, is reversibly decreased in human septic shock. *Crit. Care Med.* 22,1955–1959.

Pei, D. and Weiss, S. J. (1995). Furin-dependent intracellular activation of the human stromelysin-3 zymogen. *Nature* 375,244–247.

Peterson, J. T. (2006). The importance of estimating the therapeutic index in the development of matrix metalloproteinase inhibitors. *Cardiovasc. Res.* 69,677–687.

Phelps, D. T., Ferro, T. J., Higgins, P. J., Shankar, R., Parker, D. M. and Johnson, A. (1995). TNF-alpha induces peroxynitrite-mediated depletion of lung endothelial glutathione via protein kinase C. *Am. J. Physiol.* 269,L551–L559.

Pollock, J. S., Forstermann, U., Mitchell, J. A., Warner, T. D., Schmidt, H. H., Nakane, M. and Murad, F. (1991). Purification and characterization of particular endothelium-derived relaxing factor synthase from cultured and native bovine aortic endothelial cells. *Proc. Natl. Acad. Sci. USA* 88,10480–10484.

Pugin, J., Widmer, M. C., Kossodo, S., Liang, C. M., Preas, H. L. and Suffredini, A. F. (1999). Human neutrophils secrete gelatinase B in vitro and in vivo in response to endotoxin and proinflammatory mediators. *Am. J. Respir. Cell. Mol. Biol.* 20,458–464.

Rackow, E. C. and Astiz, M. E. (1991). Pathophysiology and treatment of septic shock. *JAMA* 266,548–554.

Radomski, M. W. (1989). Vascular endothelium in the processes of hemostasis and thrombosis. The role of prostacyclin and EDRF. *Acta Physiol. Pol.* 40(Suppl. 33), 97–109.

Radomski, M. W., Palmer, R. M. and Moncada, S. (1987). Endogenous nitric oxide inhibits human platelet adhesion to vascular endothelium. *Lancet* 2,1057–1058.

Radomski, M. W., Palmer, R. M. and Moncada, S. (1990). Glucocorticoids inhibit the expression of an inducible, but not the constitutive, nitric oxide synthase in vascular endothelial cells. *Proc. Natl. Acad. Sci. USA* 87,10043–10047.

Rajagopalan, S., Meng, X. P., Ramasamy, S., Harrison, D. G. and Galis, Z. S. (1996). Reactive oxygen species produced by macrophage-derived foam cells regulate the activity of vascular matrix metalloproteinases in vitro. Implications for atherosclerotic plaque stability. *J. Clin. Invest.* 98,2572–2579.

Reaume, A. G., Elliott, J. L., Hoffman, E. K., Kowall, N. W., Ferrante, R. J., Siwek, D. F., Wilcox, H. M., Flood, D. G., Beal, M. F., Brown, R. H., Jr., Scott, R. W. and Snider, W. D. (1996). Motor neurons in Cu/Zn superoxide dismutase-deficient mice develop normally but exhibit enhanced cell death after axonal injury. *Nat. Genet.* 13,43–47.

394

Rees, D. D., Cellek, S., Palmer, R. M. and Moncada, S. (1990). Dexamethasone prevents the induction by endotoxin of a nitric oxide synthase and the associated effects on vascular tone: An insight into endotoxin shock. *Biochem. Biophys. Res. Commun.* 173,541–547.

Remick, D. G. (2003). Cytokine therapeutics for the treatment of sepsis: Why has nothing worked? *Curr. Pharm. Des.* 9, 75–82.

Riedemann, N. C., Guo, R. F. and Ward, P. A. (2003). The enigma of sepsis. *J. Clin. Invest.* 112,460–467.

Rivers, E. P., McIntyre, L., Morro, D. C. and Rivers, K. K. (2005). Early and innovative interventions for severe sepsis and septic shock: Taking advantage of a window of opportunity. *CMAJ* 173,1054–1065.

Rosenberg, R. D. and Aird, W. C. (1999). Vascular-bed-specific hemostasis and hyper-coagulable states. *N. Engl. J. Med.* 340,1555–1564.

Salvemini, D. and Cuzzocrea, S. (2002). Oxidative stress in septic shock and disseminated intravascular coagulation. *Free Radic. Biol. Med.* 33,1173–1185.

Sato, H., Kinoshita, T., Takino, T., Nakayama, K. and Seiki, M. (1996). Activation of a recombinant membrane type 1-matrix metalloproteinase (MT1-MMP) by furin and its interaction with tissue inhibitor of metalloproteinases (TIMP)-2. *FEBS Lett.* 393,101–104.

Sawada, K., Kondo, T., Chang, J., Inokuchi, T. and Aoyagi, S. (1997). Distribution and neuropeptide content of nitric oxide synthase-containing nerve fibers in arteries and conduction system of the rat heart. *Acta Anat. (Basel)* 160,239–247.

Sawicki, G., Salas, E., Murat, J., Miszta-Lane, H. and Radomski, M. W. (1997). Release of gelatinase A during platelet activation mediates aggregation. *Nature* 386,616–619.

Schulz, R., Nava, E. and Moncada, S. (1992). Induction and potential biological relevance of a Ca^{2+}-independent nitric oxide synthase in the myocardium. *Br. J. Pharmacol.* 105,575–580.

Schulz, R., Panas, D. L., Catena, R., Moncada, S., Olley, P. M. and Lopaschuk, G. D. (1995). The role of nitric oxide in cardiac depression induced by interleukin-1β and tumour necrosis factor-α. *Br. J. Pharmacol.* 114,27–34.

Schulz, R., Smith, J. A., Lewis, M. J. and Moncada, S. (1991). Nitric oxide synthase in cultured endocardial cells of the pig. *Br. J. Pharmacol.* 104,21–24.

Schulze, C. J., Wang, W., Suarez-Pinzon, W. L., Sawicka, J., Sawicki, G. and Schulz, R. (2003). Imbalance between tissue inhibitor of metalloproteinase-4 and matrix metal-loproteinases during acute myocardial ischemia-reperfusion injury. *Circulation* 107,2487–2492.

Shipley, J. M., Wesselschmidt, R. L., Kobayashi, D. K., Ley, T. J. and Shapiro, S. D. (1996). Metalloelastase is required for macrophage-mediated proteolysis and matrix invasion in mice. *Proc. Natl. Acad. Sci. USA* 93,3942–3946.

Singal, P. K., Dhillon, K. S., Beamish, R. E., Kapur, N. and Dhalla, N. S. (1982). Myocardial cell damage and cardiovascular changes due to i.v. infusion of adreno-chrome in rats. *Br. J. Exp. Pathol.* 63,167–176.

Smith, J. A., Radomski, M. W., Schulz, R., Moncada, S. and Lewis, M. J. (1993). Porcine ventricular endocardial cells in culture express the inducible form of nitric oxide synthase. *Br. J. Pharmacol.* 108,1107–1110.

Sorrentino, R., d'Emmanuele di Villa Bianca, R., Lippolis, L., Sorrentino, L., Autore, G. and Pinto, A. (1999). Involvement of ATP-sensitive potassium channels in a model of a delayed vascular hyporeactivity induced by lipopolysaccharide in rats. *Br. J. Pharmacol.* 127,1447–1453.

Steinberg, J., Halter, J., Schiller, H., Gatto, L., Carney, D., Lee, H. M., Golub, L. and Nieman, G. (2005). Chemically modified tetracycline prevents the development of septic shock and acute respiratory distress syndrome in a clinically applicable porcine model. *Shock* 24,348–356.

Stoclet, J. C., Martinez, M. C., Ohlmann, P., Chasserot, S., Schott, C., Kleschyov, A. L., Schneider, F. and Andriantsitohaina, R. (1999). Induction of nitric oxide synthase and dual effects of nitric oxide and cyclooxygenase products in regulation of arterial contraction in human septic shock. *Circulation* 100,107–112.

Strongin, A. Y., Collier, I., Bannikox, G., Marmer, B. L., Grant, G. A. and Goldberg, G. I. (1995). Mechanism of cell surface activation of 72-kDa type IV collagenase. Isolation of the activated form of the membrane metalloprotease. *J. Biol. Chem.* 270, 5331–5338.

Suzuki, K., Enghild, J. J., Morodomi, T., Salvesen, G. and Nagase, H. (1990). Mechanisms of activation of tissue procollagenase by matrix metalloproteinase 3 (stromelysin). *Biochemistry* 29,10261–10270.

Szabo, C. (1999). Nitric oxide, peroxynitrite and poly (ADP-ribose) synthase: Biochemistry and pathophysiological implications. In *Pathophysiology and Clinical Application of Nitric Oxide* (ed. G. M. Rubanyi), pp. 69–98, Harwood Academic, Amsterdam.

Szabo, C., Zingarelli, B. and Salzman, A. L. (1996). Role of poly-ADP ribosyltransferase activation in the vascular contractile and energetic failure elicited by exogenous and endogenous nitric oxide and peroxynitrite. *Circ. Res.* 78,1051–1063.

Tasatargil, A., Dalaklioglu, S. and Sadan, G. (2005). Inhibition of poly(ADP-ribose) polymerase prevents vascular hyporesponsiveness induced by lipopolysaccharide in isolated rat aorta. *Pharmacol. Res.* 51,581–586.

Thiemermann, C. (1997). Nitric oxide and septic shock. *Gen. Pharmacol.* 29,159–166.

Titheradge, M. A. (1999). Nitric oxide in septic shock. *Biochim. Biophys. Acta* 1411, 437–455.

Toda, N. and Okamura, T. (2003). The pharmacology of nitric oxide in the peripheral nervous system of blood vessels. *Pharmacol. Rev.* 55,271–324.

Van der Vliet, A., Smith, D., O'Neill, C. A., Kaur, H., Darley-Usmar, V., Cross, C. E. and Halliwell, B. (1994). Interactions of peroxynitrite with human plasma and its constituents: Oxidative damage and antioxidant depletion. *Biochem. J.* 303(1), 95–301.

Van Wart, H. E. and Birkedal-Hansen, H. (1990). The cysteine switch: A principle of regulation of metalloproteinase activity with potential applicability to the entire matrix metalloproteinase gene family. *Proc. Natl. Acad. Sci. USA* 87,5578–5582.

Vasselon, T. and Detmers, P. A. (2002). Toll receptors: A central element in innate immune responses. *Infect. Immun.* 70,1033–1041.

Vila, E. and Salaices, M. (2005). Cytokines and vascular reactivity in resistance arteries. *Am. J. Physiol. Heart Circ. Physiol.* 288,H1016–H1021.

Visse, R. and Nagase, H. (2003). Matrix metalloproteinases and tissue inhibitors of metalloproteinases: Structure, function, and biochemistry. *Circ. Res.* 92,827–839.

396

Wakabayushi, G., Gelfand, J. A., Burke, J. F., Thompson, R. C. and Dinarello, C. A. (1991). A specific receptor antagonist for interleukin-1 prevents *Escherichia coli*-induced shock in rabbits. *FASEB J.* 5,338–343.

Walley, K. R., Hebert, P. C., Wakai, Y., Wilcox, P. G., Road, J. D. and Cooper, D. J. (1994). Decrease in left ventricular contractility after tumor necrosis factor-alpha infusion in dogs. *J. Appl. Physiol.* 76,1060–1067.

Wang, J. E., Dahle, M. K., McDonald, M., Foster, S. J., Aasen, A. O. and Thiemermann, C. (2003). Peptidoglycan and lipoteichoic acid in Gram-positive bacterial sepsis: Receptors, signal transduction, biological effects, and synergism. *Shock* 20,402–414.

Whittle, B. J. (1995). Nitric oxide in physiology and pathology. *Histochem. J.* 27,727–737.

Woessner, J. F. (1998). The matrix metalloproteinase family. In *Matrix Metalloproteinases* (eds W. C. Parks and R. P. Mecham), pp. 1–14, Academic Press, San Diego.

Xia, Y., Roman, L. J., Masters, B. S. and Zweier, J. L. (1998). Inducible nitric-oxide synthase generates superoxide from the reductase domain. *J. Biol. Chem.* 273, 22635–22639.

Xie, B., Dong, Z. and Fidler, I. J. (1994). Regulatory mechanisms for the expression of type IV collagenases/gelatinases in murine macrophages. *J. Immunol.* 152,3637–3644.

Xu, K. Y., Huso, D. L., Dawson, T. M., Bredt, D. S. and Becker, L. C. (1999). Nitric oxide synthase in cardiac sarcoplasmic reticulum. *Proc. Natl. Acad. Sci. USA* 96, 657–662.

Yamakura, F., Taka, H., Fujimura, T. and Murayama, K. (1998). Inactivation of human manganese-superoxide dismutase by peroxynitrite is caused by exclusive nitration of tyrosine 34 to 3-nitrotyrosine. *J. Biol. Chem.* 273,14085–14089.

Yates, J. C., Beamish, R. E. and Dhalla, N. S. (1981). Ventricular dysfunction and necrosis produced by adrenochrome metabolite of epinephrine: Relation to pathogenesis of catecholamine cardiomyopathy. *Am. Heart J.* 102,210–221.

Young, D. A., Phillips, B. W., Lundy, C., Nuttall, R. K., Hogan, A., Schultz, G. A., Leco, K. J., Clark, I. M. and Edwards, D. R. (2002). Identification of an initiator-like element essential for the expression of the tissue inhibitor of metalloproteinases-4 (Timp-4) gene. *Biochem. J.* 364,89–99.

Zhang, K., McQuibban, G. A., Silva, C., Butler, G. S., Johnston, J. B., Holden, J., Clark-Lewis, I., Overall, C. M. and Power, C. (2003). HIV-induced metalloproteinase processing of the chemokine stromal cell derived factor-1 causes neurodegeneration. *Nat. Neurosci.* 6,1064–1071.

Index of authors

Wang, Z. 318
Wang, Z.Q. 23
Wansbrough-Jones, M. 18
Ward, M. 11, 233–235, 276–278, 300
Ward, P.A. 139, 371
Warner, T.D. 375
Warr, C.G. 78
Wasserman, S.L. 113
Watanabe, S. 48
Watanabe, T. 348–349, 352
Watanabe, Y. 155
Waters, E. 72
Watson, W.H. 49, 136
Weave, D.F. 249
Weaver, L. 276
Webb, A. 218–219, 222, 296
Weber, E.T. 22
Wedel, B. 66
Wee, J. 383
Wegener, C. 96
Wei, C.C. 23
Wei, S. 232, 248, 254
Weidner, M.G. 374
Weigl, K. 371
Weil, M.H. 374
Weinstein, B.M. 285
Weinstein, R.B. 319
Weir, M.R. 170
Weisfeldt, M.L. 221
Weiss, B.L. 235–236, 238, 246
Weiss, J.N. 222
Weiss, S.J. 381
Weiss, S.W. 344
Weisse, M. 132
Weitzberg, E. 17–19, 200, 208
Welch, G.N. 214
Welch, W.J. 86
Weller, R. 19
Welshhans, K. 46
Wen, P.H. 345

Wendehenne, D. 28
Wendland, B. 232
Wenzel, B. 117–118
Wenzel, I. 233, 248, 255, 259
Werner, E.R. 14, 25, 155, 190
Werner-Felmayer, G. 14, 25
Weruaga, E. 111
Wesselschmidt, R.L. 381
West, K.A. 131
Westerfield, M. 241–242
Westermann, B. 297
Westfall, T.C. 373, 379
Westphal, H. 340
Wetzel, R.C. 318
Wexler, R.R. 324
Whalen, E.J. 135
White, A.R. 47
White, K.A. 153
White, W. 110
Whitehead, K.J. 249
Whiteman, M. 129, 131
Whitlow, M. 156
Whittle, B.J. 377
Whittlei, B.J.R. 294
Wicher, D. 111, 119–120
Widmer, M.C. 384
Wiegertjes, G.F. 11, 276–277
Wieland, K. 341, 350–351, 356
Wiklund, N.P. 19
Wilcox, C.S. 86
Wilcox, H.M. 378
Wilcox, P.G. 371
Wildemann, B. 119–120
Wildhirt, S.M. 379
Wilks, M. 18–19
Williams, D.L. 18
Williams, E.M. 207
Williams, G.T. 138
Williams, L. 109–110, 112–114, 116–118, 120, 182

Subject index

448

450